Lakes of New York State

VOLUME II: Ecology of the Lakes of Western New York

A view of Chautauqua Lake, New York. Courtesy of the Chautauqua Institution/Gordon Mahan.

Lakes of New York State

VOLUME II

ECOLOGY OF THE LAKES OF WESTERN NEW YORK

EDITED BY

Jay A. Bloomfield

New York State Department of Environmental Conservation
Albany, New York

ACADEMIC PRESS New York San Francisco London 1978

A Subsidiary of Harcourt Brace Jovanovich, Publishers

ACADEMIC PRESS, INC.
111 Fifth Avenue, New York, New York 10003

United Kingdom Edition published by
ACADEMIC PRESS, INC. (LONDON) LTD.
24/28 Oval Road, London NW1 7DX

Library of Congress Cataloging in Publication Data

Main entry under title:

Ecology of the Lakes of Western New York.

(Lakes of New York State ; v. 2)
Includes bibliographies and index.
CONTENTS: Mayer, J. R. and others, Chautauqua Lake.--Bannister, T. T. and Bubeck, R. C. Irondequoit Bay.--Murphy, C. B., Jr. Onondaga Lake. [etc.]
1. Lakes--New York (State) 2. Eutrophication--New York (State) 3. Water--Pollution--New York (State) I. Bloomfield, Jay A. II. Series.
CB1625.N7L34 vol. 2 574.5'2632'09747s [574.5'2332'09747]
ISBN 0-12-107302-5 (v. 2) 77-28315

PRINTED IN THE UNITED STATES OF AMERICA

Contents

List of Contributors

Numbers in parentheses indicate the pages on which the authors' contributions begin.

THOMAS T. BANNISTER (105), Department of Biology, University of Rochester, Rochester, New York

W. M. BARNARD (1), Department of Geology, State University College at Fredonia, Fredonia, New York

ROBERT C. BUBECK (105), Chesapeake Bay Program, Annapolis Field Office, U.S. Environmental Protection Agency, Annapolis, Maryland

M. D. CLADY* (367), Department of Natural Resources, New York State College of Agriculture and Life Sciences, Cornell University, Ithaca, New York

T. A. ERLANDSON (1), Environmental Resources Center, State University College at Fredonia, Fredonia, New York

J. L. FORNEY (367), Department of Natural Resources, New York State College of Agriculture and Life Sciences, Cornell University, Ithaca, New York

J. R. LUENSMAN (1), Chautauqua County Department of Planning and Development, Mayville, New York

J. R. MAYER (1), Environmental Resources Center, State University College at Fredonia, Fredonia, New York

W. J. METZGER (1), Department of Geology, State University College at Fredonia, Fredonia, New York

E. L. MILLS (367), Department of Natural Resources, New York State College of Agriculture and Life Sciences, Cornell University, Ithaca, New York

* Present address: Oklahoma Cooperative Fishery Research Unit, Oklahoma State University, Stillwater, Oklahoma 74074.

CORNELIUS B. MURPHY, JR. (223), O'Brien & Gere Engineers, Inc., Syracuse, New York

S. A. NICHOLSON (1), School of Natural Resources, The University of the South Pacific, Suya, Fiji

W. R. SCHAFFNER (367), Department of Natural Resources, New York State College of Agriculture and Life Sciences, Cornell University, Ithaca, New York

R. T. SMITH (1), Environmental Resources Center, State University College at Fredonia, Fredonia, New York

T. A. STORCH (1), Environmental Resources Center, State University College at Fredonia, Fredonia, New York

Foreword

Between 1926 and 1938 the New York State Department of Conservation published a series of biological surveys of the river drainage basins in the State. The studies carried out and the resulting reports must be considered classic biological inventories and, even today, represent the best set of biological information on the surface waters of New York State.

Anyone working in the State quickly realizes their value. Their influence on management of these waters and on the scope and direction of subsequent studies has been and continues to be so prominent that one wonders what would have been done had they not existed. Rarely does a study begin or is a report published on any lake or stream in the State that does not utilize these as basic references.

However, even as the last of the surveys was completed in 1938, the "desirability of up-to-date revisions of former surveys" was recognized. Unfortunately, this has not been done, and an examination of the original surveys reveals this need for updating. Since the original surveys, conditions in some systems have changed, the state of the art of measuring and interpreting data has advanced, and numerous studies provide a wealth of additional information.

Because of the demonstrated values of these biological inventories and the increased interest in lake management, the Department of Environmental Conservation commissioned the preparation of this treatise on lakes in New York State. The contributions were to be inventory reports modeled after the original biological surveys, but were to be expanded to cover the physical, chemical, and biological state of the lake and its drainage basins. Preparation of the contribution was not to include carrying out of any special field studies but was to be based on existing data and information. Fortunately, in New York State, for most of the important lakes, a great amount of data has been collected and reported as part of studies directed at one or another aspect of the lake. Additionally, the State has a reservoir of resident limnologists who by virtue of their location and interest were uniquely capable of authoring these articles.

The purpose of these articles may be explicitly stated as follows:

1. To provide in a unified manner an authoritative set of current data from which lake management decisions can be made.

2. To begin to meet the requirements of the 1970 Federal water pollution control legislation which requires "that the State shall prepare an identification and classification according to the trophic condition of all publicly-owned freshwater lakes in such State."

3. To provide a uniform set of data from which comparative lake studies could be made.

4. To provide a set of uniform baseline data on lakes from which future changes can be measured.

5. To provide for students a set of real data that will bring to life their classroom experience.

6. To determine where gaps in our knowledge of New York State lakes exist, and provide a basis for investment in future studies and research.

In an age when there is an increased awareness of the value of the environment, when environmental laws and regulations are multiplying at an unprecedented rate, when environmental legal proceedings are commonplace, and when environmental impact statements have become a way of life, these volumes should prove useful.

The idea of this treatise series was originally conceived in 1972 and, after many setbacks, the project commenced in 1974. The leadership, assistance, and patience of so many people were instrumental in their completion that space prevents a listing of their names and contributions. However, I feel compelled to mention the commitment of Mr. Eugene Seebald, Director of the Division of Pure Waters, who provided steadfast support for the effort, even when short-term needs suggested utilization of the required resources in other areas.

Leo J. Hetling
Albany, New York

Preface

The four lakes covered in this volume seem to have little in common except geographic location. One, Chautauqua Lake, is famous for its muskellunge fishery, while Onondaga Lake has been closed to all fishing since 1970 because of high levels of mercury in its fish. Oneida Lake is the largest lake wholly in New York State, while Irondequoit Bay is technically not a lake, but an embayment of Lake Ontario, cut off from the main lake by a narrow sandspit.

These superficial contrasts tend to obscure the general similarities among the lakes; all are productive north temperate lakes, glacial in origin and significantly influenced by man's activities. Each lake has been studied intensively in order to develop a reference point for future scientific studies and management plans.

In "Oneida Lake," E. L. Mills and co-workers review the historical data base of a large (206.7 km^2) and relatively shallow (maximum depth 16.8 m, mean depth 6.8 m) naturally productive lake. Oneida Lake is a major recreational resource and is used extensively for boating and sport fishing. Since the early 1900's, however, changes have occurred in almost every major chemical constituent and biotic component. The lake is presently managed as a warm water fishery, a strategy well suited to Oneida Lake's present eutrophic condition.

Irondequoit Bay is a scenic lake situated in a major metropolitan area, Rochester. Despite large inputs of municipal wastes to its tributary streams, the bay has great recreational potential and is heavily used by fishermen and bathers. Bannister and Bubeck have examined four topics in detail: (1) factors regulating phytoplankton growth and the potential impact of the diversion of municipal wastes, (2) the effect of diversion on rooted aquatic vegetation, (3) the probable consequences of deepening the bay's outlet to Lake Ontario, and (4) the effects of deicing salt runoff.

Mayer and seven coauthors categorize Chautauqua Lake as a productive fishery and recreational resource. The word "Chautauqua" describes the lake succinctly as it is said to be the Seneca Indian phrase for "bag tied in the middle." The lake is shallow and warm with extensive rooted and planktonic vegetation. The lake is famous for its Chautauqua muskellunge fishery (*Esox masquinongy ohioensis* Kirtland) which has been its most prized attribute for at least 100 years. Concern for this resource has led to a research and management program centered around the muskellunge. The average annual yield of muskellunge has been estimated at about 30 tonnes, and over 80% of the total young of the year muskellunge in the lake are reared at a special local hatchery which has been propagating muskellunge since 1904.

Murphy reports about a lake so abused by the activities of man, it has ceased to become much more than a receptacle for waste products. Onondaga Lake is an 11.7 km^2 highly eutrophic water body situated at the northern edge of the city of Syracuse. The initial onslaught came after 1812, when salt and brine deposits along the southern shore of the lake were extracted. The lake also served as a link in the Erie Barge Canal. In 1822, to facilitate navigation, the lake level was lowered, creating a vast expanse of salt marsh and reducing the lake surface area by 20%. Since that time, municipal and industrial wastes have been discharged, often with no treatment, directly to the lake. Onondaga Lake gained notoriety when in 1970, New York State prohibited fishing due to high concentrations of mercury in the lake's fish. A vigorous discharge abatement program was subsequently initiated to treat or divert municipal and industrial wastes. The levels of mercury in fish flesh and indications of hypereutrophic conditions have been on the decline since 1970, but contaminated bottom sediment has been hindering recovery of the lake.

In summary, the four contributions to this volume make interesting case studies for the student or professional person interested in the impact of the activities of mankind on lakes. I hope publication of this volume contributes at least slightly to an historical perspective of limnological research in New York State.

Jay A. Bloomfield

Contents of Other Volumes

Volume I. Ecology of the Finger Lakes

Volume III. Ecology of the Lakes of East-Central New York (tentative)

Chautauqua Lake—Watershed and Lake Basins

J. R. Mayer, W. M. Barnard, W. J. Metzger, T. A. Storch, T. A. Erlandson, J. R. Luensman, S. A. Nicholson, and R. T. Smith

ISBN 0-12-107302-5

HISTORY

The following brief history of Chautauqua Lake is drawn in part from factual materials developed by Dilley (1891), McMahon (1964), Meyer (1921), Powers (1924), Turner (1849), Warren (1878), and Young (1875).

Lying centrally within Chautauqua County, Chautauqua Lake is the largest and most important lake in the southern tier of western New York State. First discovered by the French explorer LaSalle in 1679, the lake serves as the headwaters for the Chadakoin, Connewango, Allegany, Ohio, and Mississippi River System. It is therefore a part of a major land and water route connecting Canada and Lake Erie on the north with the Mississippi River Valley. Because of its strategic geography, the history of the lake and its surrounding watershed reflects a rich cultural, political, social, and economic heritage.

Chautauqua Lake was formed in an elongate southeasterly draining valley during the deglaciation of western New York between 10,000 and 12,000 years ago (Muller, 1963). Water was first impounded behind a morainal complex near the present city of Jamestown, New York. As the ice front retreated northward and the size of the lake grew, a second moraine was formed in the Bemus Point-Stow, New York, area blocking the valley and forming the north shore of the lake. Continued glacial recession formed a second lake to the north. These two lakes were quickly joined by stream action eroding a channel through the moraine at Bemus Point-Stow. As the ice retreated to the north and over the local divide into the Lake Erie basin, the present size of the upper lake was established.

The name "Chautauqua" is said to be a Seneca Indian phrase meaning "bag tied in the middle"—a reference to the shape of the lake's two basins connected by the narrows at Bemus Point-Stow. The lake, which lies between Mayville and Jamestown, has a surface configuration much like the finger lakes of central New York. Chautauqua's waters are relatively warm, shallow, and eutrophic; therefore it resembles the western finger lakes much more than the eastern finger lakes. Chautauqua's waters produce good fishing, a fact known to the Indians of the region. It is not surprising, therefore, that the Chautauqua Lake area became important to the Senecas and to the other Iroquois Nations.

Early English settlement began near Chautauqua Lake in 1811 when James Prendergast built a dam across the Chadakoin River. In succeeding years Prendergast built a gristmill and several sawmills. The city of Jamestown, named for Prendergast, gradually gained a reputation first for lumbering and then for the furniture industry which flourished there. The abundance of high-quality virgin timber that covered the upper Allegheny Valley and the Chautauqua Lake region was probably the most important factor in the settling of the area. The relative ease of transporting lumber from Jamestown to Pittsburgh via the Connewango and Allegheny Rivers, together with water power provided by the lake for running sawmills, formed the basis for an extensive lumbering business that was to flourish for over 50 years.

The softwoods like hemlock, white pine, balsam, and spruce were sawed into lumber and rafted down the river. Hardwoods such as maple, oak, birch, ash, hickory, chestnut, and cherry were burned for their ashes. The lye which was produced from the leaching of water through these ashes was turned into valuable potash and pearl ash—one of the few cash commodities at the time. Pearl ash from the asheries of Chautauqua County were sent to England by way of Montreal and to New York by wagon.

The stripping of the lake watershed's hardwood and softwood timber also served to clear the land for settlers who found the soils well suited to producing hay and grain. This led to the development of beef and dairy farming. Between 1830 and 1860, the Chautauqua region became well known for its fine steers which supplied beef to New York, Boston, and Philadelphia. Milk was processed into butter and cheese.

Indian and early English civilization near Chautauqua Lake probably marks the beginning of the lake's cultural eutrophication. Deforestation during the 1800's undoubtedly had its effect on the lake. Soil erosion following lumbering and burning of the trees contributed sediment and nutrients to the tributaries and the lake basins. The trees had served to utilize available nutrients thus minimizing their transport to the lake environment. Shading by trees helped to produce cooler creek waters. The overall result was to change the nature of the Chautauqua Lake watershed, thereby changing the ecology of the lake itself.

The Chautauqua Lake region has long been known as a recreational, summer resort area. During the last 100 years extensive development of summer homes has occurred along the lake perimeter. In addition, the lake began to experience intensive use by tourists, some of whom lived nearby during the summer and others who traveled to the area by railroad and later by automobile. The steamboat era on Chautauqua Lake reflects the demand which existed in the late nineteenth and early twentieth century to use the lake to its fullest advantage. It is perhaps the world's best muskellunge lake; a State fish hatchery at Prendergast Point carries out an intensive muskellunge stocking program which serves Chautauqua and other lakes. In addition to Chautauqua's prolific fishery, the lake provides an ideal setting for swimming, power and sail boating, and water skiing. Chautauqua Institution, a widely known cultural and educational center, and Allegheny State Park at Long Point, attract many summer residents and visitors each year.

The history of Chautauqua Lake, which spans almost 300 years since its discovery, is therefore a reflection of early American civilization and of the impact of cultural change on a unique region.

DRAINAGE BASIN

Geologic Setting

Chautauqua Lake and its related watershed are located in Chautauqua County, New York, along the axis of a prominently glaciated valley in the Allegheny Plateau. The area is one of moderate relief ranging from approximately 550 m on the divides to 399 m at the outlet of Chautauqua Lake. The surficial deposits in this area have been studied extensively by Muller (1963) and are predominately Kent ground moraine deposits ranging in thickness between 3 and 30 m. Several end or ressional moraines cross the basin and play an important role in establishing the existence and nature of the lake. Fig. 1 is a surficial geology map of the basin.

At the south end of the lake, in the vicinity of Jamestown, New York, the glacial valley is blocked by the Clymer ressional moraine and associated stratified drift deposits which impond the waters of the lake. The Chadakoin River, which is the outlet for the lake, presently flows in a valley cut through these deposits. A second morainal complex, the Findley ressional moraine, is located in the bemus Point-Stow area. Here, morainal deposits and stream gravels form a constriction in the lake which generally subdivides the lake into the upper and lower basins.

Exposures of glacial sediment along the shoreline of the lake consist primarily of three types: Findley ressional moraine, Kent ground moraine, and recent stream gravel, sand, and silt. There are no bedrock exposures along the shore. However, on the northeastern shore between Big Inlet Creek and

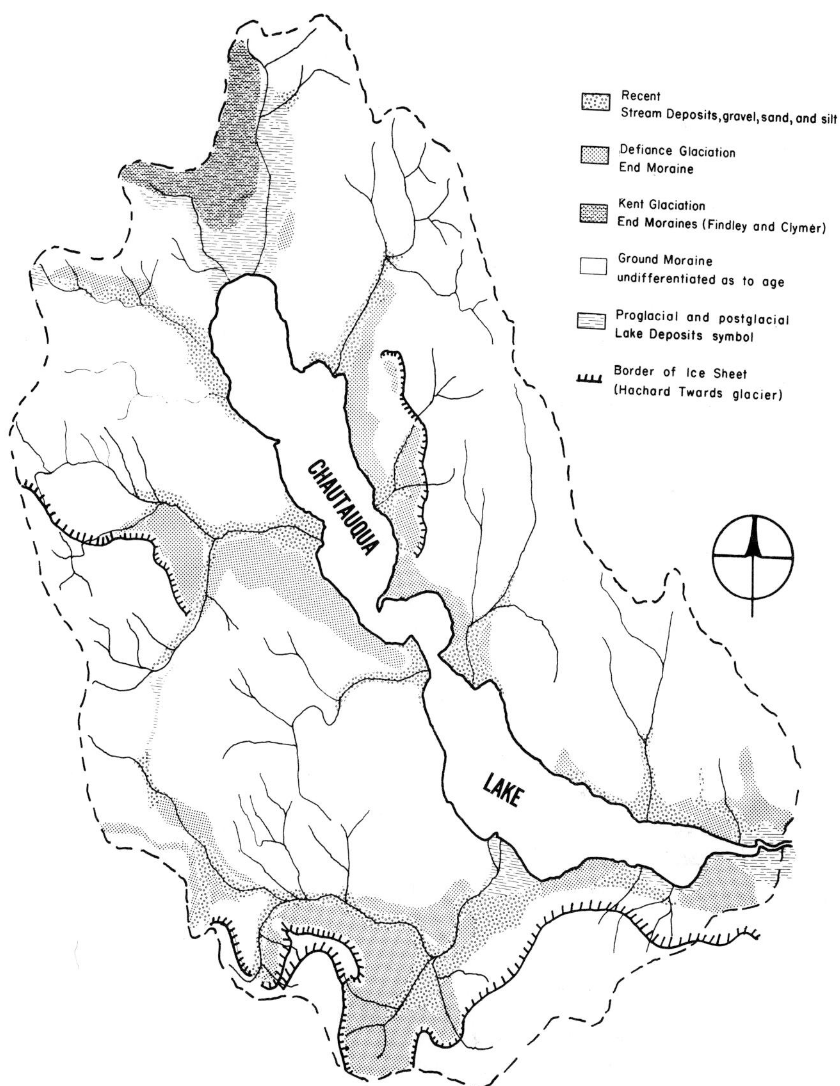

Fig. 1. Chautauqua Lake drainage basin: surficial geology. After Muller (1963).

Point Chautauqua, bedrock exists very close to the surface and many large slabs of siltstone can be observed on the bottom in the nearshore area.

These glacial deposits are underlain by a thick sequence of interbedded shale and siltstone of Upper Devonian age (Tesmer, 1963). The units consist of the Dexterville and Ellicott members of the Chadakoin Formation. Limited exposure of these rocks, especially the Ellicott member, is found in the drainage basin along several tributary streams, typically in the headwaters. Extensive exposures of both units are found along the Chadakoin River in the city of Jamestown. Despite the restricted number of outcrops in the basin, significant quantities of local bedrock are found as pebbles and cobbles in the overlying glacial drift. These materials are frequently found in great abundance in modern stream gravels where they have been concentrated by the action of running water.

Soil Distribution

For the purposes of this report, the soils of the Chautauqua Lake basin have been divided into five major groups, based on similar characteristics of drainage and agricultural potential. The distribution of these five groups is indicated in Fig. 2. A listing of the soil types which are included in each group can be found in Table 1.

The majority of the upland areas is underlain by Erie and Langford soils, both of which are developed on glacial end and ground moraine deposits. The Erie soils are found on the higher portions of the divides while the Langford soils are found in the areas of intermediate agricultural potential.

The deposits classified as Chenango soils are developed in areas underlain by sand and gravel deposits of glacial or fluvial origin or more rarely on areas of ground moraine. They are typically found in areas of low relief at

TABLE 1

Soils of the Chautauqua Lake Drainage Basin

Erie soils	Erie silt loam
Langford soils	Langford silt loam
	Lordstown shale loam
Chenango soils	Chenango silt loam
	Chenango fine sandy loam
	Chenango gravelly loam
Wooster soils	Wooster silt loam
	Wooster gravelly loam
	Wooster fine sandy loam
Mudland	Muck
	Papakating fine sandy loam

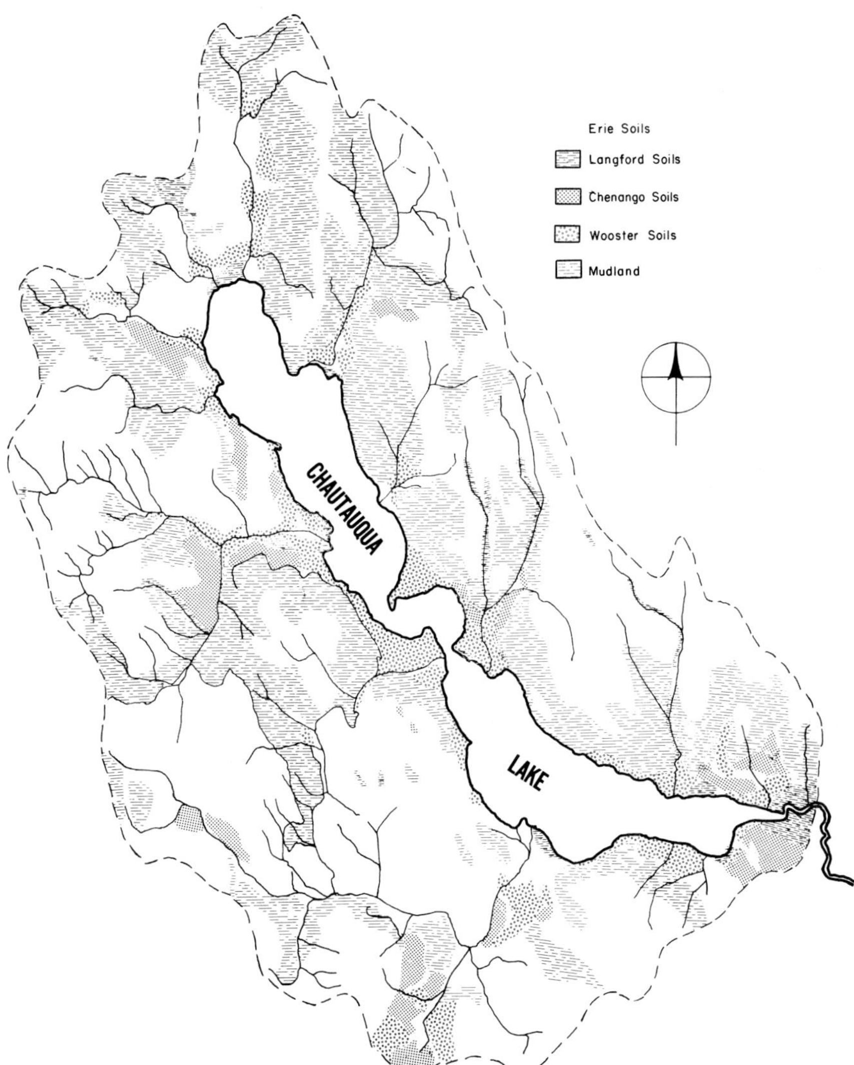

Fig. 2. Chautauqua Lake drainage basin: soil map. Modified from Howe (1936).

the lower elevations in the basin. All are well drained and have excellent agricultural potential.

The Wooster soils are associated with clay-rich glacial deposits found in areas of very low relief. They are found both at low elevations and at high elevations on the drainage divides. They are typically poorly drained and have low agricultural potentials.

The soils mapped as mudlands include all areas underlain by impervious clay deposits which are found in areas of low relief and which additionally have perennially high water tables. They are found along the lower reaches of several streams on their deltas and are extensively developed in the lowlands adjacent to the outlet of the lake at Celoron.

Population

The summer population of the Chautauqua Lake Watershed, estimated at 35,000 persons, is reached for possibly one or two weeks in the month of July or August. This estimate is made up of two elements: a year-round census and an estimate of the temporary resident capacity. Based on 1960 to 1970 population growth patterns for Chautauqua County, it is felt that the 1970 census data also represent a 1975 description or estimate of the population.

Year-Round or Census Population: 1970

Through the use of the 1970 Census of Population data and 1971 aerial photography, it is estimated that the Chautauqua Lake 1970 year-round watershed population was 23,010 persons. Year-round watershed population can be described in the following manner: While the watershed area contains parts of the towns of Portland, Stockton, and Sherman plus parts of Busti, Chautauqua, Ellery, Ellicott, Harmony, and North Harmony, no population numbers from the first four townships were used, all of the adjustment being made into the latter six townships. This allows for the use

TABLE 2

1970 Census: Estimated Year-Round Population of the Chautauqua Lake Watershed

Town or enumeration district	Population
All of the town of Chautauqua	4,341
All of the town of Ellery	4,594
All of the town of North Harmony	2,264
Half of Enumeration District 162 of Harmony	596
Enumeration Districts 156 and 157 of Busti	1,196
Nine-tenths of Enumeration District 158, town of Busti	1,115
All of the village of Lakewood	3,864
Enumeration Districts 100, 103–105, town of Ellicott	3,063
Three-fourths of Enumeration District 108, town of Ellicott	521
All of the village of Celoron	1,456
	23,010

TABLE 3

Estimated Temporary Resident Population Capacity of Chautauqua Lake Watershed, 1970

Description of units	Population
Second homes ("summer cottages"):	6,556
Trailer parks or campsites:	3,504
Hotels and motels:	1,527
Children's overnight camps:	1,110
	12,687

of existing census data in making descriptions of an existing population. Table 2 lists the enumeration districts and town population numbers used to describe the estimated year-round population of the Chautauqua Lake watershed.

Temporary Resident Population

The temporary resident population of the Chautauqua Lake watershed has been developed through an estimate of the capacity of various facilities located within the watershed. In many instances these facilities are used by persons who reside there year-round. Therefore, some capacities are notably absent from the capacity list (Lakewood, Bemus Point, Chautauqua Institution, and Mayville beaches, for example).

A listing of temporary residences from the Chautauqua County Department of Health was reviewed and backdated to 1970. This provides a part of the temporary resident capacity or population. The 1970 Census of Housing by enumeration district (using the same district and governmental units as for Year-Round Census of Population 1970) was used to count vacant seasonal and migratory units; the Census of Housing noted 2014 "summer cottages." Based on 1965 and 1970 surveys of second homes it is estimated that the average family size for a "summer cottage" is 3.25 persons (Luensman, 1973). The temporary resident population capacity of the Chautauqua Lake Watershed is shown in Table 3.

Therefore, the estimated temporary population capacity within the watershed, when combined with the estimated year-round capacity, would suggest that on a given day during the month of July or August the watershed population of Chautauqua Lake may reach over 35,000 persons.

Population Distribution

An estimated 60% of the year-round population and 95% of the temporary resident population may be found within 300 m of the shores of

Chautauqua Lake. Depending on the specific location, the pattern of location may exceed eight to ten families per acre. If a mile-wide belt were drawn about the lake, 80 to 85% of all the year-round population would be included. The majority of the watershed area has a population density of 0–50 persons per square mile. This low density is particularly true of the upper reaches of the watershed in the towns of Chautauqua, North Harmony, and Ellery.

Land Use

There are several land use records that may be used to describe activities within the Chautauqua Lake Watershed. Probably the best of these records are in the aerial photography taken in 1938–1939, 1956, and 1966 by the United States Department of Agriculture; 1961 and 1971 photos by the Chautauqua County government; and 1968–1969 photos taken by the State of New York for the state-wide Land Use and Natural Resources Inventory (LUNR) program.

Several land use investigations have been made of part or all of the area within the watershed. However, the criteria used in category identification have not been of the detail or accuracy desired, the LUNR program being a case in point.

Partial Land Use Data 1968

Through an Urban Planning Assistance "701 Program," a "wind-shield" survey was completed in the summer of 1968 (June through August) for the towns of Chautauqua, Busti, Ellery, North Harmony, and Ellicott. The data for these areas were developed on a town or village base. Table 4 is an extract from these municipalities' plan reports. It is believed that, based upon the categories shown, an approximation of the character of land use in the Chautauqua Lake watershed is obtained.

Partial Land Use 1938–1939 Compared to 1971

A second partial land use study was completed more recently (Chautauqua Lake Studies, 1972). The purpose of the study was to examine patterns of land use in the Chautauqua Lake Watershed and to make observations of changes in the patterns. The earliest available aerial photography was from 1938–1939; the latest was 1971 photography completed for the County Planning Department.

The 11 largest tributary subbasins of the Chautauqua Lake watershed were analyzed. A breakdown of the watershed as studied is given in Table 5. The results of the investigation are reported in greater detail for the 11 largest subbasins of the watershed in Table 6.

TABLE 4

Summation of 1968 Land Use in the Towns and Villages of the Chautauqua Lake Watershed[a,b,c]

Land use	Busti	Lakewood	Chautauqua	Mayville	Ellery	Bemus Point	Ellicott	Celoron	North Harmony
Family residents	763.3	161.0	665.4	94.1	581.8	22.4	504.1	31.9	586.6
Multifamily	4.7	3.4	6.1	11.0	1.0	3.7	6.6	1.1	1.5
Retail–wholesale	12.3	7.9	20.7	6.8	9.7	.6	32.0	.6	5.4
Service									
Commercial	10.2	6.9	34.8	7.0	2.2	2.2	32.2	5.1	4.7
Industrial	60.8	3.4	1.8	10.2	2.8	—	73.6	2.4	—
Transportation, communication, and utilities	326.0	87.6	405.0	54.4	281.4	13.6	462.0	29.5	240.7
Public	0.9	3.6	20.3	21.3	26.3	8.4	47.5	2.6	8.3
Semipublic	28.2	7.3	40.8	8.7	127.2	6.7	80.3	1.7	12.2
Cultural recreation	338.0	13.6	1,108.7	11.7	166.8	31.7	117.5	3.6	1,133.6
Agricultural	4,636	—	5,814.2	72.6	3,223.4	—	2,778.5	—	3,372.8
Vacant or open	6,041.7	222.1	9,011.0	212.8	7,981.1	23.9	3,422.1	102.7	5,755.2
Total	12,222.7	516.8	17,128.8	510.6	12,403.7	113.2	7,556.4	181.2	11,121.0

[a] This program was not undertaken in the town of Harmony.
[b] Whole town or village data, not just watershed data.
[c] Area given in hectares.

TABLE 5

Chautauqua Lake Watershed Land Use Study (1938–1939 Compared with 1971): Area Statistics

	Area (hectares)[a]
Major subbasins (analyzed)	31,472
Minor subbasins (not analyzed)	
Outside routes 17 and 17J	7,204
Inside routes 17 and 17J	2,764
Total subbasins	41,440
Chautauqua Lake	5,310
Total watershed	46,750

[a] One hectare = 2.471 acres.

TABLE 6

Chautauqua Lake Watershed Land Use Study[a]

	1938–1939		1971		
Land Use categories	Hectares[b]	Percent distribution	Hectares	Percent distribution	Percent change
Total forest	12,520	39.78	17,578	55.86	+40.38
Forestlands	9,220	29.29	13,995	44.47	+51.77
Forestbrush	3,265	10.38	2,787	8.86	−14.64
Plantations	35	0.11	796	2.53	+2134.09
Total farming	18,210	57.86	12,835	40.78	−29.53
Rotation	11,155	35.44	8,620	27.39	−22.72
Inactive farming	3,285	10.44	2,617	8.32	−20.38
Pasture	3,142	9.98	1,137	3,61	−63.81
Orchard	231	0.73	75	0.24	−67.54
Farmsted	397	1.27	386	1.22	−3.05
Total miscellaneous	741	2.37	1,058	3.35	+42.56
Roads	647	2.06	659	2.09	+1.69
Residential	45	0.14	171	0.54	+276.79
Commercial	33	0.11	79	0.25	+139.02
Water	5	0.02	98	0.31	+1761.54
Recreation	0	0	16	0.05	+[c]
Gravel pits	3	0.01	22	0.07	+700.00
Urban	8	0.03	13	0.04	+60.00
Total	31,471		31,471		

[a] Chautauqua Lake Studies, 1972.
[b] Hectares rounded to nearest whole.
[c] Indicates percent change cannot be calculated.

Vegetation of the Chautauqua Lake Watershed

Prior to European settlement the watershed was covered with dense forest. In general the major species, sugar and red maple (*Acer saccharum* and *A. rubrum*), beech (*Fagus grandifolia*), red oak (*Quercus rubra*), white ash (*Fraxinus americana*), elm (*Ulmus americana*), black cherry (*Prunus serotma*), and hemlock (*Tsuga canadensis*), were the same as those of present-day forests. But the numbers of other trees has changed markedly; e.g., chestnut (*Castanca dentata*), once locally abundant, has been virtually eliminated and black locust (*Robinia pseudoacacia*), native further west, has colonized many roadside areas.

The relative abundance of the various tree species in a particular forest is, of course, a function of many variables, e.g., site type, past management history, etc. In the watershed many of the dominant trees are very widespread but red maple and elm are most common in wetter areas, sugar maple and beech in the uplands, and hemlock in ravines and along streams. Aspen (*Populus tremuloriles*) is prominent in recently disturbed areas.

A wide variety of preforest successional communities occur in the watershed. Their composition and density varies with previous site activities, site properties, and past invasion regime. These communities are of three main types: (1) old fields, (2) old field–shrub mosaics, and (1) and/or (2) with invading trees. Old fields may be dominated by native or nonnative grasses or forbs, particularly goldenrod (*Solidago* spp.). Common invading shrubs include sumas (*Rhus* spp.), haw (*Crategus* spp.), blackberries (*Rubus* spp.), dogwood (*Cornus* spp.), and viburnum (*Viburnum* spp.). Many mixtures of the above shrubs and herbs exist with or without invading trees, but goldenrod, goldenrod–grass, and goldenrod–ash are some of the more common ones.

Perhaps the most significant change in the vegetation since presettlement has been in biomass, succesional status, and cycling qualities. Braun (1950) included the watershed in the Allegheny section of the hemlock–white pine–northern hardwoods forest formation. Kuchler (1964) considered all but a small portion of the watershed, a patch of beech maple at the head of the lake, to the northern hardwoods (*Acer–Betula–Fagus–Tsuga*). Presettlement vegetation, predominately unbroken forest cover with many more larger trees, and a more constant and higher year to year biomass, had presumably tighter nutrient cycles than replacing vegetation, which is more open, more frequently disrupted, and much lower in biomass. The contrasts in pre- and postsettlement vegetation are, of course, sharper for herb and shrub replacement communities, but generally also hold for young forests as well.

Unfortunately the quantitative effects of these and recent vegetation transformations on the lake are not specifically known because of the absence of appropriate studies. Periodic attempts that have been made to monitor stream discharge have ignored the well established regulatory effects of vegetation on hydrologic and nutrient cycles. Therefore, little can be said quantitatively about the effects of watershed vegetation on the lake except that exports from contemporary types—young, recently disturbed, and of low biomass—may be expected to exceed those from their presettlement predecessors, i.e., older, less disturbed, higher biomass systems.

Socioeconomic Status

The year-round population of the Chautauqua Lake Watershed may be described from data found in the 1970 Census of Population. Average family income varies from township to township within the watershed from a low of $8392 in the town of Harmony to $10,597 in the town of Ellicott. Table 7 illustrates the count of families by income group from the 1970 Census of Population.

Most of the wage earners of the watershed work outside the watershed. The majority of employed persons work in the Jamestown-Falconer business and industrial complex located immediately to the east of the outlet of Chautauqua Lake (see Table 8 for details of commuting patterns for employment).

An analysis of the employment information provided in the 1970 Census of Population indicates that, while the major land use of the watershed may be farming and forest lands, only 364 persons out of 12,512 employed at the time of the Census were directly involved with farming. About one-third of the employed work force were craftsmen, foremen, and operatives. It may be assumed that these persons, with the exception of the town of Chautauqua, worked outside of the watershed. Table 9 shows the employed work force of the watershed, and the age distribution of the Chautauqua Lake Watershed year-round population is provided in Table 10.

Point Sources of Pollutants

The principle point sources of pollution have been identified for the drainage basin and are grouped into three major categories: (1) municipal wastewater disposal plants, (2) solid waste disposal sites, and (3) industrial sources. The locations of each specific point source are indicated in Fig. 3. Additionally, there are many diffuse sources of pollution related to domestic sewage treatment installations on lakefront properties, but they are considered later in the chapter on the ecosystem. Although the city of Jamestown, New York, is the largest population center in the area, it does not lie within the Chautauqua Lake watershed and does not contribute any

TABLE 7

Count of Families by Family Income 1970 Census[a]

Income (dollars)	Busti	Chautauqua	Ellery	Ellicott	Harmony	North Harmony	Total
0–2,999	239	107	93	226	31	61	757
3,000–4,999	170	88	82	210	44	80	674
5,000–6,999	289	223	113	313	140	77	1155
7,000–9,999	554	254	375	787	152	186	2308
10,000–14,999	631	314	372	837	107	125	2386
15,000–24,999	317	96	157	323	33	95	1021
25,000+	78	34	22	105	4	19	262
Total	2278	1116	1214	2801	511	643	8563

[a] Town-wide data.

TABLE 8

Place of Employment versus Population of Chautauqua Lake Watershed[a]

Place of employment	Place of residence					
	Busti	Chautauqua	Ellery	Ellicott	Harmony	North Harmony
In watershed						
Busti	20.0%	—	6.7%	3.4.%	4.8%	12.5%
Chautauqua	—	64.7%	1.7	—	—	3.6
Ellery	—	0.8	18.1	—	—	—
Ellicott	1.8	—	0.9	4.2	2.4	—
North Harmony	—	—	—	—	4.8	10.7
Harmony	—	—	—	—	9.5	—
Out of watershed						
Jamestown	60.0	8.4	49.5	59.6	50.0	57.1
Falconer	5.5	1.7	76	26.0	9.5	8.9
Westfield	6.4	13.4	—	—	—	—
Panama	—	—	—	—	11.9	—
Other	6.5	9.8	10.1	7.0	4.8	7.2

[a] Whole town data, not just watershed population. 1970 Shopping and Employment Survey: Chautauqua County Department of Planning.

TABLE 9

1970 Census Count of Employed Population 16 Years Old and Over by Occupation in the Chautauqua Lake Watershed[a]

Profession	Busti	Chautauqua	Ellery	Ellicott	Harmony	North Harmony	Total
Professional, technical, managers, and administrators	708	415	401	814	120	152	2,610
Sales, clerical, and kindred workers	831	312	378	1,072	186	131	2,910
Craftsmen, foremen, and operatives, except transport	1,112	524	472	1,397	281	306	4,092
Transport equipment operatives	129	99	67	133	43	57	528
Laborers, except farm	90	96	57	102	22	61	428
Farmers, farm managers, farm laborers, and farm foremen	51	115	61	52	44	41	364
Service workers including private household workers	382	228	289	509	65	107	1,580
Total	3,303	1,789	1,725	4,079	761	855	12,512

[a] Whole town data, not just watershed population.

TABLE 10

1970 Census: Estimated Year-Round Population by Age in the Chautauqua Lake Watershed[a]

	Municipality							
Age group	Busti	Chautauqua	Ellery	Ellicott	Harmony	North Harmony	Total	Percent total
Under 5	415	326	351	339	53	174	1,658	7.2
5–14	1,180	891	1,024	882	162	506	4,645	20.2
15–24	920	630	700	711	79	321	3,361	14.6
25–34	640	477	498	496	85	252	2,448	10.6
35–44	799	469	581	600	71	223	2,743	11.9
45–54	810	495	572	735	58	315	2,985	13.0
55–64	666	479	456	626	43	250	2,520	10.9
65 and over	765	574	412	667	45	223	2,686	11.6
Total	6,195	4,341	4,594	5,056	596	2,264	23,046	100.0

[a] Enumeration District data were used.

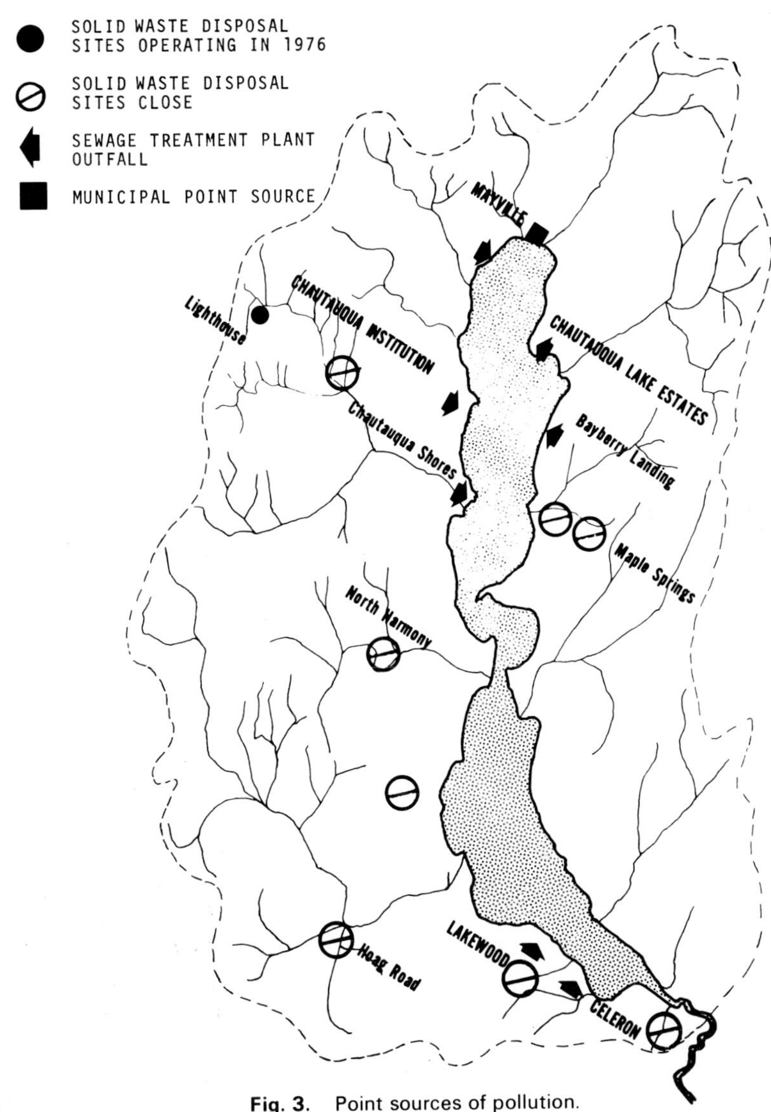

Fig. 3. Point sources of pollution.

pollution to the lake. All of Jamestown's municipal and industrial effluents are discharged into the Chadakoin River or Cassadaga Creek downstream from the lake.

The sewage treatment plants for the village of Mayville and the Chautauqua Institution outfall into the upper basin of the lake (see Table 11). The small population centers at Chautauqua Shores and Bayberry Landing each have primary sewage treatment plants and Chautauqua Lake

TABLE 11

Point Sources of Pollutants

Source	Receiving water	Treatment	Daily flow (m^3)	Annual P (kg)
Chautauqua Institute STP[a]	Upper lake	Primary	1385	1973.2[b]
Mayville STP[a]	Mud Creek (upper lake)	Primary	635	1018.3[b]
Chautauqua Lake Estates	Upper lake	Tertiary	33	107.0[b]
Chautauqua Shores	Upper lake	Primary	121	36.2[b]
Bayberry Landing	Upper lake	Primary	3	8.3[b]
Chautauqua Malted Plant	Little Inlet (upper lake)	Primary	254	5014.3
Lakewood STP[a]	Lower lake	Primary	3175	3444.4[b]
Celoron STP[a]	Chadakoin River	Primary	197[c]	250.2[c]

[a] STP, sewage treatment plant.

[b] Annual phosphorus based on an average per capita yield of 0.834 kg P/year (see discussion in text).

[c] Daily flow and annual phosphorus represents 5% of the total outflow from this plant which refluxes back into the lake (see discussion in text).

Estates operates a tertiary wastewater treatment plant; each of these installations also contributes effluent to the upper basin.

The lower lake basin receives wastewater directly from the primary treatment plant in Lakewood. A similar plant located in Celoron discharges its wastes into the Chadakoin River near the southern end of the lake. Because of wave and current action, an estimated 5% of this effluent is refluxed back into the lower lake while 95% flows directly down the river and out of the basin. There are no small population centers operating private waste water treatment plants in the lower basin.

There is presently only one solid waste disposal site operating in the watershed. It is located in the Lighthouse Creek subbasin and is being operated by Chautauqua County as a sanitary landfill. There are numerous abandoned waste disposal sites which were also indicated in Fig. 3. All of these areas were operating as recently as 1970, but each has been closed as part of a comprehensive county-wide solid waste management system. These abandoned landfills are possible point sources of pollution, as has been demonstrated by Barnard *et al.* (1972) who found that the sites on Lighthouse and Maple Springs Creeks contribute high concentrations of calcium, magnesium, potassium, and sodium to their respective streams.

The only significant industrial point sources of pollution in the basin are located in Mayville, New York, where they measurably effect the water quality of Little Inlet Creek. Little Inlet has been shown by Barnard *et al.* (1972, 1973) and Hopke *et al.* (1972) to have extremely high phosphate and higher than normal sodium and chloride concentrations at the mouth of the creek. With the exception of the high phosphate values, this pollution most likely comes from a wide variety of sources, including an abandoned municipal dump and a stockpile of highway salt located in the subbasin. The major source of phosphate is the wastewater treatment plant serving the Chautauqua Malted Milk Plant. the outfall from this plant contained an average of 54 mg/liter phosphorus and contributed approximately 40% of the total phosphorus load for the entire upper basin. It is of interest that, because of a corporate reorganization, this plant has been closed since January, 1976.

HYDROLOGY

Climatology

There are no official long-term weather observation stations within the Chautauqua Lake Watershed. General observations indicate that the two nearest stations do not reflect but only approximate conditions within the watershed. From the Westfield weather observation station to the northerly edge of the Chautauqua Lake Watershed is a horizontal distance of 5.47 km. The Jamestown weather station has been located at two places: once 3.06 km outside of the watershed, it is now 9.50 km from the watershed.

Because of topographic conditions from the northern edge of the watershed to the Bemus Point–Long Point area of the lake, it is locally observed that there is a distinct weather difference. The upper portion of the watershed is in a "snow belt" or lake effect (Lake Erie) area which has different conditions than those found from Bemus Point down lake to the Jamestown weather station. Conversely, the Lake Erie moderating effect which is reflected in the information of the Westfield weather station stops at the northern edge of the watershed. Unfortunately, the only careful recording of climate over time is from the two weather stations noted. There is no long-term record of hours of sunshine, cloud cover, or ice cover on Chautauqua Lake.

Precipitation

Mean precipitation recorded at the Westfield weather station (see Table 12A) is 40.57 in.; that recorded in Jamestown is 42.35 in. (see Table 12B). Westfield's records from 1941 through 1972 show the lowest monthly

TABLE 12A

Total Precipitation for Westfield (1941–1972)[a,b]

Year	Month												Annual
	Jan.	Feb.	Mar.	Apr.	May	Jun.	Jul.	Aug.	Sep.	Oct.	Nov.	Dec.	
1941	—	—	—	—	—	—	3.60	2.25	2.44	3.26	2.68	2.07	—
1942	0.76	2.31	5.10	2.79	5.04	2.36	6.85	8.47	6.01	3.78	5.33	3.97	52.77
1943	1.11	1.51	2.03	4.26	4.84	2.08	4.13	5.27	5.32	4.86	2.47	1.96	39.84
1944	1.36	2.23	2.81	4.56	3.60	5.74	1.42	2.64	6.36	3.37	3.14	2.92	40.15
1945	1.54	2.68	4.82	3.41	2.43	5.00	3.41	3.34	8.62	7.17	3.64	2.11	48.17
1946	1.29	2.82	2.37	1.84	6.30	3.87	4.67	2.12	1.90	4.54	3.80	3.20	38.72
1947	4.35	0.83	1.29	7.21	6.54	4.92	5.62	3.97	5.41	0.35	3.21	2.51	46.21
1948	2.02	2.57	5.04	4.57	3.35	5.37	3.14	2.55	1.06	5.98	3.64	2.04	41.33
1949	2.86	1.60	3.30	2.98	3.28	1.48	4.23	4.66	3.90	1.89	4.02	2.57	36.77
1950	4.84	5.17	3.47	3.75	2.72	2.72	5.98	2.91	3.54	4.07	8.09	1.72	48.98
1951	2.37	3.26	4.11	4.52	3.20	4.31	2.63	0.52	3.03	2.66	3.74	2.59	36.94
1952	3.45	2.28	2.18	2.09	5.11	1.54	1.69	4.76	5.38	1.10	2.31	2.45	34.34
1953	2.79	0.87	4.22	2.51	4.72	2.12	2.91	3.54	5.09	0.87	2.93	3.55	36.12
1954	2.46	1.81	4.42	6.84	1.59	4.52	1.98	1.95	2.57	9.28	2.84	4.37	44.63
1955	1.76	2.08	2.57	3.41	2.26	1.60	2.09	6.33	2.54	6.70	3.92	1.82	37.08
1956	1.65	3.28	3.47	4.06	4.79	1.03	5.94	7.89	3.70	1.57	4.10	2.66	44.14
1957	2.93	1.45	1.57	5.15[c]	4.14	5.62	1.49	3.02	5.90	1.90	2.42	2.29	37.88
1958	2.85	1.38	1.35	3.42	3.51	6.01	5.10	2.47	6.05	2.12	3.62	1.62	39.50

1959	4.74	2.82	3.56	3.04	2.40	3.79	3.09	2.27	3.66	7.38	3.86	3.71	44.32
1960	3.14	3.04[c]	0.71	1.22	3.84	2.94	1.33	0.97	1.39	2.21	1.91	1.39	24.09
1961	0.71	4.11	2.15	8.63	3.79	4.19	2.92	4.62	2.68	4.15	2.54	2.12	42.61
1962	2.64	1.77	1.36	2.19	2.28	3.23	2.83	3.63	6.01	4.13	2.52	2.61	35.26
1963	2.01	1.05	2.34	3.65	2.18	1.83	2.89	3.01	2.43	0.89	7.28	2.36	31.92
1964	2.18	2.24	3.99	5.18	3.34	2.36	3.04	4.28	3.41	3.04	3.43	4.11	40.60
1965	4.58	2.79	4.10	2.46	1.92	3.38	3.13	4.26	2.30	5.67	5.11	2.65	42.35
1966	2.46	1.77	4.31	5.30	3.07	3.48	2.69	4.30	6.48	2.30	7.78	4.19	48.13
1967	1.44	1.50	1.75	4.32	4.27	2.98	5.22	3.59	7.21	4.01	5.89	2.43	44.61
1968	2.55	1.15	2.53	3.32	4.21	3.33	1.75	3.38	4.15	3.71	4.79	4.30	39.17
1969	3.60	0.65	1.94	6.70	6.62	6.20	4.51	1.05	2.99	5.15	3.56	3.22	46.19
1970	1.84	2.49	1.49	3.08	5.19	2.24	6.52	5.84	11.78	5.79	5.77	3.02	55.25
1971	1.29	2.40	2.02	2.34	2.80	3.32	4.98	4.43	2.70	3.35	5.67	4.67	39.97
1972	2.18	2.94	4.39	3.22	4.25	9.35	3.27	2.69	8.04	3.98	4.60	3.94	52.83
Means[d]	2.49	2.27	2.99	4.03	3.80	3.52	3.48	3.60	4.17	3.74	3.81	2.67	40.57
D[e]	2.89	2.60	2.86	2.91	3.98	3.47	4.17	3.48	3.46	3.97	3.28	3.03	40.10
F[f]	24	22	25	25	25	23	24	24	23	24	24	24	—

[a] Source: Climatological Data, New York, Annual Summaries, U.S. Department of Commerce, Weather Bureau.

[b] All information shown in inches. To convert inches to centimeters, multiply by 2.54.

[c] Amount is wholly or partially estimated.

[d] Means for the period of 1941 through 1969.

[e] D: means for period prior to 1931.

[f] F: number of years used to obtain means in D.

TABLE 12B

Total Precipitation for Jamestown (1926–1972)[a,b]

Year	Month												Annual
	Jan.	Feb.	Mar.	Apr.	May	Jun.	Jul.	Aug.	Sep.	Oct.	Nov.	Dec.	
1926	1.31	2.60	2.45	3.50	0.89	4.79	3.11	4.17	4.92	5.87	4.39	4.41	42.41
1927	5.17	3.70	2.35	2.97	5.44	3.63	5.72	2.54	0.81	3.35	10.07	4.99	50.74
1928	3.10	2.98	2.83	3.80	1.34	11.00	4.60	2.36	2.13	5.21	4.79	1.64	45.78
1929	5.84	2.04	3.64	6.86	6.20	1.89	6.43	1.04	1.65	4.48	4.71	3.12	47.90
1930	4.13	2.86	3.90	3.52	3.68	2.95	1.87	2.03	3.74	1.68	3.77	2.97	37.10
1931	1.97	2.06	2.67	3.96	4.10	4.63	3.53	2.22	3.45	3.16	3.81	3.84	39.40
1932	6.25	3.00	5.26	3.12	3.85	3.37	5.00	3.68	2.47	4.43	4.30	3.93	48.66
1933	1.52	2.73	4.48	2.18	4.76	7.10	3.44	2.82	4.22	2.32	6.89	4.82	47.28
1934	4.30	2.05	3.71	2.64	1.43	2.15	0.87	3.11	3.91	2.54	2.97	4.57	34.25
1935	4.11	3.40	1.61	1.72	2.20	6.36	7.55	2.37	5.15	2.64	2.20	6.72	46.03
1936	3.38	3.43	5.54	3.17	2.18	2.23	1.63	3.34	1.58	4.15	6.03	4.42	41.08
1937	7.08	3.14	3.10	7.07	2.51	6.33	4.97	4.01	2.25	4.96	3.90	5.06	54.38
1938	2.75	4.61	3.04	4.94	2.63	2.53	8.50	4.34	9.20	1.40	3.81	3.44	51.19
1939	4.11	5.26	4.65	3.25	1.57	2.79	4.89	2.44	5.20	3.68	1.65	5.38	44.87
1940	5.14	3.89	3.18	4.45	4.10	6.57	1.89	5.38	5.49	1.66	3.84	4.00	49.59
1941	3.00	1.70	1.93	1.72	1.69	3.96	2.82	2.26	2.54	5.07	1.32	2.17	30.18
1942	1.58	3.28	5.71	1.94	6.00	3.70	5.83	2.05	4.32	4.71	4.27	5.17	48.56
1943	3.41	3.43	2.84	5.88	6.67	3.26	3.12	5.04	1.17	4.35	3.52	2.23	44.92
1944	1.87	3.72	3.79	4.85	3.94	4.65	2.84	2.14	6.03	1.79	3.03	6.36	45.01
1945	3.24	3.86	5.10	3.69	2.70	4.26	2.61	3.36	6.30	5.40	3.84	3.57	47.93
1946	1.74	2.94	2.45	1.61	5.04	4.70	3.10	1.30	0.80	3.46	3.57	3.43	34.14
1947	4.84	1.73	3.87	8.07	6.69	4.83	5.42	1.49	3.23	0.63	3.45	1.94	46.19
1948	2.39	2.46	5.55	4.14	3.66	5.17	3.60	2.47	0.93	4.40	2.74	2.63	40.14
1949	3.47	1.94	2.80	2.83	4.25	1.57	4.64	3.78	2.89	1.32	3.39	2.36	35.24

1950	6.39	4.18	4.25	4.02	2.23	2.37	5.47	2.91	5.40	2.90	7.67	2.63	51.07
1951	2.77	3.18	4.08	2.43	2.99	4.35	4.55	1.49	4.03	1.69	4.91	4.65	41.12
1952	3.80	2.82	2.30	1.88	3.62	1.38	3.08	3.37	2.50	2.05	2.78	2.53	32.11
1953	3.19	1.35	3.55	2.09	6.83	1.31	3.41	1.93	3.87	0.84	2.00	2.84	33.21
1954	2.06	2.00	4.40	6.01	2.14	2.68	1.66	1.76	1.90	7.79	2.10	4.25	38.75
1955	1.90	2.02	4.06	3.30	2.59	2.81	1.65	5.93	2.09	6.96	4.66	2.62	40.59
1956	1.57	3.89	5.49	4.10	4.35	4.44	7.09	5.00	2.81	1.59	2.99	3.20	46.52
1957	3.41	1.61	1.66	5.65	1.85	6.49	2.06	1.23	3.14	1.27	3.25	3.76	35.18
1958	3.28	1.51	1.58	4.40	3.02	6.02	5.86	3.20	5.93	2.26	2.91	2.37	42.34
1959	5.71	3.12	3.31	3.99	1.91	4.55	3.82	2.35	0.42	6.54	4.90	3.38	44.00
1960	2.93	4.42	1.18	1.61	4.86	3.89	2.24	1.91	NA[c]	NA[c]	NA[c]	NA[c]	NA[c]
1961	0.93	4.49	1.94	7.39	4.06	3.83	2.16	4.10	1.27	3.28	3.31	3.39[d]	40.15
1962	3.16	2.62	1.27	3.79	2.63	5.06	2.34	1.15	4.08	3.72	2.29	3.88	—
1963	1.13	1.32	2.85	4.04	2.06	1.70	3.53	3.01	1.97	0.33	6.69	3.91	32.54
1964	2.05	1.38	4.48	4.07	2.59	3.43	5.64	4.32	2.04	2.15	1.74	4.01	37.90
1965	4.66	2.72	3.84	2.40	3.13	3.62	1.95	3.86	2.74	3.52	4.51	3.04[d]	39.99
1966	3.71	1.66	4.12	4.06	3.55	3.63	1.55	3.54	3.95	2.07	4.55	4.15	40.54
1967	1.10	2.08	1.90	4.55	3.44	5.12	7.77	8.83	5.18	3.01	5.31	2.16	50.45
1968	2.61	1.17	1.84	1.80	4.49	3.78	2.70	4.60	3.56	5.77	4.71	5.89	42.92
1969	2.93	1.87	2.07	6.02	4.22	5.57	5.26	2.03	2.20	4.21	4.55	4.81	45.74
1970	1.74	2.35	1.60	3.07	3.61	3.56	5.24	4.18	6.44	4.69[d]	7.33	3.85	47.67[d]
1971	2.50	2.37	2.30	1.26	2.17	2.17	2.57	3.46	2.35	2.68	3.99	4.00	31.62
1972	2.53	3.51[d]	3.80	2.54	4.05	8.63	2.99	3.62	4.22	4.34	4.79	4.61	49.63
Mean[e]	3.29	2.79	3.33	3.85	3.48	4.10	3.91	3.09	3.33	3.36	4.00	3.82	42.35

[a] Source: Climatological Data, New York, Annual Summaries, U.S. Department of Commerce Weather Bureau.
[b] All information shown in inches. To convert inches to centimeters, multiply by 2.54.
[c] NA, information not available.
[d] Amount is wholly or partially estimated.
[e] Mean for 1926–1969.

precipitation of 0.71 in. in January of 1961 with a maximum of 11.78 in. in September of 1970. Mean monthly averages range from 2.27 to 4.17 in., with 1970 being the wettest year of record (55.25 in.) and 1963 the driest year (31.92 in.) as observed at the Westfield weather station. Jamestown's records from 1926 through 1972 show the lowest monthly precipitation at 0.80 in. in September of 1946 with a maximum of 11.00 in. in June of 1928. Mean monthly averages range from 2.79 to 4.10 in., with 1937 being the wettest year (54.38 in.) of record and 1971 the driest year (31.62 in.) as observed at the Jamestown weather station.

Temperature

Temperature extremes and freeze data are available from 1953 through 1972 for the two weather stations. At Westfield (see Table 13A), the longest growing season (days with minimum temperatures above 0°C) was 212 days in 1955, while the average growing season was 185 days. The highest temperature in Westfield was 35°C, recorded on September 3, 1953; the lowest temperature was −21.7°C on February 2, 1961 and on February 4, 1970. At Jamestown (see Table 13B), the longest growing season was 200 days in 1955, while the average growing season was 144 days. The highest temperature at Jamestown was 36.7°C on September 2, 1953; the lowest temperature was −30.6°C on February 2, 1961.

Stream Flow

Since the summer of 1972, an extensive study of the streams tributary to Chautauqua Lake has been made (Metzger *et al.*, 1972, 1973; Bogden *et al.*, 1974). Eleven streams were selected for monitoring and staff gauges were installed on each. Through the cooperation of the Water Resources Division of the United States Geological Survey in Ithaca, New York, a continuous recorder was installed on Ball Creek at one observation site. During the summer months of the ensuing years, each stream was visited at least five days each week and the staff gauge readings taken. Arrangements were made with volunteers so that staff gauges at Dewittville and Prendergast Creeks were read daily throughout the year. Rating curves were prepared for each stream based on at least ten discharge measurements made over as wide a range of flow conditions as possible. These rating curves include observations through the range of normal variations in discharge but do not include the highest flows associated with floods.

The stream flow data reported in Table 14 were determined in the following manner. A hydrograph of each stream was constructed from data taken from staff gauge readings. Where data were sparse, estimations of flow were made by extrapolation from the records of the nearest stream where records

TABLE 13A

Temperature Extremes and Freeze Data for Westfield (1953–1972)[a]

					Last spring minimum of				First fall minimum of				Number of days between dates	
Year	Highest temperature (°F)	Date	Lowest temperature (°F)	Date	16°F or below temperature (°F)	Date	32°F or below temperature (°F)	Date	32°F or below temperature (°F)	Date	16°F or below temperature (°F)	Date[b]	16°F or below	32°F or below
1953	95	9–3	9	2–2	24	3–10	30	4–21	28	11–5	24	11–6	241	198
1954	90	7–14	1	1–11	16	4–3	32	5–4	32	10–31	15	12–20	261	180
1955	91	7–5	0	2–4	11	3–8	28	4–8	32	11–6	10	11–28	265	212
1956	91	7–1	9	3–25	9	3–25	32	5–24	24	11–10	16	12–29	279	170
1957	90	6–18	−6	1–17	13	3–4	32	5–3	32	10–26	NA[c]	NA	NA	176
1958	86	7–5	−2	2–17	15	2–21	30	4–26	30	11–23	13	11–29	281	211
1959	94	9–9	−2	2–2	15	3–23	30	4–22	29	10–21	16	11–17	239	182
1960	90	9–7	−2	12–24	10	3–26	29	4–19	32	10–24	15	12–8	257	188
1961	87	9–13	−7	2–2	16	3–18	32	5–2	30	11–9	13	12–14	271	191
1962	91	7–8	−4	2–11	14	3–5	31	4–20	31	10–25	15	12–10	280	188
1963	91	6–28	15	1–24	14	3–11	32	5–2	31	11–25	12	12–24	278	207
1964	92	7–28	5	2–22	16	4–1	30	4–11	32	10–7	16	12–18	261	179
1965	89	8–17	−5	1–17	13	4–4	32	4–24	29	10–28	7	10–20	260	187
1966	92	7–3	−5	1–26	12	3–9	31	5–11	31	10–18	13	12–2	268	160
1967	90	6–16	−6	2–13	0	5–13	31	5–13	30	10–29	15	12–26	282	169
1968	92	7–18	−6	1–12	14	3–14	29	5–17	27	11–10	14	12–9	270	187
1969	91	6–27	0	2–15	15	3–12	31	5–1	32	10–22	16	12–2	265	174
1970	87	9–24	−7	2–4	14	3–16	28	5–7	31	11–8	16	11–24	253	185
1971	92	6–28	−3	2–2	14	3–31	31	4–27	24	11–7	16	12–2	237	194
1972	90	7–22	−2	1–16	14	4–8	29	4–28	32	10–10	16	12–7	243	165

[a] Source: Climatological Data, New York, Annual Summaries, U.S. Department of Commerce, Weather Bureau.
[b] 1953 figures based on 24°F rather than 16°F.
[c] NA, information not available.

TABLE 13B

Temperature Extremes and Freeze Data for Jamestown (1953–1972)[a]

					Last spring minimum of				First fall minimum of				Number of days between dates	
Year	Highest temperature (°F)	Date	Lowest temperature (°F)	Date	16°F or below temperature (°F)	Date	32°F or below temperature (°F)	Date	32°F or below temperature (°F)	Date	16°F or below temperature (°F)	Date	16°F or below	32°F or below
1953	98	9–2	7	2–2	24	4–19	30	4–29	27	10–8	19	11–6	201	162
1954	93	7–14	3	2–13	14	4–4	32	5–5	30	10–7	11	12–6	246	155
1955	95	7–27	−8	2–4	16	3–27	24	4–8	32	10–25	15	11–28	246	200
1956	92	6–14	3	1–26	7	3–25	28	5–25	30	10–11	15	11–23	243	139
1957	94	6–17	−11	1–15	14	3–6	31	5–17	27	9–27	8	12–5	274	133
1958	90	7–2	−4	2–17	16	4–9	29	5–10	29	10–2	16	11–29	234	145
1959	93	9–9	−4	2–20	13	3–29	28	5–16	29	9–17	11	11–18	234	124
1960	NA[c]	NA	NA	NA	10	3–26	30	5–2	NA	NA	NA	NA	NA	NA
S.P[b]	NA	NA	−15	12–28	NA	NA	NA	NA	29	10–1	16	11–8	NA	NA
1961	88	9–3	−23	2–2	11	3–18	32	5–30	32	9–18	16	11–11	238	111
1962	90	9–1	NA	NA	15	4–3	28	5–10	32	10–21	NA	none	NA	164
1963	90	7–26	−20	2–17	12	3–15	26	5–24	30	9–23	12	12–5	265	122
1964	91	7–29	−5	2–12	16	4–4	32	5–30	31	9–13	13	11–22	232	106
1965	89	8–17	−17	1–17	14	4–4	32	4–29	28	10–5	16	10–29	208	159
1966	NA	NA	−12	1–29	13	3–28	31	5–12	32	9–16	13	11–20	237	127
1967	90	6–15	−10	3–19	−9	3–20	31	5–24	31	10–13	13	11–15	240	142
1968	90	6–9	−13	1–12	12	3–25	27	5–8	32	10–22	11	12–9	259	167
1969	91	6–28	−6	2–16	11	4–1	30	5–27	30	10–16	12	11–22	235	142
1970	87	8–2	−8	2–14	16	3–30	26	5–8	NA	NA	14	11–24	239	NA
1971	92	6–29	−12	1–19	16	4–1	31	5–15	30	11–14	12	11–9	222	173
1972	87	7–20	−14	2–9	13	4–10	28	6–2	25	10–10	16	11–30	234	120

[a] Source: Climatological Data, New York, Annual Summaries, U.S. Department of Commerce, Weather Bureau.
[b] SP, sewage plant.
[c] NA, information not available.

TABLE 14

Annual Water Balance Data for the Chautauqua Lake Watershed (July 1973–June 1974)

	Area (hectares)[a]	Rainfall ($m^3 \times 10^6$)	Total discharge ($m^3 \times 10^6$)	Precipitation yield[b]	Runoff (%)[c]
Chautauqua Lake	5,324	55.35	55.35[d]	100	22.0
Big Inlet Basin	2,699	28.13	5.60	20	2.2
Little Inlet Basin	515	5.37	2.80[e]	52	1.1
Mud Creek Basin	1,406	14.66	10.27	70	4.1
Lighthouse Basin	1,067	11.12	3.49	31	1.4
Prendergast Basin	6,055	63.13	31.23	49	12.5
Ball Basin	2,484	25.89	15.77	61	6.3
Goose Basin	7,612	79.35	50.06	63	20.1
Dutch Hollow Basin	1,621	16.90	3.75	22	1.5
Bemus Basin	3,131	32.65	7.47	23	3.0
Maple Springs Basin	1,197	12.47	4.99	40	2.0
Dewittville Basin	3,686	38.42	14.06	37	5.6
Other basins	9,969	103.92	41.57[f]	40	16.7
Sewage plants	—	—	3.05[g]	—	1.2
Total drainage basin	46,766	487.4	249.5	51	

[a] Based on data from Chautauqua Lake Studies, 1972, pp. 28–29.
[b] Percentage of rainfall on the basin that was accounted for in the stream flow.
[c] Percentage of total discharge for entire basin that was measured as stream flow in the basin.
[d] Estimate only.
[e] Corrected to include the entire basin.
[f] Estimated as 40% of the total precipitation.
[g] Calculated from daily averages supplied by the sewage treatment plants.

were complete or from Ball Creek where a continuous record was available. Daily flow estimates were made by visual averaging on the hydrographs. Estimates of peak flows during floods were always made on the conservative side as the rating curves were not verified at high flow conditions. The average daily flow in cubic meters was recorded and totaled for each month. The discharge for each stream listed in Table 14 is the sum of the monthly flows taken during the 12 months from July 1973 through June 1974.

Significant additional inflows of water derive from the sewage treatment plants at Mayville, Celoron, Lakewood, and the Chautauqua Institute. Daily flow data were taken for these installations and are recorded in Table 14.

In order to determine the total input of water to the lake watershed system, an estimation of total precipitation for the basin was made. Rainfall data for Jamestown, New York, reported by the National Oceanic and Atmospheric Administration (NOAA, 1973a, 1974) were assumed to be representative for the entire basin. During the 12-month study period, 104 cm of the precipitation were recorded. This compares well with the average of 103.4 cm which has been measured from 1941–1970 (NOAA, 1973b). The total precipitation is estimated to be 487.4×10^6 m^3. This water was divided proportionally among the subbasins on the basis of area as shown in Table 14. The total input of water to the lake by surface streams and sewage treatment plants during the study year was 249.5×10^6 m^3. Thus 51% of the total rainfall came off the watershed as direct or delayed runoff. The individual subbasins vary greatly in percentage yield of precipitation runoff ranging from 20% for Big Inlet Creek to 70% for Mud Creek. The factors which influence these differences are not easily evaluated. The nature and distribution of the soil types in the subbasins are remarkably similar. The relative steepness of the slopes is also similar. It is possible that the apparent variations may simply reflect the fact that precipitation totals are calculations based upon the average precipitation falling equally on every point in the basin. Those basins with apparent low yields of runoff may not have received the average amount of rainfall; those with high yields may have had more than average precipitation. Since a uniform precipitation pattern over an area as large as the Chautauqua basin is highly unlikely, this may provide the most reasonable explanation of the observed variations in percentage of runoff yield.

Annual inflow of water to the lake from the tributaries, sewage treatment plants (STP's), and ground water sources should essentially equal the outflow, assuming no net gain or loss of water in the lake. During the study year the inflow from the streams and STP's totaled 249.5×10^6 m^3 while the Chadakoin River outflow totaled 297.0×10^6 m^3. (This is somewhat lower than the 38 year average reported by the United States Geological

Survey: 301.9 $\times$ 10^6 m^3.) This means that during the study year (July 1973–June 1974) there was a net loss of water equal to 47.5 $\times$ 10^6 m^3. This difference can be accounted for primarily in two ways. First, estimates of stream discharge during floods were intentionally conservative. It is possible that a significant portion of this unaccounted for water may enter the lake during flood events. The second source of water most certainly is groundwater flow. According to Crain (1966), there is little evidence of a hydrologic connection between the lake and the recent and glacial sediments in the Chautauqua valley. If this conclusion is correct, then additions of ground water to the lake should be small. However, if a significant portion of the 47.5 $\times$ 10^6 m^3 shortfall does not occur during floods, then Crain's conclusion may require some modification.

The net balance of water in the basin as a whole may be set forth in the following way. The total rainfall on the basin during the study year was equivalent to 487.5 $\times$ 10^6 m^3. The Chadakoin River discharged only 297.0 $\times$ 10^6 m^3, indicating that there were 190.4 $\times$ 10^6 m^3 unaccounted for in surface runoff. There are two primary processes which affect this difference. The first of these is groundwater flow. Crain (1966) emphasized that there is no continuous subsurface aquifer in the Chautauqua valley as is known to be present in adjacent valleys. However, there may be some transmission of groundwater from the areas around Jamestown or Lakewood to adjacent areas which are out of the basin.

The second process is evaporation and is considered to be the most important in this case. The entire 190.4 $\times$ 10^6 m^3 shortfall can be accounted for by an average of 40.6 cm of evaporation-transpiration in the basin. There have been no studies of evaporation rates in the Chautauqua lake watershed. Climatological data published by NOAA include pan evaporation rates as determined at Lockport, New York, which can be taken as being fairly representative of this area. During the four years 1971–1974 pan evaporation measured at Lockport stations ranged from 45.7 cm to 74.8 cm. Considering that the actual evaporation/transpiration rate may be only one-half of this value, it is reasonable to assume that 40.6 cm of evaporation-transpiration did occur during the year.

In summary, of the 487.4 $\times$ 10^6 m^3 of precipitation falling on the Chautauqua Lake basin, 51% (249.5 $\times$ 10^6 m^3) is accounted for as surface runoff, 10% (47.5 $\times$ 10^6 m^3) as floods and groundwater inflow to the lake, and 39% (190.4 $\times$ 10^6 m^3) as water lost through evaporation and groundwater flow out of the basin.

Groundwater

The groundwater resources of the western portion of Chautauqua County have been studied by Crain (1966) as part of a program describing the groundwater resources of the Jamestown, New York, area. He reports that the water supplies of the Chautauqua valley area appear to be related to glacial sand and gravel deposits with limited lateral extent. There is no evidence of a continuous aquifer such as is found in the adjacent Cassadaga valley. Most of the thick valley fill deposits which exceed 123 m in the Hartfield area are very fine grained, have low permeability, and are not capable of producing significant amounts of water.

Tests on monitoring wells in the Mayville and Lakewood areas have been conducted and transmissibilities of the local aquifers were calculated to range between 200 and 700 m^3 per meter (Crain, 1966). Recharge is thought to occur through local surficial deposits in most cases. However, wells in the Lakewood area produce from a pumping water level 10 to 15 m below the level of Chautauqua lake and locally some recharge may be derived from the Lake.

There are presently no large water supplies developed in the deposits in the Chautauqua valley. Crain (1966) concludes that because (1) the productive deposits are local in extent, (2) most of the valley fill is too fine grained to be a major source of water, and (3) rapid lateral variations in size and type of material seem to characterize these deposits, no large-scale water supplies can be developed in the area. He further concludes that most water-bearing units which are presently productive do not freely interconnect with Chautauqua Lake and therefore limit their future potential.

PHYSICAL LIMNOLOGY

Lake Morphometry and Bathymetry

Chautauqua Lake consists of a northern basin 14.8 km (9.20 miles) long with a mean depth of 7.8 m (26 ft). The southern basin is 13.1 km (8.14 miles) long with a mean depth of 3.5 m (11 ft). The average width of the lake is 1.9 km (1.2 miles). While the surface area of the northern lake (2856 ha; 7071 acres) is only 15.7% greater than the southern lake (2468 ha; 6110 acres), the volume of the northern basin (2.23×10^8 m^3) accounts for 72% of the total lake volume. The southern basin volume is 0.87×10^8 m^3 (Chautauqua Lake Studies, 1975). These data and related lake morphometry and bathymetry appear in Table 15.

Some of this information differs significantly from that reported previously (New York State Conservation Department, 1938). For example, the volume of the northern lake basin is 20% less than that calculated in 1938. Similarly the volume of the southern basin appears to be 33% less. These differences reflect the availability now of a more accurate

TABLE 15

Chautauqua Lake Morphometry

	Northern basin	Southern basin
Length	14.8 km (9.20 miles)	13.1 km (8.14 miles)
Width		
Greatest	3.5 km (2.2 miles)	3.2 km (2.0 miles)
Mean	2.0 km (1.2 miles)	1.9 km (1.2 miles)
Depth		
Greatest	23 m (75 ft)	6.0 m (20 ft)
Mean	7.8 m (26 ft)	3.5 m (11 ft)
Surface area	2856 ha (7071 acres)	2468 ha (6110 acres)
Volume	2.23×10^8 m^3	0.87×10^8 m^3
Length of shore line	26.4 km	27.4 km
Hydraulic retention	526 days	105 days
Latitude	N 42° 10′	
Longitude	W 79° 24′	
Drainage basin area	467.5 km^2 (180.5 miles^2)	
Elevation above sea level	399 m (1308 ft)	

bathymetric map of Chautauqua Lake prepared during the summer of 1975 (Lake Erie Environmental Studies, State University College at Fredonia). Figure 4 represents this map in 1-m contour intervals.

Very significant differences between the northern and southern lake basins are obvious from the bathymetry. The southern basin is rather shallow, averaging 3.5 m in depth. It has been observed to have approximately 25–28 m of postglacial sediments deposited over a layer of outwash gravel seen in the cores at the Bemus Point-Stow proposed bridge crossing site (see Fig. 4). The surface of this basin is a smooth, flat, and relatively featureless plain.

The northern basin can be divided into two areas of quite different character. The northernmost area is shallow, with the bottom sloping gently to the southeast down the axis of the lake. Water depths range from less than 1 m on the northern shore to about 12 m off Point Chautauqua. The surface is relatively featureless. The southern two-thirds of the northern lake is considerably different from the other areas. It is characterized by much deeper water and a more variable topography. It is in this lake region that numerous kettles, locally called "deep holes," occur. The deepest of these is the 23 m hole near Point Chautauqua.

While Chautauqua Lake as a whole rarely stratifies, the typical deep hole environment becomes stratified during summer periods and exhibits a true

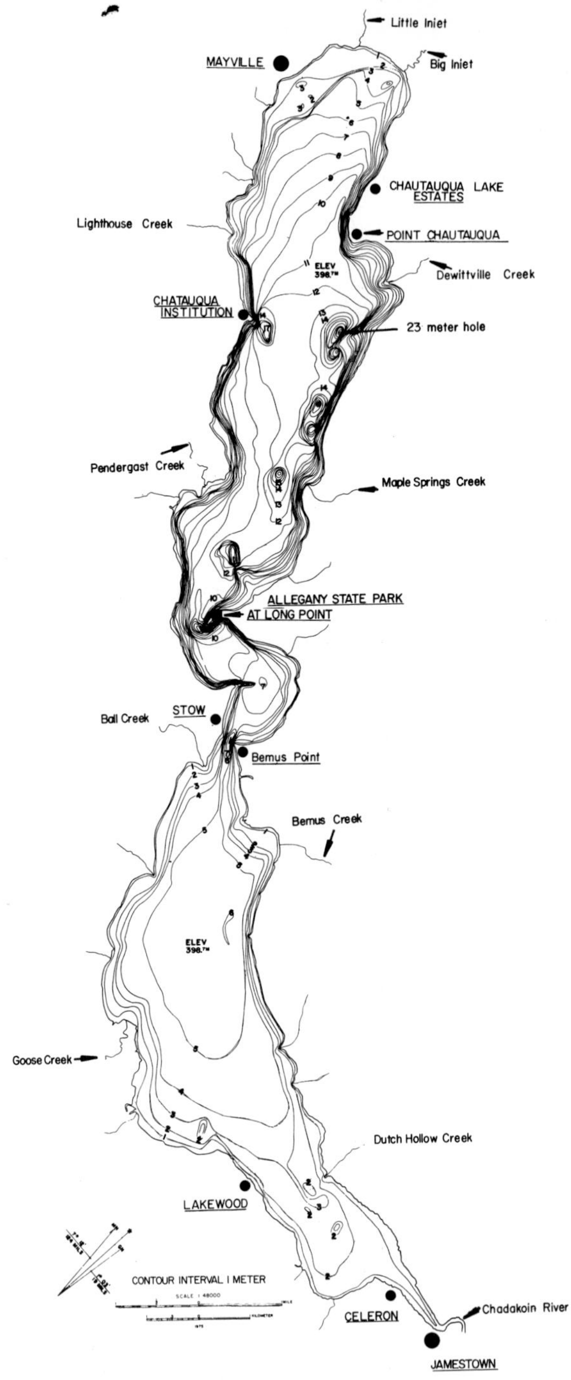

Fig. 4. Chautauqua Lake bathymetric map.

thermocline. Only the lower portion of these holes develop summer anoxic states; the rest of the lake retains relatively high levels of dissolved oxygen.

The differences among these three subdivisions cannot be attributed to differences in postglacial sedimentation alone. Each of the subareas is fed by a significant number of tributary streams which drain comparable watershed areas with similar soils and meteorological events. It is therefore important to recognize the differences in sedimentation which occurred during the deglaciation events in the basin.

The southern lake appears to have been extensively filled with outwash deposits at the time the moraine complex at Bemus Point-Stow was forming. The glacially deepened valley was nearly completely filled with these deposits. Recent sediments have added to these layers so that only about 3.5 m of water remain. The northern lake was not so extensively filled. The outwash deposits associated with an end moraine system north of the lake extended into the area now occupied by Chautauqua Lake. But the supply of sediments was not sufficient to completely fill the basin. The southern two-thirds of the northern lake is characterized by a thin layer of modern sediments which do not mask the glacial features formed in the area as the ice retreated to the north.

We have, therefore, concluded that the differences in the characteristics among the three subareas of the lake have been primarily inherited from the period immediately following the deglaciation of the area. These characteristics have been modified by postglacial sedimentation most significantly in the shallow portions of the lake while the deep southern portion of the northern lake seems to have been effected very little.

Recent Lake Sediments

Areal Distribution

The recent sediments of Chautauqua Lake have been extensively studied by Clute (1973) and Crowley *et al.* (1972). In order to determine the areal distribution, texture, sorting, and composition of the sediments, 92 grab samples were taken and analyzed. These samples have been categorized using the classification of Shepard (1954). The areal distribution of the six subclasses recognized in the lake are shown in Fig. 5. The majority of the sediments are, and the greatest area in the lake is covered by, sediments in the clayey silt range. This type of sediment dominates the central portion of all areas of the lake. The near shore areas and deltas are characterized by the coarsest sediments which are typically sand, silty sand, or sandy silt. Two rather large areas in the lake are predominantly sand; one of these is adjacent to the shore at the northern end of the lake at Mayville, the other

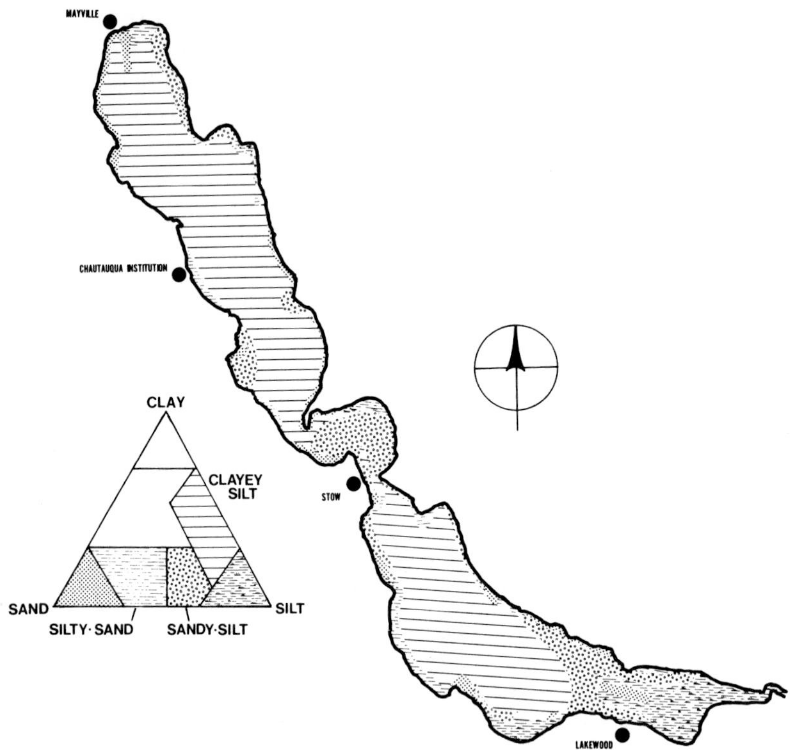

Fig. 5. Distribution of modern sediments in Chautauqua Lake.

is found near the center at the extreme southern part of the lake near Lakewood and Celoron.

Texture

Textural studies of each lake sample were made by Clute (1973) and Clute *et al.* (1974). A rather wide variation in sorting was found, ranging from well-sorted samples (phi range 0.35–0.50) to very poorly sorted samples (phi range 2.00–4.00). However, most of the sediments fall into the poorly sorted class (phi range 1.00–2.00) including most of the delta samples as well as those taken from the center of the lake basin.

Determination of the skewness of the grain size distribution reveals that most samples are positively skewed (0.1–0.3) indicating a greater average percentage of fine material in each sample. Negatively skewed samples were found in the areas of the deltas and some other near-shore areas indicating either a significant supply of coarse materials to these areas or an effective winnowing away of the fine grain sizes owing to current or wave action.

Mineral Composition

The minerological composition of the sediments in the lake varies with grain size as may be expected. The clay-sized fractions, analyzed by x-ray diffraction techniques, are composed predominantly of the clay mineral illite. Iron chlorite comprises an important but subordinate percentage of every sample. X-ray patterns typically indicate a small but persistent percentage of quartz in each clay sized sample. The coarser sized particles found in all other samples are almost exclusively quartz with minor amounts of feldspar, chert, jasper, flint, and some lithic fragments. A study of the heavy minerals suite from each area has not been undertaken.

Organic Carbon

The organic content of the sediments was determined by the methods of Jackson (1958; see Crowley *et al.*, 1972, Clute, 1973). The average value of the organic content in Chautauqua Lake sediments is 8.9% and ranged between 0.5% and 18.4%. There is an inverse relationship between increasing amounts of organic matter and decreasing grain size in the sediments. Clute (1973) suggests that there is a stronger correlation between the mean grain size of the sediments and organic carbon than between the percentage of clay mineral and organic carbon. The organic materials, therefore, do not necessarily have a strong affinity for the clay minerals. The distribution suggests more strongly that the effective diameter of the organic matter is in the proper size range to be concentrated with the generally finer sediments regardless of their mineralogical composition.

Iron–Manganese Nodules

Iron–manganese nodules have been observed at a single location in the upper lake near sample Site 3 (Fig. 6). Samples were recovered both with the Peterson dredge and by scuba divers (Clute, 1973; Clute *et al.*, 1974). The iron–manganese concentrations occur as discrete discoidal accumulations less than 1 cm in thickness and also as an irregular crust 1 to 3 cm thick. *In situ* observations revealed that the concentrations occur approximately 1 cm below the sediment surface. X-ray diffraction analysis demonstrates that the material is amorphous to x-rays. Neutron activation analysis of a single sample yielded data which suggest an iron content of approximately 58% (Clute, 1973). This value is considered to be very high when compared to other iron–manganese nodules from freshwater lakes.

Subbottom Sediments

Seismic Profiles

In the previous section, the character of the surface sediments in the lake was outlined. Two additional types of data have been collected and analyzed

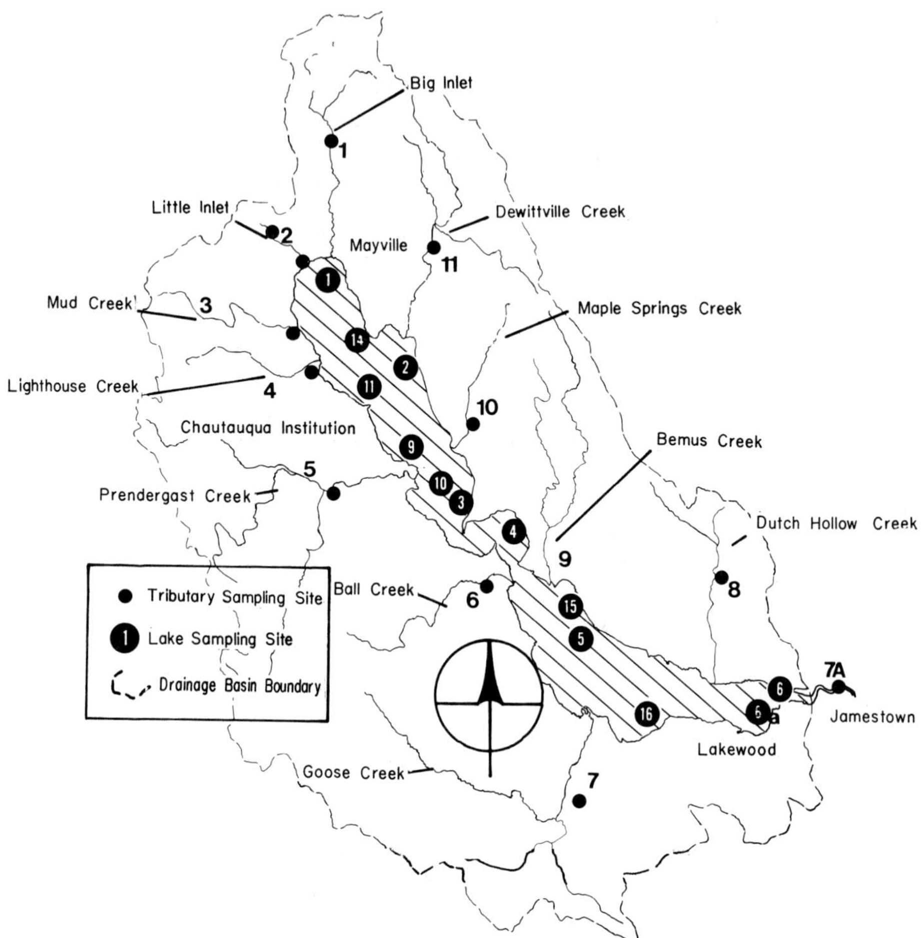

Fig. 6. Chautauqua Lake drainage basin.

to give a more complete description of the lake sediments. The first type of data comes from high resolution, broad frequency seismic profiling using an E. G. & G. Uniboom System (Metzger, 1974). The interpretation of profiles run during the summer of 1974 reveals that in most areas of the upper lake signal penetration was typically 15 to 20 m. In areas where thick, water-saturated clayey silt occurs at the surface there was effectively no penetration.

Of the many features which can be seen on these records, two are worthy of special mention. The first, illustrated in Fig. 7, shows a uniboom profile and interpretation of the 23-m depression of the upper lake. The profile clearly shows the truncated layers of glacial sediment along the steep north side of the feature which is essentially a 15-m cliff. A layer of recent lake sediment can be seen to drape over similar glacial sediments on the south side. The feature is interpreted to be a kettle hole which formed in the glacial sediments and which has remained unfilled with recent sediments. It is most probable that the numerous depressions in the upper lake shown on the bathymetic map (see Fig. 4) are kettle holes which are inherited from the deglaciation events of this area.

The second type of feature is illustrated in Fig. 8. The small hill located at the bottom of the depression which protrudes up through the surrounding recent lake sediments has been identified on several profiles and appears to be a discontinuous ridge which extends across the lake from Long Point toward Prendergast Point (see Fig. 4). This feature is a ridge of glacial sediments associated with the end moraine complex of the Bemus Point area. The depressions are most probably kettle holes.

Core Profile—N.Y. Department of Transportation

As part of the sediment sampling program, a limited number of cores were taken by State University College at Fredonia geologists during the summer of 1972 using a benthos corer from a boat and a hand-held piston corer operated by scuba divers. The maximum effective length of these cores was approximately 1 m and little new sedimentologic information was gained from them. However, the New York State Department of Transportation, while studying a potential lake bridge site, undertook an extensive deep coring program in the Stow–Bemus Point area. Lake sediments were cored to an average depth of 35 m and to a maximum depth of over

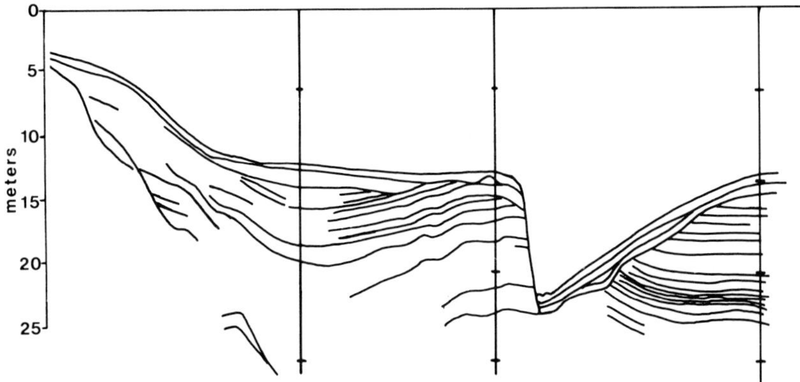

Fig. 7. Geological interpretation of the uniboom profile of the deepest depression in the northern basin. Note especially the truncation of the subhorizontal glacial deposits and steep clifflike eastern side of the feature. Bar scale on left in meters. Scale on right is subdivided into 24-ft intervals.

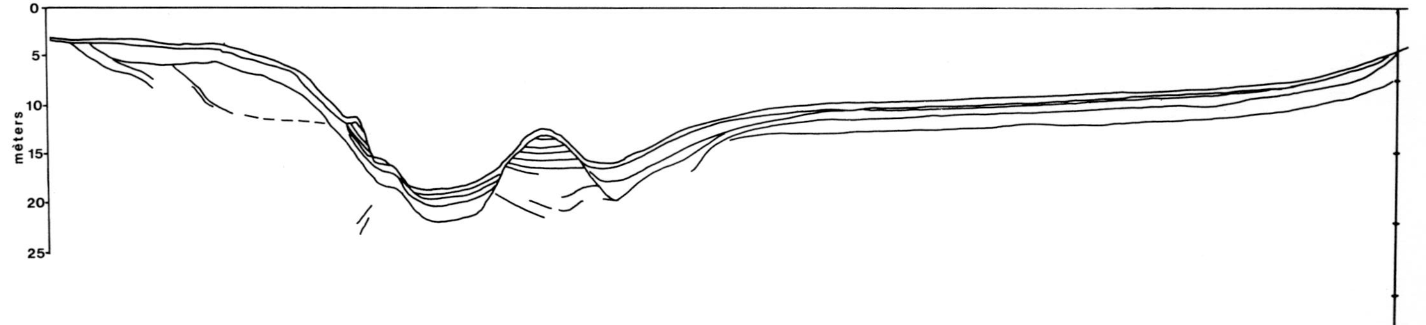

Fig. 8. Geological interpretation of a uniboom profile record taken between Long Point (left) and Prendergast Point (right). Record reveals a glacial moraine on the floor of the lake which is seen as the hump at the bottom of the depression. Bar scale on left in meters. Scale on right is subdivided into 24-ft intervals.

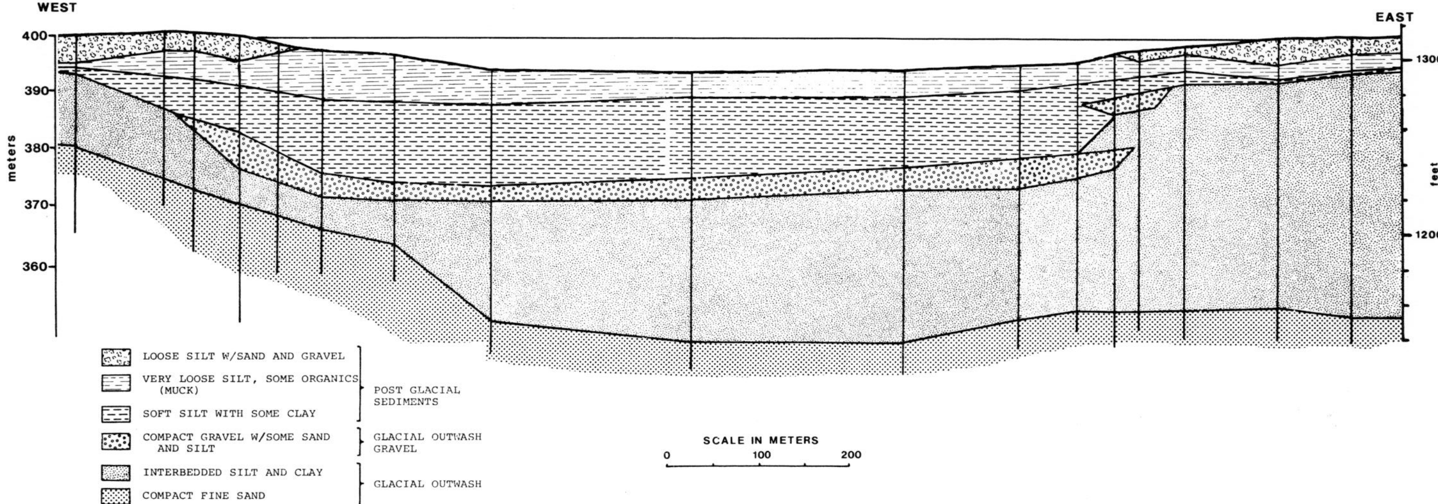

Fig. 9. Geologic cross section prepared from descriptions of cores at the proposed Southern Tier Expressway bridge site between Stow, New York (west), and Bemus Point, New York (east).

70 m. A geological interpretation of these sediments was made using a core split of one of the cores as a guide; the interpretation is illustrated in the cross section in Fig. 9.

Long-Term Sedimentation Rate

An attempt at determining the long-term rate of sedimentation has been made using DOT Core Data. The layer of material labeled glacial gravel in Fig. 8 has been interpreted to have formed during the glacial event associated with the formation of the end moraine complex now seen in the Long Point State Park area. The gravels are thought to represent an outwash deposit associated with that moraine. The sediments above the gravel are therefore considered to have accumulated since that glacial event. There are no radiocarbon dates of the deglacial events in Chautauqua

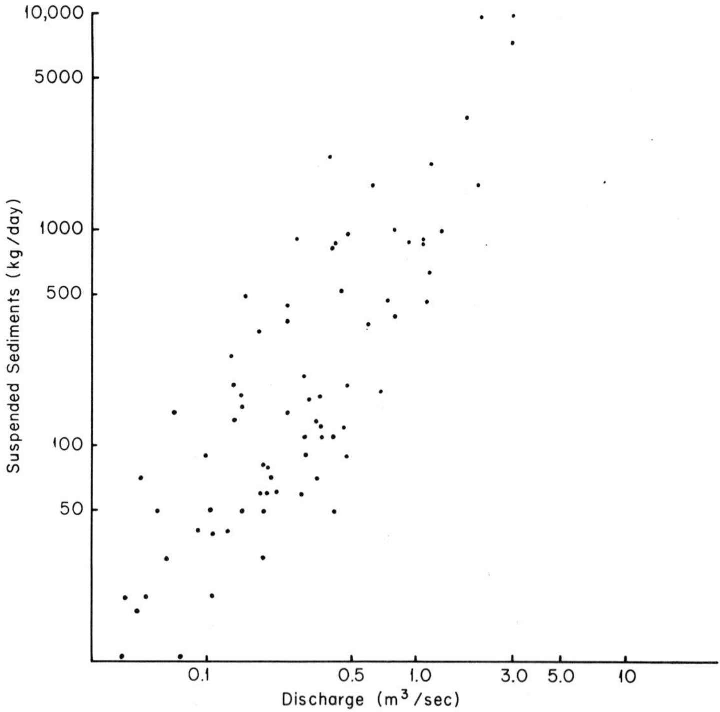

Fig. 10. Diagram in which the measured suspended sediment load is plotted against the discharge of the streams. The data are a composite of all the tributary streams in the basin and represents flows representative of an entire year.

County proper. However, by correlating the deposits in the Chautauqua Lake Basin to others where dates have been made, it was found that these deglacial events probably occurred between 10,000 and 12,000 years ago. An examination of the cross section (Fig. 9) reveals that the postglacial sediments range in thickness between 25 and 28 m. If a thickness of 28 m is taken and the span of time since deglaciation began is assumed to be 10,000 years, the annual rate of sedimentation is approximately 0.28 cm/year. Using the same average thickness and the longer period of 12,000 years, the sedimentation rate is 0.23 cm/year.

Modern Sedimentation Rates

Studies of suspended sediment loads have been made on each of the 11 main tributaries to the lake. Sediment yields are typical for streams with similar discharge and basin characteristics. The results of approximately 70 suspended sediment determinations are recorded in Fig. 10. The graph shows the variations in sediment yield with changing conditions of streamflow.

Several generalizations can be made from these determinations. Under normal baseflow conditions, the streams typically yield less than 100 kg/day. Higher flows produce sediment yields as high as 10,000 kg/day. Observations indicate that the same flow rates result in very different sediment yields depending on the degree of weathering and preparation of the surficial material. The highest sediment production comes from stream flows associated with thunderstorms during the late spring and early summer months.

CHEMICAL LIMNOLOGY

The Lake

Introduction

Chautauqua Lake is mesotrophic in its northern basin and eutrophic in its southern basin. This status is reflected in the lake's phytoplankton and macrophyte production: there is also good correlation between trophic state and nutrient loading from natural and cultural sources. Lake chemistry is largely determined by expected levels of bicarbonate, carbonate, and major and trace elements. It would appear, however, that the impact of municipal and industrial effluents, of watershed runoff, and of past aquatic plant control measures has changed lake chemistry to a degree. These changes have influenced the community structure, diversity, and related biological characteristics of lake algae and rooted aquatic plant communities.

Hydrogen Ion Concentration (pH), Alkalinity, and Dissolved Oxygen

While the lake pH may normally fall within a range of 6.5 to 8.5 (Ruttner, 1961), highly alkaline conditions often prevail during summer periods in lower Chautauqua Lake. Typical lower-lake surface pH maxima observed in recent years are: 10.6 (1971), 9.0 (1972), 10.2 (1973), and 9.4 (1974) (Chautauqua Lake Studies, 1971–1974). The maximum pH during 1937, 9.4, was found in the lower lake's surface water (New York State Conservation Department, 1938). These relatively high pH values appear to

TABLE 16

Comparison of Chautauqua Lake Temperature and Dissolved Oxygen Data for June

	June									
	1937		1971		1972		1973		1974	
Depth (m)	Temperature (°C)	Dissolved oxygen	Temperature (°C)	Dissolved oxygen	Temperature (°C)	Dissolved oxygen	Temperature (°C)	Dissolved oxygen	Temperature (°C)	Dissolved oxygen
Site I										
0	20.0	7.8	23.1	8.8	19.2	8.5	23.0	8.0	18.2	8.6
3	18.9	7.4	22.3	9.0	18.2	8.3	22.8	7.5	17.5	7.2
6	18.9	7.2	21.6	9.1	18.1	8.3	22.6	7.0	17.3	3.2
8	16.1	3.8	—	—	17.9	4.9	20.4	5.5	—	—
Site 2										
0	20.0	7.6	23.5	8.4	20.5	8.3	24.4	8.0	18.5	6.6
3	19.4	7.4	22.8	8.2	18.4	8.5	22.9	7.8	18.2	6.0
6	18.9	7.2	22.5	8.4	18.3	8.2	22.8	7.8	18.2	5.9
9	17.2	6.6	17.0	2.3	18.2	8.1	18.2	6.0	18.1	5.7
12	13.9	2.3	14.8	0.5	16.6	6.9	15.9	3.4	17.4	3.2
15	12.8	2.6	14.2	0.5	16.1	4.1	15.5	2.8	16.1	1.6
18	12.8	2.3	14.0	—	12.9	0.5	15.4	—	12.3	—
21	12.8	2.2	13.8	—	12.7	0.2	15.4	—	12.2	—
Site 3										
0	—	—	23.8	7.0	17.8	8.9	23.8	6.8	—	—
3	19.4	7.7	23.0	5.5	17.6	8.9	23.3	6.6	—	—
6	18.3	7.1	20.0	2.1	17.1	8.3	23.0	6.5	—	—
11	13.9	2.6	15.5	0.2	16.3	2.7	16.6	3.0	—	—
Site 4										
0	—	—	24.6	9.7	18.5	8.9	26.6	8.2	20.0	8.1
3	21.1	7.6	23.8	7.6	17.4	8.8	23.3	7.6	19.7	7.9
6	20.6	7.4	22.3	1.3	—	8.8	23.1	7.0	19.4	7.0
Site 5										
0	20.0	7.6	24.5	9.6	19.3	9.6	26.2	7.0	19.6	7.0
3	20.0	7.2	23.5	9.0	18.6	9.8	23.7	6.5	19.4	6.5
5	19.4	6.8	21.5	1.8	17.0	—	23.6	6.0	19.3	3.2
Site 6										
0	20.6	8.0	25.5	10.9	17.4	9.9	26.7	9.1	21.0	9.1
2.5	20.0	7.5	24.5	11.4	16.4	10.1	25.4	10	20.8	9.5

[a] From Faigenbaum (1938).
[b] From Chautauqua Lake Studies (1971–1974).

be related to intensive algal and macrophyte primary production which depletes lake water carbon dioxide while contributing hydroxide ion (Ruttner, 1961).

Total alkalinity was reported in 1937 to range between 38 and 56 ppm $CaCO_3$ (Tressler *et al.*, 1940). Corresponding ranges observed in recent years are: 43.5–60.7 (1971), 39.1–57.9 (1972), 37.7–64.6 (1973), and 38.0–68.4 (1974) (Chautauqua Lake Studies, 1971–1974).

Dissolved oxygen and temperature data reported in 1937 together with corresponding information obtained during the period 1971–1974 are shown in Table 16. The more recent data were collected at lake stations close to

and August of 1937[a] and 1971–1974[b]

	August									
	1937		1971		1972		1973		1974	
Depth (m)	Temperature (°C)	Dissolved oxygen	Temperature (°C)	Dissolved oxygen	Temperature (°C)	Dissolved oxygen	Temperature (°C)	Dissolved oxygen	Temperature (°C)	Dissolved oxygen
Site 1										
0	24.4	8.3	21.8	7.6	20.9	8.0	23.4	8.0	22.2	8.5
3	23.3	8.2	21.5	7.4	20.8	8.4	23.4	8.0	22.2	9.1
6	22.2	7.3	20.5	4.6	20.6	8.0	22.3	7.5	22.1	8.9
8	18.9	5.0	—	—	20.6	—	22.8	5.0	21.9	5.0
Site 2										
0	—	—	21.0	7.2	20.7	8.0	24.4	8.5	23.2	9.0
3	23.9	8.7	21.0	6.8	20.6	8.1	23.6	8.5	22.0	8.6
6	23.9	8.5	21.0	6.6	20.5	6.6	23.6	8.5	22.0	8.5
9	18.9	2.0	20.5	4.0	20.3	2.7	21.3	3.0	21.9	8.4
12	—	—	—	—	—	—	—	—	—	—
15	16.1	1.3	15.7	0.0	16.6	0.5	15.6	0.5	17.8	0.5
18	—	—	—	—	—	—	—	—	—	—
21	15.6	1.5	15.1	—	13.7	0.6	15.6	—	12.4	—
Site 3										
0	—	—	21.5	8.3	20.5	8.3	24.7	9.0	—	—
3	23.9	8.6	21.5	8.0	20.5	6.8	24.3	9.0	—	—
6	22.2	5.6	21.5	8.0	20.4	4.6	23.5	7.0	—	—
11	17.8	1.5	19.5	0.2	20.9	1.4	20.6	2.5	—	—
Site 4										
0	23.9	9.6	21.8	10.1	23.0	8.3	25.3	8.5	22.0	8.8
3	23.9	9.4	21.8	10.1	22.9	7.7	24.4	8.5	21.8	8.0
6	21.1	5.2	21.5	6.2	—	0.3	23.3	8.0	21.7	7.8
Site 5										
0	23.9	10	22.1	8.7	22.7	8.9	25.0	8.5	22.5	7.6
3	23.9	9.4	21.8	7.5	22.7	7.6	23.6	8.5	21.8	7.2
5	21.7	0.7	21.2	0.4	22.6	0.2	23.6	7.0	21.8	6.7
Site 6										
0	—	—	23.0	12.5	21.2	10.4	26.3	10	22.3	9.6
2.5	23.3	5.6	22.5	0.4	21.2	0.3	25.0	6.0	21.8	9.0

the 1937 stations. Comparisons between temperature and dissolved oxygen data for the more recent years relative to 1937 have failed to disclose any significant differences. It would appear that the contemporary lake is capable of achieving and maintaining dissolved oxygen levels at least equal to levels observed in 1937. This is due primarily to the fact that the lake is relatively shallow and does not develop a stable thermocline (Tressler and Bere, 1938; Chautauqua Lake Studies, 1971–1974). Oxygen depletion is found only in the lake's "deep holes" where true thermoclines develop each summer.

Turbidity

In 1937 Tressler and Bere (1938) noted that Chautaugua Lake transparency, measured by Secchi disk, is related to suspended organic matter. Mayer *et al.* (1974) compared seasonal Secchi disk data obtained off Maple Springs in 1937 with 1971 measurements and found increased turbidity (see Table 17). Acciardi *et al.* (1972) compared both 1971 and 1972 Secchi disk information with 1937 data and reached the same conclusion. It was also demonstrated that the lower lake generally displays more turbidity than the upper lake on any given day; observations of 1.0 m or less were found to be common in Burtis Bay.

Temperature

Owing to its shallow depth, Chautauqua Lake is known to become relatively warm during the summer period. Chautauqua's thermal characteristics define its fish community and their populations; the lake abundantly produces muskellunge, yellow perch, bass, walleye, bullheads, and carp.

In 1938 the published results of a biological survey of Chautauqua Lake included lake temperature observations made during the summer of 1937

TABLE 17

Chautauqua Lake Turbidity[a]

Year	Early July	Mid-July	Early August	Mid-August	Mid-September
1937	3.3[b]	3.9	3.3	2.3	2.6
1971	2.3	2.0	2.6	2.6	1.3
1972	3.0	2.5	2.5	2.4	2.3
1973	3.5	1.6	1.5	2.2	—
1974	2.3	2.3	1.9	1.3	—

[a] Average Secchi disk readings at Maple Springs (Site 3) in 1937 and from 1971 to 1974 (Mayer *et al.*, 1974).

[b] Measurements given in meters.

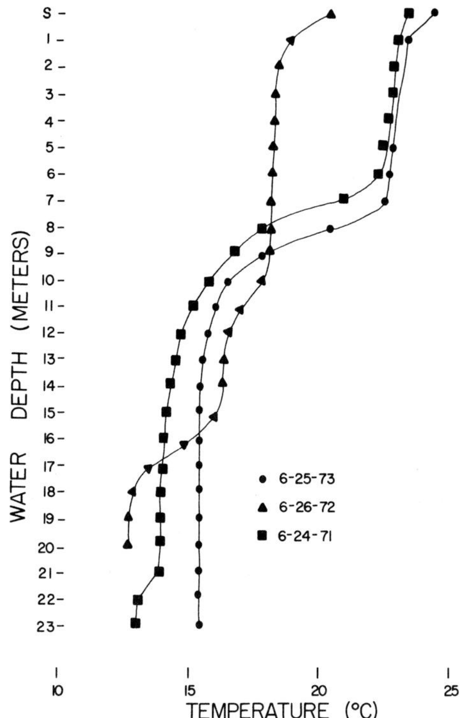

Fig. 11. Temperature profiles of Chautauqua Lake's 23-m deep hole in June 1971, 1972, and 1973.

(New York State Conservation Department, 1938). Vertical temperature profiles, obtained at two-week intervals at a station near Maple Springs, together with thermal data at other lake stations, were reported by Tressler and Bere (1938). The principal observation was that a mid-summer lake bottom temperature maximum of 16.0°C was found at the Maple Springs station in 14 m of water. This was regarded as being rather high. The surface temperature maximum at the same station in 1937 was 24.0°C, while an upper lake maximum of 26.0°C was recorded in Bemus Bay on July 13. No persistent thermocline was found in Chautauqua Lake during the summer of 1937 except in the 23-m "deep hole" region. Transient thermal discontinuities in most of the lake areas were easily disturbed by wind action (Tressler and Bere, 1938).

Beginning in 1971 and continuing on an annual basis thereafter, vertical profiles of temperature data were obtained weekly at several lake stations. (Chautauqua Lake Studies, 1971–1974). In general, the more recent observations parallel the temperature data recorded in 1937. Figure 11, which shows typical summer temperature profiles measured in the 23-m hole in 1971 and 1973, also shows a less well-defined thermal stratification in 1972 due to the influence of Hurricane Agnes on June 20, 1972. Presumably similar thermal stratification also exists in the other deep holes of the upper lake basin.

The influence of major weather systems on lake temperatures was observed by Acciardi *et al.* (1972); the heavy rainfall produced by Hurricane Agnes in June 1972 lowered the average surface water temperature by 5.4°C. As a result the lake's surface water in June 1972 was 4.4°C cooler than in June 1971.

Plant Nutrients

Chautauqua Lake's eutrophic state may be assessed in several ways. As a fishery, these waters were early recognized as an abundant resource (see History). Biologists carrying out a survey of the Allegheny and Chemung watersheds in 1937 reported that "Few lakes in New York State equal Chautauqua Lake in the large number of fish produced" (New York State Conservation Department, 1938). The same survey revealed that intensive algal and macrophyte growth dominate the lake much of the year.

These observations indicate that the lake is rich in nutrients. It is not yet clear, however, to what degree this is a reflection of natural events within

TABLE 18

Summary of Phosphorus Data Obtained at Four Experimental Stations in Chautauqua Lake between April 15, 1975 and September 30, 1975[a]

	Unfiltered		Filtered		
	Ortho-P[b]	Total P[b]	Ortho-P[b]	Total P[b]	
Site 9					
Range	2–23	8–48	<1–19	3–32	Upper lake
Mean	8.2	21.9	5.4	11.1	
Site 14					
Range	2–16	7–48	<1–14	2–45	
Mean	7.5	20.9	5.2	9.8	
Site 15					
Range	3–35	14–101	<1–25	3–42	Lower lake
Mean	15.5	43.1	7.7	17.9	
Site 16					
Range	2–27	9–87	0–13	2–67	
Mean	14.2	43.3	6.1	18.0	

[a] From Chautauqua Lake Studies (1975).
[b] Data given as μg/liter.

TABLE 19

1973 Levels of Nitrogen Compounds in Chautauqua Lake[a]

Nitrogen compound	Range (μg/liter)	Average (μg/liter)	Number of analyses
Nitrate	10–908	269	60
N (ammonia)	0–861	117	40
N (organic)	0–448	115	40

[a] From Chautauqua Lake Studies (1973).

the lake watershed and to what extent cultural developments and activity have accelerated, increasing levels of added nutrients.

Analyses for phosphorus were undertaken during 1975 at four experimental stations where phytoplankton studies were being performed concurrently (Chautauqua Lake Studies, 1975). Stations 9 and 14 are upper lake sites while 15 and 16 are lower lake sites (see Fig. 6). The range and mean values for unfiltered orthophosphate and total P and for filtered orthophosphate and total P are given in Table 18. The analyses were performed using the Environmental Protection Agency method modified to include reduction of arsenate to arsenite (Johnson, 1971). Based on these data, unfiltered total P monitored between April and October 1975 averaged 21.4 μg/liter P in the upper lake and 43.2 μg/liter in the lower lake. Lake levels of nitrogen compounds determined in 1973 are reported in Table 19 (Chautauqua Lake Studies, 1973).

Major and Trace Elements

During the years 1971–1974 benchmark measurements were made jointly by the State University College at Fredonia and Jamestown Community College to define the contemporary state of Chautauqua Lake (Chautauqua Lake Studies, 1971, 1972, 1973, 1974). Major and trace element concentrations determined by atomic absorption analyses of lake water are summarized in Table 20. Of special interest is the relatively high concentration of dissolved arsenic first observed by Lis and Hopke in 1971 (Lis and Hopke, 1973). Arsenic levels in North American waters average between 2 and 4 μg/liter. Chautauqua Lake's average of between 15 and 24 μg/liter arsenic is apparently the consequence of large-scale use of sodium arsenite as a lake weed herbicide from 1955 to 1963. Recent studies have shown that lake sediment arsenic levels range from 0.5 to 59 mg/liter (Ruppert *et al.*, 1974). The release of arsenic from the sediments may mimic phosphorus release which is known to be controlled by redox conditions overlying the lake bottom (Gumerman, 1970; Hayes and Phillips, 1959; Shukla *et al.*,

TABLE 20

Concentrations of Major and Trace Chemical Elements in Chautauqua Lake[a,b]

Element	Level	1971		1972		1973		1974	
		Range	Mean[c]	Range	Mean[d]	Range	Mean[e]	Range	Mean[f]
Sodium	mg/liter	—	3.03	1.3–10.5	3.70	2.7–17	3.84	2.4–12.2	3.50
Potassium	mg/liter	—	—	0.7–1.7	1.03	0.6–1.55	0.97	0.5–2.1	1.26
Calcium	mg/liter	—	21.1	12.9–24.8	19.2	14.6–29.5	21.1	16.8–23.4	20.1
Magnesium	mg/liter	—	4.26	2.9–8.5	4.9	3.3–8.0	4.3	2.7–4.6	3.6
Iron	μg/liter	5–120	31	<1–247	24	1–542	38	—	—
Manganese	μg/liter	—	50	<1–568	62.4	6.3–1488	99.9	—	—
Copper	μg/liter	—	—	<1–16.0	4.8	<1–12.5	1.2	—	—
Zinc	μg/liter	—	10	<1–29.4	9.6	—	—	—	—
Chromium	μg/liter	—	—	—	—	0.1–1.2	0.35	—	—
Arsenic	μg/liter	3.5–43.4	15.1	10–45	24	—	—	—	—
Lead	μg/liter	—	—	<1–10.6	3.1	—	—	—	—
Cadmium	μg/liter	—	—	<1–3.8	1.6	0.1–6.0	1.5	—	—

[a] Source: Chautauqua Lake Studies (1971–1974).
[b] Determined by atomic absorption analysis.
[c] Number of analyses = 180.
[d] Number of analyses = 140.
[e] Number of analyses = 125.
[f] Number of analyses = 125.

1971). The observed concentrations of dissolved arsenic in Chautauqua Lake are within established limits for this element in drinking water (New York State Department of Health, 1970).

A preliminary study of arsenic in fish was conducted in 1972 on lake sucker, bass, and walleye pike (Hopke, *et al.,* 1972). Excluding viscera, the average level of arsenic found was 0.255 μg/gram.

Also of particular interest is the finding that dissolved manganese occurs in concentrations as high as 1.49 mg/liter. Maximum levels of this element appear in the lake's "deep hole" (23 m) near Point Chautauqua. This observation most likely reflects a natural characteristic of the lake's geochemistry.

Special chemical studies of the herbicide Ortho Diquat, a product of the Chevron Chemical Company, were carried out in 1973 and 1974 (Chautauqua Lake Studies, 1973, 1974). This herbicide has been used since 1965 to control Chautauqua Lake macrophytes, especially in the lower lake. In 1972, for example, 171 hectares of lower lake weeds were sprayed during early summer. This represents 16.7% of the total weedbed area in the lower basin based on a 1938 weed map (New York State Conservation Department, 1938). In general it was found that Diquat kills macrophytes more or less completely within about one week but does not diminish rapid regrowth. Unused Diquat combines rapidly with macrophytes, suspended matter, and lake sediments. Within about 48 hours Diquat levels decrease to analytically undetectible levels (<1 μg/liter). Furthermore, no Diquat was found north of the lake's narrows between Bemus Point and Stow during periods of lower-lake herbicide application (Chautauqua Lake Studies, unpublished data).

Tributaries

Stream water is principally a mixture of overland flow (rain and snow melt) containing low concentrations of dissolved constituents and more highly mineralized groundwater. Its chemical quality is thus determined by the minerals, gases, and other constituents available for solution by overland flow and groundwater and the relative proportion contributed by each to the stream discharge.

Regionalization

Except for a narrow 2- to 7-mile wide belt along the Lake Erie shore and two small areas in the southeastern part of the county, Chautauqua County, including the Chautauqua Lake basin, lies in a region whose stream water quality is characterized as moderately hard and of the calcium bicarbonate type (Fig. 12). Table 21 lists the ranges of measurement of the properties of

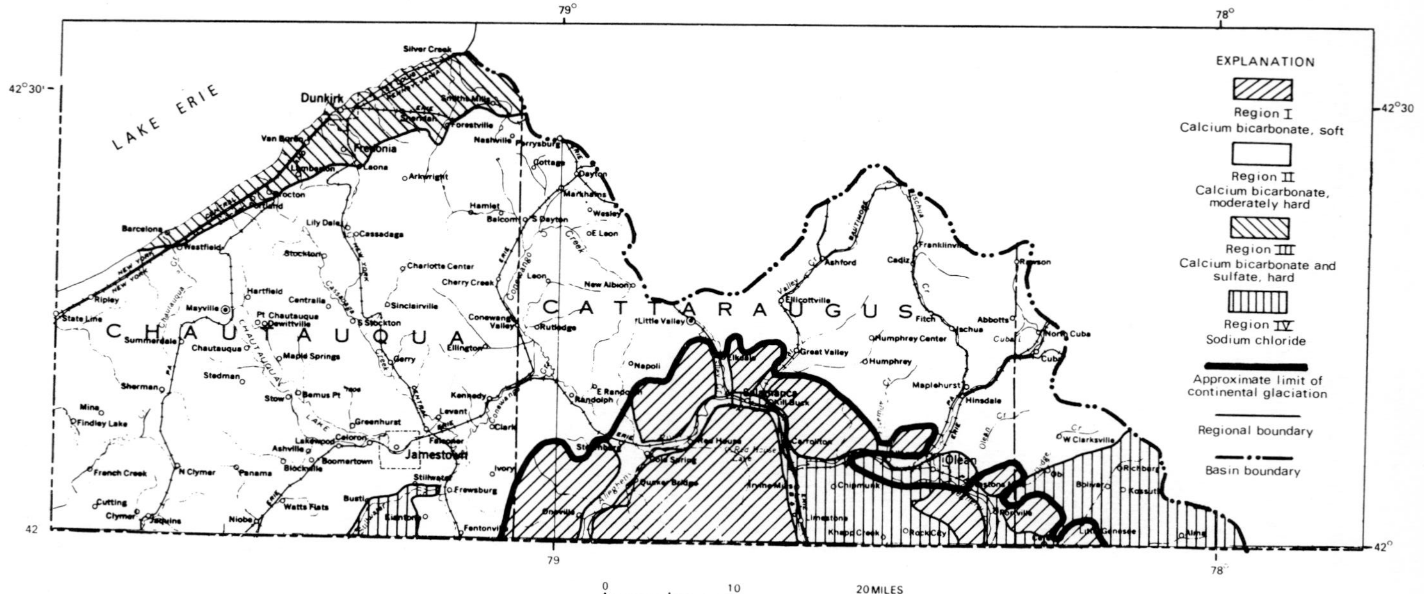

Fig. 12. Regionalization of streamflow quality. From Frimpter (1973).

TABLE 21

Range of Measurement of Property or Dissolved Constituent in Region II[a,b]

Constituent or property	Range (mg/liter)[c]
Bicarbonate (HCO_3)	43–191
Calcium (Ca)	18–59
Magnesium (Mg)	2.7–11
Chloride (Cl)	2.5–61
Dissolved solids (residue on evaporation)	105–241
Fluoride (F)	0–0.2
Hardness as $CaCO_3$	61–180
pH, units	6.8–8.1
Iron (Fe)	0.00–0.71
Manganese (Mn)	0.00–0.35
Nitrate (NO_3)	0–13
Silica (SiO_2)	1.1–8.3
Sodium (Na)	1.6–40
Potassium (K)	0.6–2.1
Specific conductance, μmhos/cm at 25°C	129–400
Sulfate (SO_4)	12–52

[a] Source: Modified from Frimpter (1973), Table 6.

[b] Excludes measurements at very low and very high flows.

[c] All measurements are mg/liter except pH and specific conductance.

dissolved constituents (excluding measurements at very low and very high flows) for stream water in this region. The water quality reflects the relatively insoluble nature of the bedrock and the greater solubility of fragments of glacially transported carbonate rocks now incorporated in the soils and other unconsolidated sediments (Frimpter, 1973).

Chemical Data

Chemical analyses of stream water from the Chautauqua Lake watershed are included in Barnard *et al.* (1972, 1973, 1974b), Hopke *et al.* (1972, 1973), Acciardi *et al.* (1972), Frimpter (1973), and the U.S. Environmental Protection Agency (1974).

From 1967 data Frimpter (1973) tabulates one analysis each of a water sample from Dewittville, Bemus, and Goose Creeks and three water samples from Prendergast Creek.

From 1972 to the present time, 11 tributaries, draining over 70% of the land area of Chautauqua Lake's 467.5 km² (180.5 square miles) watershed, have been intensively investigated by personnel associated with the Lake Erie Environmental Studies program at State University College at Fredonia. Table 22 lists concentration ranges of constituents determined in these streams from samples collected near their mouths during parts of 1972 and 1973; the ranges are in excellent agreement with those tabulated for the region (see Table 21). Except for higher concentrations of sodium, chloride, and phosphate found in Little Inlet, which are attributable pri-

TABLE 22

Concentration Ranges of Constituents in Stream Waters of Eleven Chautauqua Lake Tributaries

Constituent	Range	Unit of concentration
I. Ranges for samples: mid-March to mid-August, 1972[a]		
Alkalinity	18–130	mg/liter $CaCO_3$
Cadmium	<1–4	μg/liter
Calcium	6.7–45	mg/liter
Chloride	<0.5–9.8[b]	mg/liter
Chromium	<1–4	μg/liter
Copper	<1–38	μg/liter
Iron	10–850	μg/liter
Lead	<1–13	μg/liter
Magnesium	0.5–8.3	mg/liter
Nitrate	0.2–12.7	mg/liter
Phosphate	0–270[c]	μg/liter
Potassium	0.5–3.2	mg/liter
Sodium	1.3–7.6[d]	mg/liter
Sulfate	9.1–28	mg/liter
Zinc	<1–42	μg/liter
II. Ranges: late May to mid-August, 1973[e]		
Dissolved oxygen	0–14.6	mg/liter
pH, standard	6.1–9.2	units
Arsenic	11.4–20.3	μg/liter
Ammonia nitrogen	0–280[f]	μg/liter
Organic nitrogen	0–575[g]	μg/liter

[a] From Barnard *et al.* (1972), Hopke *et al.* (1972), and Acciardi *et al.* (1972).

[b] Excludes Little Inlet where range was 15–47 mg/liter.

[c] Excludes Little Inlet where range was 1700–9800 μg/liter.

[d] Excludes Little Inlet where maximum was 42 mg/liter.

[e] From Hopke *et al.* (1973).

[f] For June only and excludes Little Inlet with range 631–3100 μg/liter.

[g] For June only and excludes Little Inlet.

marily to industrial waste and leachate from stockpiles of highway salts, the chemistry of the 11 tributaries is not markedly dissimilar. This is evident from the averaged values for each tributary listed in Table 23; analytical data were obtained on samples collected biweekly from late May to mid-August, 1974.

Two studies have been undertaken to establish nutrient loading of phosphorus and nitrogen by Chautauqua Lake's tributaries. Analytical data obtained on near-surface grab samples collected between November 1972 and October 1973 near the mouths of eight tributaries by the U.S. Environmental Protection Agency are presented in Table 24. Data obtained on phosphorus in samples taken periodically between February 1975 and January 1976 from 11 tributaries and the lake's outlet by personnel of the Lake Erie Environmental Studies program are given in Table 25. The actual amount of nutrient loading by the tributaries is discussed in a later section.

Analytical data for samples collected throughout the reaches of the individual tributaries are given in Barnard *et al.* (1972, 1973, 1974b). These include the major cations (calcium, magnesium, potassium, and sodium) in 11 creeks (Barnard *et al.*, 1972); the major cations, iron, manganese, and specific conductance in 11 tributaries with chloride for five tributaries at the northwest end of Chautauqua Lake (Barnard *et al.*, 1973); and the major cations, cadmium, chromium, copper, iron, manganese, chloride, and specific conductance for Lighthouse, Prendergast, and Ball Creeks watersheds only (Barnard *et al.*, 1974b). Most of these determinations were comparable to the regional and local ranges previously established; however, there were point sources (such as sanitary landfill sites in the Lighthouse and Maple Springs watersheds) where values were many times higher than those typically found.

Correlations

Correlations among specific conductance, discharge, and chemical constituents have been established on both regional and local levels. For the Allegheny River basin Frimpter (1973) has shown high degrees of correlation for linear relationships between dissolved solids and (individually) specific conductance, bicarbonate, chloride, sulfate, and calcium. For the Chautauqua Lake basin, Barnard *et al.* (1973, 1974a,b, 1975) quantified relationships involving specific conductance, discharge, and the concentrations of various elements in water from 11 major tributary watersheds.

Precipitation

Precipitation is the principal source of both ground and surface water in Chautauqua County. Published chemical analyses of precipitation in the

TABLE 23

Average Determinations for Water Samples Collected Biweekly from May 30 to August 8, 1974

Tributary	L_s (μmhos)	Ca (mg/liter)	Cd (μg/liter)	Cr (μg/liter)	Cu (μg/liter)	Fe (μg/liter)	K (mg/liter)	Mg (mg/liter)	Mn (μg/liter)	Na (mg/liter)
Big Inlet	265.3	35.2	<1	<1	<2.6	466.7	1.4	7.3	129.5	3.9
Little Inlet (Route 17)	320.7	37.3	—	—	—	577.5	2.0	7.4	—	7.9
Little Inlet (Whallon Rd)	333.3	32.2	<1	1.8	<3	833.3	2.8	6.0	132.7	23.8
Mud	271.2	37.8	<1	<1	<1.8	345.0	0.8	8.3	107.0	2.7
Lighthouse	245.2	34.8	<1	<1	<2.0	190.0	1.2	6.2	58.2	3.0
Prendergast	279.3	35.1	<1	<1	<1	81.7	0.9	6.7	36.0	2.0
Ball	280.3	41.7	<1	<1.2	<1.8	126.7	1.2	6.9	94.8	2.5
Goose	263.2	37.3	<1	<1.5	<1.8	262.0	1.2	7.4	97.0	3.6
Dutch Hollow	241.2	34.0	<1.4	<1	<1	1675.0	3.3	5.2	<39.2	4.5
Bemus	176.2	25.8	<1	<1	<4.8	182.5	0.9	4.2	<27.0	2.2
Maple Springs	268.8	40.3	<1	<1.8	<2.2	218.3	1.1	6.7	56.3	2.3
Dewittville	286.2	41.5	<1.5	<1	<2.0	90.0	1.1	7.0	54.2	2.8
Averages	272.3	36.4	<1.1	<1.1	<2.1	415.9	1.5	6.7	<78.3	5.2

TABLE 24

Range of Nutrient Concentrations for Samples from Selected Tributaries (November 1972–October 1973)[a]

	Total $NO_2 + NO_3$ (mg/liter)	Total Kjel N (mg/liter)	Total NH_3–N (mg/liter)	Dissolved ortho-P (mg/liter)	Total P (mg/liter)
Little Inlet	0.018–0.450	0.340–8.40	0.031–2.500	0.330–5.880	0.065–7.00
Big Inlet	0.016–0.650	0.240–2.40	0.017–0.120	<0.005–0.039	0.015–0.06
Wing Creek	0.240–0.660	<0.100–1.18	0.008–0.099	<0.005–0.056	0.010–0.17
Ball Creek	0.092–0.700	0.140–1.25	<0.005–0.052	<0.005–0.017	0.010–0.15
Goose Creek	0.033–0.570	0.140–1.68	0.014–0.069	<0.005–0.039	0.010–0.11
Bemus Creek	0.130–0.550	0.130–0.77	<0.005–0.220	<0.005–0.013[b]	<0.005–0.02[b]
Dutch Hollow	<0.010–0.750	0.100–1.80	<0.005–0.072	<0.005–0.015	0.010–0.11
Dewittville	0.019–1.14	<0.100–2.73	<0.005–0.078	<0.005–0.015	<0.005–0.11

[a] Source: U.S. Environmental Protection Agency (1974).
[b] These data do not include phosphorus determinations for samples with very high nitrogen values.

TABLE 25

Chautauqua Lake Tributaries Phosphate Data (February 1, 1975–January 31, 1976)[a,b]

	Mean	Number determined	Standard deviation
Big Inlet			
Ortho	9.03	38	4.92
Total	29.91	37	14.79
Little Inlet (at Route 17)			
Ortho	10.18	39	7.86
Total	35.16	38	21.94
Little Inlet (at Whallen Rd.)			
Ortho	1132.0	39	844.3
Total	1512.0	31	976.7
Mud Creek			
Ortho	8.33	39	5.13
Total	26.39	38	15.38
Lighthouse Creek			
Ortho	4.29	38	5.11
Total	14.89	36	10.26
Prendergast Creek			
Ortho	3.97	39	3.00
Total	19.39	38	18.24
Ball Creek			
Ortho	6.74	39	4.13
Total	28.68	38	26.78
Goose Creek			
Ortho	11.28	39	18.20
Total	34.00	38	29.88
Dutch Hollow Creek			
Ortho	10.69	39	8.43
Total	33.77	39	30.59
Bemus Creek			
Ortho	4.36	39	2.50
Total	15.79	39	10.65
Maple Springs Creek			
Ortho	5.05	39	3.29
Total	21.92	36	24.94
Dewittville Creek			
Ortho	4.19	38	2.64
Total	28.32	37	30.42
Chadakoin River (outlet)			
Ortho	8.34	38	7.24
Total	50.05	38	32.46

[a] Source: Chautauqua Lake Studies (unpublished).
[b] Concentrations given in μg/liter.

immediate area of Chautauqua Lake are virtually nonexistent. However, records of chemical analyses of monthly composite precipitation samples collected at the National Weather Service Station in Allegany State Park, some 50 to 60 km east and "down-weather" from the Chautauqua area, have been available since August 1965 and are published annually (U.S. Department of the Interior Geological Survey, 1970–1974). Analytical and statistical data for the 5-year period October 1969 to September 1974 are presented in Table 26.

Groundwater

Groundwater in the Allegheny River basin, of which Chautauqua Lake is a part, has its principal source in precipitation that has percolated downward through the soil, dissolving minerals and carbon dioxide gas produced by bacteria in the soil. Regionally, the Allegheny River basin in New York can be divided into three areas based upon general dissolved solids content (Fig. 13, from Frimpter, 1974).

Because the Chautauqua Lake region is blanketed by glacially transported material containing abundant carbonate (limestone and dolostone) rock fragments, the shallow groundwater is somewhat harder and more mineralized than that of nearby nonglaciated areas. The rocks underlying the entire area are not very soluble and therefore do not produce highly mineralized water. Deeper layers of rock, however, contain salty connate water—water trapped in the sediments when they were deposited in a former sea. This salty water has subsequently been flushed out of the shallow rock layers by the circulation of relatively fresh water derived from precipitation (Frimpter, 1973).

In contrast to the abundance of chemical data available for streams in the Chautauqua Lake area, published chemical data for groundwater are essentially limited to Crain (1966); no additional data for the immediate area are provided by Frimpter (1974).

The quality of groundwater in southeastern Chautauqua County, which includes the Chautauqua Lake region, is of two distinctly different types—fresh water and mineralized water. The fresh water, originating as precipitation, typically contains less than 10 mg/liter chloride and 100 to 300 mg/liter dissolved solids. The mineralized water contains over 1000 mg/liter chloride and several thousand mg/liter dissolved solids, making it unfit for nearly all purposes. This water probably originated as the result of the deep circulation of fresh groundwater which dissolved portions of deep salt beds or as the result of ascending connate saline waters. As the mineralized water moves upward along fractures and faults in the rock, it mixes with fresh groundwater (Crain, 1966).

TABLE 26

Chemical and Physical Determinations on Monthly Composite Precipitation Samples

Period of collection		Calcium (mg/liter)	Magnesium (mg/liter)	Sodium (mg/liter)	Potassium (mg/liter)	Bicar. bonate (mg/liter)
From	To					
10-01-69	11-01-69	0.5	0.06	0.6	0.4	0.0
11-01-69	12-01-69	0.2	0.06	0.3	0.0	0.0
01-01-70	02-01-70	1.8	0.28	1.0	0.3	0.0
02-01-70	03-01-70	1.7	0.30	2.2	0.6	0.0
03-01-70	04-01-70	1.8	0.25	1.9	0.1	0.0
04-01-70	05-01-70	1.0	0.21	1.2	0.1	0.0
05-01-70	06-01-70	1.0	0.11	0.4	0.0	0.0
06-01-70	07-01-70	1.5	0.11	0.4	0.2	0.0
07-01-70	08-01-70	0.9	0.00	0.6	0.3	0.0
08-01-70	09-01-70	0.5	0.10	0.2	0.0	0.0
09-01-70	10-01-70	1.0	0.04	0.1	0.1	0.0
10-01-70	11-01-70	1.0	0.10	0.1	0.3	0.0
11-01-70	12-01-70	0.6	0.16	0.5	0.1	0.0
12-01-70	01-01-71	1.0	0.11	0.2	0.0	0.0
01-01-71	02-01-71	1.2	0.16	0.7	0.1	0.0
02-01-71	03-01-71	0.6	0.07	0.2	0.0	0.0
03-01-71	04-01-71	1.7	0.30	0.7	0.2	2.0
04-01-71	04-31-71	4.5	0.77	0.2	0.2	5.0
04-31-71	06-01-71	2.0	0.32	0.3	0.2	0.0
06-01-71	07-01-71	2.0	0.30	0.3	1.8	0.0
07-01-71	08-01-71	3.6	0.41	5.0	3.2	6.0
08-01-71	09-01-71	1.4	0.51	2.0	0.6	0.0
09-01-71	10-01-71	0.4	0.10	0.9	0.1	0.0
10-01-71	11-01-71	1.7	0.54	4.6	0.7	12.0
11-01-71	12-01-71	1.0	0.22	7.4	1.1	10.0
12-01-71	01-01-72	1.0	0.21	0.7	0.6	0.0
02-01-72	02-29-72	1.9	0.41	1.2	0.6	0.0
03-01-72	04-01-72	6.6	1.00	2.1	0.7	0.0
04-01-72	05-01-72	15.0	—	—	—	0.0
05-01-72	06-01-72	1.0	0.20	0.4	0.3	0.0
06-01-72	07-01-72	0.2	0.10	0.2	0.0	0.0
07-01-72	08-01-72	0.9	0.20	0.6	0.2	0.0
08-01-72	09-01-72	0.8	0.91	3.0	1.3	7.0
10-01-72	11-01-72	—	0.12	1.7	0.3	0.0
11-01-72	12-31-72	0.9	0.17	1.2	0.6	0.0
12-31-72	01-31-73	—	—	3.0	4.5	0.0
03-01-73	04-30-73	—	—	—	—	—
06-01-73	07-01-73	0.4	0.70	0.5	0.0	0.0
07-01-73	08-01-73	0.5	0.11	0.1	0.0	0.0
08-01-73	09-01-73	0.0	0.68	0.2	0.0	8.0
10-01-73	11-01-73	0.5	0.1	0.2	0.1	0.0
11-01-73	12-01-73	1.0	0.2	0.7	0.3	0.0
12-01-73	01-01-74	0.3	0.1	0.1	0.1	0.0
01-01-74	02-01-74	0.5	0.2	0.0	0.1	0.0
02-01-74	03-01-74	1.2	0.2	0.2	0.1	0.0
03-01-74	04-01-74	0.8	0.1	0.1	0.1	0.0
04-01-74	05-01-74	1.0	0.3	0.2	0.2	0.0
05-01-74	05-31-74	0.7	0.2	0.1	0.5	0.0
05-31-74	07-01-74	0.3	0.0	0.0	0.2	0.0
07-01-74	08-01-74	0.5	0.1	0.0	0.0	0.0
08-01-74	09-30-74	6.5	2.1	0.0	0.0	0.0
Number of samples		48	48	49	49	50
Mean		1.6	0.29	1.0	0.4	1.0
Standard deviation		2.40	0.35	1.45	0.81	2.76

[a] Source: U.S. Department of the Interior Geological Survey (1970–1974).

[b] Total phosphate.

[c] Specific conductance.

Collected at Allegany State Park[a]

Sulfate (mg/liter)	Chloride (mg/liter)	Ammonia as N (mg/liter)	Nitrate as N (mg/liter)	Phosphate[b] (mg/liter)	Conductance[c] (micromhos)	pH
5.00	1.35	0.08	0.52	0.020	40	4.30
3.50	0.05	0.17	0.52	0.010	32	4.40
11.00	1.00	1.71	4.07	0.080	186	3.50
5.80	1.70	0.74	1.74	0.300	42	4.60
6.00	0.85	0.62	1.15	0.010	46	4.40
6.30	0.60	0.19	0.73	0.040	32	4.50
6.00	0.40	0.23	0.52	0.050	47	4.30
6.50	0.50	0.04	0.37	0.020	37	4.40
6.30	2.00	0.08	0.37	0.130	37	4.35
2.00	0.60	0.12	0.41	0.020	45	4.10
5.50	0.45	0.04	0.50	0.040	50	4.20
3.85	0.05	0.04	0.00	0.040	24	4.50
4.90	1.30	0.00	0.45	0.040	38	4.50
6.00	0.30	0.19	0.72	0.060	54	4.10
5.20	0.00	0.23	0.61	0.060	36	4.50
5.40	0.25	0.08	0.02	0.070	31	4.40
5.90	0.50	0.12	0.00	—	20	6.00
16.00	1.00	0.54	0.04	0.200	39	6.20
8.50	0.65	0.62	—	0.060	63	4.20
11.00	1.80	1.09	1.20	0.090	81	4.10
15.00	8.00	0.23	0.01	0.640	62	6.00
9.00	2.50	0.70	0.49	0.120	58	4.50
3.30	0.70	0.23	0.00	0.060	22	4.70
5.70	1.80	0.12	0.10	—	39	6.50
7.70	3.00	0.40	0.19	—	52	6.60
7.60	1.50	0.54	0.16	—	47	4.50
13.00	1.30	0.62	0.47	—	97	4.00
31.00	2.50	1.23	0.85	—	178	3.00
60.00	10.00	4.81	1.62	—	236	4.30
7.30	0.80	0.31	0.22	—	55	4.10
2.10	0.10	0.05	0.03	—	17	4.50
11.00	0.70	0.03	0.18	—	75	3.90
14.00	1.00	2.95	0.14	—	54	6.20
6.00	1.90	0.18	0.45	—	230	4.70
12.00	1.70	1.80	1.12	—	75	4.10
—	—	—	—	—	212	3.85
0.00	5.70	6.00	1.35	—	252	—
6.40	0.20	0.48	0.65	—	55	4.00
8.00	0.20	0.64	0.85	—	72	3.80
6.20	0.40	0.12	0.69	—	67	8.20
4.5	0.0	0.55	0.22	—	42	4.1
4.8	0.6	0.75	0.67	—	39	4.2
2.8	0.1	0.47	0.16	0.00	13	4.2
5.2	0.3	0.61	0.19	—	41	4.1
8.3	1.5	0.76	0.09	—	77	3.7
4.7	0.2	0.59	0.11	—	38	4.1
5.0	0.4	0.60	0.14	0.061	39	6.5
7.0	0.4	0.51	0.09	0.123	54	4.0
5.6	0.3	0.39	0.09	0.061	35	4.1
7.4	0.2	0.00	0.26	0.031	71	—
3.4	1.2	0.36	0.04	0.031	45	4.0
50	50	50	49	28	51	49
8.29	1.29	0.68	0.52	0.09	67.2	4.59
8.84	1.89	1.12	0.67	0.12	58.0	0.94

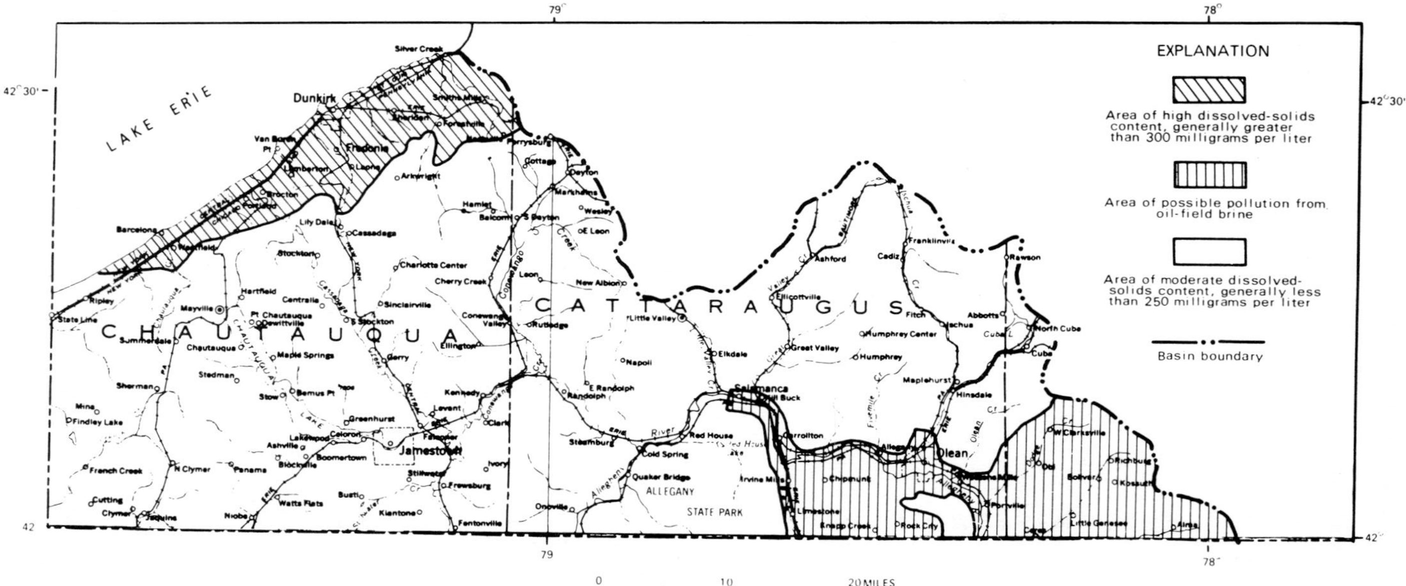

Fig. 13. Dissolved solids content of groundwater, Allegheny River basin, New York. From Frimpter (1973).

TABLE 27

Comparison of Selected Parameter Ranges for Groundwater from Southeastern Chautauqua County and the Immediate Chautauqua Lake Area[a]

	Southeastern Chautauqua County		Chautauqua Lake area	
Parameter	Number of samples	Range	Number of samples	Range
Dissolved solids (mg/liter)	112	90–4440	14	142–261
Chloride (mg/liter)	119	0.0–2420	16	1.4–120
Iron (mg/liter)	80	0.01–3.4	9	0.01–0.91
Hardness (mg/liter $CaCO_3$)	116	46–1090	16	74–268
pH	43	6.9–8.6	11	7.2–7.7
Specific conductance (μmhos/cm at 25°C)	107	145–7560	14	229–440

[a] Source: Crain (1966).

Crain (1966) tabulates chemical analyses for water from 119 wells and springs, of which 16 are located in the Chautauqua Lake watershed and the adjacent western side of the city of Jamestown. The ranges of the dissolved solids, chlorides, hardness, iron, specific conductance, and pH for the 119 sites in southeastern Chautauqua County and for the 16 sites in the immediate Chautauqua Lake area are given in Table 27. The relatively narrow ranges for the Chautauqua Lake area as compared to those for the more spacious southeastern part of the county are due principally to the fact that all the wells in the former were producing fresh water (possibly with limited mixing of mineralized water) from unconsolidated sands and gravels and not from bedrock.

A comparison of the dissolved solids, hardness, chloride and iron in well waters from southeastern Chautauqua County for wells ending in shallow (<45 m) and deep (>45 m) unconsolidated deposits and in bedrock is given in Table 28.

Gaseous hydrogen sulfide is common in the groundwater throughout the area. It is objectionable because it imparts the characteristic odor of rotten eggs and makes water acidic and corrosive. No quantitative determinations of hydrogen sulfide are available, but the gas tends to occur in the bedrock aquifer or in unconsolidated aquifers confined under silt and clay. Methane is also widespread.

In summary, groundwater for the Chautauqua Lake area in unconsolidated deposits less than 45 m below surface tends to be somewhat hard, but very low in chloride, iron, and dissolved solids, and is usually satisfactory

TABLE 28

Comparison of Constituents for Wells Ending in Shallow and Deep Unconsolidated Deposits and in Bedrock in Southeastern Chautauqua County[a]

Constituent	Unconsolidated deposits <45 m	Unconsolidated deposits >45 m	Bedrock
Iron (mg/liter)	75% < 0.31	70% < 0.31	64% > 0.3
Chloride (mg/liter)	80% < 10	>10	35% > 100 15% > 250
Dissolved solids (mg/liter)	70% + <201	relatively low	55% > 200 25% > 500
Hardness	hard	moderately hard	hard

[a] Source: Crain (1966).

for most purposes. At the southeastern edge of the lake, in the area adjacent to the Chadakoin River, the groundwater may be chemically polluted, however.

Throughout southeastern Chautauqua County, wells in bedrock in the uplands yield water of good quality, but those ending in bedrock in the valleys tend to be poorer because of high chloride, dissolved solids, and/or iron content. Many wells in bedrock also contain hydrogen sulfide and methane. The possibility of encountering water high in chloride, dissolved solids, hydrogen sulfide, and methane also increases with the depth of the well (Crain, 1966).

BIOLOGICAL LIMNOLOGY

Phytoplankton

Documented studies of the phytoplankton in Chautauqua Lake deal mainly with phytoplankton types and numbers. Lists of the various phytoplankters have been made by Tressler and Bere (1938), Giebner (1951), DeShong and Wood (1973), and the U.S. Environmental Protection Agency (1974). A list of the common phytoplankton genera compiled from the published reports of these investigators and the work of Storch (unpublished) is presented in Table 29. Historically, reports of Chautauqua Lake phytoplankton indicate that the lake is highly productive, but that it is a "diatom lake" with blue-green algae occurring in large numbers during the summer months.

In the first study of Chautauqua lake phytoplankton in the mid-1930's, diatoms were observed to be the dominant phytoplankters throughout the entire year (Tressler and Bere, 1938; Tressler *et al.*, 1940). *Stephanodiscus* was the dominant winter diatom. *Melosira, Fragilaria, Asterionella, Tabellaria,* and *Cyclotella* were found in great numbers during the spring, summer, and fall. *Coelosphaerium, Aphanizomenon,* and *Gloeotrichia* were found in great abundance during the blue-green algae blooms in July, August, and September. The largest numbers of blue-greens occurred in the

TABLE 29

Common Chautauqua Lake Phytoplankton Genera

Myxophyceae	*Closterium*	*Tetradesmus*
Anabaena	*Coccomyxa*	*Tetraedron*
Anacystis	*Coelastrum*	*Tetrastrum*
Aphanizomenon	*Cosmarium*	*Trochiscia*
Aphanocapsa	*Crucigenia*	*Ulothrix*
Aphanothece	*Desmidium*	*Volvox*
Calothrix	*Dictyosphaerium*	*Westella*
Chroococcus	*Dimorphococcus*	Bacillariophyceae
Coelosphaerium	*Elakatothrix*	*Achnanthes*
Dactylococcopsis	*Errerella*	*Amphora*
Gloeocapsa	*Euastrum*	*Asterionella*
Gloeotrichia	*Eudorina*	*Cocconeis*
Gomphosphaeria	*Geminella*	*Cyclotella*
Holopedium	*Gloeocystis*	*Cymbella*
Merismopedia	*Golenkinia*	*Epithemia*
Microcystis	*Gonatozygon*	*Eunotia*
Nostoc	*Gonium*	*Fragilaria*
Oscillatoria	*Hydrodictyon*	*Gomphonema*
Pleurocapsa	*Kirchneriella*	*Gyrosigma*
Plectonema	*Lagerheimia*	*Melosira*
Tetrapedia	*Micractinium*	*Navicula*
Tolypothrix	*Mougeotia*	*Nitzschia*
Chrysophyceae	*Myrmecia*	*Pinnularia*
Dinobryon	*Nephrocytium*	*Stauroneis*
Mallomonas	*Oocystis*	*Stephanodiscus*
Ochromonas	*Palmellococcus*	*Surirella*
Rizochrysis	*Pandorina*	*Synedra*
Synura	*Pediastrum*	*Tabellaria*
Uroglena	*Phacus*	Cryptophyceae
Chlorophyceae	*Phacotus*	*Cryptomonas*
Actinastrum	*Pleodorina*	*Rhodomonas*
Ankistrodesmus	*Quadrigula*	Dinophyceae
Apiocystis	*Scenedesmus*	*Ceratium*
Arthrodesmus	*Schroederia*	*Gymnodinium*
Botryococcus	*Selenastrum*	*Peridinium*
Characium	*Sphaerocystis*	Euglenophyceae
Chlamydomonas	*Spondylosium*	*Euglena*
Chlorella	*Staurastrum*	*Lepocinclis*
Chlorogonium	*Stylosphaeridium*	*Trachelomonas*

southern basin. *Ankistrodesmus* and *Scenedesmus* were some of the more abundant green algae observed in the lake during the first study.

More recent studies conducted in the 1970's indicate that the phytoplankton communities differ from those observed in the mid-1930's. The most striking difference is that Chautauqua can no longer be considered a "diatom lake" (Fig. 14). During the winter and early spring, diatoms (especially *Asterionella, Fragilaria,* and *Synedra*) are abundant. However, during this same period, phytoflagellates (*Cryptomonas, Dinobryon,* and *Rhodomonas*) often occur in greater numbers than the diatoms (T. Storch, unpublished). Throughout the summer, from mid-June to mid-September, the blue-green algae are dominant (DeShong and Wood, 1973; Flanders and Sobon, 1972; Nicholson and Rosenthal, 1973; T. Storch, unpublished). In the early fall, blue-green algae and diatoms are most abundant. However, by late fall (mid-November), phytoflagellates and diatoms again dominate the phytoplankton community.

The most common blue-green algae observed in Chautauqua Lake during the 1970's are *Anabaena, Aphanizomenon,* and *Gloeotrichia. Asterionella, Nitzschia, Melosira,* and *Stephanodiscus* are some of the common summer

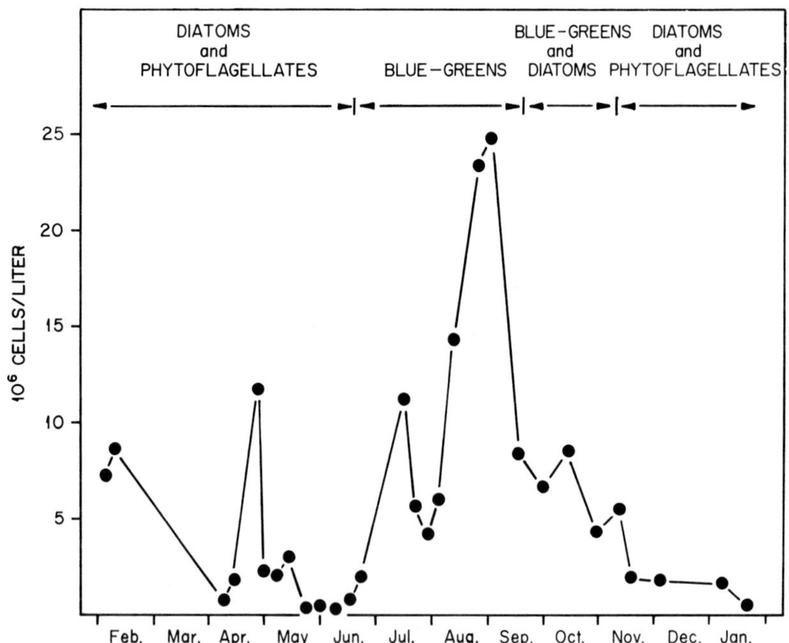

Fig. 14. Phytoplankton cell numbers at 0 m in the northern basin of Chautauqua Lake (February 1975 through January 1976).

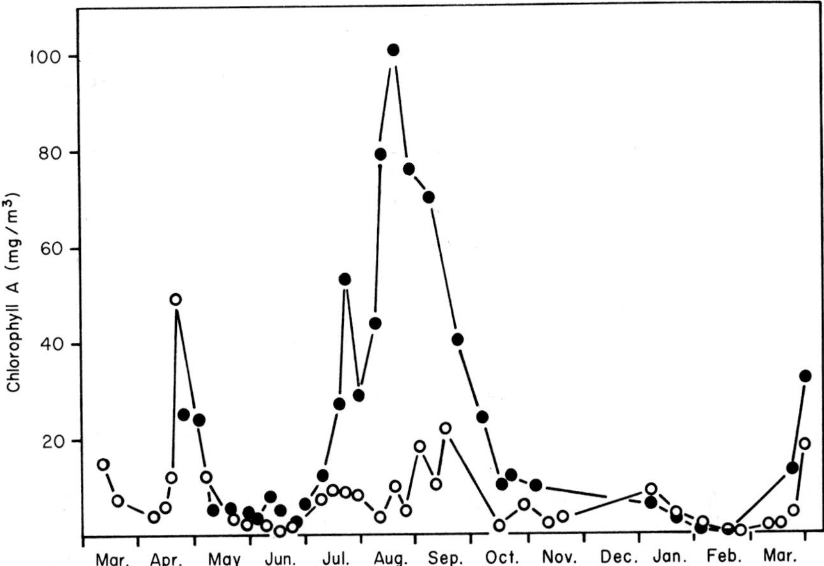

Fig. 15. Concentration of phytoplankton chlorophyll *a* at 0 m in the northern (○) and southern (●) basins of Chautauqua Lake (March 1975 through March 1976).

diatoms. Green algae have never been observed to occur in large numbers. However, a wide variety of green algae are present during the summer; these include such forms as *Actinastrum, Ankistrodesmus, Oocystis,* and *Scenedesmus* (DeShong and Wood, 1973; T. Storch, unpublished).

Algal chlorophyll analyses were conducted in 1972 (Flanders and Sobon, 1972; U.S. Environmental Protection Agency, 1974) and in 1975 and 1976 (T. Storch, unpublished). In 1975, surface chlorophyll *a* values ranged from approximately 0.1 to 101.5 mg/m^3 with lowest concentrations occurring in mid-winter and late spring and the highest values in early spring and summer (Fig. 15). The surface chlorophyll *a* concentrations are generally representative of the chlorophyll values throughout the vertical column of water (T. Storch, unpublished). This is apparently a result of the shallow nature of the lake and the consequent frequent mixing due to wind activity. The chlorophyll concentration, however, is distributed on a horizontal basis with the highest values generally occurring in the southern basin, especially during the summer (Fig. 15). Measurements of algal photosynthesis, numbers, and biomass also indicate that phytoplankton densities are usually greatest in the southern basin (Flanders and Sobon, 1972; DeShong and Wood, 1973; T. Storch, unpublished).

Algal primary production measurements have been conducted since 1973 in Chautauqua Lake (DeShong and Wood, 1973; Storch *et al.*, 1974a,b; T. Storch, unpublished). The daily net rate of photosynthesis, as measured by the radioactive ^{14}C method, ranges between 0.08 and 3.37 gm m^{-2} day^{-1}. The photosynthetic rates are highest following the melting of the ice cover in April and during the summer blue-green algae bloom (July, August, and September). Lowest rates occur in late fall and winter (November through March) and in late spring (late May and early June). The low photosynthetic rates in May and June are associated with a large die-off of the algae resulting in extremely low algal biomass and chlorophyll concentrations. Maximum Secchi disk transparency (sometimes greater than 8 m) occurs during this same period.

The distribution of algal photosynthesis with depth in Chautauqua lake shows a pattern characteristic of photosynthesis in eutrophic lakes: (1) photoinhibition at the surface, (2) maximum rate of photosynthesis between 0 and 2 m, and (3) decrease in photosynthesis with depth below the depth of maximum photosynthesis. The maximum depth where photosynthesis occurs is usually between 6 and 8 m. A detailed 2-year study of primary production at one site in the northern basin indicated that within the 24-hour day in which carbon is photosynthetically fixed by the algae, on the average, 36% is respired, 7% is released as extracellular dissolved organic carbon, and 57% remains as particulate carbon in the algae cells (T. Storch, unpublished).

Storch *et al.* (1974b) measured algal photosynthesis in the littoral and pelagic zones of the northern basin. Their study revealed that between June and November, algal photosynthesis per volume of water was, on the average, 25% higher in the littoral zone than in the pelagic zone. A comparison of algal photosynthesis in the southern and northern basins of Chautauqua Lake over a 12-month period indicated that algal photosynthesis per square meter surface area is about 35% greater in the southern basin (T. Storch, unpublished). This is somewhat misleading, however, because the trophogenic layer of the northern basin is more extensive. The mean depth of the northern basin (7.8 m) is approximately twice as great as that of the southern basin (3.5 m). When algal photosynthesis is determined per unit volume of lake water in the trophogenic layer of both basins, the annual rate of photosynthesis is more than 300% greater in the southern basin than in the northern basin.

Preliminary information is available on algal growth-limiting nutrients in Chautauqua Lake. The first analysis was conducted in 1972 by the U.S. Environmental Protection Agency as part of their National Eutrophication Survey (1974). This algal assay, using the test organism *Selenastrum capricornutum,* indicated that Chautauqua Lake primary production was controlled by nitrogen when the assay was conducted in October 1972. The nitrogen to phosphorus ratios, which are usually less than 9 to 1 in Chautauqua Lake, also suggest that the lake is nitrogen limited. However, further studies conducted in 1975, using indigenous algal communities, indicated that nitrogen may not always be the limiting nutrient (T. Storch, unpublished). In these studies, additions of phosphorus and nitrogen in combination always had the greatest stimulatory effect on the natural algal communities. However, depending upon both the time of year in which the assay was run and the sites within the lake where samples were taken for the assay, sometimes phosphorus was observed to be limiting and on other occasions nitrogen was the limiting nutrient. These apparent oscillations between nitrogen and phosphorus as the limiting nutrient correlate both with periodic fluctuations in the indigenous concentration of these nutrients and with changes in the composition of the algal community.

With respect to the response of algae to nutrients from specific nutrient loading sources to the lake, algal growth is enhanced by the input of treated sewage to the lake from the four primary wastewater treatment plants and the single tertiary treatment plant discharging into the lake (Flanders and Storch, 1974; T. Storch, unpublished). All tributaries, with the exception of Little Inlet, have been shown to have no significant effects on the growth of natural communities of phytoplankton in the lake (T. Storch, unpublished). Little Inlet, however, which is the receiving water for industrial wastes in the Mayville area, has a large stimulatory effect on the growth of algae similar to that of sewage from the wastewater treatment plants.

Zooplankton

There have been two major studies of zooplankton of Chautauqua Lake. Tressler and Bere (1938) collected plankton samples during their limnological survey conducted in 1937, and T. A. Erlandson (unpublished) sampled zooplankton at two northern basin and two southern basin stations during the summer of 1975. Although Erlandson did not classify below the order level, Tressler and Bere identified the major groups according to genus, reporting 5 genera of copepods, 15 genera each of cladocerans and rotifers, and 11 genera of protozoans. The most commonly observed genera were

Copepoda: *Cyclops* and *Diaptomus*
Rotifera: *Anuraea*
Protozoa: *Ceratium* and *Glenodinium*
Cladocera: *Bosmina* and *Ceriodaphnia*

Tressler and Bere (1938) reported a total zooplankton density at a single station ranging from 162 to 545 plankters per liter, reaching a peak of 545 per liter in early August. The recent data of Erlandson (Table 30) show a four-station range of from 172 to 2674 plankters per liter, with a four-station mean peak of 1785 per liter in late June; no August samples were taken.

A comparison of these data show that the density of zooplankton may have increased by as much as five times since 1937. Table 30 also shows that the density at the northern basin stations exceeded that of the southern basin stations during five of the seven weeks sampled. The sampling period was too brief to show any definite seasonal trends, although there appears to be a decline in density from mid-June through mid-July, with a late July increase.

Table 31 presents both the density in number per liter and the importance in terms of percent of the sample collected for the major zooplankton taxa in the 1975 survey (T. A. Erlandson, unpublished). Considering the entire sampling period, the rotifers were by far the most important component, with a seasonal mean importance value of 88.03%, followed by copepods (7.19%), cladocerans (3.50%), and ostracods (1.23%). However, over the course of the summer, the relative importance of the Rotifera declined, while that of the other three groups increased. A comparison of these data with similar data extracted from the report of Tressler and Bere (1938) indicates that since 1937 rotifers have become a more important component of the planktonic fauna, accounting for 88% in 1975 compared with 41% in 1937.

Bacteria

Introduction

The purpose of conducting selected bacteriological studies of Chautauqua Lake was to make a contribution toward a comprehensive data bank which would describe current biological, chemical, and physical parameters of the lake. In addition, it is to be hoped that the above-mentioned data bank will ultimately lead to a comprehensive management plan for Chautauqua Lake.

Bacteriological studies, which began in 1972 and continued during the summers of 1973 and 1974, concentrated on total coliform counts, fecal coliforms, and fecal streptococci. An additional dimension was added in 1973 and 1974 which dealt with investigations for the *Salmonella* group.

Twenty-eight sites were sampled twice each week from June until mid-August of the 3 years mentioned. Some of the sites were public and private bathing areas, tributaries to the lake, some in mid-lake and others at certain locations selected because of human and agricultural activities in the proximity. Testing for the *Salmonella* group was limited to ten sites which were sampled weekly.

TABLE 30

Total Numbers of Zooplankton in the Chautauqua Lake Northern and Southern Basins (1975)[a]

	Southern basin stations				Northern basin stations			
Date	1[b]	2[c]	Total	$\bar{x}$	3[d]	4[e]	Total	$\bar{x}$
June 16	1306.25	742.62	2048.87	1024.44	NS	NS	—	—
June 23	992.45	1516.37	2508.82	1254.41	2258.88	NS	—	2258.88[g]
June 30	NS[f]	2296.69	—	2296.69[g]	799.96	2258.17	3058.13	1529.07
July 7	720.99	664.32	1385.31	692.66	2674.80	814.16	3488.96	1744.40
July 14	NS	302.12	—	302.12[g]	837.99	546.30	1384.29	692.15
July 21	259.28	172.82	432.10	216.05	428.47	289.12	717.59	358.80
July 28	894.35	1189.80	2084.15	1042.08	427.64	454.13	881.77	440.89

[a] Source: T. A. Erlandson *et al.* (unpublished).
[b] Station 1, Maple Bay.
[c] Station 2, Arnold's Bay.
[d] Station 3, Chautauqua Bay.
[e] Station 4, Point Chautauqua.
[f] NS, no sample.
[g] Not a true mean.

TABLE 31

Density and Total Percentage of Major Zooplankton Taxa (1975)[a]

Date	Total	Ostracoda		Cladocera		Copepoda		Rotifera	
		Number[b]	Percentage[c]	Number	Percentage	Number	Percentage	Number	Percentage
June 16	512.22	4.41	0.86	3.85	0.75	25.43	4.96	478.53	93.42
June 23	794.62	1.45	0.18	2.42	0.30	16.23	2.04	774.52	97.47
June 30	894.73	0.49	0.05	6.00	0.67	20.67	2.31	867.57	96.96
July 7	609.28	1.65	0.27	10.35	1.70	50.13	8.23	547.15	89.80
July 14	270.25	2.29	0.85	16.16	5.98	39.74	14.70	212.06	78.47
July 21	143.71	5.16	3.59	14.06	9.78	14.22	9.89	109.77	76.38
July 28	370.75	10.38	2.8	19.67	5.31	30.42	8.20	310.29	83.69
Mean	513.65	3.69	1.23	10.36	3.50	28.12	7.19	471.41	88.03

[a] Source: T. A. Erlandson *et al.* (unpublished).
[b] Number per liter.
[c] Percent of total.

The Coliform Group

Total coliform tests were conducted primarily to ascertain the direction of the studies and the intensity of investigations at each site. Generally speaking, all mid-lake sites showed few if any coliforms present. The sites located on tributaries, along with lake shoreline, and at the specially selected sites did, however, reveal the presence of the coliform group.

Although the presence of coliform bacteria indicates that a water sample may be polluted, it does not necessarily prove that the contamination had a fecal source. It was for this reason that additional testing was begun to differentiate between fecal and nonfecal strains of coliforms. It is obvious that such testing can be a valuable tool for checking for recent and potentially dangerous pollution of water.

The Millipore technique (Geldreich *et al.*, 1965) was used to obtain coliform data at or near the mouths of the principal lake tributaries. Data obtained during the years 1972–1974 indicated the presence of significant fecal coliform counts at the following test sites on Chautauqua Lake:

1. Big Inlet Creek: >200/100 ml
2. Little Inlet Creek: >200/100 ml
3. Lighthouse Creek: >200/100 ml
4. Stow Ferry: >200/100 ml
5. Dutch Hollow Creek: >200/100 ml
6. Goose Creek: <200/100 ml
7. Boatlanding (Jamestown): <200/100 ml
8. Burtis Bay: <200/100 ml
9. Bemus Bay: <200/100 ml
10. Dewittville Creek: <200/100 ml
11. Bemus Creek: <200/100 ml

The significance of the fecal coliform counts, while indicating the presence of fecal contamination, draws added attention when the sites with counts greater than 200/100 ml are compared with the data on *Salmonella* found below.

Fecal Streptococci

The positive finding of fecal streptococci indicates the presence of warm-blooded animal pollution from a variety of mammals and birds. Because fecal streptococci do not occur in pure water or virgin soil and because they do not multiply in water, a decision was made to include a study to determine their incidence and to make comparisons of fecal coliform to fecal streptococci ratios. Strains of *Streptococcus fecalis* are not considered pathogenic but have been known to occur with endocarditis. *Streptococcus bovis* and *Streptococcus equinus* from cows and horses were identified along with *S. fecalis*.

The presence of these three strains of streptococci in untreated water is an indicator of fecal pollution by warm-blooded animals. Accordingly, the results obtained between 1972–1974 on Chautauqua Lake indicated a high correlation between the incidence of fecal streptococci and fecal coliforms with the streptococci being found in all locations except the midlake sites.

The literature describes data produced elsewhere indicating that when the ratios of fecal streptococci to fecal coliforms are greater than one, the presence of fecal pollution from human sources is indicated. Similarly, ratios less than one are indicative of nonhuman sources of fecal contamination. The ratios obtained as part of these studies are greater than 1.0 for many sites, but no clear pattern was shown to exist. Further studies are necessary before any definitive statements can be made.

The Salmonella Group

Beginning in 1973 and continuing in 1974, selected studies were made on the *Salmonella* group in and around Chautauqua Lake. Using serological techniques, the *Salmonella* group was detected at several locations and efforts were made to provide for species identification. All samples tested for *Salmonella typhosa* proved negative. Positive identifications were obtained, however, for *Salmonella choleraesius* in Big Inlet Creek and *Salmonella typhimurium* in Big Inlet Creek, Little Inlet Creek, Lighthouse Creek, and at the Stow Ferry. *Salmonella paratyphi* was detected at all of the above locations in addition to Dutch Hollow Creek.

The reader's attention is called to the fact that the rate of isolation for *Salmonella* was greatest at those locations where the fecal coliform counts were greater than 200/100 ml.

Summary

Bathing water standards have been suggested at 200 fecal coliforms or less per 100 ml. Coliform are an acceptable indicator of the presence of fecal material which may contain other pathogenic bacteria. The standard of 200/100 ml for fecal coliforms is therefore probably reasonable.

One important aspect not studied is dose contact. In the development of a short- and long-term management plan for Chautauqua Lake, dose contact studies should be made. As human encounters with water increase through recreational contact, consumption of water, and other ways, the more certain we must be of the dose in the water. It would seem that these selected bacteriological studies offer a small but important segment of information which should be useful in the development of public health standards and in the design and operation of public works facilities for the treatment of both drinking water and domestic sewage.

Macroinvertebrates

There have been two major studies of macroinvertebrates of Chautauqua Lake. Townes (1938) included quantitative information at the generic and specific levels, and T. A. Erlandson (unpublished) obtained quantitative data primarily at the family level from dredge and macrophyte samples. Most of Townes' data are from 174 dredge samples taken primarily from Arnold's Bay in the lake's southern basin, but include insects collected as adults from the vegetation along the shore and assumed to be developed from immature stages inhabiting the lake itself. Erlandson's data were obtained from 75 dredge samples taken along the length of the lake in 1972, as well as from extensive macrophyte sampling conducted in several areas around the lake from 1972 through 1974. Table 32 presents a list of taxa compiled from both studies. A total of 109 genera were positively recorded, and several other genera were not identified. Of the 109 genera identified, 62 were insects, representing the largest single class. Excluding the leeches and chironomids, which were not identified to genus by Erlandson, a total of 92 genera were considered in both reports. Of these 92 genera, 35 were reported in both studies, 39 by Townes only, and 18 by Erlandson only. Thus, using the same genera considered, there were 73 genera reported for the lake in 1938, and 53 genera collected in the 1970's. Although Townes and Arlandson used different sites and collection methods, these data indicate that a decrease in macroinvertebrate diversity occurred between the 1930's and the 1970's. The decrease is especially notable for mites and insects which may indicate deteriorating water and/or habitat quality.

Both Townes and Erlandson obtained data on macroinvertebrate density. Townes included the density for each species collected and reported an average of 39 gm of invertebrate per square meter. According to Townes, weed beds contain 60 to 100 gm of invertebrates per square meter, while the weedless bottom supports 10 to 50 gm per square meter. Erlandson compared the benthos of the lake's two major basins by taking 40 dredge samples along the length of the northern basin and 35 along the length of the southern basin. Data from both the 1938 and 1972 surveys are summarized for the major taxa in Table 33.

Considering the ranges in the number of individuals per square meter, it is evident that between the 1930's and the 1970's there has been a notable decrease in the density of snails, oligochaetes, and chironomids, possibly indicating a deterioration in water and/or habitat quality. Further, Table 33

TABLE 32

Macroinvertebrate Taxa Collected from Chautauqua Lake

Phylum, Class, Order	Family	Genus and number of species[a]
Cnidaria		
Hydrozoa		
Hydroida	Hydridae	*Hydra* (?)[b]
Platyhelminthes		
Turbellaria		
Tricladida	Planariidae	*Dugesia* (?)[b]
Aschelminthes		
Aphasmidia		
Enoplida	Dorylaimidae	*Dorylaimus* (?)[c]
Tardigrada[4] (unclassified)		*Limnomermis* (?)[c]
Ectoprocta		
Gymnolaemata		
Ctenostomata	Paludicellidae	*Paludicella* (1)[c]
Phylactolaemata	Plumatellidae	*Plumatella* (?)[b]
Plumatellina	Cristatellidae	*Cristatella* (1)[d]
	Lophopodidae	*Pectinatella* (1)[b]
Mollusca		
Gastropoda	Planorbidae	*Gyraulus* (3)[b]
		Helisoma (4)[b]
	Physidae	*Physa* (1)[b]
	Lymnaeidae	*Lymnaea* (2)[c]
	Hydrobiidae	*Amnicola* (1)[b]
		Bithynia (1)[c]
	Valvatidae	*Valvata* (2)[b]
	Viviparidae	*Viviparus* (1)[d]
		Campeloma (1)[b]
	Ancylidae	*Ferrissia* (1)[b]
Pelecypoda	Unionidae	*Anodonta* (1)[b]
		Elliptio (1)[c]
		Lampsilis (2)[b]
		Ptychobranchus (1)[c]
	Sphaeriidae	*Musculium* (2)[b]
		Pisidium (1)[b]
		Sphaerium (2)[b]
Annelida		
Oligochaeta	Tubificidae	*Branchiura* (?)[d]
		Stylaria (2)[b]
		Limnodrilus (1)[b]
	Naididae	*Naidium* (1)[b]
	Lumbriculidae[d]	unidentified
	(plus a number of unidentified forms)	

TABLE 32 *(Continued)*

Phylum, Class, Order	Family	Genus and number of species[a]
Hirudinea[e]	Glossiphoniidae	*Glossiphonia* (2)
		Helobdella (1)
		Placobdella (1)
	Erpobdellidae	*Erpobdella* (1)
	Hirudidae	*Haemopis* (1)
	Piscicolidae[b]	unidentified
Arthropoda		
Crustacea		
Isopoda	Asellidae	*Asellus* (1)[b]
Amphipoda	Gammaridae	*Gammarus* (1)[b]
Decapoda	Talitridae	*Hyalella* (1)[b]
	Astacidae	*Cambarus* (1)[b]
Arachnida		
Acarina	Limnocharidae	*Limnochares* (?)[d]
	Lebertiidae	*Frontipoda* (1)[c]
	Limnesiidae	*Limnesia* (?)[b]
	Hygrobatidae	*Hygrobates* (?)[c]
	Unionicolidae	*Koenikea* (?)[c]
		Unionicola (?)[c]
	Pionidae	*Piona* (?)[c]
	Arrenuridae	*Arrenurus* (?)[c]
	Oribatidae	*Notaspis* (?)[c]
Insecta		
Collembola	Sminthuridae	*Sminthurides* (1)[d]
Ephemeroptera	Ephemeridae	*Hexagenia* (1)[d]
	Heptageniidae	*Stenonema* (3)[d]
	Baetidae	*Choroterpes* (1)[d]
		Caenis (3)[b]
		Callibaetis (1)[d]
Odonata	Aeschnidae	*Basiaeschna* (1)[c]
		Aeschna (1)[c]
		Anax (1)[c]
	Cordulegasteridae	*Didymops* (1)[c]
		Epicordulia (1)[c]
	Libellulidae	*Libellula* (2)[c]
		Leucorrhinia (1)[c]
		Sympetrum (1)[c]
	Lestidae	*Lestes* (1)[c]
	Coenagrionidae	*Amphiagrion* (1)[c]
		Enallagma (7)[b]
		Ischnura (2)[c]

(Continued)

TABLE 32 *(Continued)*

Phylum, Class, Order	Family	Genus and number of species[a]
Hemiptera	Gerridae	*Gerris* (1)[d]
	Notonectidae[d]	unidentified
	Pleidae	*Plea* (1)[c]
	Nepidae	*Ranatra* (1)[b]
	Corixidae[b]	unidentified
Megaloptera	Sialidae	*Sialis* (1)[d]
Coleoptera	Dytiscidae	*Bidessus* (1)[b]
	Gyrinidae	*Byrinus* (1)[d]
	Hydrophilidae	*Berosus* (1)[b]
	Psephenidae	*Psephenus* (1)[c]
	Elmidae	*Elmis* (1)[b]
	Haliplidae[d]	unidentified
	Curculionidae	*Hyperodes* (1)[d]
	Chrysomilidae	*Donacia* (1)[d]
Arthropoda		
Insecta		
Trichoptera	Hydroptilidae	*Agraylea* (1)[c]
		Orthotrichia (1)[c]
		Hydroptila (1)[c]
		Oxythira (1)[c]
	Polycentropidae[c]	unidentified
	Limnephilidae	*Pycnopsyche*[d]
	Molannidae	*Molanna* (2)[b]
	Leptoceridae	*Leptocerus* (1)[b]
		Leptocella (2)[c]
		Oecetis (3)[b]
		Setodes (1)[c]
		Mystacides (2)[c]
		Triaenodes (1)[c]
	Helicopsychidae	*Helicopsyche* (1)[c]
	Glossosomatidae[d]	unidentified
Lepidoptera	Pyralidae	*Nymphula* (1)[b]
Diptera	Culicidae	*Chaoborus* (2)[b]
	Ceratopogonidae	*Culicoides* (1)[d]
		Palpomyia (1)[c]
		Probezzia (1)[c]
	Chironomidae[f]	*Pentaneura* (2)
		Tanypus (2)
		Procladius (1)
		Clinotanypus (1)
		Orthocladius (1)
		Trichocladius (1)
		Cricotopus (2)
		Coryoneura (1)

TABLE 32 *(Continued)*

Phylum, Class, Order	Family	Genus and number of species[a]
		Tanytarsus (10)
		Pseudochironomus (1)
		Pentapedilum (1)
		Chironomus (34)
	Scopeumatidae	*Hydromyza* (1)[c]
	Psychodidae	*Psychoda* (1)[d]
	Tabanidae	*Chrysops* (1)[d]

[a] The number in parenthesis indicates the number of species identified for that genus.
[b] The genus was recorded by both Townes (1938) and T. A. Erlandson *et al.* (unpublished).
[c] The genus was recorded by Townes (1938) only.
[d] The genus was recorded by T. A. Erlandson *et al.* (unpublished) only.
[e] No generic determinations of leeches were made by Erlandson *et al.*
[f] No chironomid generic determinations were made by Erlandson *et al.*

shows that there are significant differences in the macroinvertebrate densities of the northern and southern basins, with the southern basin having consistently lower density values for all taxa considered. The cause of these differences is not known, but they may be related to chemical herbicide treatment of the lake, especially the southern basin, since the 1940's. Erlandson (unpublished) showed that there were no significant correlations between benthos densities and depth, as well as various bottom sediment parameters, namely, percent of sand, silt, organic matter, and clay.

In addition to the dredge samples taken by Erlandson, a vast quantity of phytomacrofauna data were obtained from 1972 to 1974 only some of which is reported (Erlandson *et al.,* 1974). These data, obtained from macrophyte samples taken from untreated areas as well as areas treated with the herbicide Ortho Diquat and areas mechanically harvested, have been only partially analyzed. Ultimately, such studies will yield information on the effects of herbicides and harvesting on macroinvertebrate populations. The macrophyte/invertebrate data that have been analyzed show that platyhelminthes, molluscs, annelids, and arthropods are most frequently found in association with plants, as follows:

Platyhelminthes: 1 family
Molluscs: 8 families
Annelids: 5 families
Arthropods: 39 families

TABLE 33

Comparison of the Density of Benthic Macroinvertebrates in 1937[a] and in the Northern and Southern Basins in 1972[b]

Taxon	1937	Northern basin (1972)		Southern basin (1972)	
	Range (no./m^2)	Range (no./m^2)	$\bar{x}/m^2$	Range (no./m^2)	$\bar{x}/m^2$
Hydra	50–100	—	—	—	—
Dugesia	50–9000	—	—	—	—
Gastropoda	10–6500	0–435	99.2	0–359	54.0
Large Pelacypoda	1–10	—	—	—	—
Sphaerndae	25–800	0–1323	165.9	0–208	18.9
Oligochaeta	50–13,000	0–718	151.2	0–473	112.3
Hirudinea	50–750	0–737	28.4	0–302	14.0
Isopoda	50–3000	—	—	—	—
Ephemeroptera naiads	10–250	—	—	—	—
Coleoptera	0–225	—	—	—	—
Trichoptera larvae	10–2800	—	—	—	—
Chaoboridae larvae	25–1000	0–1285	173.9	0–95	9.2
Chironomidae larvae	25–5000	0–1323	362.9	0–1285	223.6

[a] Source: Townes (1938).
[b] Source: Erlandson *et al.* (unpublished).

The number of invertebrates per gram dry weight of plant material varies greatly, depending on site and plant species, ranging from 5.71 to 1623, with a mean of 234.2. Normally, the most abundant invertebrates associated with macrophytes are snails, amphipods, and chironomids.

Aquatic Vascular Plants

A large variety of rooted vascular plants, or aquatic macrophytes, occurs in Chautauqua Lake. McVaugh (1938) reported that several species, including forms of *Potamogeton* and *Scirpus,* were found in various parts of the lake. But, as floristic studies of the 1970's are still in progress (S. A. Nicholson, unpublished, 1973), accurate comparisons between the current flora and that of 1937 cannot readily be made. Also, accurate characterization of contemporary communities must await thorough analyses of accumulated biomass data.

Species Distributions

Acciardi *et al.* (1973) discussed distributions of several macrophytes over depth and substrate gradients. Subsequently, Levey *et al.* (1973) quantified sediment size parameters and related them to macrophyte abundance. It is believed that there are statistically significant correlations between macrophyte species abundance (biomass) and sediment size parameters. Furthermore, it is hypothesized that water activity or turbulence vs. texture per se, is of key importance in regulating macrophyte distributions.

Zonation patterns of several macrophyte species across depth have been presented by Nicholson and Aroyo (1973) for an undisturbed transect. In discussing these patterns in more detail, Nicholson and Aroyo offered evidence that biotic factors, e.g., interspecific competition, not merely light, influenced depthward limits of many species. Recent species removal experiments (S. A. Nicholson, unpublished) modeled after J. L. Harper's competition studies in terrestrial communities will better define the nature and importance of inspecific competition in regulating population distributions of macrophytes (S. A. Nicholson, unpublished).

Density

Macrophytes occur throughout the lake from shore to 3–4 m, but are generally concentrated between 1.3–2.6 m. There are, however, large spatial and horizontal variations. For example, biomass is greater on silty sites than on sandy sites in shallow water (1.5 m) but not deeper (Nicholson *et al.,* 1975). Biomass also tends to increase with enrichment and/or disturbance, but is not necessarily greater at stream mouths and bays than elsewhere as some have stated. Biomass may reach 300+ and 700 gm/m², dry weight above ground, in rich submergents and emergents, respectively (Nicholson, unpublished). Inclusion of below-ground biomass may increase the above values by 10–20% (Nicholson and Best, 1974) and 100–200% (Nicholson and Aroyo, 1975), respectively. Since 15–25% of submergents' biomass is ash (S. A. Nicholson and L. W. Post, unpublished), peak organic standing crop may be 225–255 gm/m² or more above ground in these communities.

The above root:shoot ratio pattern tends toward an increasing value of this ratio from submergents shoreward where all vegetation zones are present. But leaf area index tends to be constant at 5, across the hydrosere (Nicholson and Best, 1974), because shoreward decreases in leaf thinness (cm²/gm) cancel out increases in above-ground biomass. However, the leaf area index of chronically enriched/disturbed communities can be much greater, e.g., 10–15 in early season *P. crispus* (Nicholson and Best, 1974), and 20–30 in late season *Myriophyllum* (Nicholson *et al.*, 1974).

Predictive dimension–weight regression equations as have been derived for individual (Nicholson, 1974a) and mixed tree populations (Monk *et al.*, 1970) have been developed for some macrophytes. For example, Nicholson *et al.* (1974) found good length–weight regression fits for *Myriophyllum* sp. but wide-diameter–weight scatter.

Production

Macrophyte productivity varies temporally and spatially in accordance with biomass trends discussed previously. Maximum above-ground production rates in submergents, as estimated by harvest methods, approximate 1–3 gm m^{-2} day^{-1} in some communities in May (Nicholson *et al.*, 1973). Total minimum estimated production for macrophyte communities from May through August is on the order of 100–450+ gm/m² above ground plus about 10–20% below ground additional.

However, true macrophyte net production is higher than the above ranges since losses due to mortality, shedding, grazing, decay, and DOM exudates have not as yet been included (S. A. Nicholson, unpublished). Grazing has generally been found to be small in macrophytes, approximating 5–10% of net production (Westlake, 1965) and angiosperm communities in general. DOM losses, though measurable, are also probably quite low, as are losses to decay in healthy, intact communities. Although frequent biomass harvests as were obtained in many communities should minimize losses to mortality, its potential importance cannot be neglected. Cold season growth by semievergreen forms must also be considered as another addition to annual totals.

Fish

Chautauqua Lake is generally inhabited by fish characteristic of shallow eutrophic lakes in the midwestern section of the United States. The lake is situated in the Ohio Basin and its fish fauna resembles those of the Mississippi region (Evermann and Goldsborough, 1902; Greeley, 1938). In this respect, the fish of Chautauqua Lake are distinct from those of many lakes in the State of New York which drain into the Atlantic Ocean and are characterized by native Atlantic drainage fish.

As a habitat for fish, Chautauqua Lake is generally restricted to warm water species. Because of its shallow morphometry, the lake is essentially homothermal and provides little habitat for cold-water species. The only cold-water areas during the summer months are the eight deep kettle holes in the northern basin. However, these deep holes become anaerobic during the summer and during periods of prolonged ice cover in the winter (Tressler and Bere, 1938; T. Storch, unpublished). Except for these deep holes, which comprise a very small fraction of the total lake area, the lake maintains oxygen concentrations at or near saturation levels throughout the entire year. Thus, summer and winter oxygen concentrations impose no apparent stress on indigenous fish species of Chautauqua Lake.

One of the first known lists of Chautauqua Lake fish was published in 1902 by Evermann and Goldsborough. They reported 32 species and indicated that some of the common fish of the lake included muskellunge (*Esox masquinongy ohioensis* Kirtland), bluegill (*Lepomis macrochirus* Rafinesque), largemouth bass (*Micropterus salmoides* lacepede), brown bullhead (*Ictalurus nebulosus nebulosus* Le Sueur), and black bullhead (*Ictalurus melas* Rafinesque). Carp (*Cyprinus carpio* Linnaeus), which was introduced into the lake prior to 1900, was also reported to be common. In a later study conducted in 1937, five of the species observed in 1902 were not found (Greeley, 1938; Odell and Senning, 1938): paddlefish (*Polydon spathula* Walbaum), shortnosed gar (*Lepisosteus platostomus* Rafinesque), bowfin (*Amia calva* Linnaeus), black bullhead (*Ictalurus melas* Rafinesque), and red fin sucker (*Moxostoma aureolum*). However, 11 new species were found in 1937, including an important game fish, calico bass (*Pomoxis nigromaculatus* Le Sueur).

The most recent list of Chautauqua Lake fish species was compiled in 1975 by D. Bimber (unpublished) (Table 34). Through an extensive review of the literature and of annual records kept by the New York State Department of Environmental Conservation, Bimber noted that a number of new species are now present which were not found during the 1937 survey. Included is a new game fish, walleye (*Stizostedeon vitreum vitreum* Mitchill), which has become important in the lake's sports fishery.

With respect to fish numbers, Chautauqua Lake is a very productive lake (Odell and Senning, 1938; Mooradian and Shepherd, 1973). In comparison with other lakes in New York State, Odell and Senning (1938) reported that few lakes in the state equaled Chautauqua Lake in the large number of fish produced. This is in part a result of the dense macrophytic flora (McVaugh, 1938; Anderson, 1973) and the abundant invertebrate fauna (Townes, 1938) which exist throughout much of the lake basin. In addition, the large numbers and diversified types of fish in the lake play an important role in the overall size of the fish population. Many of the fish species are of little value for human consumption or for game fish. However, with the exception of a few types, the group of nongame fish is made up of small species which are

TABLE 34

List of Chautauqua Lake Fish Species[a]

Common name	Scientific name[b]
Paddlefish	*Polyodon spathula* Walbaum
Bowfin	*Amia calva* Linnaeus
Longnose gar	*Lepisosteus osseus* Linnaeus
Shortnose gar	*L. platostomus* Rafinesque
Spotted gar	*L. oculatus* Winchell
Common sucker	*Catostomus commersoni* Lacepede
Northern hog sucker	*Hypentelium nigricans* Le Sueur
Silver red horse	*Moxostoma anisurum* Rafinesque
Carp	*Cyprinus carpio* Linnaeus
Goldfish	*Carassius auratus* Linnaeus
Golden shiner	*Notemigonus crysoleucas* Mitchill
Redside dace	*Clinostomus elongatus* Kirtland
Creek chub	*Semotilus atromaculatus atromaculatus* Mitchill
Western blacknosed dace	*Rhinichthys atratulus meleagris* Agassiz
Mimic shiner	*Notropis volucellus volucellus* Cope
Blacknose minnow	*N. heterolepis* Eigenmann and Eigenmann
Blackchin minnow	*N. heterodon* Cope
Spottail shiner	*N. hudsonius hudsonius* Clinton
Western satin fin minnow	*N. spilopterous* Cope
Emerald shiner	*N. atherinoides* Rafinesque
Central common shiner	*N. cornutus crysecephalus* Rafinesque
Fathead minnow	*Pimephales promelas* Rafinesque
Bluntnose minnow	*Pimephales notatus* Rafinesque
Stoneroller minnow	*Campostoma anomalum* Rafinesque
Barred mad tom	*Noturus muirus* Jordan
Brown bullhead	*Ictalurus nebulosus nebulosus* Le Sueur
Black bullhead	*I. melas* Rafinesque
Mad tom	*Schilbeodes marginatus*
Muskellunge	*Esox masquinongy ohioensis* Kirtland

TABLE 34 *(Continued)*

Common name	Scientific name[b]
Grass pickerel	*E. americanus vermiculatus* Le Sueur
Banded killifish	*Fundulus diaphanus* Le Sueur
Brook silversides	*Labidesthes sicculus* Cope
White bass	*Morone chrysops* Rafinesque
Largemouth bass	*Micropterus salmoides* Lacepede
Smallmouth bass	*M. dolomieui* Lacepede
Bluegill	*Lepomis macrochirus* Rafinesque
Pumpkinseed	*L. gibbosus* Linnaeus
Rock bass	*Ambloplites rupestris* Rafinesque
Calico bass	*Pomoxis nigromaculatus* Le Sueur
White crappie	*P. annularis* Rafinesque
Western johnny darter	*Etheostoma nigrum nigrum* Rafinesque
Bluntnose darter	*E. chlorosomum* Hay
Rainbow darter	*E. caeruleum* Storer
Iowa darter	*E. exile* Girard
Striped fantail darter	*E. flabellare lineolatum* Agassiz
Log perch	*Percina caprodes* Rafinesque
Blackside darter	*P. maculata* Girard
Yellow perch	*Perca flavescens* Mitchill
Walleye	*Stizostedeon vitreum vitreum* Mitchill
Mottled sculpin	*Cottus bairdi* Girard
Brown trout	*Salmo trutta* Linnaeus
Rainbow trout	*S. gairdneri* Richardson
Brook trout	*Salvelinus fontinalis* Mitchill
Cisco	*Coregonus artedi huronicus* Koelz
Redfin sucker	*Moxostoma aureolum*

[a] Source: D. Bimber (unpublished).
[b] Nomenclature after Eddy (1969).

essential forage for the game species (Greeley, 1938). Despite large numbers of predatory fish, forage fish remain plentiful (Odell and Senning, 1938).

The lake has a great abundance of pan fish such as pumpkinseed (*Lepomis gibbosus* Linnaeus), bluegill (*Lepomis macrochirus* Rafinesque) and calico bass (*Pomoxis nigromaculatus* Le Sueur). During a 5-day creel census in the summer of 1937, Moore *et al.* (1938) reported that of the legal fish caught, 84% were pan fish. Since its introduction into the lake in 1915, the calico bass has become one of the most abundant fish (Odell and Senning, 1938) and supports a large sports fishery, especially in the spring.

Due to the great abundance of fish in Chautauqua Lake, there is strong competition, especially between small fish (Odell and Senning, 1938). For example, Odell and Senning (1938) reported that the yellow perch (*Perca flavescens* Mitchill) are stunted and show good growth only during the first year. Pumpkinseed (*Lepomis gibbosus* Linnaeus) and bluegill (*Lepomis macrochirus* Rafinesque) grow well only for the first 2 years. Odell and Senning (1938) hypothesized that the period of slow growth occurs when young fish reach the life stage where they become dependent on other fish for a large part of their diet. The older and larger fish, on the other hand, are fewer in number, and the competition for prey among them is not as intense. Thus, large fish such as muskellunge (*Esox masquinongy ohioensis* Kirtland) and largemouth bass (*Micropterus salmoides lacepede*) grow well after their first few years.

Apparently, Chautauqua Lake has never supported significant commercial fisheries. Except for the taking of bullheads by hook and line for local market use (Greeley, 1938) and the exporting of some muskellunge (Levy, 1962), commercial fishing has been limited to programs for controlling less desirable fish such as longnose gar (*Lepisosteus osseus* Linnaeus) and shortnose gar (*Lepisosteus platostomus* Rafinesque) in the 1890's (Evermann and Goldsborough, 1902) and carp (*Cyprinus carpio* Linnaeus) in the 1900's (Odell and Senning, 1938; Mottley, 1938). On the other hand, Chautauqua Lake has been well known since the late 1800's for its sports fisheries. The 1937 study by the State of New York Conservation Department revealed that the lake is capable of maintaining remarkably diversified game fish fauna. Probably the most sought after and prized game fish in the lake is muskellunge, followed by small and largemouth bass (Moore *et al.*, 1938; Heacox, 1946; Mooradian and Shepherd, 1973). In fact, heavy sports fishing pressure is believed to have played a major role in keeping down the numbers of large-mouth bass and muskellunge for the past several decades in Chautauqua Lake (Odell and Senning, 1938; Mooradian and Shepherd, 1973).

The Chautauqua muskellunge (*Esox masquinongy ohioensis* Kirtland), a native subspecies of the upper Allegheny River drainage system, is found in greatest abundance in Chautauqua Lake (Heacox, 1946). It has been the most prized and important game fish in Chautauqua Lake for at least 100 years (Evermann and Goldsborough, 1902; Mooradian and Shepherd, 1973). As a result of the importance and the heavy sports fishing pressure placed on the muskellunge, most of the fisheries management and piscine research conducted on Chautauqua Lake have centered around the muskellunge. In 1890, the New York State Fish Commission began hatching muskellunge and stocking them in Chautauqua Lake (Evermann and Goldsborough, 1902). Since 1904, a hatchery has existed on Chautauqua Lake for the purpose of propagating muskellunge (Heacox, 1946). This hatchery has played a crucial role in muskellunge fishery on the lake.

Between 1920 and 1940, the muskellunge population in the lake underwent a drastic decrease as a result of four major factors (Heacox,

1946): (1) increased fishing pressure; (2) use of improved fishing tackle; (3) direct and indirect competition between the carp, calico bass, and the muskellunge for food and habitat; and (4) removal of immature female muskellunge from the breeding population. New programs and regulations implemented by New York State since the 1940's have helped to reestablish a larger muskellunge population. For example, between 1962 and 1971, the annual yield of muskellunge to the sports fisheries ranged between 20 and 40 tons with an annual average in excess of 30 tons (Mooradian and Shepherd, 1973). The role that stocking plays in the muskellunge fisheries is underscored by the results of a study conducted between 1968 and 1971 indicating that over 80% of the total young-of-the-year muskellunge in the lake are hatchery reared (Mooradian and Shepherd, 1973).

Presently, there is concern about the possibility of interspecific competition pressure between the walleye (*Stizostedeon vitreum vitreum* Mitchill) and the muskellunge in Chautauqua. The walleye was unintentionally introduced into the lake just within the last few decades; yet today the walleye contributes significantly to sports fishery in Chautauqua Lake (D. Bimber, unpublished). Interestingly, decreases in the angler catch of yellow perch (*Perca flavescens* Mitchill) and muskellunge in the mid-1970's correlates with increases in the catch of walleye (J. R. Luensman, personal communications). A large study conducted by the New York State Department of Environmental Conservation in 1976 was undertaken to provide definitive information concerning the suspected impact of the walleye on the muskellunge population.

DISCUSSION

Biota

Chautauqua Lake is inhabited by fauna and flora characteristic of productive northern temperate lakes. Diatom genera, such as *Asterionella, Fragilaria, Melosira, Stephanodiscus,* and *Synedra,* well known in productive temperate lakes of the world (Hutchinson, 1967), are abundant in Chautauqua lake at least at certain times of the year. In the summer, the phytoplankton communities are dominated by blue-green algae: *Anabaena, Aphanizomenon,* and *Gloeotrichia.* Phytoplankton annual gross productivity ranges from approximately 275 gm carbon m^{-2} $year^{-1}$ in the northern basin to 375 gm C m^{-2} $year^{-1}$ in the southern basin (T. Storch, unpublished). These productivity rates indicate that Chautauqua Lake is a mesotrophic-eutrophic lake, based on the general classification system where mesotrophic lakes have an annual productivity of between 100 and 300 gm C

m^{-2} $year^{-1}$ and eutrophic lakes have a productivity in excess of 300 gm C m^{-2} $year^{-1}$ (Wetzel, 1975).

In a comparison of 26 New York lakes studied by the U.S. Environmental Protection Agency in 1972, 19 of the lakes had less mean phytoplankton chlorophyll *a* and 17 had greater Secchi disk transparencies than Chautauqua Lake (USEPA, 1974). The relationship between Secchi disk transparency and chlorophyll *a* for both basins of Chautauqua lake (T. Storch, unpublished) and other lakes in the New York Area is presented in Fig. 16. Relative to the New York lakes presented in Fig. 16, the northern basin of Chautauqua Lake is intermediate with respect to chlorophyll *a* and

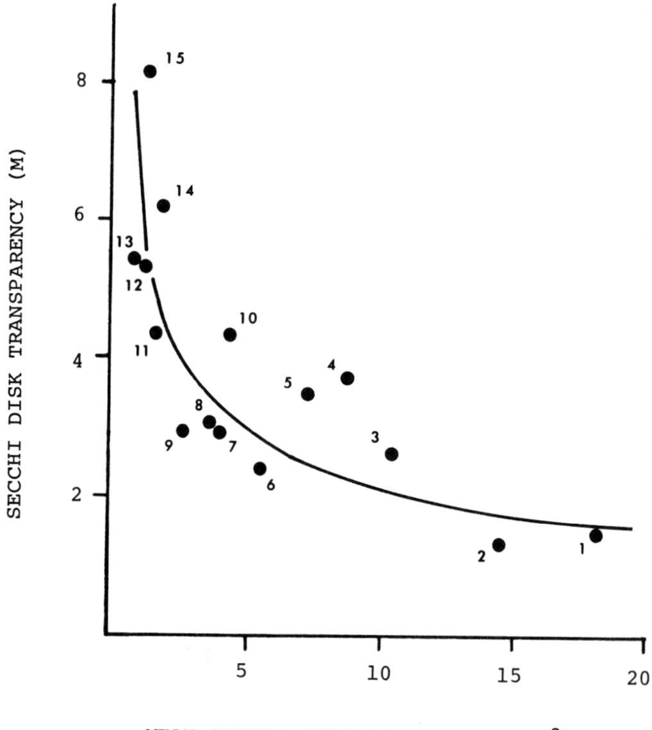

Fig. 16. Mean annual chlorophyll *a* concentration in Chautauqua Lake versus Secchi disk transparency in Chautauqua Lake and other temperate lakes. (1) Chautauqua Lake southern basin 1975–1976; (2) Lake Erie western basin 1970–1971; (3) Oneida Lake 1975; (4) Conesus Lake 1972; (5) Chautauqua Lake northern basin 1975–1976; (6) Hemlock Lake 1972; (7) Hemlock Lake 1973; (8) Owasco Lake 1973; (9) Owasco Lake 1972; (10) Conesus Lake 1973; (11) Muskoka Lake 1969–1970; (12) Skaneateles Lake 1972; (13) Skaneateles Lake 1973; (14) Rosseau Lake 1969–1970; (15) Joseph Lake 1969–1970.

transparency. The southern basin, however, has the highest chlorophyll *a* concentration and the lowest Secchi disk transparency.

Studies conducted to determine the limiting nutrient in Chautauqua Lake suggest that algal growth is nitrogen and phosphorus limited (USEPA, 1974; T. Storch, unpublished). In the northern basin, the limiting nutrient appears to oscillate between phosphorus and nitrogen; phosphorus is limiting in the spring and summer and nitrogen is limiting in the fall (T. Storch, unpublished). In the southern basin, phosphorus is apparently the only nutrient controlling the growth of the algae at all times of the year (T. Storch, unpublished).

The species composition of the invertebrate communities are characteristic of a mesotrophic lake. In deeper waters, the benthos is dominated by species commonly known to tolerate low oxygen concentrations, primarily sphaerid clams, oligochaetes, and chironomids. The fauna of shallow waters are more diverse and include organisms associated with less productive lakes: valvatid and viviparid snails, amphipods, mites, and numerous insects. These shallow water fauna are usually associated with varied and dense macrophytic communities consisting of rooted aquatic plants such as *Najas, Myriophyllum, Ceratophyllum, Vallisneria,* and *Heteranthera.*

Based on changes in the biota, it is apparent that the lake has become more productive over the past four decades. Phytoplankton communities, previously dominated by diatoms throughout the entire year in 1937 (Tressler and Bere, 1938), now consist mainly of blue-green algae during the summer and diatom–phytoflagellate populations in late fall, winter, and early spring (T. Storch, unpublished). A comparison of the lake's macroinvertebrate fauna during the 1970's (T. A. Erlandson, unpublished) with that of the 1930's (Townes, 1938) indicates that their diversity has decreased, and that many taxa, especially those associated with less productive waters, are either lacking or are no longer present in significant numbers. Notable among these organisms are various dragon flies, damsel flies, and caddis flies. Consistent with the macroinvertebrate evidence, a comparison of the zooplankton survey in 1937 (Tressler and Bere, 1938) with that of 1975 (T. A. Erlandson, unpublished) shows that rotifers characteristic of eutrophic waters have become a more important component of the planktonic fauna during the past 40 years.

The decrease in the diversity of the macroinvertebrate fauna is paralleled by a decrease in the diversity of the macrophytic flora. The number of commonly occurring macrophytic species in the northern basin has decreased from 18 to only 8 between 1937 and 1973 (Anderson, 1973). In addition, the macrophytic flora has become generally denser in many areas. During the approximately 40-year period in which these changes in the littoral zone biota have occurred, numerous aquatic macrophyte control programs have been conducted using herbicides and mechanical harvesters (Nicholson, 1974b). However, cause and effect relationships between these control processes and the large-scale changes in the biota have not been definitively established.

Nutrient Loading

Chautauqua Lake was one of 26 lakes in New York State included in the National Eutrophication Survey initiated in 1972 by the U.S. Environmental Protection Agency (USEPA, 1974). That study attempted to estimate the phosphorus and nitrogen loading of the lake from analyses of lake water, grab samples collected near the mouths of eight tributaries, and monthly effluent samples provided by three wastewater treatment plants. Unfortunately the EPA estimates contain significant errors and their conclusion that Chautauqua Lake experiences a net annual nutrient loss has been seriously questioned.

The Environmental Resources Center (ERC) at State University College at Fredonia undertook a year-long study of the phosphorus budget of Chautauqua Lake from February 1975 through January 1976. Analytical and statistical data for determinations of dissolved orthophosphate and of total unfiltered phosphate on samples taken periodically from 11 tributaries and the lake's outlet were presented above in Table 25. Sources and magnitudes of phosphorus loading estimated by ERC are outlined in Table 35. Estimates are based on the following considerations:

Inputs

a. Tributaries. The phosphorus contribution is based on an estimated total discharge (40-year average) for each tributary, corrected for an increased 1975 discharge and the mean total unfiltered phosphorus concentration of 36 to 39 determinations on each stream made from February 1975 through January 1976. Little Inlet received in addition 9.28×10^4 m^3 of industrial processed water containing 54.04 mg/liter P.

b. Minor Tributaries and Immediate Drainage. The phosphorus contribution is based on an estimated total discharge (40-year average) corrected for an increased 1975 discharge and the mean total unfiltered phosphorus concentration found in the 11 major tributaries (less the industrial input in Little Inlet). Of the total calculated for the entire lake, 49.1% and 50.9% were allotted to the upper and lower basins, respectively, corresponding to the relative lengths of their shorelines.

c. Known Municipal Sewage-Treatment Plants. The phosphorus contribution is based on 0.834 kg P cap^{-1} $year^{-1}$, the mean value

TABLE 35

Annual Total Phosphorus Loading in Chautauqua Lake[a]

	kg P/year	Percentage of total in basin
Upper basin		
Inputs		
a. Tributaries		
Big Inlet	244.0	1.97
Little Inlet	5,154.5	41.59
Mud Creek	394.9	3.19
Lighthouse Creek	75.7	0.61
Prendergast Creek	882.2	7.12
Maple Springs Creek	159.4	1.29
Dewittville Creek	580.1	4.68
b. Minor tributaries and immediate drainage (nonpoint load)	779.1	6.29
c. Known municipal sewage treatment plants		
Chautauqua Institute	1,973.2	15.92
Mayville	1,018.3	8.22
d. Condominia	107.0	0.86
e. Septic tanks	172.9	1.40
f. Known industrial (only known included in Little Inlet)		
g. Direct precipitation	852.5	6.88
Total	12,393.8	100.0
Output (to lower basin)	3,810.5	
Net annual phosphorus gain	8,583.3	
Lower basin		
Inputs		
a. Tributaries		
Ball Creek	658.9	5.15
Goose Creek	2,479.7	19.37
Dutch Hollow	184.5	1.44
Bemus	171.8	1.34
b. Minor tributaries and immediate drainage (nonpoint load)	807.7	6.31
c. Known municipal sewage treatment plants		
Lakewood	3,444.4	26.90
Celoron (5%)	250.2	1.95
d. Condominia—none		
e. Septic tanks	259.7	2.03
f. Known industrial—none		
g. Direct precipitation	736.6	5.75
h. Input from upper basin	3,810.5	29.76
Total	12,804.0	100.0
Output (Chadakoin River outlet)	13,433.9	
Net annual phosphorus gain (entire lake)	7,953.4 kg P	

[a] Source: Chautauqua Lake Studies (1976).

determined from analysis of samples from the four sewage treatment plants and the population served. The Celoron plant discharges into the Chadakoin River, not Chautauqua Lake, but some effluent, here assumed to be 5%, may drift back into the lower part of the lake's southern basin.

d. Condominia. The phosphorus contribution is based on 0.834 kg P cap^{-1} $year^{-1}$ and the estimated population.

e. Septic Tanks. The phosphorus contribution is based on 0.417 kg P cap^{-1} $year^{-1}$, half that where effluent is discharged into the lake to allow for possible retention of phosphates in leach beds and soil. The population is estimated by calculating the average number of occupants in near-shore structures only and their average time of residence.

f. Known Industrial Effluent. The only known industrial effluent containing phosphorus is received by Little Inlet and the amount has been included under the contribution by that tributary.

g. Direct Precipitation. The phosphorus contribution is calculated from a 40-year average of 1.04 m precipitation falling annually on the surface of each basin and a mean of 0.088 mg/liter PO_4 determined in 28 samples collected at Allegany State Park from October 1969 through September 1974.

h. Output from Upper Basin to Lower Basin. Phosphorus loading is based on northern basin average retention time and known concentrations of total phosphorus.

Output

The only lake outlet is the Chadakoin River which is gauged at Falconer, New York, downstream from the storm sewer outlet of Jamestown and the wastewater discharge from the Celoron sewage treatment plant. Correcting for these sources, the total streamflow for the year is 362.1×10^6 m^3. The mean concentration of phosphorus found in the Chadakoin River downstream from the Celoron sewage treatment plant was 50.04 μg/liter in a total flow of 363.4×10^6 m^3. This resulted in 18,187.8 kg of phosphorus. Subtracting 95% of the Celoron sewage phosphorus that is contributed to the river and not to the lake, the total net output of phosphorus was 13,433.9 kg.

When the data from the recently concluded phosphorus loading study (see Table 35) are applied to the phosphorus load–water load (mean depth ÷ retention time) relationship formulated by Vollenweider (1973), Chautauqua Lake falls into the mesotrophic range on the phosphorus loading vs.

water loading plot (Fig. 17). Interestingly, Chautauqua Lake is situated in the same part of the plot as other lakes in the New York State area. However, the lake has both the lowest phosphorus load and water load relative to lakes in New York State represented in Fig. 17.

The phosphorus loading–chlorophyll *a* relationship presented by Bachmann and Jones (1974; Jones and Bachmann, 1975) provides a means of predicting the benefits of reducing phosphorus loading to Chautauqua Lake (Fig. 18). According to the work of Edmondson (1970) and Bachmann and Jones (1974), annual phosphorus loading in lakes must be reduced below 0.02 mg $liter^{-1}$ $year^{-1}$ in order to significantly reduce algal biomass to the point (10 mg/m^3 chlorophyll *a* or less) where there is a noticeable improvement in water transparency. Based on the regression line in Fig. 18, in order to achieve such a reduction in Chautauqua Lake, phosphorus input must be reduced by approximately 50% in the northern basin and by approximately 80% in the southern basin. In light of the sources of phosphorus loading to

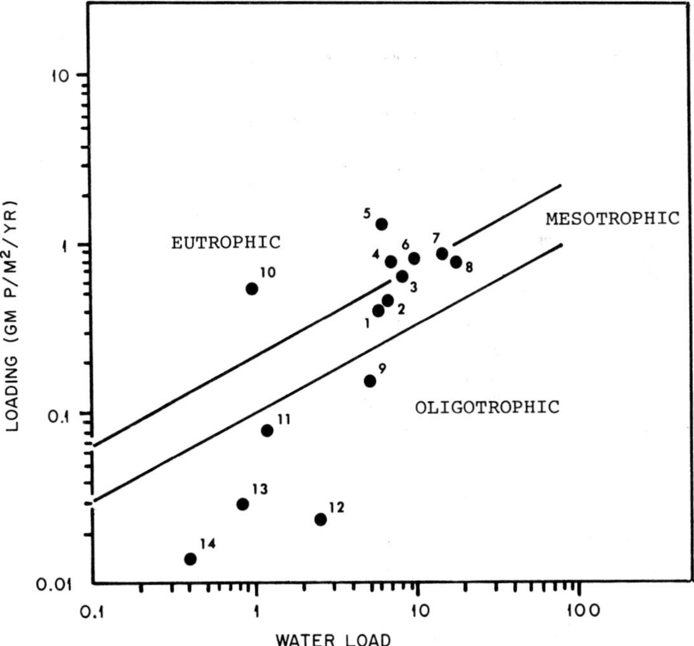

Fig. 17. Annual phosphorus loading versus water loading (mean depth/water retention time) for Chautauqua Lake and other lakes. (1) Chautauqua Lake; (2) Hemlock Lake; (3) Conesus Lake; (4) Lake Ontario; (5) Lake Erie; (6) Oneida Lake; (7) Owasco Lake; (8) Muskoka Lake; (9) Lake Rosseau; (10) Lake Mendota; (11) Lake Joseph; (12) Skaneateles Lake; (13) Lake Superior; (14) Lake Tahoe.

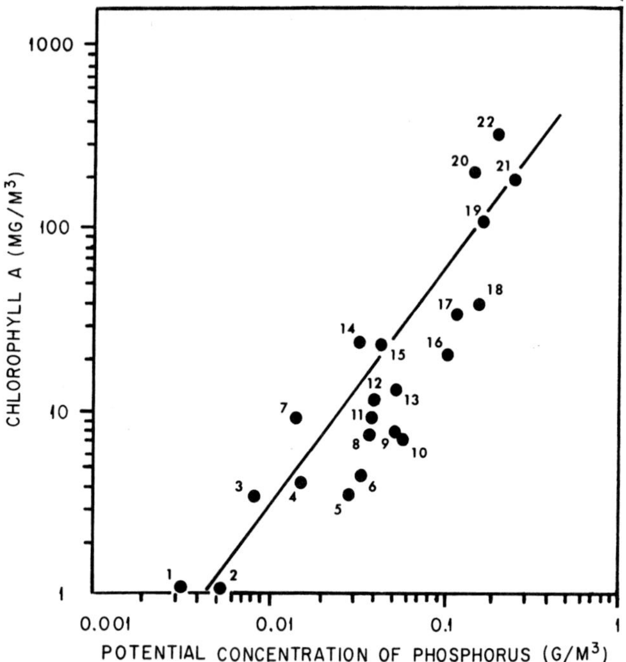

Fig. 18. Mean summer (July–August) chlorophyll *a* concentration versus potential phosphorus concentration for Chautauqua Lake and other lakes. From Bachmann and Jones (1974). (1) Clear Lake, Canada; (2) Skaneateles Lake; (3) Lake West Okoboji, Iowa, 1972; (4) Lake West Okoboji, Iowa 1973; (5) Owasco Lake; (6) Hemlock Lake; (7) Lake Washington, Washington 1967–1970; (8) Lake Sammamish, Washington 1970; (9) Chautauqua Lake, northern basin; (10) Conesus Lake; (11) Lake Sabasticook, Maine; (12) Oneida Lake; (13) Lake Sammamish, Washington, 1965; (14) Lake Washington, Washington 1957–1966; (15) Lower Lake Minnetonka, Minnesota; (16) Lake Mendota, Wisconsin; (17) Lake 227, Canada; (18) Chautauqua Lake southern basin; (19) Lake East Okoboji, Iowa, 1973; (20) Lower Gar Lake, Iowa, 1973; (21) Lake East Okoboji, Iowa, 1971; (22) Lower Gar Lake, Iowa, 1971.

Chautauqua Lake (see Table 35), it is apparent that this degree of phosphorus reduction cannot be readily achieved solely by eliminating phosphorus input from cultural wastes discharged by the sewage treatment plants.

With regard to the benefits of nutrient control in the northern basin, the recent (January 1976) elimination of the large point source of phosphorus on Little Inlet (Chautauqua Malted Milk Company) (see Table 35) and the proposed improvement of the treatment facilities at the Chautauqua Institution sewage treatment plant should lead to a visible improvement in the transparency of the water in the northern basin. This should effect a decrease of approximately 50% in the annual phosphorus input to the northern basin, where the average 1975 summer (July–August) algal chlorophyll *a* concentration was already less than 10 mg/m^3.

In the southern basin, however, the amounts and sources of phosphorus loading appear to dictate that an 80% reduction in the phosphorus load cannot be readily achieved (see Table 35). Thus, while the implementation of "realistic" nutrient reduction programs in the southern basin will lead to decreases in the algal biomass, these programs apparently will not visibly improve water transparency in the southern basin.

SUMMARY

Chautauqua Lake lies centrally within Chautauqua County in western New York State and is known as a productive fishery and recreational resource. The lake was formed during deglaciation of the region between 10,000 and 12,000 years ago and is now the headwaters for the Chadakoin, Connewango, Allegany, Ohio, and Mississippi River System. Divided near its middle by a narrow passage, Chautauqua Lake is comprised of two basins, a northern basin lying between Mayville and Bemus Point, and a more shallow southern basin between Bemus Point and Jamestown.

The northern basin is 14.8 km long, and has a 3.5-km mean width, a 7.8-m mean depth, and a shoreline 16.4 km long. The southern basin is 13.1 km long, and has a mean width of 1.9 km, mean depth of 3.5 m, and a shoreline length of 17.0 km. While the southern lake bottom is flat and featureless, the northern basin has several kettles or deep holes, the deepest of which is 23 m off Point Chautauqua. Chautauqua Lake lies at 399 m above sea level and serves as the catch basin for the surrounding 467.5 km^2 watershed area.

Owing to its shape and generally shallow character, Chautauqua Lake rarely stratifies—the exceptions being the kettle hole depressions in the northern basin. The 23-m deep hole near Point Chautauqua, for example, develops a true thermocline at about 10 m from June through August. Only the lower regions of the lake's deep hole environments become anoxic during the summer period; the rest of the lake displays well-oxygenated waters at all times probably as a result of frequent wind-driven mixing.

The majority of the lake basins' upland area is underlain by Erie and Langford soils both of which are developed on glacial end and ground moraine deposits. Chenango soils are developed in areas underlain by sand and gravel deposits of glacial or fluvial origin. The Wooster soils are associated with clay-rich glacial deposits found in very low relief areas.

Soils mapped as Mudlands include all regions underlain by impervious clay deposits in areas of low relief and perennially high water tables.

Following early English settlement near Jamestown in 1811, the Chautauqua Lake region population growth was spurred by watershed lumbering, potash and pearl ash production, and beef and dairy farming. But the region has become known best for its cultural and recreational resources—the former centered at Chautauqua Institution. Over the past 100 years there has been extensive development of summer homes near the lake. In addition the lake has become the object of more intensive use by tourists and by year-round residents as well. It is estimated that year-round lake watershed population currently approximates 23,000 people while summer population reaches a maximum of 35,000 people.

Cultural development around the lake has led to a variety of point sources of pollution. With the exception of Jamestown, which is not within the lake watershed proper, six primary and one tertiary treatment plants outfall to the lake directly or indirectly by way of a tributary. One sanitary landfill is presently operational within the watershed; several abandoned landfills also exist. A few important industrial point sources in Mayville affect the water quality of Little Inlet and therefore of the lake also. These include a milk processing plant, recently closed down, which typically contributed high levels of phosphorus to Little Inlet; an abandoned municipal dump and a stockpile of highway salt also pollute this tributary.

The chemistry of the lake's waters has been shown to be very typical of mesotrophic-eutrophic lakes. Levels of major and minor elements, inorganic carbonates and bicarbonates, pH, alkalinity, and nutrients are within expected ranges. The only exceptions are the relatively high pH values observed during late summer and early fall and the high concentrations of dissolved arsenic (15–24 μg/liter) due apparently to release of arsenical herbidical chemicals used in the 1950's and now associated with lake sediments. Total alkalinity averages 51 ppm $CaCO_3$ which is almost the same as that reported in 1937. While lake turbidity during the summer months at Maple Springs in 1937 averaged 3.2 m (Secchi Disk), turbidity was significantly greater from 1971 through 1974 during which Secchi disk readings averaged between 2.0 in 1974 and 2.6 m in 1972.

Chemical analyses for phosphorus carried out in 1975 showed that levels of unfiltered total phosphorus averaged 21.4 μg/liter P in the northern basin between April 15 and September 30; in the southern basin, unfiltered total phosphorus averaged 43.2 μg/liter P during the same period. The production of lake algae and weeds parallels the availability of nutrients. In 1975, surface chlorophyll *a* ranged from 0.1 to 101.5 mg/m^3 with lowest values occurring in mid-winter and late spring and the highest values in early spring and summer. Algal primary production, measured by the radioactive

^{14}C method, ranges between 0.08 and 3.37 gm^{-2} day^{-1}. Photosynthetic rates are highest following the melting of the ice cover in April and during the summer blue-green algae bloom (July, August, and September). Algal production appears to be phosphorus limited in Chautauqua lake except for the fall when northern basin algae are nitrogen limited.

Lake macrophytes appear to reflect the impact of cultural change affecting water quality. Comparisons between contemporary macrophyte community structure and diversity and that of 1937 show a significant decline in diversity and the emergence of a small number of dominant pest varieties, especially water milfoil (*Myriophyllum exalbescens*) and curly leaf pondweed (*Potamogeton crispus*). Macrophyte production approximates 1–6 gm m^{-2} day^{-1} from May through August.

Lake tributary water quality is characterized as moderately hard and of the calcium bicarbonate type. Except for higher concentrations of sodium, chloride, and phosphate found in Little Inlet, which are attributable primarily to industrial waste and leachate from highway salt stockpiles, the chemistry of the 11 principal tributaries to Chautauqua Lake is not significantly dissimilar from that of other streams in this region. Average levels of total phosphorus in the lake tributaries range between 14.9 and 1512 μg/liter P, the latter value reflecting very high concentrations of phosphorus observed in Little Inlet. If Little Inlet is excluded, the range of tributary total phosphorus is 14.9–35.2 μg/liter.

A comparison of zooplankton density observed during the summer of 1975 with that reported in 1937 shows that the density may have increased by as much as five times. Recent data document a range of from 172 to 2674 plankters/liter in late June, 1975. An investigation of benthic macroinvertebrates between 1972 and 1974 disclosed that, while 73 genera were identified in 1937, only 53 genera could be identified during 1972–1974. This decrease in macroinvertebrate diversity parallels observed losses in diversity among Chautauqua Lake macrophytes.

Special studies of 28 lake and tributary sites for coliform bacteria from 1972 through 1974 resulted in fecal coliform counts greater than 200 organisms/100 ml at five sites; rates of isolation for *Salmonella* were greatest at these same sites.

The abundant production of a variety of fish species is well known in Chautauqua Lake. One of the first species lists developed in 1902 documented the occurrence of 32 species and identified the major species as muskellunge, bluegill, large mouth bass, brown bullhead, black bullhead, and carp. In 1937 it was found that 11 new species could be identified while five species observed in 1902 could not be. A more recent species list discloses the presence of an even larger number of species than previously compiled, including an important new game fish, walleye. The Chautauqua muskellunge, a native subspecies of the upper Allegheny River drainage system, is found in greatest abundance in Chautauqua Lake and has been the most highly prized and important game fish in the lake for at least 100 years.

When compared to 26 other New York lakes studied by the U.S. Environmental Protection Agency in 1972, 19 of the lakes had less mean planktonic chlorophyll *a* and 17 had greater mean Secchi disk transparancies than Chautauqua Lake. Recent phosphorus loading data, when studied in terms of the lake's mean depth and the retention time of each lake basin, indicate that both lake basins are mesotrophic.

REFERENCES

Acciardi, F., Best, D., DeShong, R. L., Fiedler, A., Flanders, R., Schultz, C., Sobon, B., and Ver, C. (1972). Limnological measurements: Chautauqua Lake. *In* "Chautauqua Lake Studies," (J. R. Mayer, ed.) pp. 190–197. State University College, Fredonia, New York.

Acciardi, F., Nicholson, S. A., and DeShong, R. (1973). Environmental factors related to macrophyte species distributions. *In* "Chautauqua Lake Studies," (J. R. Mayer, ed.) pp. 133–137. State University College, Fredonia, New York.

Anderson, D. (1973). Macrophyte study of upper Chautauqua Lake. *In* "Chautauqua Lake Studies," (J. R. Mayer, ed.) pp. 351–365. State University College, Fredonia, New York.

Bachmann, R. W., and Jones, J. R. (1974). Phosphorus inputs and algal blooms in lakes. *Iowa State J. Res.* **49,** 155–160.

Barnard, W. M., Pazdersky, G., Salerno, M., and Schneider, H. I. (1972). Geochemical data and studies. *In* "Chautauqua Lake Studies," (J. R. Mayer, ed.) pp. 97–113. State University College, Fredonia, New York.

Barnard, W. M., Schneider, H. I., and Robinson, S. L. (1973). Geochemistry of tributaties of Chautauqua Lake, Chautauqua County, N.Y. Part II. *In* "Chautauqua Lake Studies," (J. R. Mayer, ed.) pp. 142–213. State University College, Fredonia, New York.

Barnard, W. M., Schneider, H. I., and Robinson, S. L. (1974a). Relationships of specific conductance to chemical constituents in the Chautauqua Lake watershed, Chautauqua County, N.Y. *Geo. Soc. Am., Abstr. Programs, Northeast. Sect., 9th Annu. Meet.* Abstract, pp. 1–4.

Barnard, W. M., Suib, S. L., Leetaru, H. E., and Campbell, M. A. (1974b). Geochemistry of tributaries of Chautauqua Lake, Chautauqua County, N.Y. Part III. Investigation in 1974. *In* "Chautauqua Lake Report 1974," Vol. 1, pp. 16–25 and Vol. 2, pp. 74–116. State University College, Fredonia, New York.

Barnard, W. M., Suib, S. L., Leetaru, H. E., and Campbell, M. A. (1975). Relationships among discharge, specific conductance, and chemical constituents in tributaries of Chautauqua Lake, Chautauqua County, N.Y. *Geol. Soc. Am., Abstr. Programs, Northeast. Sect., 10th Annu. Meet.*, Abstract, pp. 24–25.

Bogden, J., Anderson, R., and Metzger, W. (1974). Rating curves for the tributaries of Chautauqua Lake. *In* "Chautauqua Lake Studies," (J. R. Mayer, ed.) Vol. 1, p 37 and Vol. 2 pp. 117–128. State University College, Fredonia, New York.

Braun, E. L. (1950). "Desiduous Forests of Eastern North America." Blakiston, Philadelphia, Pennsylvania.

Chautauqua Lake Studies. (1971). A preliminary report on Chautauqua Lake. State University College, Fredonia, New York, and Jamestown Community College, Jamestown, New York.

Chautauqua Lake Studies. (1972). State University College, Fredonia, New York, and Jamestown Community College, Jamestown, New York.

Chautauqua Lake Studies. (1973). State University College, Fredonia, New York, and Jamestown Community College, Jamestown, New York.

Chautauqua Lake Studies. (1974). State University College, Fredonia, New York, and Jamestown Community College, Jamestown, New York.

Chautauqua Lake Studies. (1975). State University College, Fredonia, New York.

Chautauqua Lake Studies. (1976). State University College, Fredonia, New York.

Clute, P. R. (1973). Chautauqua Lake sediments. Master's Thesis, State University College, Fredonia, New York (unpublished).

Clute, P. R., Crowley, D. J., and Metzger, W. (1974). Grain-size parameters of Chautauqua lake sediments. *Geol. Soc. Am., Abstr. Programs* Abstract, Vol. 5, No. 1, p. 13.

Crain, L. J. (1966). Ground-water resources of the Jamestown area, New York with emphasis on the hydrology of the major stream valleys. *N.Y., Water Resour. Comm., Bull.* **58,** 1–167.

Crowley, D. J., Metzger, W. J., Clute, P. R., Flis, J., and Mittlefehldt, D. (1972). Geological studies of Chautauqua Lake. *In* "Chautauqua Lake Studies," (J. R. Mayer, ed.) pp. 198–210. State University College, Fredonia, New York.

DeShong, R., and Wood, K. (1973). Chautauqua Lake phytoplankton. *In* "Chautauqua Lake Studies," (J. R. Mayer, ed.) pp. 232–275. State University College, Fredonia, New York.

Dilly, B. F. (1891). "Biographical and Portraitcyclopedia of Chautauqua County, New York." Gresham, Philadelphia, Pennsylvania.

Eddy, S. (1969). "The Freshwater Fishes." W. C. Brown, Dubuque, Iowa.

Edmondson, W. T. (1970). Phosphorus, nitrogen and algae in Lake Washington after diversion of sewage. *Science* **169,** 690–691.

Erlandson, T. A., Anderson, D. N., Jones, W. H., and Parker, R. (1974). Effect of harvesting on habitat—invertebrates. *In* "Chautauqua Lake Studies," (J. R. Mayer, ed.) pp. 151–157. State University College, Fredonia, New York.

Evermann, B., and Goldsborough, E. (1902). "Notes on the Fishes and mollusks of Lake Chautauqua, N.Y.," Report of the Commissioner, New York State Fish Commission pp. 169–175.

Faigenbaum, H. M., (1938). Chemical Investigations of the Allegheny and Chemung Watersheds. *In* "A Biological Survey of the Allegheny and Chemung Watersheds," Suppl. to 27th Annu. Rep., New York State Dept. of Conservation, Albany.

Flanders, R., and Sobon, B. (1972). Phytoplankton survey. *In* "Chautauqua Lake Studies," pp. 177–189. State University College, Fredonia, New York.

Flanders, R., and Storch, T. (1974). The effect of primary sewage effluent from wastewater treatment plants on the growth of Chautauqua Lake algae. *In* "Chautauqua Lake Studies," (J. R. Mayer, ed.) pp. 69–80. State University College, Fredonia, New York.

Frimpter, M. H. (1973). "Chemical Quality of Streams, Allegheny River Basin and Part of the Lake Erie Basin, New York," Basin Planning Rep. ARB-3. State of New York, Department of Environmental Conservation, Albany.

Frimpter, M. H. (1974). "Ground-water Resources, Allegheny River Basin and Part of the Lake Erie Basin, New York," Basin Planning Rep. ARB-2. State of New York, Department of Environmental Conservation, Albany.

Geldreich, E. E., Clark, H. F., Huff, C. B., and Best, L. C. (1965). Fecal coliform organism medium for the membrane filter technique. *J. Am. Water Works Assoc.* **57,** 208.

Giebner, B. M. (1951). The plankton algae of the southeast and of Chautauqua Lake. *Proc. Rochester Acad. Sci.* **9,** 409–420.

Greeley, S. (1938). Fishes of the area with annotated list. *In* "A Biological Survey of the Allegheny and Chemung Watersheds," Suppl. to 27th Annu. Rep., pp. 48–73. New York State Department of Conservation, Albany.

Gumerman, R. C. (1970). Aqueous Phosphate and Lake Sediment Interaction *Proc. Conf. Great Lakes Res.* **13,** 1–673.

Hayes, F. R., and Phillips, J. E. (1959). "Radiophosphorus equilibrium with mud, plants, and bacteria under oxidized and reduced conditions." *Limnol. Oceanogr.* **3,** 1–459.

Heacox, C. (1946). The Chautauqua lake muskellunge: Research and management applied to a sport fishery. *Trans. North Am. Wildl. Conf.* **11,** 419–425.

Hopke, P. K., Chun Chan, Y., DiPalma, R., Knab, W., Lis, S., Peterson, R., and Ruppert, D. (1972). Chemical studies of Chautauqua Lake. *In* "Chautauqua Lake Studies," (J. R. Mayer, ed.) pp. 78–96. State University College, Fredonia, New York.

Hopke, P. K., Bolton, B., Colasanti, V., Cunningham, W., Foti, J., Korwin, J., Liesing, M., Palmer, L., Peterson, R., Ruppert, D., and Salerno, M. (1973). Physical and chemical limnology. *In* "Chautauqua Lake Studies," (J. R. Mayer, ed.) pp. 3–141. State University College, Fredonia, New York.

Hutchinson, G. E. (1967). "A Treatise on Limnology." Wiley, New York.

Johnson, D. L. (1971). Simultaneous determination of arsenate and phosphate in natural waters. *Environ. Sci. Technol.* **5,** 411.

Jones, J. R., and Bachmann, R. W. (1975). Algal response to nutrient inputs in some Iowa lakes. *Verh. Int. Ver. Theor. Angew. Limnol.* **19,** 904–910.

Kuchler, A. W. (1964). Potential natural vegitation of the conterminous U.S. *Am. Gogi. Soc., Spec. Publ.* No. 36.

Levey, R. A., Nicholson, S. A., and Clute, P. R. (1973). Macrophytes and sediments in Chautauqua Lake: Near-shore sediments associated with macrophyte beds. *In* "Chautauqua Lake Studies," (J. R. Mayer, ed.) pp. 322–344. State University College, Fredonia, New York.

Levy, H. (1962). "Man Against Musky, Stories of the Muskellunge, King of Fresh Water." Stackpole Company, Harrisburg, Pennsylvania.

Lis, S. A., and Hopke, P. K. (1973). Anomalous arsenic concentrations in Chautauqua Lake. *Environ. Lett.* **5,** 45.

Luensman, J. R. (1973). Personal communications based on: Second homes and their impact on the economy of Chautauqua County, Chautauqua County Department of Planning, Mayville, New York.

Luensman, J. R., Malinoski, A., and Phillips, D. (1972). Chautauqua Lake watershed land use survey. *In* "Chautauqua Lake Studies," (J. R. Mayer, ed.) pp. 29–40. State University College, Fredonia, New York.

McMahon, H. G. (1964). "Chautaugua County; A History." Henry Stewart, Inc., Buffalo, New York.

McVaugh, R. (1938). Aquatic vegetation of the Allegheny and Chemung watersheds. *In* "A Biological Survey of The Allegheny and Chemung Watersheds," Suppl. to 27th Annu. Rep., pp. 176-195. New York State Dept. of Conservation, Albany.

Mayer, J. R., Barnard, W., Hopke, P. K., Storch, T., Flanders, R., Dietrich, G., Bolton, B., Salerno, M., Alessi, C., Korwin, J., Suib, S., and Colasanti, V. (1974). Chautauqua Lake chemistry and limnology. *In* "Chautauqua Lake Studies," (J. R. Mayer, ed.) Vol. 1 pp. 8–15 and Vol. 2 pp. 1–73. State University College, Fredonia, New York.

Metzger, W. J. (1974). Sedimentation in Chautauqua Lake. *In* "Chautauqua Lake Studies," (J. R. Mayer ed.) pp. 26–36. State University College, Fredonia, New York.

Metzger, W. J., Crowley, D. J., Clute, P., Flis, J., and Mittlefehldt, D. (1972). Geological studies of Chautauqua Lake. *In* "Chautauqua Lake Studies," (J. R. Mayer, ed.) pp. 198–210 (plus 17 figures). State University College, Fredonia, New York.

Metzger, W. J., Anderson, R., Levey, R., Mittlefehldt, D., and Ostrye, T. (1973). Geological studies of Chautauqua Lake. *In* "Chautauqua Lake Studies," (J. R. Mayer, ed.) pp. 214–231. State University College, Fredonia, New York.

Meyer, F. T. (1921). "A Book of Information About Chautauqua Lake." Lindstrom & Meyer, Jamestown, New York.

Monk, C. D., Child, G. I., and Nicholson, S. A. (1970). Biomass, litter and leaf surface area estimates of an oak-hickory forest. *Oikos* **21,** 138–141.

Mooradian, S., and Shepherd, W. (1973). Management of muskellunge in Chautauqua Lake. *N.Y. Fish Game J.* **20,** 152–157.

Moore, E., and Staff Members. (1938). A creel census at Chautauqua Lake, New York. *Trans. Am. Fish. Soc.* **67,** 130–138.

Mottley, C. (1938). Carp control studies with special reference to Chautauqua Lake. *In* "A Biological Survey of the Allegheny and Chemung Watersheds," Suppl. to 27th Annu. Rep., pp. 226–235. New York State Dept. of Conservatism, Albany.

Muller, E. H. (1963). Geology of Chautauqua County, New York, Part II. Pleistocene geology. *N.Y. State Mus. Sci. Serv., Bull.* **392.**

National Oceanic and Atmospheric Administration (NOAA). (1973a). "Monthly Averages of Temperature and Precipitation for State Climatic Divisions, 1941–1970." Environmental Data Service.

National Oceanic and Atmospheric Administration, (NOAA). (1973b). "Monthly Normals of Temperature, Precipitation, and Heating and Cooling degree Days, 1941–1970." Environmental Data Service.

National Oceanic and Atmospheric Administration (NOAA). (1974). "Climatological Data," Annual Summary. Environmental Data Service.

New York State Conservation Department. (1938). "A Biological Survey of the Allegheny and Chemung Watersheds," Suppl. to 27th Annu. Rep. NYSCD, Albany.

New York State Department of Health. (1970). "Drinking Water Standards in New York State." NYSDH, Albany.

Nicholson, S. A. (1973). (unpublished).

Nicholson, S. A. (1974a). Macrophytes in Chautauqua Lake. *In* "Chautauqua Lake Studies," (J. R. Mayer, ed.) pp. 89–106. State University College, Fredonia, New York.

Nicholson, S. A. (1974b). Effects of harvesting on macrophyte communities in Chautauqua Lake. *In* "Chautauqua Lake Studies," (J. R. Mayer, ed.) pp. 139–150. State University College, Fredonia, New York.

Nicholson, S. A. (1975). Foliage biomass in temperate tree species. *Am. Midl. Nat.* **93,** 44–52.

Nicholson, S. A., and Aroyo, B. (1973). A case study in hydrach zonation. *In* "Chautauqua Lake Studies," (J. R. Mayer, ed.) pp. 285–303. State University College Fredonia, New York.

Nicholson, S. A., and Aroyo, B. (1975). A case study in hydrach zonation. *Vegetatio* **30,** 207–212.

Nicholson, S. A., and Best, D. G. (1974). Root: shoot and leaf area relationships of macrophyte communities in Chautauqua Lake, New York. *Bull Torrey Bot. Club* **101,** 96–100.

Nicholson, S. A., and Rosenthal, B. (1973). Midsummer near-shore phytoplankton communities of Chautauqua Lake. *In* "Chautauqua Lake Studies," (J. R. Mayer, ed.) pp. 255–275. State University College, Fredonia, New York.

Nicholson, S. A., Acciardi, F., and DeShong, R. (1973). Spatial and temporal distribution of macrophytes in Chautauqua Lake. *Bull. Ecol. Soc. Am.* **54,** 17.

Nicholson, S. A., Bakker, L. J., and Setari, C. (1974). Dimension analysis and leaf and stem surface area estimates for *Myriophyllum* Sp. *In* "Chautauqua Lake Studies," Lake Erie Environ. Stud., (J. R. Mayer, ed.) pp. 108–120. State University College, Fredonia, New York.

Nicholson, S. A., Levey, R. A., and Clute, P. R. (1975). Macrophyte-sediment relationships in Chautauqua Lake. *Verh. Int. Ver. Theor. angew. Limnol.* **19,** 2758–2764.

Odell, T., and Senning, W. (1938). Lakes and ponds of the Allegheny and Chemung watersheds. *In* "A Biological Survey of the Allegheny and Chemung Watersheds," Suppl. to 27th Annu. Rep., pp. 74–101. New York State Dept. of Conservation, Albany.

Powers, M. (1924). "Yehsennohwehs." Roycraft Shops, East Aurora, New York.

Ruppert, D. F., Hopke, P. K., Clute, P., Metzger, W., and Crowley, D. (1974). Arsenic concentration and distribution in Chautauqua Lake sediments. *J. Radioanal. Chem.* **23,** 159–169.

Ruttner, F. (1961). "Fundamentals of Limnology" (transl. by D. G. Frey and F. E. J. Fry) 3rd ed, Univ. of Toronto Press, Toronto.

Shepard, F. P. (1954). Nomenclature based upon sand-silt ratios. *J. Sediment. Petrol.* **24,** 151–158.

Shukla, S. S., Syers, J. K., Williams, J. D. H., Armstrong, D. E., and Harris, R. F. (1971). *Soil Sci. Soc. Am., Proc.* **35,** 1–244.

Storch, T., Dietrich, G., Flanders, R., and Priznar, F. (1974a). Algal primary production in littoral and pelagic areas of the northern basin of Chautauqua Lake. *In* "Chautauqua Lake Studies," (J. R. Mayer, ed.) pp. 39–607. State University College, Fredonia, New York.

Storch, T., Dietrich, G., Flanders, R., and Priznar, F. (1974b). Phytoplankton photosynthesis in a harvested and an unharvested littoral site in Chautauqua Lake. *In* "Chautauqua Lake Studies," (J. R. Mayer, ed.) pp. 158–168. State University College, Fredonia, New York.

Tesmer, I. H. (1963). Geology of Chautauqua County, New York. Part I. Stratigraphy and paleontology (Upper Devonian). *N.Y. State Mus. Sci. Serv., Bull.* **391.**

Townes, H. (1938). Studies of the food organisms of fish. *In* "A Biological Survey of the Allegheny and Chemung Watersheds, Suppl. to 27th Annu. Rep., pp. 162–173. New York State Dept. of Conservation, Albany.

Tressler, W. L., and Bere, R. (1938). A Limnological study of Chautauqua Lake. *In* "A Biological Survey of the Allegheny and Chemung Watersheds," Suppl. to 27th Annu. Rep., pp. 196–210. New York State Department of Conservation, Albany.

Tressler, W. L., Wagner, L. G., and Bere, R. (1940). A limnological study of Chautauqua Lake. II. Seasonal variations. *Trans. Am. Microsc. Soc.* **59,** 12–30.

Turner, O. (1849). "Pioneer History of the Holland Purchase of Western New York." Jewett, Thomas & Co., Buffalo, New York.

U.S. Department of the Interior Geological Survey. (1970–1974). "Water Resources Data for New York," Part 2: Water Quality Records. U.S. Environmental Protection Agency, Washington, D.C.

U.S. Environmental Protection Agency, National Eutrophication Survey. (1974). "Report on Chautauqua Lake, Chautauqua County New York," EPA Reg. II, Working Pap. No. 155, pp. 1–38. USEPA, Washington, D.C.

Vollenweider, R. A. (1973). Input-output models. *Schweiz. Arch. Hydrol.*

Warren, R. M. (1878). "Chautauqua Sketches. Descriptive History." Otis, Buffalo, New York.

Westlake, D. F. (1965). Basic data for investigations of the productivity of aquatic microphytes. *In* "Primary Production in Aquatic Environments" C. Goldman, (ed.), Univ. of California Press, Berkeley. pp. 231–247.

Wetzel, R. G. (1975). "Limnology." Saunders, Philadelphia, Pennsylvania.

Young, A. W. (1875). "History of Chautauqua County, New York, from its First Settlement to the Present Time; With Numerous Biographical and Family Sketches." Matthews & Warren, Buffalo, New York.

Limnology of Irondequoit Bay, Monroe County, New York

Thomas T. Bannister and Robert C. Bubeck

ISBN 0-12-107302-5

INTRODUCTION

Irondequoit Bay is a lake 6.7 km in length and 1 km in width located 6 km northeast of the center of the city of Rochester. At the north end, it is separated from Lake Ontario by a sandbar, and its principal tributary, Irondequoit Creek, which drains a 395-km^2 watershed, enters at the south end. Situated in a major metropolitan area, and bounded by steep, wooded hillsides, the bay is an extraordinarily valuable asset of the community. Its scenic value enhances neighboring real estate, its hillsides have great potential as public parks, and, despite large sewage inputs, it is heavily used by boaters and swimmers.

The earliest historical document is a map of the bay recently located in Paris (Dardenne, 1974) prepared about 1684 by the French who navigated the Bay in campaigns against the Senecas. White settlement of the area began about 1800; clearance of forests for farms must have brought sediment as the first pollutant. By the end of the nineteenth century, the bay had become a popular recreation spot, and the shore was dotted with hotels and cottages. Trolley lines provided transportation from Rochester. Human activities on and around the bay in the nineteenth century are recorded in local histories (Thompson, 1956; Sassaman, 1970). Of interest are mentions of commercial fishing, cutting of ice, and the harvest of "flags" (presumably cattails for cooperages).

The eastern half of Rochester drains into Irondequoit Bay via small streams. By 1889, Densmore, Hobbie, and Thomas Creeks were severely polluted by the sewage of an estimated 35,000 persons. Kuichling (1889) noted ". . . the serious defilement of a number of small brooks, which has given rise to many law suits for damages wherein the plaintiffs have been successful and which under our present laws may be permanently enjoined." He further observed that Irondequoit Bay ". . . is one of the most accessible and popular pleasure resorts for the entire community, and the limited amount of partially clarified sewage which now finds its way therein has already produced an appreciable pollution of its waters and the atmosphere in the vicinity of the mouths of the streams into which some of the large sewers now empty." To correct these conditions, Kuichling recommended construction of the Rochester East Side Interceptor. This was built and in operation by 1896 (Kuichling 1907). To what extent the condition of the bay improved is unclear.

In 1912, Whipple (1913) was clearly impressed by the abundance of algae and aquatic vegetation. He wrote "The bay supports an immense growth of aquatic vegetation. There are many acres of flags which grow in such abundance that they are annually harvested. In addition to these flags, there are thick growths of *Chara*, Anacharis, *Ceratophyllum*, Potamogeton, *Lemna*, etc. In places these are so thick as to form an almost impenetrable mass. These water weeds are overgrown with various kind of filamentous algae, and they also harbor large numbers of crustacea and other animal forms. The biological changes that take place in this bay are, therefore, intense, and have an important bearing upon the quality of the water itself. When these growths decay their odor is sometimes very objectionable." These observations together with high counts of phytoplankton (see Phytoplankton Growth in the Bay, p. 173) indicate a highly eutrophic state in 1912.

Sparse evidence suggests that in the 1920's, when suburban growth commenced in townships of the watershed, water quality as judged by clarity, fishing, and species and extent of aquatic weeds began to deteriorate rapidly. By 1939, when the bay was studied extensively (New York State Department of Conservation, 1940; Tressler *et al.*, 1953), the deterioration of the bay and the lower reaches of Irondequoit Creek was well advanced. Water quality surveys by the New York State Department of Health (1955, 1964), the Federal Water Pollution Control Administration (1955, 1965, 1968), and the Monroe County Department of Health (unpublished data, 1965) show that deterioration continued up to the mid-1960's. At this time and continuing to the present, 10–20% of the water entering the bay was raw sewage or treatment plant effluent. Coliform levels frequently reached high levels in the south end of the bay, and very dense phytoplankton blooms occurred from May to October. In recent years, the bay has been frequently likened to a sewage oxidation pond.

Beginning in 1964, in a series of over 20 reports on the water quality of Irondequoit Bay and its tributaries, the Rochester Committee for Scientific Information (1964–1972) brought to public attention the extreme eutrophication and possible health hazards of waters in some parts of the bay. Aroused public concern led to the creation of the Monroe County Pure Waters Agency which undertook construction of improved waste treatment facilities (Monroe County Pure Waters Agency, 1969). Beginning in late 1977, sanitary sewage of the Irondequoit Bay watershed will be transported to the new Van Lare treatment plant, from which effluent will be discharged into Lake Ontario.

In recent years, public interest in the bay has focused on a number of issues. With regard to water quality, the heavy algal blooms, the possible hazards of sewage contamination, and the effects of road deicing salt runoff have been concerns. The potential of the Bay for fishing presumably remains of interest also, although the number of anglers is now small.

Property owners and others along the shore were concerned with flooding during the springs of 1971 and 1972 (when Lake Ontario levels were very high) and especially in June 1972 during Hurricane Agnes. High water caused not only flooding, but also shoreline erosion, undercutting and slumping of the silt bluffs, increased sediment loads of tributaries, and extension of stream deltas. Landfills, particularly at Empire Boulevard and in the wetlands just to the south of the boulevard, have encroached upon bay and wetland areas appreciably. Public planning of future development of the bay area has been actively pursued and has had to deal with the conflicting aims of real estate, commercial, and public interests, and with the problem of reconciling heavy public use with maintenance of woodlands and stabilization of the hillsides. An issue first arising in 1939, and periodically exciting intense interest since then, is the possibility of deepening the outlet into Lake Ontario and removing the low highway and railroad bridges so that pleasure boats can pass freely. It has been argued that deepening the outlet might improve water quality in the bay through increased mixing with Lake Ontario water.

Relatively extensive limnological studies of Irondequoit Bay have been conducted only twice. One was in 1939–1940 as part of the "Biological Survey of the Lake Ontario Watershed" (New York State Department of Conservation, 1940; Tressler *et al.*, 1953). The second was in 1969–1972 when studies were carried out by three independent groups. During this period, Bubeck, Bannister, Diment, and others (see Acknowledgments), at the University of Rochester, studied the physical, chemical, and phytoplankton limnology of the bay; only aspects dealing with road deicing salt runoff have been previously published (Bubeck *et al.*, 1971; Bubeck, 1972; Diment *et al.*, 1973, 1974). During this same period, Forest and associates

(1973, 1975; Forest, 1976), at the State University College at Geneseo, studied the rooted aquatic vegetation of the bay. During the summer of 1972, K. Harbison and a group of University of Rochester students studied the chemistry and other aspects of both the bay and Irondequoit Creek. A report of the creek study is on file (Thorne, 1973). A brief report of the study of the bay is also available (Harbison, 1973).

The purpose of this chapter is to compile and review available limnological information about the bay. Results of earlier studies will be summarized briefly and unpublished findings of recent studies will be presented in detail. The available information provides a basis for discussion of several topics: (1) factors regulating phytoplankton growth and the probable effects of waste diversion, (2) the effects of deicing salt runoff, (3) the probable effect of deepening the outlet to Lake Ontario, and (4) the probable effect of diversion on rooted aquatic vegetation. The diversion soon to commence constitutes a large-scale experiment in lake management; some recommendations are offered for exploitation of this experiment to generate new, fundamental, limnological knowledge.

DRAINAGE BASIN

Geology

The geology of Irondequoit Bay, its watershed, and adjacent areas has been described in articles by Fairchild (1894, 1895, 1896a,b, 1906, 1909, 1913, 1919, 1928, 1930, 1932, 1935), Chadwick (1917), Legette *et al.* (1935), and Grossman and Yarger (1953). Concise summaries were presented by Tressler *et al.* (1953) and Bubeck (1972).

Monroe County is thickly blanketed by glacial drift deposited on sedimentary strata. The latter dip about 7.6 m/km to the south. The strata were laid down during Ordovician, Silurian, and Devonian periods when shallow seas covered the area. Names and thicknesses of older strata exposed in the Genesee River gorge are, from bottom to top, Queenston shale (213 m), Grimsby sandstone (17 m), the Clinton group (24 m) of shales and limestones, Rochester shale (26 m), and Lockport dolomite (52 m). Outcroppings of the latter occur south of the head of Irondequoit Bay. Southward, bedrock is comprised of progressively younger strata: Vernon shale (98 m) just south of the Barge Canal, Camillus shale and dolomite (with inclusions of salt and gypsum) in the latitude of Scottsville and Fishers (168 m), Bertie limestone (26 m) near Mendon and Victor, Onondaga limestone (46 m) along the Monroe–Ontario County line, and the Hamilton group (18 m) of shales and limestones in the vicinity of Hop-

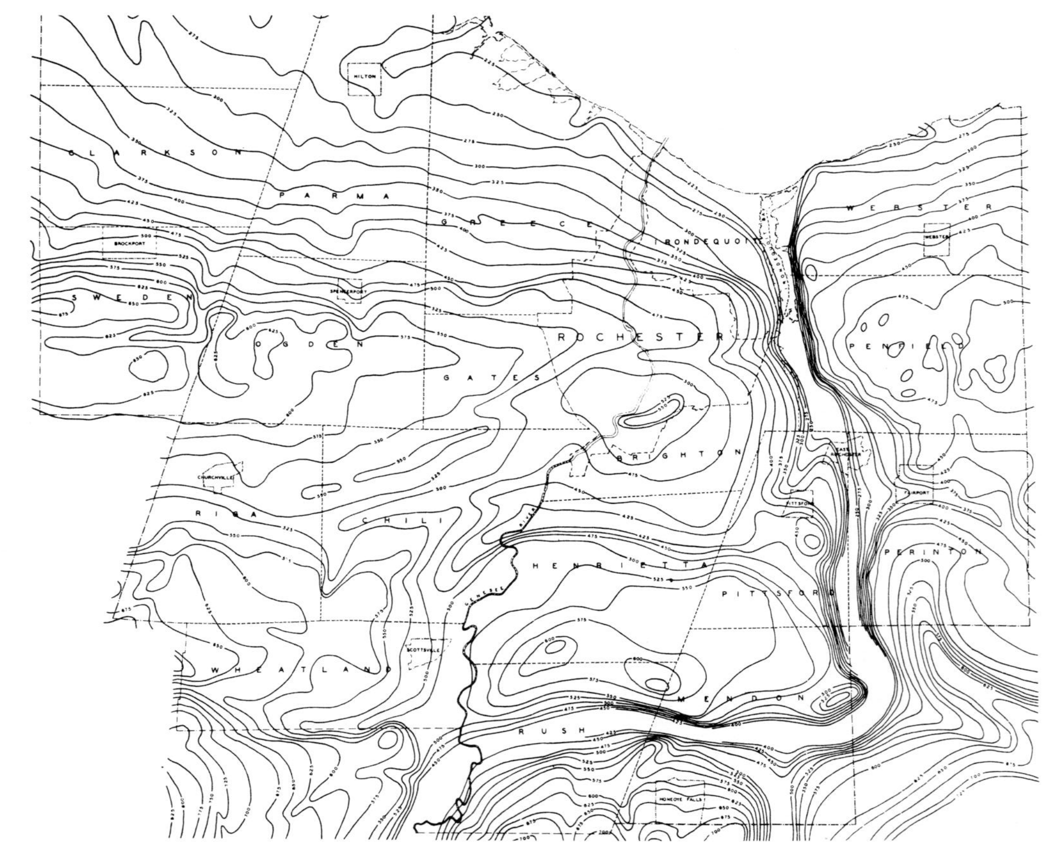

Fig. 1. Bedrock elevations above sea level (in feet). The valley of the preglacial Genesee River is evident; present day-Irondequoit Bay and Irondequoit Creek lie over the old valley. From Leggett *et al.* (1935).

per Hill where Irondequoit Creek has its source. A bedrock map of the area is available (New York State Museum and Science Service, 1970).

Early in the Tertiary Period, stream flow in the region was reversed from a generally southerly to a northerly course, and the northward-flowing Genesee River became a tributary of the great Ontarian River which eroded the present basin of Lake Ontario. The valley developed by the Genesee during this period is clearly defined by the contours of bedrock (Fig. 1), which show that the valley proceeded eastward under Rush and Mendon townships, turned abruptly in the vicinity of Fishers, and then followed a northward course toward the Ontarian River. Irondequoit Bay and the town of East Rochester lie directly over the buried valley of the Old Genesee River.

During the Pleistocene, Lake Ontario and northern New York were covered by a glacial ice sheet which deposited a thick layer of glacial drift throughout the region and completely filled the old Genesee Valley, except for the lower valley over which Irondequoit Bay now lies. South of Bushnell's Basin, in the region of the upper valley, moraines, eskers, kames, drumlins, and kettle lakes are characteristic features. The recession of the ice sheet occurred in stages marked by lakes of different elevations between the ice sheet to the north and the high ground to the south. During this time, the incompletely filled lower valley of the old Genesee (from Penfield northward) was an embayment of Lakes Dawson and Iroquois and great thicknesses of silts and clays accumulated. By the Lake Emmons period, the embayment was closed off leaving a lake in the lower valley; later, as the waters of the Ontario basin continued to recede and became an inland extension of the sea, the lake in the lower valley probably disappeared. Finally, as a result of elevation of the St. Lawrence region, the old Ontarian basin filled with fresh water to become Lake Ontario and Irondequoit Bay was reformed in the old valley of the Genesee. In recent times, a sandbar formed separating Irondequoit Bay from Lake Ontario; thus the Bay became a separate lake.

The drainage systems of the area were greatly altered by the ice sheet. Glacial fill blocked the Genesee from resuming its old course to the east in the vicinity of Rush. Instead it took a new course to the north, eventually excavating the Rochester canyon. To the east, over the buried valley of the old Genesee, developed the Irondequoit Creek system, draining northward into Irondequoit Bay.

Description of the Irondequoit Bay Drainage Basin

Figure 2 shows the boundary of the basin and the coverage by the topographic maps reproduced in Fig. 3. The latter were prepared by the New

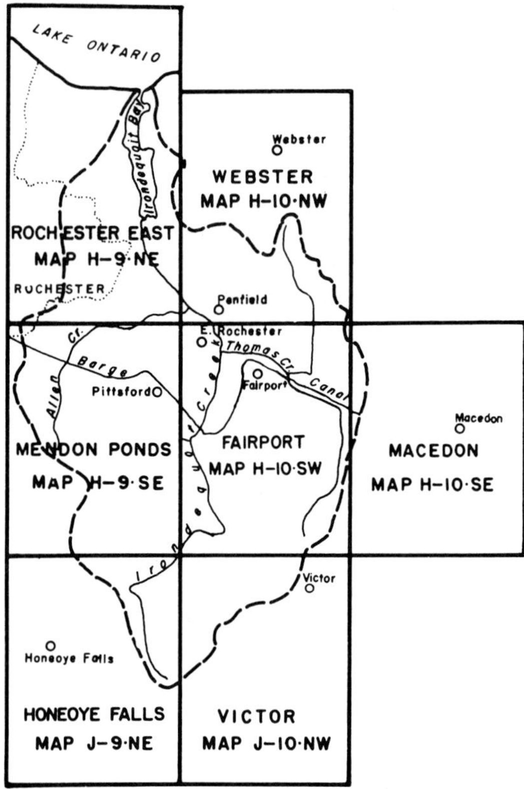

Fig. 2. Outline of Irondequoit Bay watershed and coverage by U.S. Geological Survey quadrangles. From New York State Department of Health (1955).

York State Department of Health (1955) from Geological Survey topographic maps (1:24,000) and show the streams of the basin. The streams and other bodies are identified by index numbers assigned by the New York State Department of Conservation (1940) and revised by the New York State Department of Health (1955). From the sandbar on Lake Ontario to the southernmost point, the north–south distance across the basin is about 22 miles (35.4 km); the maximum east–west width is about 13 miles (20.9 km). The total area of the basin including Irondequoit Bay is about 178 square miles (460 km^2). Elevations range from 246 ft (75 m) (mean level of Lake Ontario and Irondequoit Bay) to about 1100 ft (335 m) atop Hopper Hill at the southwest corner of Victor Township in Ontario County. The basin includes portions of the city of Rochester and the townships of Irondequoit, Webster, Brighton, Henrietta, Penfield, Pittsford, Perinton, and Mendon in Monroe County; portions of the townships of Victor and

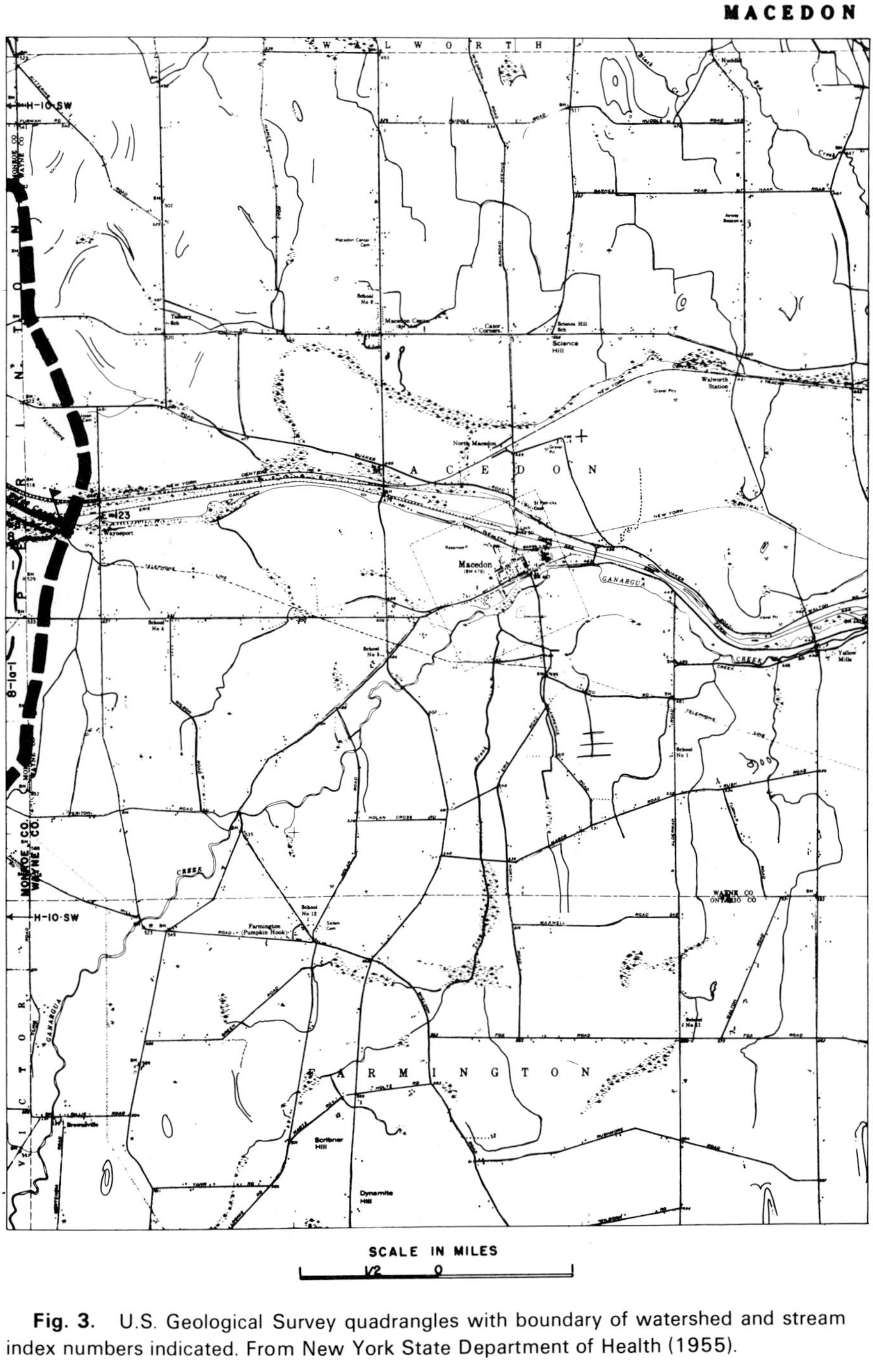

MACEDON

SCALE IN MILES

1/2 0

Fig. 3. U.S. Geological Survey quadrangles with boundary of watershed and stream index numbers indicated. From New York State Department of Health (1955).

ROCHESTER EAST

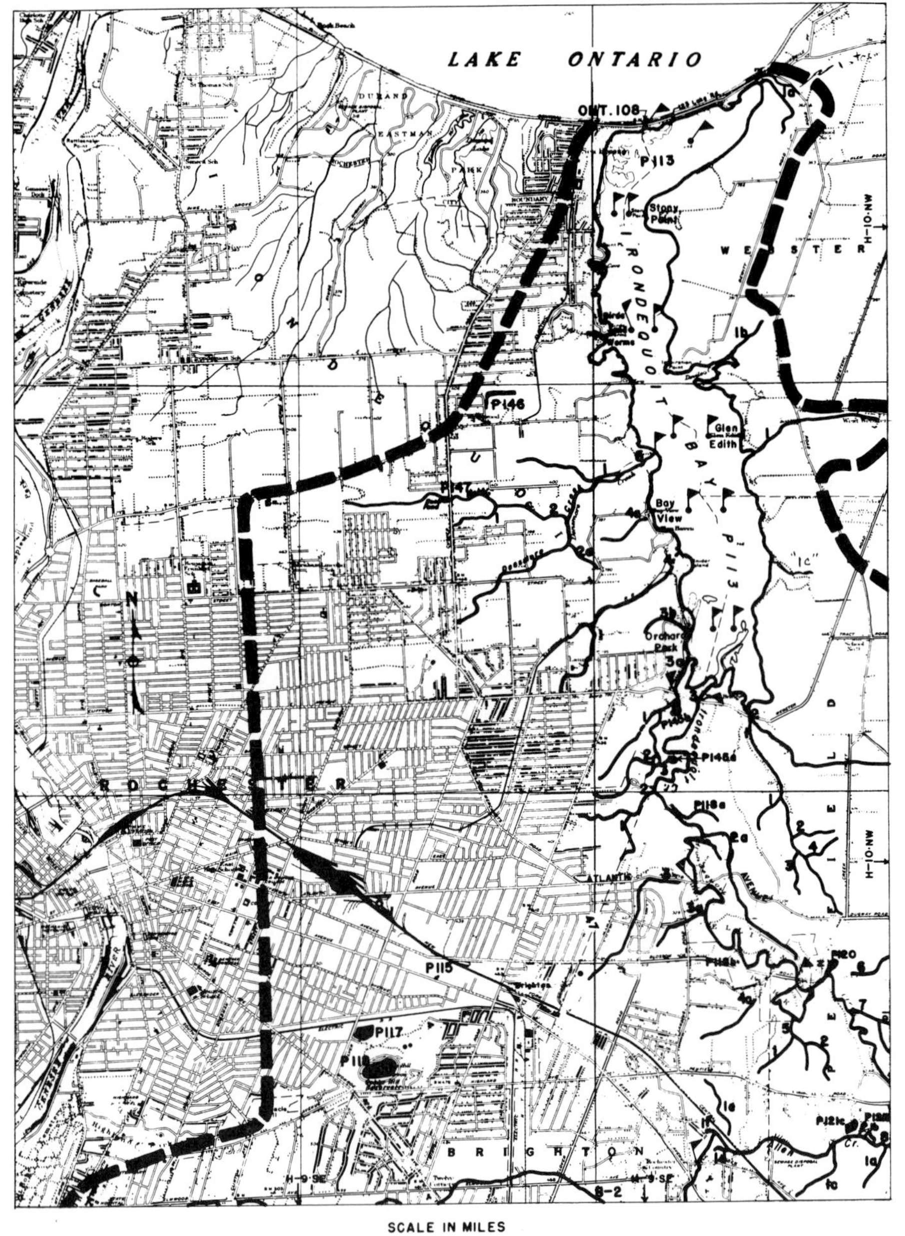

SCALE IN MILES

1/2 0

Fig. 3. (*Continued*)

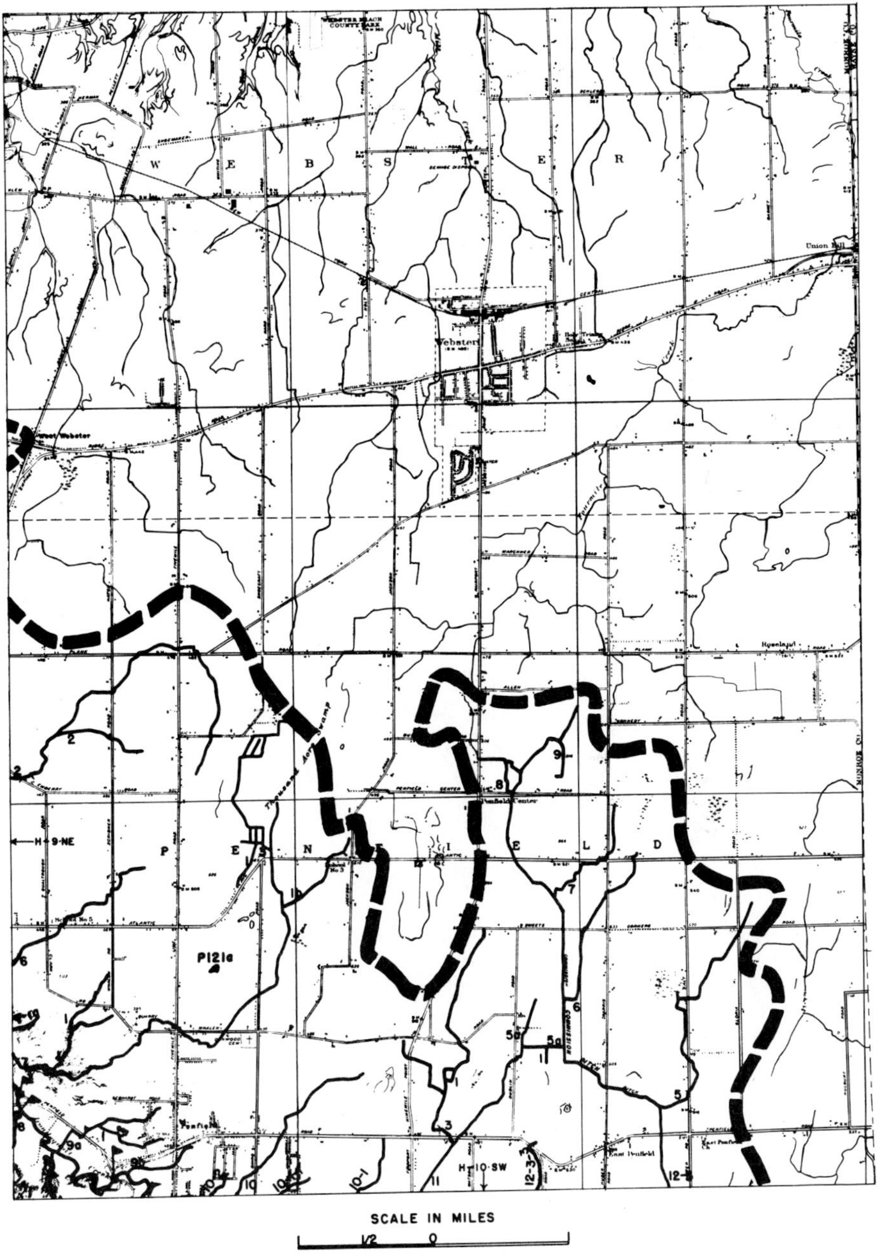

Fig. 3. *(Continued)*

MENDON PONDS

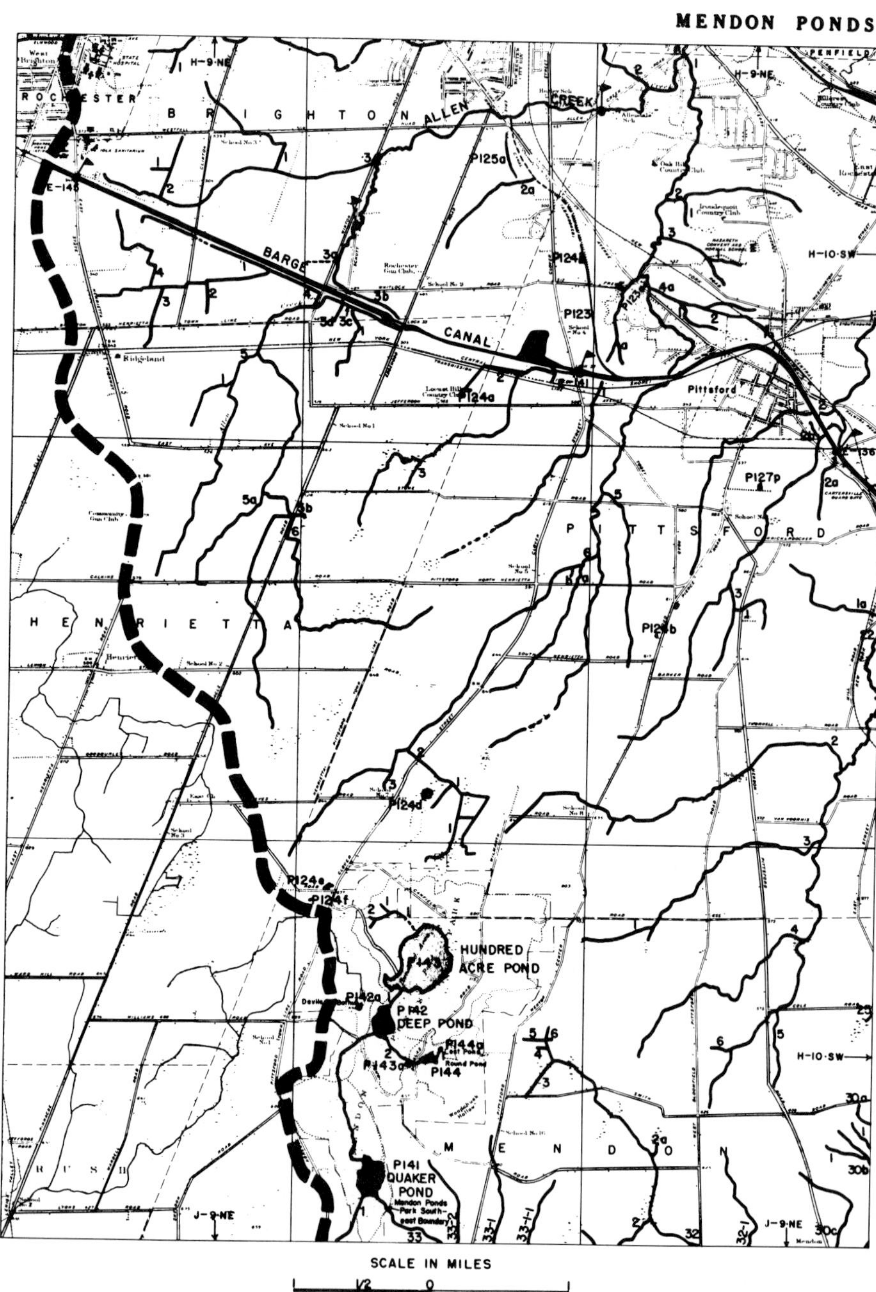

Fig. 3. (*Continued*)

FAIRPORT

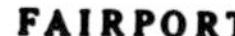

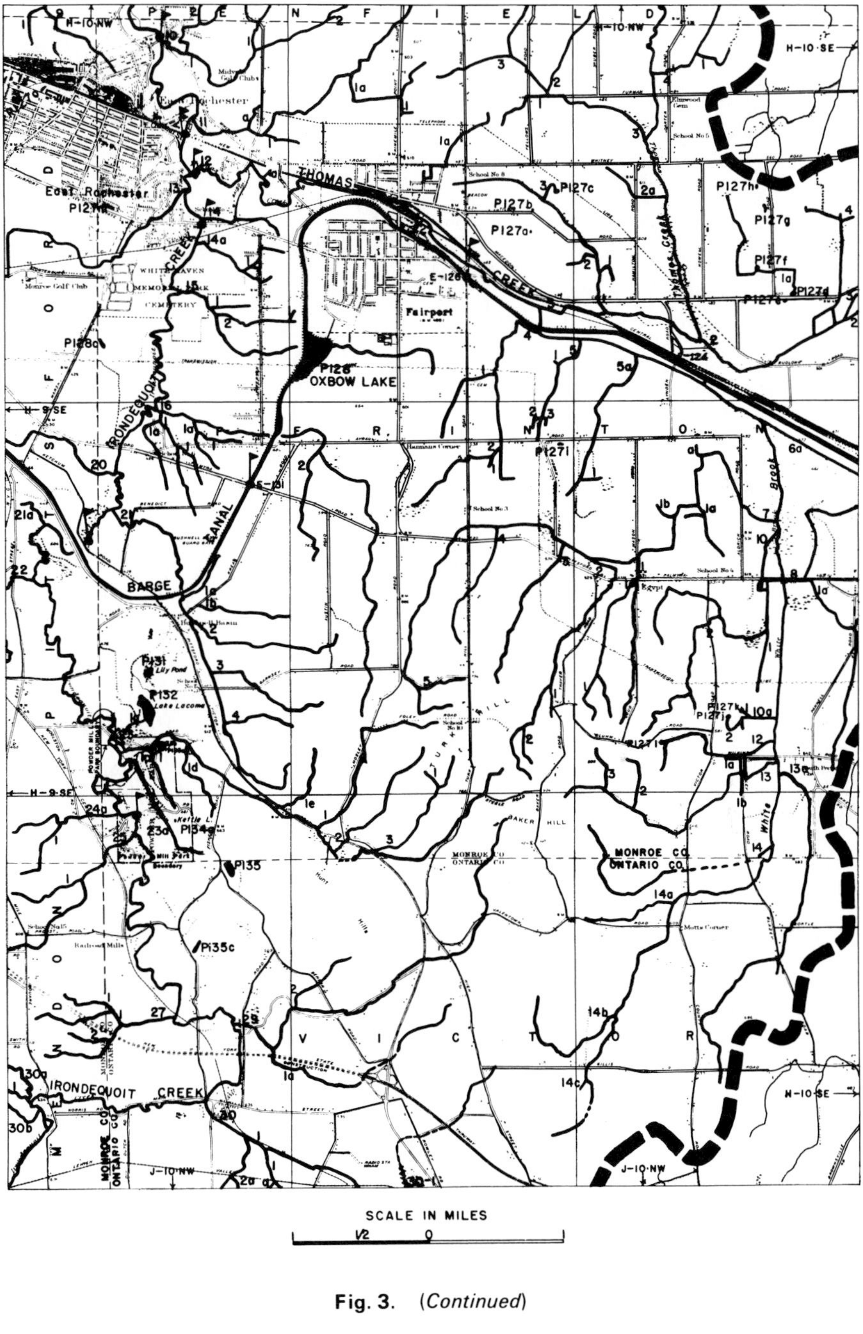

SCALE IN MILES

1/2 0 1

Fig. 3. *(Continued)*

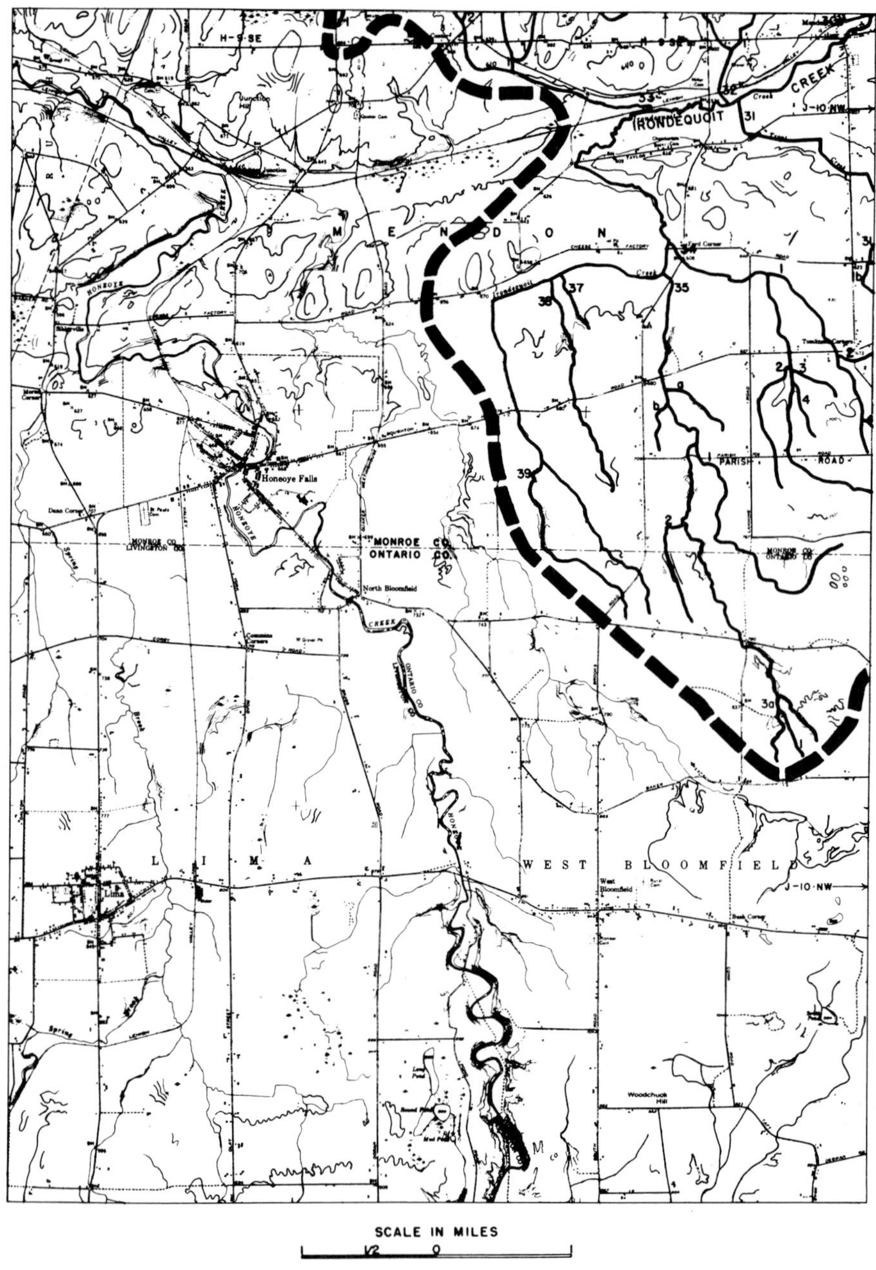

Fig. 3. *(Continued)*

VICTOR

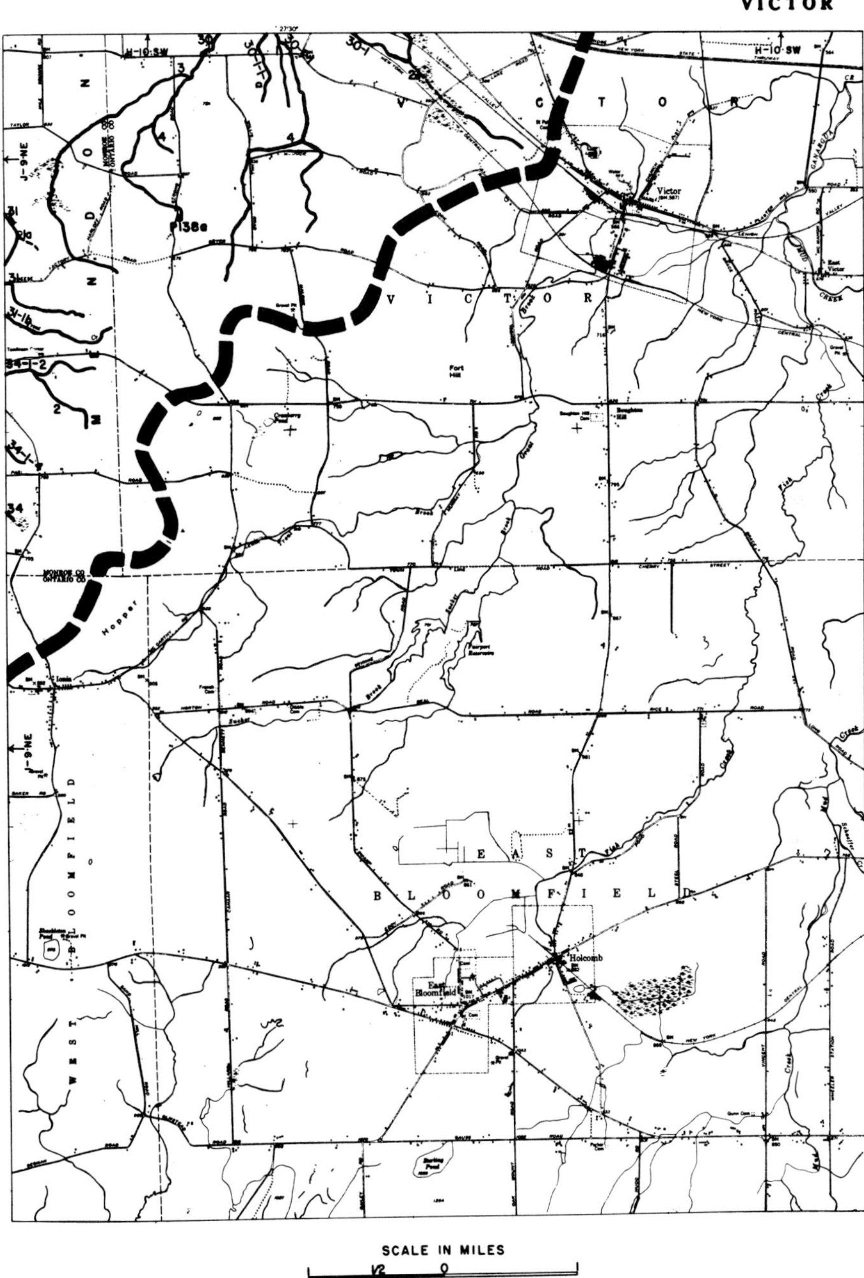

Fig. 3. *(Continued)*

West Bloomfield in Ontario County; and a very small area in Macedon Township in Wayne County. Irondequoit Bay and Irondequoit Creek form the north–south backbone of the basin. In describing the basin, it is convenient to divide this backbone into six sectors, along which are staged 11 subbasins.

Bay–Wetlands Sector

This sector extends from the outlet of the bay at the sandbar on Lake Ontario south 9.0 km (by a straight line) to Irondequoit Creek at Browncroft Boulevard. This distance combines the length of the bay (6.9 km) ending at Empire Boulevard, and the wetlands lying between Empire and Browncroft. Four subbasins are staged along this sector. The first is the bay proper (length, 6.7 km; average width, about 1.0 km; area, 6.7 km^2). The north end is bounded by a sandbar 2.1 km long and about 90 m wide, upon which lie a single track railroad, a highway, and numerous cottages. The outlet of the bay is a channel about 10 m wide and 1.5 m deep. Boat passage between the bay and Lake Ontario is severely restricted by low bridge clearance (about 1.2 m) and by shifting sandy shoals on the Lake Ontario side. Possible opening of the bay has been studied since the late 1930's (U.S. Army Corps of Engineers, 1957, 1968, 1969, 1971). From the time of French expeditions against the Senecas (1687) until the mid-nineteenth century, the outlet was larger, and small vessels sailed down the bay to dock at The Landing near Tryon Park. Along most of its perimeter, the bay is confined by steep, wooded hillsides which rise 30 to 45 m over horizontal distances of 0.2 to 0.5 km. In a number of locations, the shoreline is bounded by nearly vertical silt bluffs, up to 15 m high, which are bare of vegetation. High water levels, as in the springs of 1972, 1973, and 1974, undercut these bluffs and cause them to slump. South of the mouth of Irondequoit Creek at Empire Boulevard, a marshy flood plain varying from 0.16 to 0.8 km in width extends to Browncroft Boulevard. Large stands of cattails occur. Near Browncroft Boulevard, the Stappenbeck Rendering Company maintains a waste treatment lagoon which discharges to Irondequoit Creek (Thorne, 1973). Along the entire sector, high, nearly level ground, at an elevation of 120–140 m, occurs to both east and west. The valley seldom exceeds 1.6 m in width.

The Webster–Northwest Penfield subbasin is a strip along the east side of the bay and wetlands, about 24.5 km^2 in area. It is drained by small streams (stream index numbers 113-1a, 1b, 1, 1c, 2; see Fig. 3) which have cut steep ravines. A small sewage treatment plant (Penfield Northwest) discharges about 0.05 m^3/sec into the bay. The high ground is largely occupied by suburban residences. The third subbasin of 12.9 km^2 lies in eastern Irondequoit Township. It is drained by four small streams (113-3b, 4, 4a, and

5) of which the last, Densmore Creek, is the longest. The area is almost entirely developed with suburban housing. An Irondequoit sewage treatment plant discharges 0.03 m^3/sec to the bay via Densmore Creek. The fourth subbasin is an area of 26.1 km^2 comprising a large part of the city of Rochester east of the Genesee River. In principle, combined sanitary and storm runoff from the area is conducted by the East Side Interceptor to the Eastman–Durand sewage treatment plant on Lake Ontario; discharges at the overflows into Densmore Creek and the small tributary (113-3-2) of Irondequoit Creek should occur rarely. In fact, the capacity of the Interceptor is exceeded (Black and Veatch, Consulting Engineers 1969; O'Brien and Gere Engineers, Inc., 1976), and overflows are frequent and significant.

Lower Irondequoit Creek Sector

This sector extends from Browncroft Boulevard to Old Penfield Road 3.4 km by straight line. In this sector, Irondequoit Creek meanders through a level but generally dry flood plain 0.2 to 0.8 km in width. Creek elevation increases about 0.48 m along the 5.3-km-long creek bed. Ellison Park lies midway along the sector and includes portions of the level flood plain and steep, wooded slopes forming the sides of the valley. Drainage in the sector is by several small streams of a Brighton–Penfield subbasin (area 22.8 km^2) lying mainly to the east of Irondequoit Creek. A Brighton sewage treatment plant (Rich's Dugway) discharges 0.06 m^3/sec to the creek.

Middle Irondequoit Creek Sector

This sector extends from Penfield Road 3.7 km southeast to the mouth of Thomas Creek. Just upstream from Penfield Road, the broad floodplain ends, and Irondequoit Creek becomes narrowly confined in a valley 0.2 to 0.5 km wide with steep, wooded slopes 15 to 25 m high. The creek gradient is large: about 32 m in 7.2 km along the creek bed; indeed, between Panorama Plaza and the Linden Street bridge, the gradient is 25 m in 2.1 km, and rapids and falls occur in this section. Along this sector, there is drainage from three subbasins. First, Allen's Creek (113-3-8), which enters Irondequoit Creek just above Penfield Road, drains 83.0 km^2 lying to the west in Brighton, Henrietta, and Pittsford Townships. About 3.2 km upstream from the mouth of Allen's Creek, a Brighton sewage treatment plant (Allen's Creek) discharges 0.3 m^3/sec. Secondly, Thomas Creek (113-3-12) drains 82.7 km^2 lying to the east in Penfield, Perinton, Victor, and Macedon Townships. At 2.4 and 4.8 km upstream on Thomas Creek, two sewage treatment plants (Fairport and Thomas Creek) discharge about 0.05 m^3/sec. Waste discharges of food-processing companies occur (Plymouth Rock Provision Co. on Thomas Creek and Comstock Foods on White Brook). Thirdly, an East Rochester–Penfield subbasin of 23.3 km^2 drains via several small streams into Irondequoit Creek. In this subbasin, Penfield Central and East Rochester sewage treatment plants discharge about 0.08 m^3/sec. Welkley Dairy discharges wastes to Irondequoit Creek. Until ceasing operation in 1968, Lawless Brothers Paper Mill, Inc., at East Rochester, discharged about 0.06 m^3/sec of waste into Irondequoit Creek.

Upper Irondequoit Creek Sector

This sector extends from Thomas Creek 3.7 km south southwest to a station just upstream from the Barge Canal, under which Irondequoit Creek passes in culverts. The creek valley is 0.16 to 0.32 km wide and the banks decline gradually to about 10 m high. The gradient is small: about 8 m in 5.9 km by creek bed. The Pittsford–Bushnell's Basin subbasin (37.7 km^2) drains via several streams into this sector. A small sewage-treatment plant in Perinton (Crystall Springs) discharges about 0.02 m^3/sec. A Pittsford sewage treatment plant (Jefferson Heights) discharges about 0.02 m^3/sec into the Barge Canal during the navigation season and into Irondequoit Creek during winter.

Powder Mill Sector

This extends from the Barge Canal 6.6 km south-southeast to the hamlet of Fishers in the township of Victor. Streams from both east and west drain the Powder Mill subbasin of 50.7 km^2. In this sector, Irondequoit Creek threads a torturous course through marshes and drumlins and rises 25 m in 13.3 km by creek bed. There are no sewage discharges in this or the last sector.

Headwaters Sector

This sector extends from Fishers 10.3 km south-southwest to the source of Irondequoit Creek on the west side of Hopper Hill (elevation 335 m). The headwaters subbasin of 90.5 km^2 drains the area of Mendon Ponds Park to the northwest, an area toward Victor Village to the southeast, and the southern area extending into West Bloomfield. The elevation of the creek at Fishers is 144 m. It rises to 172 m at Mendon Village, 3.2 km to the southwest, and follows a marshy valley westward about 1.6 km, then turns south and climbs for several kilometers through rising ground to heights of about 275 m on the slopes of Hopper Hill. The headwaters subbasin is devoted to farms; only about 10% of the area (the steeper hillsides) is forested.

Barge Canal

The canal traverses the Irondequoit Creek basin and modifies its hydrology mainly by augmenting flow in Irondequoit Creek and some tributaries. In Monroe County, east of the Genesee River, the canal is filled in April with Genesee River water and drained in November, partially into the Irondequoit Creek basin. Flow in the canal is from west to east, and water levels are 156 m at the Genesee River, 148 m east of Lock 33 at Edgewood Road (at the northwest corner of Henrietta Township), and 140 m east of Lock 32 at Clover Street (west of Pittsford Village). The 140-m level extends into Wayne County. Discharges from the canal occur at several locations: (1) into Allen Creek (113-3-8) east of Winton Road by two siphons (2 in., 8 in.) which operate continuously during the navigation season; (2) into West Brook (113-3-8-1), a tributary of Allen Creek, by three siphons (2 in., 4 in., 6 in.), at a point between Monroe Avenue and Clover Street—only the largest siphon is thought to be operable and used irregularly; (3) into Irondequoit Creek 0.7 miles west of Bushnell Basin, by three Tainter Gates used to drain the canal in November; (4) into Thomas Creek in Fairport, by two fixed weirs used for level control and by four Tainter Gates used for draining in November; and (5) into White Brook (113-3-12-3), a tributary of Thomas Creek, east of Fairport, by a weir used for level control. Nearly all streams that cross the canal pass beneath the canal in culverts. One small stream (E-141-1) drains about 5.2 km^2 within the Allen Creek basin into the Canal. The Canal receives about 0.17 m^3/sec of effluent from the Henrietta, Pittsford Village, and Jefferson Heights sewage-treatment plants.

Soils

Of the total area of Monroe County (2060 km^2), the Soil Survey of Monroe County (1973) classifies 6.8% as urban and 3.8% as alluvial, marsh, cut and fill, made land, muck, pits and quarries, rock land, or lake beaches. The remaining 89.4% is covered by soils derived from glacial and lacustrine deposits. According to the Soil Survey, four major soil groups and eight soil associations (numbered as in the Soil Survey) occur within the Irondequoit Bay drainage basin.

Group I

Associations dominated by soils formed in glacial till include (1) the Madrid-Massena Association, with deep, well- to poorly drained soils and moderately coarse- to medium-fine textured subsoil; (2) the Ontario-Hilton Association, with deep, well- to moderately well-drained soils and medium- to moderately fine-textured subsoil; and (3) the Lima-Honeoye-Ontario Association, with well- to moderately well-drained soils and medium- to moderately fine-textured subsoil over limestone bedrock.

Group II

These associations are dominated by soils formed in gravelly or sandy glacial water deposits. They follow the preglacial Genesee valley and are noted for sand and gravel pits as well as mixed farming. They include (1) the Palmyra-Wampsville-Eel Association, with deep, excessively to well-drained soils with medium-coarse to moderately fine-textured subsoil over sand and gravel; and (2) the Colonie–Elnora–Minoa Association, with deep, excessively drained to somewhat poorly drained soils with coarse- to medium-textured subsoil over sand and is derived from sandbars, beaches, and deltas of lake plains.

Group III

These associations are dominated by soils formed in lake-laid deposits of silt and very fine clay. They include (1) the Arkport-Collamer Association, with deep, well- to moderately well-drained soils that have a coarse-textured to moderately fine-textured subsoil.

Group IV

These associations are dominated by soils formed in clayey, lake-laid deposits and include (1) the Hudson–Rhinebeck–Madalin Association, with deep, moderately well- to very poorly drained soils and fine- to moderately fine-textured subsoils; and (2) the Schoharie–Odessa–Cayuga Association, with deep, well- to somewhat poorly drained soils and fine to moderately fine subsoils.

Each of the associations comprises one or more soil series distinguished by profile characteristics, and each series is subdivided into one or more phases according to surface texture, stoniness, and slope. Altogether, 111 soil phases, forming an extremely intricate mosaic, occur in Monroe County. Within the county, about 4.2% of the area is characterized by slopes of 6–15%, 1.2% of the area by slopes of 15–25%, and 2.4% of the area by slopes exceeding 25%.

Population

Portions of the city of Rochester and nine townships lie in the drainage basin. The population of the basin is difficult to estimate accurately for three reasons: (1) boundaries of the watershed and of the census districts (the towns) never coincide; (2) the Rochester East Side Interceptor, by diverting sanitary and storm runoff to the Genesee River or to Lake Ontario, reduces the effective population of the Irondequoit Basin; (3) the Barge Canal receives municipal sewage effluents (though little surface

TABLE 1

Estimated Population of the Drainage Basin in 1970

Division	Population in 1970	Percent area in basin	Percent population in basin	Estimated population in basin
City of Rochester	296,233	30	30	100,000
Town of Brighton	35,065	75	87	31,000
Town of Henrietta	33,017	20	20	6,600
Town of Irondequoit	63,675	33	33	21,000
Town of Mendon	4,541	67	40	1,800
Town of Penfield	23,782	95	95	22,000
Town of Perinton	31,568	90	90	28,500
Town of Pittsford	25,058	98	98	25,000
Town of Victor	5,071	40	40	2,000
Town of Webster	24,739	5	10	2,400
Total population of basin				240,300
Effective population assuming complete diversion of Rochester sewage				140,000

runoff) at some points in the basin and discharges an unknown fraction into the basin at other points. In Table 1, U.S. Census data for the city and towns for 1970 are listed together with crude estimates of the percentages of area and of population within the basin. From these, the total population of the basin is estimated to have been 240,000, and the effective population, assuming complete diversion of Rochester sewage, about 140,000. Previous estimates for 1970 were 206,000 (Bubeck *et al.*, 1971) and 187,000 (Hennigan, 1970).

The historical pattern of population growth is shown in Table 2. From

TABLE 2

Historical Growth of Population

Year	In Rochester	In nine townships
1830	9,207	—
1850	36,403	24,100
1870	62,386	26,823
1890	133,896	27,257
1910	218,149	31,575
1930	328,132	59,421
1950	332,448	94,374
1970	296,233	246,516

the beginning of its settlement about 1810, the city of Rochester grew exponentially until about 1920 and reached a peak population of 332,448 in 1950. The pattern of growth was different in the townships. From 1850 to about 1900, the mainly rural populations of the townships were essentially stable at 2000 to 4000 per town. Except for Perinton, which grew slowly to about 12,000 in 1950 (as a result of industry in the villages of Fairport and East Rochester), the low populations of the towns were maintained until the onset of rapid suburban growth. Census data show that rapid growth began about 1920 in Irondequoit and Brighton, about 1950 in Penfield and Pittsford, in the mid-1950's in Henrietta and Webster, and about 1960 in Perinton. In 1970, suburbanization was just beginning in Mendon and Victor, and their populations, 4541 and 5071, respectively, were as yet only double the levels maintained from 1850 to 1950.

The pattern of growth implies that pollution of Irondequoit Bay and Creek occurred in two stages. From 1810 to 1920, pollution was primarily that associated with land clearance and the establishment of farms, and with sewage input from the east side of Rochester. (The Rochester East Side Interceptor was constructed about 1890 to abate pollution of ditches and streams leading to the bay.) From 1920 to 1970, a large additional pollution load arose due in part to the growth of the town populations within the basin from about 20,000 to at least 140,000, and in part to the rapidly increasing use of chemical fertilizers. In this 50-year period, the bay evolved from a eutrophic lake to something akin to a sewage oxidation pond.

Land Use

Urban areas of the basin include Rochester; the portions of the towns of Irondequoit, Webster, Henrietta, and Brighton lying in the basin; and the northern halves of Pittsford and Perinton. To the south, particularly southeast along Interstate 490 to Bushnell Basin and beyond, suburban tract development is rapidly advancing into agricultural areas. In the basin, areas where farming predominates include the eastern portions of Penfield and Perinton, the southern half of Pittsford, and essentially all of Mendon, Victor, and West Bloomfield. Relatively crude land-use maps for Monroe County have been published (Genesee Finger Lakes Regional Planning Board, 1969, 1970). An excellent multicolored land-use map of Monroe County has been published by the Monroe County Department of Planning (1975). Heavy and light industrial, commercial, multi- and single-family residential, park, and recreational areas are shown; however, agricultural land, idle land, and forested areas are not differentiated. The Soil Survey Monroe County, New York (1973) reproduces large-scale aerial photographs on which maps of soil type are overlaid. The aerial photographs show not more than 10% of the area forested, and that mainly in parks, stream banks, steep hillsides, and small wood lots. The rural area is devoted to mixed farming: dairying, grains, market produce, and some livestock. For land immediately adjacent to Irondequoit Bay, the Monroe County Department of Planning has prepared vegetation, land-use (present and proposed), ownership, and other maps (see, for example, Monroe County Department of Planning, 1973).

HYDROLOGY

Stream Flow

No flow rates are available for the outlet of the bay into Lake Ontario. In the Irondequoit Creek drainage basin, the only permanent gauging station has been that maintained by the U.S. Geological Survey since 1959 on Allen's Creek just upstream from the Brighton sewage-treatment plant. Recently, an additional permanent gauging station was established on Irondequoit Creek at Linden Avenue downstream from the village of East Rochester. Studies of stream flow in the basin are discussed below; those carried out prior to 1972 have also been reviewed by Bubeck (1972).

Grossman and Yarger (1953) estimated that low flows in Irondequoit Creek just above the Barge Canal exceed 0.36 m^3/sec 98% of the time. The estimate was based on U.S. Geological Survey measurements during the period of 1949–1953. The New York State Department of Health (1955) reported flows on October 26 and November 11, 1954, in m^3/sec, to have been 0.63 and 1.32 in Irondequoit Creek at Browncroft, 0.64 in Irondequoit Creek just above Thomas Creek, 0.18 and 0.23 in Thomas Creek at its mouth, and 0.24 and 0.26 in Allen's Creek at its mouth. The U.S. Geological Survey (Dunn, 1962), participating in a pollution study by the New York State Department of Health (1964), reported flows at several locations during the period of July–September 1962. From daily measurements, the low, approximate mean, and high flows (m^3/sec) were 0.34, 0.42, and 1.25 in Irondequoit Creek above the Barge Canal; 1.13, 1.42, and 3.4 in Irondequoit Creek at Browncroft; 0.23, 0.34, and 1.5 (in July–August) and 0.23, 0.51, and 3.1 (in September) in Allen's Creek above the Brighton sewage treatment plant. Less extensive measurements were made at several other locations, including Thomas Creek below the Fairport sewage treatment plant (0.23, 0.57, and 2.0 m^3/sec) and Allen's Creek at its mouth (0.37, 0.45, and 0.62 m^3/sec). Minimum seven-day flows in 10 years were estimated to be 0.28 (Irondequoit Creek just above the Barge Canal), 1.0 (Irondequoit Creek at Browncroft), 0.25 (Thomas Creek below the Fairport sewage treatment plant), and 0.23 (Allen's Creek at its mouth). Time of travel data and a list of sewage treatment plant and industrial discharges were also reported.

Hennigan (1970) summarized discharge data (for example, U.S. Department of the Interior, Geological Survey, 1968) for the Allen's Creek gauging station and meteorological data for the Rochester airport, for the period of 1961–1968. Table 3 is revised from Hennigan's. He noted that an annual mean flow of about 0.40 m^3/sec must be attributed to augmentation from the Barge Canal in 1963 and later years. Since this augmentation is limited to the months of April to November, the actual average augmentation must be about 0.60 m^3/sec. Evidently, during summer, 50% and sometimes much more of the Allen's Creek flow is attributable to the Barge Canal.

Bubeck (1972, also unpublished data, 1975) established a gauging station on Irondequoit Creek at the Browncroft Boulevard bridge and made measurements of instantaneous discharge from November 1970 through December 1972. During most of this period, measurements were made daily. His data are summarized in Table 4 (see also Fig. 4), in which maximum, minimum, and mean monthly flows are listed for the two years. Discharges ranged from a minimum of 1.15 m^3/sec in September 1971 to a maximum of 25.2 m^3/sec after snowstorms and melting in March 1972. In 1971, flows were measured infrequently during the period of low flow from June to October; as a result the best estimate of mean annual flow is the mean

TABLE 3

Precipitation and Runoff Data for Allen's Creek, 1961–1969[a]

Water year	Precipitation (in.)	Discharge (in.)	Mean annual flow (m^3/sec)
1961	27.39	8.58	0.50
1962	30.33	8.19	0.48
1963	23.16	11.21	0.65
1964	23.50	12.39	0.72
1965	21.47	10.38	0.61
1966	28.99	13.69	0.80
1967	26.63	13.17	0.77
1968	33.47	19.30	1.13
Average	26.87	12.11	0.71

[a] Revised from Hennigan (1970). Discharge data from USGS annual summary series. Precipitation from NWS annual summaries. Area of drainage basin 28.0 square miles (72.5 km^2).

TABLE 4

Flow Rates (m^3/sec) in Irondequoit Creek at Browncroft Boulevard in 1971 and 1972[a]

	1971				1972			
Month	Days	Maximum	Minimum	Mean	Days	Maximum	Minimum	Mean
Jan.	25	7.42	1.60	3.39	29	7.42	1.01	3.53
Feb.	27	21.56	1.90	5.43	26	9.94	1.43	3.44
Mar.	30	22.96	4.54	8.15	30	25.20	4.20	10.33
Apr.	29	9.69	2.77	4.76	29	14.84	4.26	6.52
May	16	12.32	2.77	4.54	29	20.16	4.62	7.64
Jun.	13	7.08	2.04	2.91	30	23.52	3.98	7.28
Jul.	13	8.54	1.65	3.30	20	4.28	2.66	3.53
Aug.	12	3.64	1.15	2.10	30	10.50	1.99	3.33
Sep.	1	1.15	1.15	1.15	29	10.08	1.82	3.05
Oct.	5	1.90	1.29	1.56	31	5.38	1.49	2.91
Nov.	30	6.38	1.43	2.37	30	17.08	2.52	7.67
Dec.	30	6.66	1.54	3.56	30	14.28	5.04	7.96
Mean annual flow estimated as								
(a) mean of all measured flows for year				4.24				5.68
(b) mean of monthly means				3.68				5.60

[a] Data of R. C. Bubeck (unpublished data, 1975). For each month, the number of days when flow was measured, and the maximum, minimum, and mean flows are listed. Mean annual flow can be estimated as the mean of all recorded flows for the year. For 1971, this estimate is too high because low flows in June–October were infrequently measured; the mean annual flow is probably more accurately estimated as the mean of the monthly means. For 1972, because of more frequent measurements, both methods of estimating mean annual flow give similar values.

of the monthly means, about 3.7 m^3/sec. For 1972, the mean annual flow was considerably higher, about 5.7 m^3/sec. Total precipitation at the Rochester airport in 1971 was 86.8 cm, about 5% higher than the long-term average of 82.7 cm. 1972 was a wetter year (total precipitation, 97.2 cm) and was remarkable especially for the very heavy rainfall (16.5 cm) in June due to Hurricane Agnes. Probably the estimated mean annual flow (3.7 m^3/sec) for 1971 is the better indication of mean flow in an average year. Previously, Bubeck (1972) estimated the mean annual flow to be about 4.0 m^3/sec, and Lozier Engineers, Inc. (1967) calculated a value of 3.6 m^3/sec, based on an assumed runoff coefficient. For the July–September period, the mean flow in Irondequoit Creek is probably about 2.4 m^3/sec in an average year. During winter (December–March) the mean flow is probably about 5.1 m^3/sec in an average year.

For the period of June–August 1972, Thorne (1973) reported average

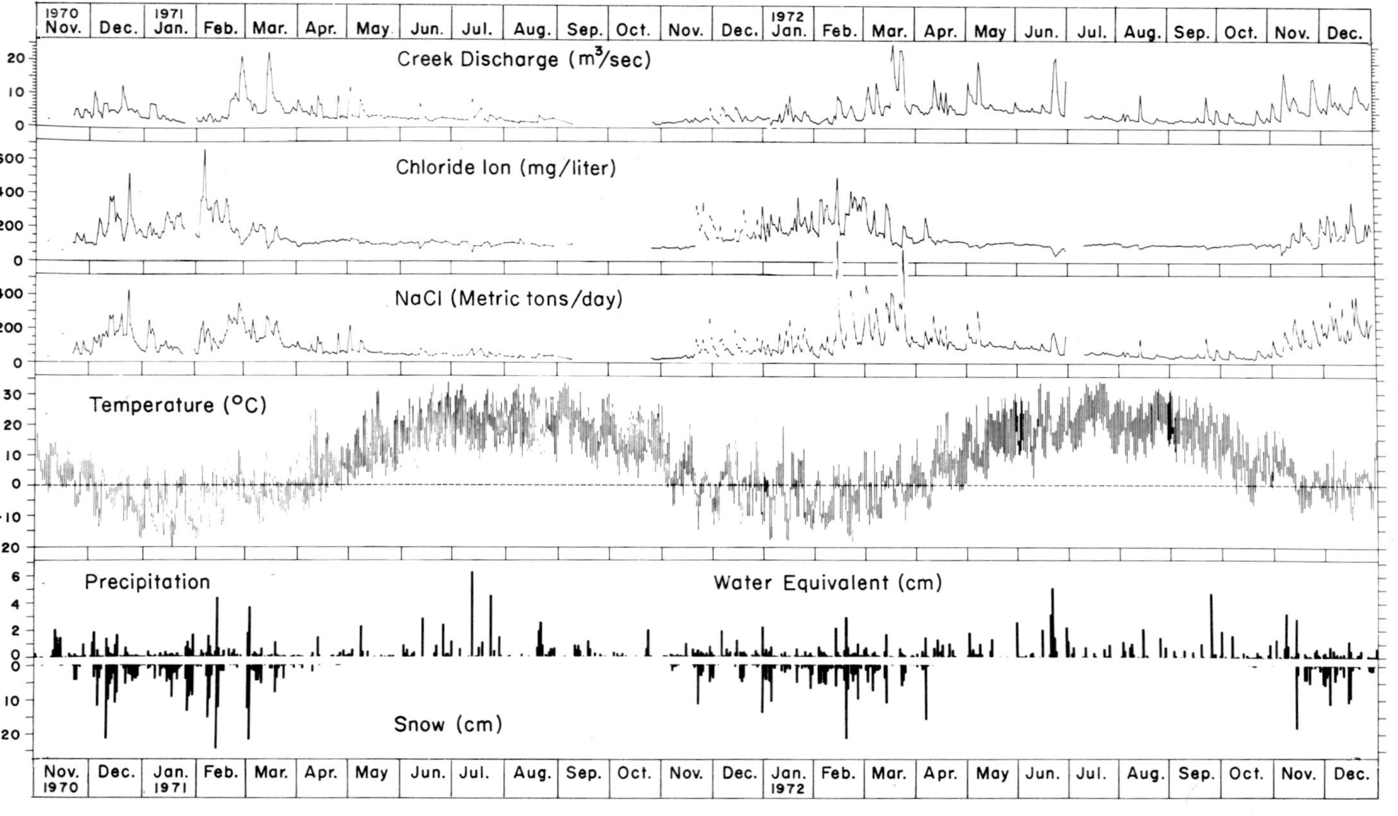

Fig. 4. Discharge, chloride ion concentration, and transport of sodium chloride in Irondequoit Creek at Browncroft Boulevard from November 1970 to December 1971. Precipitation and temperatures (maximum and minimum) recorded at the Monroe County Airport by the National Weather Service. From Diment *et al.* (1974).

flows (m^3/sec) at several locations on Irondequoit Creek: 4.1 at Browncroft, 2.4 just below the mouth of Thomas Creek, 1.1 just above the Barge Canal, and 0.84 at Fishers. The Barge Canal input was 0.06 m^3/sec. Average flows in Allen's Creek were 1.24 near the mouth and 0.12 above the Barge Canal; the Canal input was 0.37. Average flows in Thomas Creek were 0.66 near the mouth and 0.55 upstream of the Barge Canal input.

During 1975, O'Brien and Gere Engineers, Inc., began a study to determine the volume and chemical composition of discharges from interceptors at three sites on the east side of Rochester. At two of the sites (Site 28 "Norton Street Screenhouse" and Site 29 "Densmore Bypass"), the discharges enter Densmore Creek (113-5); at the third site (Site 31), the discharge is into Thomas Creek (113-3-2) at Tryon Park. At all three sites, the total discharges are made up partly of overflows from the Rochester combined sanitary–storm interceptor and partly of discharges from separate storm sewers. O'Brien and Gere Engineers, Inc., attempted to determine the discharge volume in two ways: (1) by direct measurement of flows, and (2) by computation using a hydrological model and assuming a runoff coefficient.

Direct measurements consisted of automatically recording flow rates at 15-minute intervals, during periods when the discharge exceeded about 10 MGD (0.44 m^3/sec). We have examined the complete data (O'Brien and Gere Engineers, Inc., 1975) only for the period of February to June 1975, during which flows were not recorded at Site 29. Between February 5 and June 5 (120 days), 2350 rates ranging from 10 to 300 MGD (0.44 to 13.2 m^3/sec) were recorded at the Norton Street Screenhouse (Site 28). Thus discharges greater than 10 MGD apparently occurred 22% of the time. The mean recorded rate was approximately 25 MGD. The total discharge during the period was about 600 MG, and the time average rate during the 120 days was about 5 MGD or 0.22 m^3/sec. At Tryon Park (Site 31), 507 discharge rates (mean value about 35 MGD) were recorded between January 25 and June 19, 1975. Mean flow for the period was therefore about 1.3 MGD or 0.06 m^3/sec. Although discharge at Densmore Bypass (Site 29) was not recorded, O'Brien and Gere Engineers, Inc. (1976a) indicated that the discharge was roughly comparable to that at Site 28. Then the mean total discharge at the three sites, during the 120-day period, was about 11.3 MGD or 0.50 m^3/sec. Presumably, from June to September, the mean discharge would be smaller. As a first approximation we assume mean discharges of 0.20 m^3/sec during the summer and 0.35 m^3/sec annually. These estimates would seem to be conservative, since discharges smaller than 10 MGD were not recorded. However, O'Brien and Gere Engineers, Inc. (1976a,b) believe that these values overestimate the discharge in an average year by as much as threefold. According to them, the discrepancy is due to (1) heavier than normal precipitation during February–June 1975 and (2) instrument problems resulting in spurious records of discharge.

Much lower estimates of the interceptor discharges were calculated with a hydrological model and assumed values of runoff and of a runoff-to-overflow ratio (O'Brien and Gere Engineers, Inc., 1976a,b). In one calculation, the total annual discharge was estimated as 1243 MG, of which 625 MG were from separate storm sewage and 216, 155, and 247 MG from combined storm–sanitary sewage at Sites 28, 29, and 31, respectively. The corresponding annual discharge rate was 3.4 MGD or 0.15 m^3/sec. In a second calculation, the total annual discharge was found to be 952 MG, of which 625 MG were from separate storm sewage and 327 MG from combined storm–sanitary sewage from the three sites; the mean annual discharge rate was 2.6 MGD or 0.11 m^3/sec. Evidently, the annual discharge obtained by calculation is only about ⅓ of the value (0.35 m^3/sec) based on the discharge measurements in 1975. It is believed that O'Brien and Gere Engineers, Inc., continued measurement of the discharges at least through 1975; it is hoped that a report will resolve the discrepancy. For present purposes, we assume that the high discharges based on the measured rates are more nearly correct; we will use these high values in estimating nutrient discharges in a later section (see Other Nutrient Inputs to the Bay, p. 169).

The total water input into Irondequoit Bay is estimated in Table 5. The total input should equal the input from Irondequoit Creek at Browncroft Boulevard plus that from the Bay–Wetlands Sector. The latter includes

TABLE 5

Estimated Average Water Influxes (m^3/sec) into Irondequoit Bay from Various Sources

Source	Dec.-Mar.	Jun.-Sep.	Year
Irondequoit Creek at Browncroft Boulevard	5.10	2.36	3.70
East Irondequoit and Webster–Northwest Penfield subbasins (including wetlands)			
Runoff	0.37	0.18	0.26
STP effluent	0.08	0.08	0.08
Rochester interceptor overflows	0.50	0.20	0.35
Precipitation on bay	0.17	0.17	0.17
Total	6.22	2.99	4.56
Water loading (m/day)	0.080	0.039	0.059

contributions by direct precipitation on the bay (about 0.17 m^3/sec), overflow from Rochester interceptors (0.35 m^3/sec), effluents from three small sewage treatment plants discharging directly into the Bay (0.08 m^3/sec), and runoff from the Irondequoit and Webster–Penfield subbasins. A conservative estimate of the latter is the product of the flow in Irondequoit Creek at Browncroft Boulevard (3.7 m^3/sec) less the contributions by upstream sewage treatment plants (0.5 m^3/sec) and the Barge Canal (0.5 m^3/sec), and the ratio of the area (14.5 square miles) of Irondequoit and Webster–Penfield subbasins and the area (151 square miles) of the Irondequoit Creek basin above Browncroft. The mean annual runoff from the Irondequoit and Webster–Penfield subbasins is then about 0.26 m^3/sec. Table 5 shows that the mean annual total input to the bay is about 4.56 m^3/sec, and the corresponding water loading is about 0.068 m/day. Also shown are the estimated inputs and water loadings during the December–March and June–September periods.

From the mean annual input (4.56 m^3/sec) and the volume (45.9×10^6 m^3), and neglecting evaporation, a replenishment time of about 116 days can be calculated.

Climatology

Table 6, taken from Soil Survey Monroe County (1973) concisely summarizes the climate in the basin. Irondequoit Bay normally freezes over in late December and remains ice covered until approximately the beginning of April. From the mean precipitation (0.81 m/year), it can be calculated that the flow in Irondequoit Creek at Browncroft Boulevard would be about 10.2 m^3/sec, were all precipitation converted to runoff. The actual flow, after subtracting sewage treatment plant and Barge Canal components, is about 2.8 m^3/sec. Thus runoff appears to be about 28% of precipitation.

During the years 1970–1973, Hubbard (1975) recorded the daily solar radiation at Brockport, New York, about 25 miles west of Irondequoit Bay. Table 7 lists monthly mean radiation values. Table 8 shows how daily radiation fluctuated during 1971; the radiation of 785 langleys on June 10 and of 13 langleys on December 30 were the highest and lowest, respectively, during the four years.

Groundwater Flow

No studies have been found from which groundwater inputs to Irondequoit Bay or its tributaries can be estimated. Groundwater resources of Monroe County were described by Leggette *et al.* (1935).

TABLE 6

Temperature and Precipitation Data for Rochester[a]

Month	Temperature				Precipitation				
			7 years in 10 will have					Snowfall	
						3 years in 10 will have			
	Average daily maximum (°F)	Average daily minimum (°F)	Maximum temperature equal to or higher than (°F)	Minimum temperature equal to or lower than (°F)	Average total (in.)	More than (in.)	Less than (in.)	Average total (in.)	7 years in 10 will have more than (in.)
Jan.	33	18	46	3	2.3	2.8	1.6	20	13
Feb.	33	17	50	3	2.6	2.8	2.0	21	14
Mar.	40	24	62	11	2.7	2.9	2.1	15	8
Apr.	55	36	74	26	2.8	3.2	2.2	3	1
May	67	46	84	34	2.7	3.2	2.2	[b]	[c]
Jun.	78	56	90	45	2.5	3.3	1.8	0	0
Jul.	83	61	92	51	2.8	2.8	1.9	0	0
Aug.	80	59	90	47	2.8	3.4	2.1	0	0
Sep.	73	52	88	39	2.4	2.7	1.9	0	0
Oct.	62	42	79	31	2.5	3.3	1.4	[b]	[c]
Nov.	48	33	69	22	2.6	3.0	2.0	7	4
Dec.	36	22	51	5	2.2	2.2	1.8	14	8
Year	57	39	94	−2	30.9	32.9	27.0	80	66

[a] From Soil Survey Monroe County, New York (1973).
[b] Less than half an inch.
[c] One year in 10 will have more than 1 in.

TABLE 7

Monthly Average Values of Total Daily Solar Radiation (cal/ cm² daily) at Brockport, New York (43°N 78°W)[a]

Month	Year				4-year average
	1970	1971	1972	1973	
Jan.	167	158	150	146	155
Feb.	225	195	209	217	212
Mar.	294	309	304	250	289
Apr.	444	417	430	365	414
May	459	507	485	351	450
Jun.	536	549	465	505	514
Jul.	488	563	506	533	522
Aug.	492	478	449	431	462
Sep.	311	332	326	330	325
Oct.	181	246	194	224	211
Nov.	104	130	109	—	114
Dec.	96	91	61	—	83

[a] From Hubbard (1975).

PHYSICAL LIMNOLOGY

Description of the Bay

General features of the bay are indicated on the Rochester East (photo-revised 1969) Quadrangle (Fig. 5), a recent aerial photograph (Fig. 6), and the hydrographic map (Fig. 7). The latter was published by Odell (1940) and originally drawn in 1931. Bubeck (1972) made numerous transects with a recording fathometer, and concluded that this chart was still generally accurate in describing depths of the basins. The elevation of the bay water level is the same as that of Lake Ontario. The elevation of Lake Ontario is shown on the National Oceanic and Atmospheric Administration–Lake Survey charts as the Lake Ontario Low Water Datum of 242.8 ft (74.0 m); this is the International Great Lakes Datum (1955) for Lake Ontario. United States Geological Survey topographic maps give Lake Ontario and Bay levels as 245 ft (74.7 m). Figure 8 shows the seasonal dependence of water level in recent years.

Figure 7 shows that the bay can be divided into four quarters, north to south. The northernmost quarter, situated behind the sand bar, is a shallows (1 to 1.5 m), in which rooted aquatic growth is abundant. Early maps show small islands clustered south of the outlet; on these islands and elsewhere in the northern quarter emergent vegetation (especially cattails) was once

TABLE 8

Total Daily Solar Radiation (cal/cm² daily) at Brockport, New York, during 1971[a]

	Month											
Day	Jan.	Feb.	Mar.	Apr.	May	Jun.	Jul.	Aug.	Sep.	Oct.	Nov.	Dec.
1	130	275	390	286	429	698	340	554	571	425	240	180
2	108	273	287	233	363	147	701	544	308	420	216	206
3	110	192	115	295	295	303	743	297	476	245	187	89
4	69	152	160	305	731	620	718	422	604	271	180	71
5	61	96	404	367	374	496	550	641	421	338	245	154
6	218	152	208	474	629	537	551	587	397	180	150	30
7	208	137	251	559	688	625	696	607	395	153	278	20
8	189	104	254	576	73	408	664	629	485	329	278	21
9	194	222	445	390	195	784	657	590	343	106	130	14
10	81	192	228	537	705	785	706	480	497	415	92	39
11	86	184	186	608	587	736	513	427	226	168	40	205
12	218	73	430	382	238	527	733	675	110	212	86	157
13	162	87	232	206	430	270	217	604	117	235	87	181
14	61	301	452	654	747	320	653	526	115	282	207	60
15	115	205	218	615	641	362	665	127	396	331	50	39
16	206	347	241	512	507	510	659	630	276	269	64	100
17	191	72	365	500	715	691	556	603	270	144	202	78
18	206	320	297	679	413	669	703	591	485	338	185	131
19	191	142	49	593	596	464	197	557	52	372	55	54
20	186	159	117	397	334	623	702	494	191	345	71	58
21	89	120	491	495	581	311	694	489	480	312	57	37
22	202	120	427	571	744	745	592	479	470	200	130	201
23	231	256	412	657	733	707	628	467	299	27	232	101
24	201	106	318	356	310	706	105	588	475	53	91	21
25	154	284	417	180	305	252	470	200	340	55	107	168
26	142	288	513	187	234	584	422	261	111	50	113	270
27	126	385	517	439	201	666	735	232	137	292	53	32
28	148	227	164	92	692	601	648	59	318	306	26	28
29	184	0	189	84	742	721	261	538	186	272	18	134
30	223	0	261	279	764	595	433	443	397	289	22	13
31	208	0	526	0	736	0	528	479	0	186	0	163
Days	31	28	31	30	31	30	31	31	30	31	30	31
High	231	385	526	679	764	785	743	675	604	425	278	270
Low	61	72	49	84	73	147	105	59	52	27	18	13
Mean	158	195	308	416	507	548	562	478	331	245	129	98

[a] From Hubbard (1975).

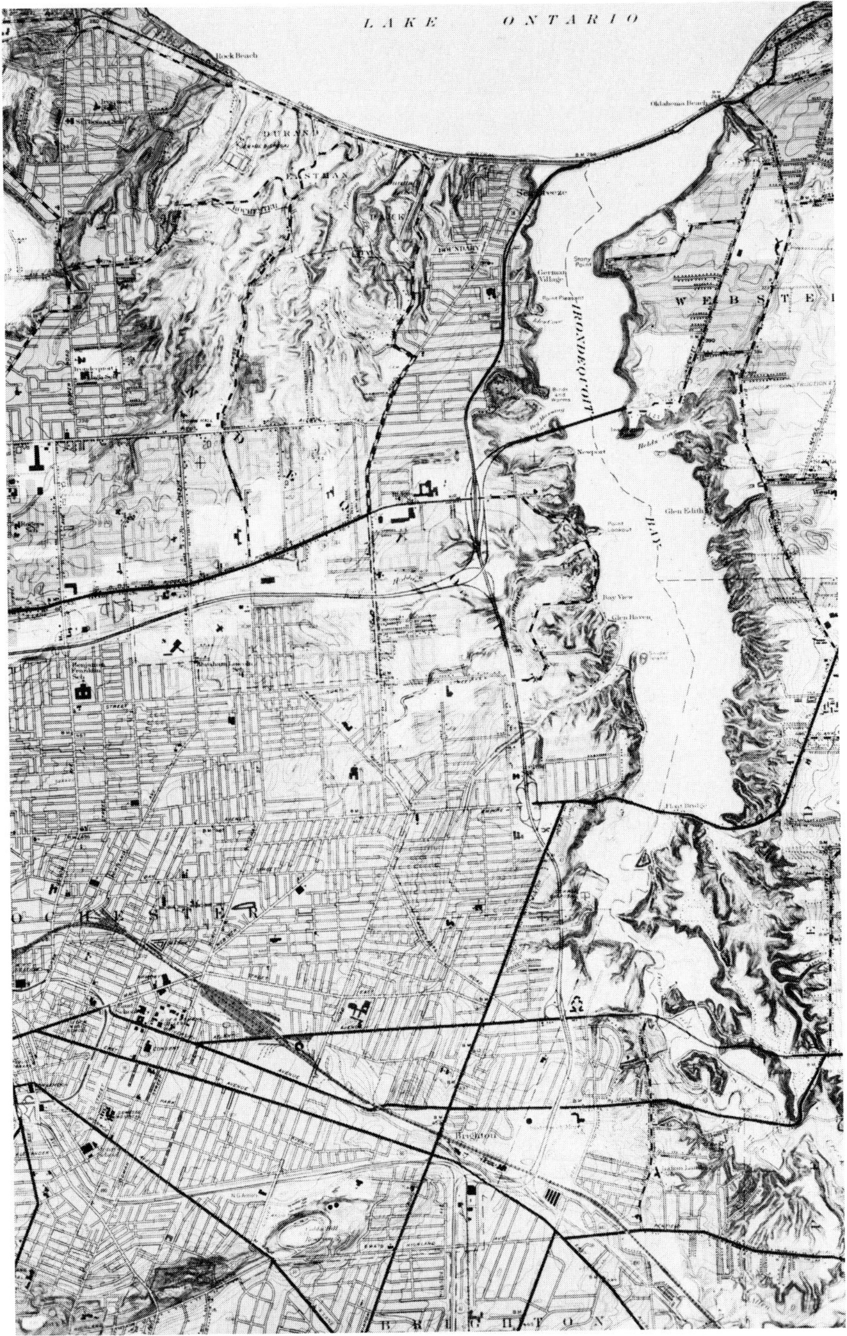

Fig. 5. Portion of U.S. Geological Survey quadrangle, Rochester East (photorevised 1969), showing Irondequoit Bay.

Fig. 6. Aerial photograph of Irondequoit Bay dated April 6, 1968.

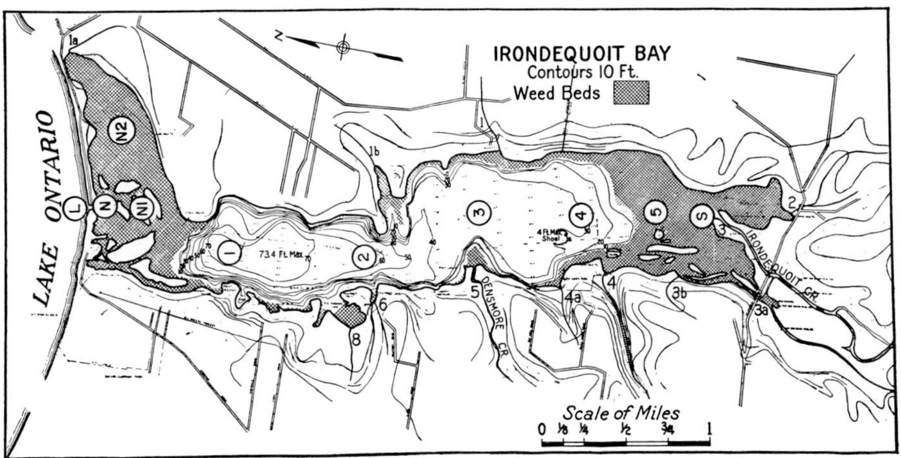

Fig. 7. Bathymetric map of Irondequoit Bay. Circled numbers indicate sampling stations of Bubeck (1972). From Odell (1940).

much more extensive. During high water levels around 1952, the cattail beds were dislodged, and large patches were dragged to shore by boaters. Only one islet now remains. The second quarter, from Stony Point to Inspiration Point, is a relatively broad, deep basin. The maximum depth reported by Bubeck (1972) was about 24 m and occurred due east of Ides Cove. Station 1 of Bubeck and others was at this approximate location. Ides Cove (or Round Hole) on the west side is a small but deep (8.8 m) cove nearly isolated by a 1-m-deep submerged sill. Bubeck (1972) studied Ides Cove in detail and showed it to be marginally meromictic. The deep basin terminates just south of the Bay Bridge, which since 1970 has traversed the narrowest point (0.4 km) from just north of Newport House on the west to Inspiration Point on the east. Depth at the middle span of the bridge is about 20 m.

The third quarter extends southward a little beyond Snider's Island and comprises a broad (up to 1.1 km) basin having a maximum depth of about 11.0 m at Bubeck's Station 3 between Glen Edith and the Densmore Creek delta. Held's (or Devils) Cove on the east side is protected by shoals and islets and, to landward, by relatively undisturbed hillsides. The southernmost quarter is a shallows of 3 m or less and supports heavy stands of rooted aquatics. The water is often turbid and sediment laden, and, as Fig. 6 indicated, a turbid plume of Irondequoit Creek sometimes extends far up the eastern shore of the Bay.

Bubeck (1972) noted that several shore features have undergone changes. The disappearance of the islets and cattail beds at the north has been pre-

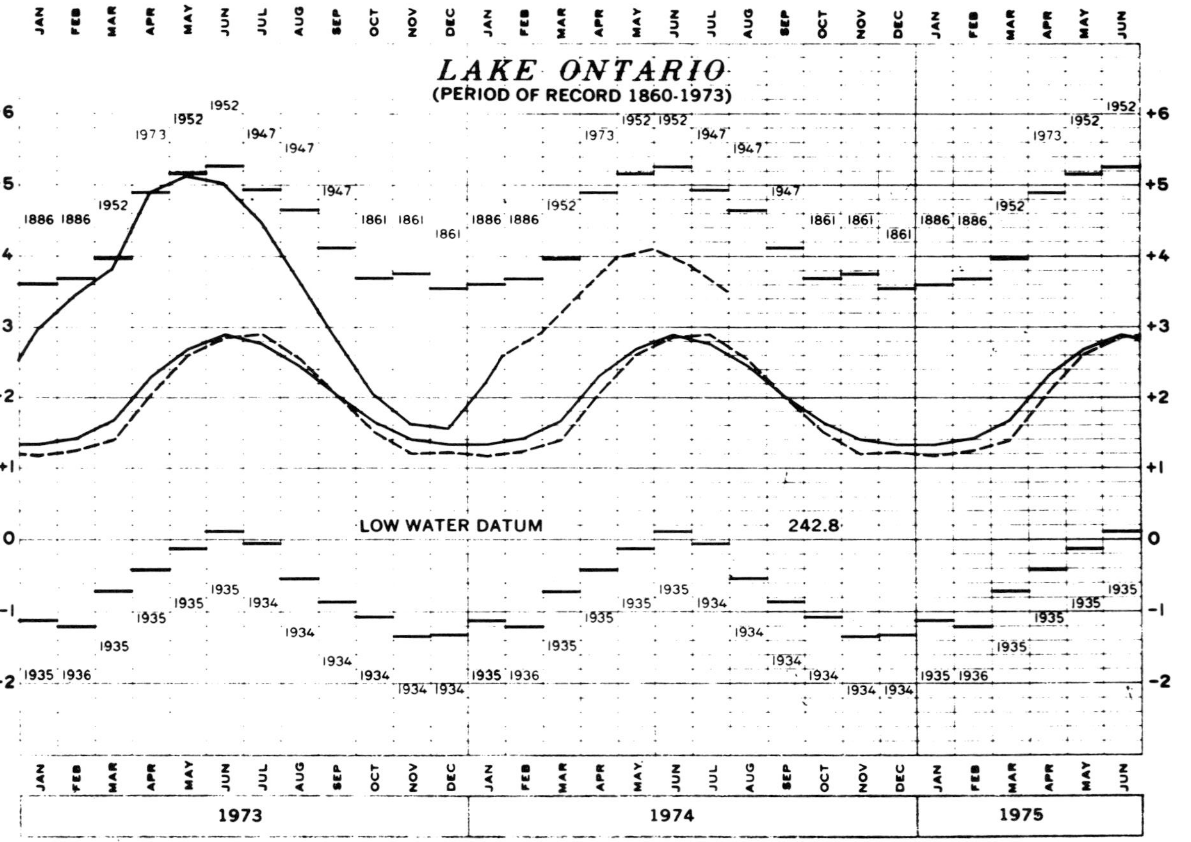

Fig. 8. Water level record (in feet) for Lake Ontario and Irondequoit Bay. Record high and low levels and date of occurrence indicated by horizontal bars. Intermediate curves show mean levels during past 10 years (dashed curve) and during entire period of record (solid curve). Upper curve shows observed levels in 1973 (solid portion) and predicted levels in 1974 (dashed portion). From U.S. Department of Commerce, NOAA, Lake Survey Center, January 1974.

viously mentioned. The location of the outlet to Lake Ontario appears to have been centrally located on the sand bar until a flood in 1853 created the present channel further west. A comparison of early maps and of 1931 and 1968 aerial photos records the considerable extension of the Densmore Creek delta, the joining of Snider's Island to the mainland, and the filling and extension of the Irondequoit Creek Delta at the south end of the bay.

Deep sediment layers underlie the bay. Prior to construction of the Bay Bridge, the New York State Department of Transportation examined 27 sediment cores. Samples of the core logs were presented by Marine Resources, Inc. (1970), and Bubeck (1972) interpreted them as follows. At the center of the middle span of the bridge, the depth of water is about 20.1 m. Below this position the upper layer, about 13.4 m thick, is "black muck underlain by grey marl (very soft)"; then a middle layer, about 28.1 m thick of "layered grey to grey brown fine sand and silt (medium dense to very dense), trace of clay," with an apparent 1.2–1.8 m thick lens of "red-brown sand, some silt and stones" near the contact with the lowest layer. The lowest layer is about 30 m thick and composed of "red brown silt, stone, and sand, trace of clay (very dense)." Bedrock of Queenston shale is reached 73.2 m below bay bottom, apparently verified by seismic measurements.

Morphometry

Bubeck (1972) reported the following dimensions of the bay: maximum length 6.7 km, greatest width 1.2 km, surface area 6.67 km^2, volume 45.9 $\times$ 10^6 m^3, maximum depth 23.8 m, mean depth 6.8 m. Tressler *et al.* (1953) reported similar values of area and mean depth, but apparently erred in calculating a volume of 57.5 $\times$ 10^6 m^3. Figures 9 and 10 show how area and volume change with depth. Approximately 52% of the area lies over shallows less than 3 m deep. Approximately 52% of the volume lies above the 6-m depth, the approximate depth of the epilimnion in summer.

Temperature

From temperature measurements at station 1, the location of greatest depth, Diment *et al.* (1974) prepared the isopleth diagrams shown in Figs. 11, 12, and 13. These show an annual cycle characterized by mixing in April and again in late November and December, with strong stratification from May through October, and with ice cover from late December through March. The fall mixing is complete and extends to the bottom. As detailed later, cold water containing in excess of 400–600 mg/liter chloride (from runoff of road deicing salt) accumulates on the bottom during winter and

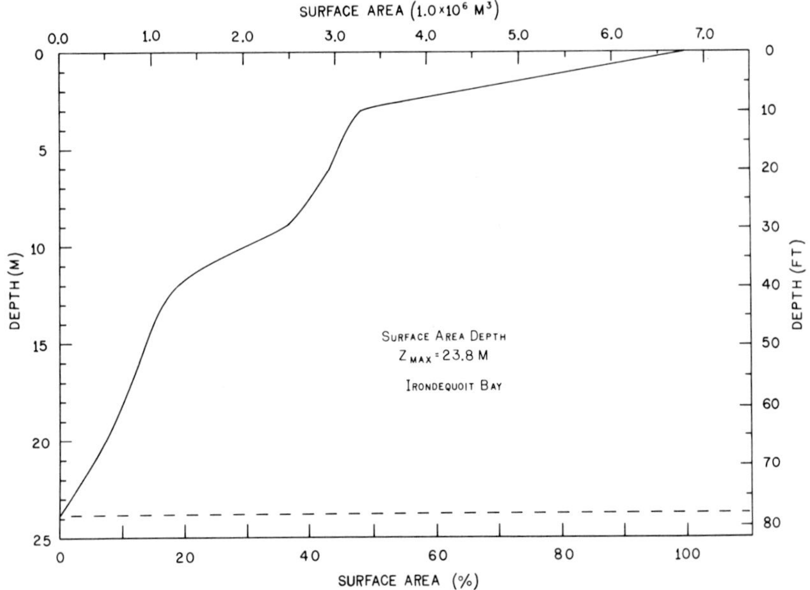

Fig. 9. Area versus depth for Irondequoit Bay. From Bubeck (1972).

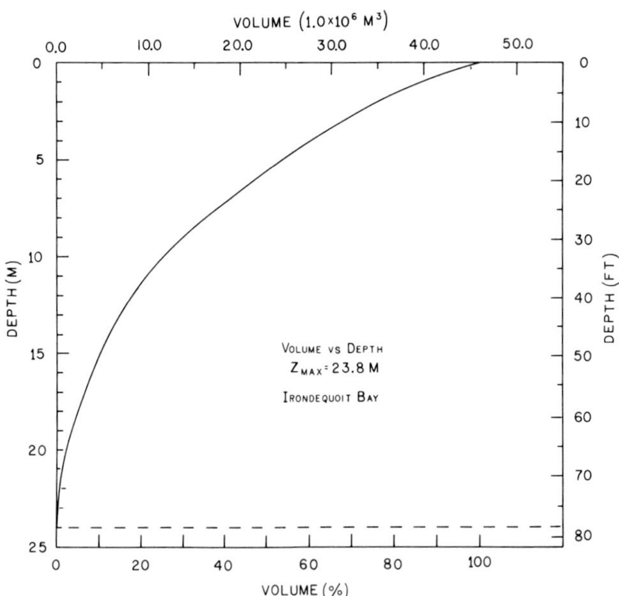

Fig. 10. Volume below depth z as a function of depth z. From Bubeck (1972).

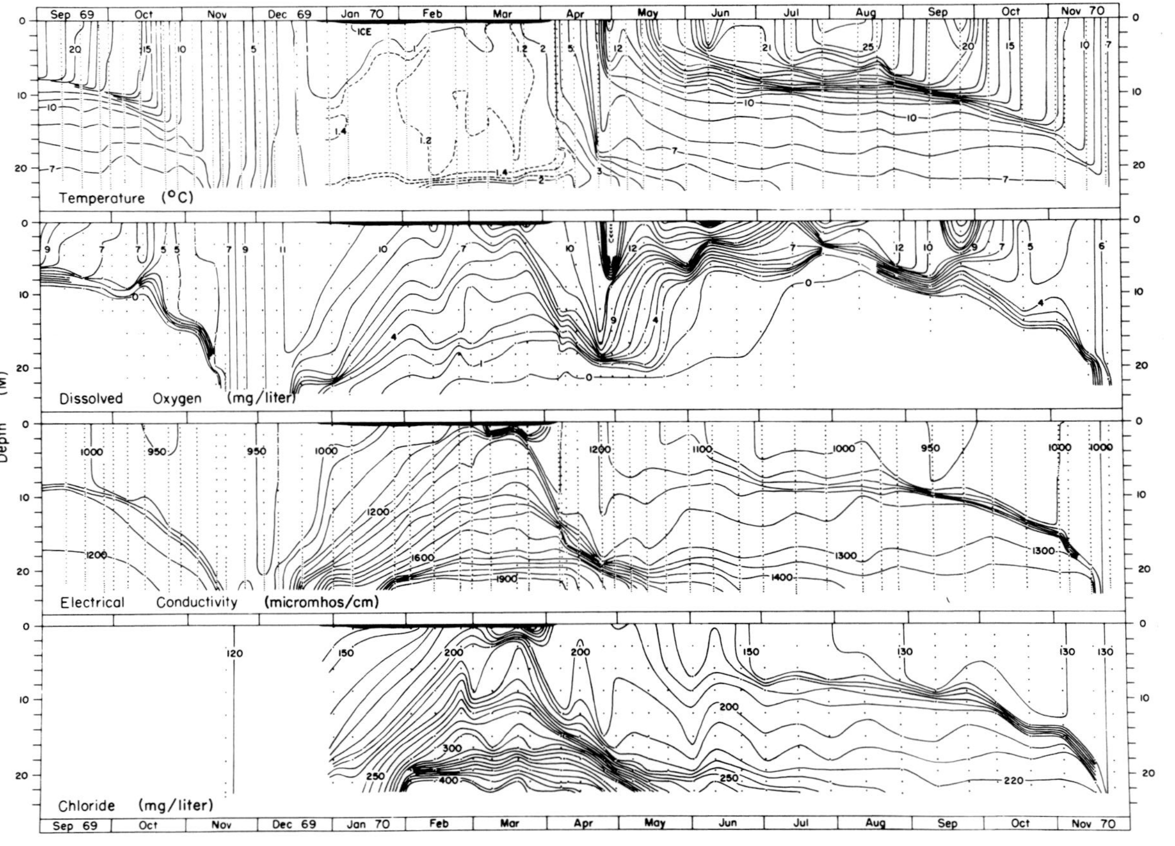

Fig. 11. Temperature, dissolved oxygen, conductivity at 25°C, and chloride isopleths, September 1969 to November 1970. From Bubeck *et al.* (1971). (Copyright 1971 by the American Association for the Advancement of Science.)

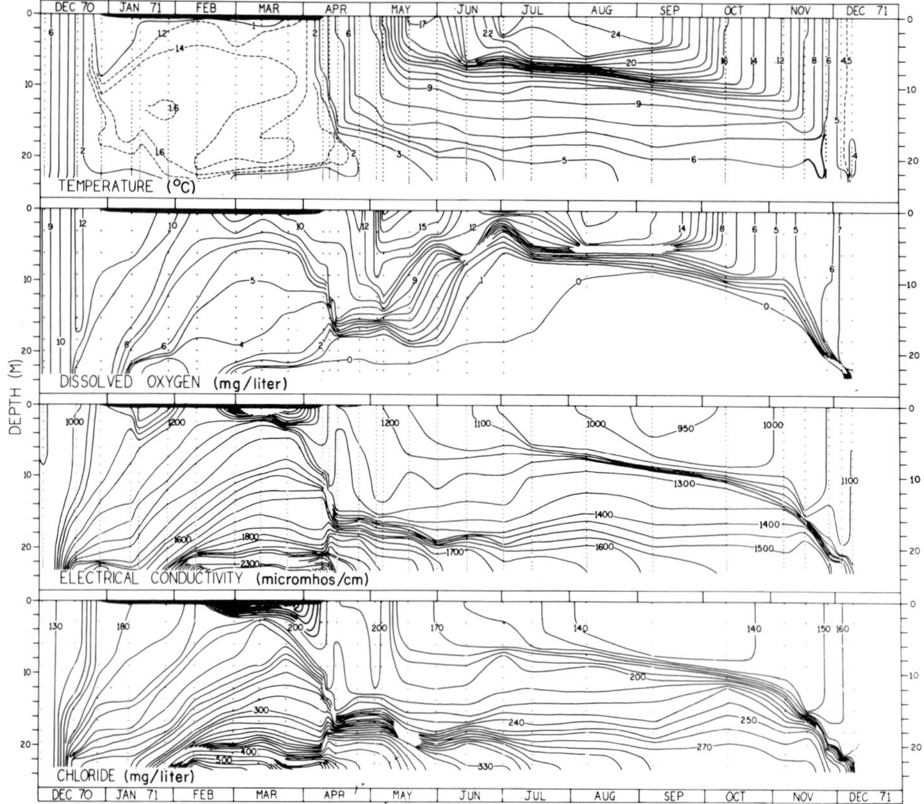

Fig. 12. Temperature, dissolved oxygen, conductivity at 25°C, and chloride isopleths, December 1970 to December 1971. From Diment *et al.* (1973). (Reprinted with the permission of the Highway Research Board, National Academy of Sciences.)

limits mixing in the spring to the upper 12–18 m. The depth of the epilimnion (i.e., the depth to the upper margin of the thermocline) is about 4 m in early June, 6 m in July, and 8 m in September. During summer the depth of the well-mixed layer generally ranges between about 5 m, when mixing is vigorous, and about 3 m at times of strong secondary stratification within the epilimnion.

Level Fluctuation; Seiching

During the ice-free seasons from November 1969 to November 1971, R. G. Austin recorded surface levels, at intervals of one to several days, using a staff gauge located at the shore of his home (R. C. Bubeck,

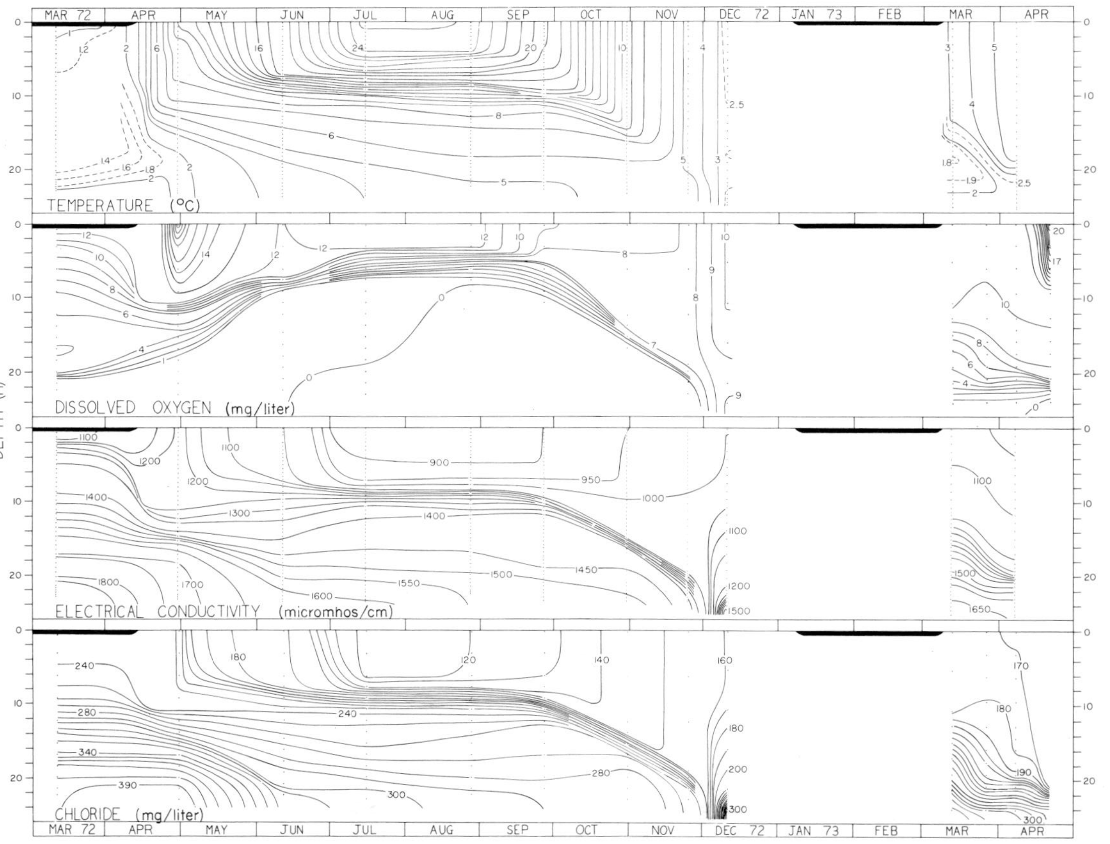

Fig. 13. Temperature, dissolved oxygen, conductivity at 25°C, and chloride isopleths, March 1972 to April 1973. From Diment *et al.* (1974). (Reprinted with the permission of the Northern Ohio Geological Society, Inc.)

unpublished data). A plot of the data showed that the level, computed as a 2-week running average, closely matched the Lake Ontario levels for the same period. The amplitude of short-term fluctuations (over a few days) reached a maximum of about 25 cm in late November 1970 (after 5 days of rain), and attained 15 to 18 cm on three other occasions. At other times, day-to-day fluctuations did not exceed about 7 cm.

During part of the period August–December 1970, the water level was continuously recorded at Austin's shore. Many of the tapes show trains of oscillations lasting several hours or more. The peak-to-peak amplitude sometimes approached 1 cm and the period was about 18 min. According to an analysis of Witten (1971), the steep drop-off near Stony Point prevents the bay from seiching with a period of about 25 min, which would be predicted from the overall length and mean depth. An 18-min period was predicted for seiching of the deep water mass lying south of Stony Point. For the shallows north of Stony Point, an independent seiche with a period of about 6 min was predicted, but was not evident in the tapes.

Horizontal Mixing

Horizontal circulation within the bay has not been carefully studied. As was shown by Fig. 6, the inflow from Irondequoit Creek is directed along the east side of the bay, and, in periods of high discharge, a plume of turbid water extends northward for a considerable distance. As a result of the input from the creek, and because of the shallow depth and the beds of cattails and other aquatics, the surface waters at the south end probably differ from those of the deep basins. Similarly, waters of the shallows at the north end may differ from those of the deep basin. Moreover, during strong northerly winds, there are occasions when Lake Ontario water flows into the bay. Bubeck (1972) showed that, at such times, dilution by Lake Ontario water reduced the conductivity and chloride content of bay water near the outlet. Since changes in the elevation of the level of the bay are generally smaller than 1% of the mean depth, it is certain that the occasional injections of Lake Ontario water cannot amount to more than about 1% of the bay volume. Thus significant dilution effects are necessarily limited to the area close to the outlet.

Over the two deep basins, from Stony Point in the north to Glen Haven in the south, available data indicate approximate horizontal homogeneity of surface waters. Bubeck (1972) measured temperature, conductivity, Secchi depth, and dissolved oxygen at stations 1, 2, 3, and 4 (Fig. 7) fortnightly from May to October 1969. At stations 1 and 2, in the deeper northern basin, the data were similar. At stations 3 (off Densmore Creek) and 4 (off Glen Haven), dissolved oxygen and turbidity were somewhat higher and the thermocline was less steep. During 1970 and 1971, Bannister carried out phytoplankton analyses of samples taken at station 1 and at the end of the dock (about 15 m from shore where depth is about 4.5 m) of the Rochester Canoe Club located on the west shore between Newport House and Point Pleasant. Species and numbers were similar at the two locations.

Chemical analyses of the surface waters at different stations on the bay have been performed by the Federal Water Pollution Control Administration (FWPCA) (1965), Monroe County Department of Health (MCDH) (unpublished data, 1965), and Harbison (1972). The FWPCA data for nine stations, sampled in May, July, and September, show approximate homogeneity from Stony Point to Glen Haven. At the mouth of the bay, most parameter values were low as a result of dilution with Lake Ontario water. At the southernmost station, near the mouth of Irondequoit Creek, silica (SiO_2), sulfate (SO_4), soluble and total phosphate (PO_4), and nitrate (NO_3) concentrations were high on one or two but not all three dates. MCDH (unpublished data, 1965) analyzed surface water on 20 dates at six stations between Stony Point and Glen Haven. At the bay outlet, soluble PO_4 and ammonium (NH_4) contents were often higher than at main bay stations. Near the mouth of Irondequoit Creek, soluble PO_4 was lower and NO_3 and NH_4 were much higher than at main bay stations.

Harbison (1972) assayed several parameters, as functions of depth, in samples taken at 15 stations on June 17, 1972. He found that, in surface water, total inorganic nitrogen ranged from 0.40 to 0.48 mg/liter at Snider Island and more southerly stations, 0.38 off Glen Haven, 0.30 at the Bay bridge, 0.19–0.27 off Stony Point, and 0.02 to 0.17 mg/liter at stations in the Northeast Arm. Thus there was a small gradient (about 0.15 mg/liter N in 4 km) between Glen Haven and Stony Point, and steeper gradients from Stony Point into the Northeast Arm and from the mouth of Irondequoit Creek to Glen Haven. Harbison also reported gradients of pH (from 9.0 in the south to 9.3 at Stony Point to 9.4–9.8 in the Northeast Arm) which are correlated with the sampling schedule (which began at 9 A.M. in the Northeast Arm and progressed to the southernmost station at 3:30 P.M.). Taken together, the FWPCA, MCDH, and Harbison data indicate that sometimes, but not always, a gradient in inorganic nitrogen exists along the axis of the Bay. Even when it exists, the gradient is small, and horizontal homogeneity is a fair approximation between Stony Point and Glen Haven. For other constituents, available data do not indicate gradients between Stony Point and Glen Haven.

Transparency

Tressler *et al.* (1953) measured Secchi disk depths monthly from August 1939 to June 1940, at station 1. The only values published were 5.8 m in June (the maximum for the 11-month period), 1.3 m in September (the minimum for the period), and 2.7 m (the average for the period). Regrettably, the value for May was not published, and no measurement was made in July. The FWPCA (1965) recorded Secchi depths at several stations on June 10 and August 25, 1965. At the outlet and at the mouth of Irondequoit Creek, the depth was 0.5 m; at other bay stations, the depths were 1.0–1.25 m in June and 1.5–2.0 m in August. During 1969–1973, Bubeck (1972; also unpublished data, 1975) recorded Secchi depths at station 1. From mid-June to early October, depths ranged between 0.4 and 1.3 m, the lowest values occurring in July. Maximum depths of 4 to 5 m were recorded under ice in January and February. All available Secchi depth data are graphed in Fig. 14. In addition to Secchi depth, Bubeck (1972) recorded horizontal light transmission. Bannister made a few measurements with a silicon-cell photometer in summer 1970 and found vertical light transmission to be about 10%/m, for the 400–700 nm spectral band. Since Secchi depths correspond approximately to the depth for 15% transmission, the photometer measurements of vertical transmission imply a Secchi depth of about 0.8 m, in close agreement with Bubeck's experimental values.

Apparently, the Secchi depth has decreased since 1939–1940. This is indicated both by the 11-month average of 2.7 m and the June depth of 5.8 m for 1939–1940, much larger values than those for 1969–1972. However, the specific conclusion that summer transparency has decreased is less certain because (1) no values are available for May and July 1940, (2) the Secchi

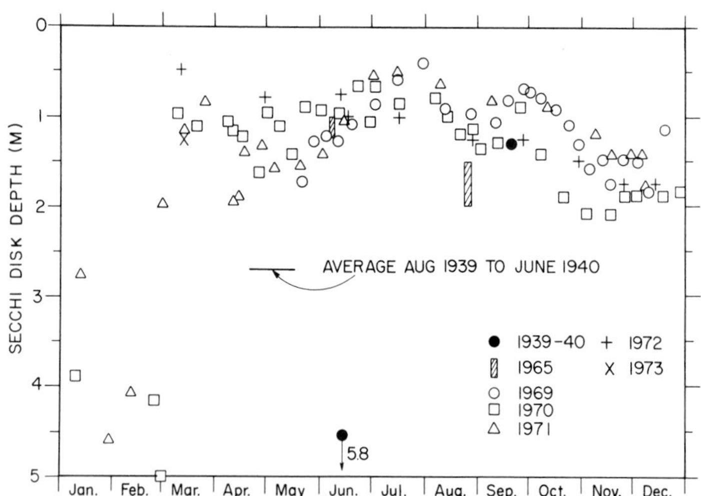

Fig. 14. Secchi disk depths at station 1, 1939–1973. Data for 1939–1940 from Tressler *et al.* (1953), for 1965 from FWPCA (1965), for 1969–1973 from Bubeck (1972, also unpublished data)

depth in June 1940 is suspiciously high, and (3) the Secchi depth in September 1939 lies within the range of the 1969–1972 values. The 1965 data, for two days only, are too few to prove any trend; probably there has been no significant change between 1965 and 1969–1973. The Secchi depth in June 1965 was similar to those for 1969–1973; the depth in August 1965 was somewhat higher than those for 1969–1973, but also higher than for September 1939.

CHEMICAL LIMNOLOGY

Chemistry of the Bay in 1969-1972

The most comprehensive set of data is that of Bubeck (1972) who determined depth profiles of physical and chemical properties, at station 1, at least monthly and usually at more frequent intervals. All data obtained up to December 1971 are recorded in the thesis of Bubeck (1972). Isopleth diagrams for temperature, conductivity, dissolved oxygen (DO), and chloride (Cl), for the period from mid-1969 to December 1972, were shown in Figs. 11, 12, and 13. These diagrams were discussed in previous articles in relation to road deicing salt runoff (Bubeck *et al.*, 1971; Diment *et al.*, 1973, 1974). Other chemical constituents were analyzed in 1970 and 1971. The frequency of chemical sampling through the water column was indicated by the vertical dots shown on the chloride isopleth diagrams of Figs. 11, 12, and 13. Isopleth diagrams for sulfate (SO_4), sulfide (HS), and alkalinity are shown in Fig. 15, for nitrate nitrogen (NO_3–N), nitrite nitrogen (NO_2–N), and ammonium nitrogen (NH_4–N) in Fig. 16, and for pH, silica (SiO_2), and soluble phosphate (PO_4) in Fig. 17. At less frequent intervals, analyses of calcium (Ca), magnesium (Mg), bromide (Br), and fluoride (F) were performed, and a few determinations of sodium (Na) and potassium (K) were also carried out. Table 9 summarizes the values of chemical and physical parameters at the surface and at 22 m, at four seasons of the year, during 1970 and 1971. Figure 18 shows time courses of temperature, pH, NH_4–N, NO_3–N, total inorganic N, SiO_2, and soluble PO_4 in surface water during 1970–1971.

Table 9 shows that Irondequoit Bay is a hardwater lake with high concentrations of Ca, Mg, Na, SO_4, Cl, and bicarbonate (HCO_3). To a fair approximation, the content of major mineral components in the epilimnion in summer may be regarded as composed of NaCl (4 m*M*), $MgSO_4$ (about 1 m*M*), and $Ca(HCO_3)_2$ (about 2.5 m*M*). The SO_4 content varies relatively little with time of year or depth, though some depletion in deep water occurs during the period of anoxia and is reflected by accumulation of HS. The NaCl content changes dramatically during the year and with depth,

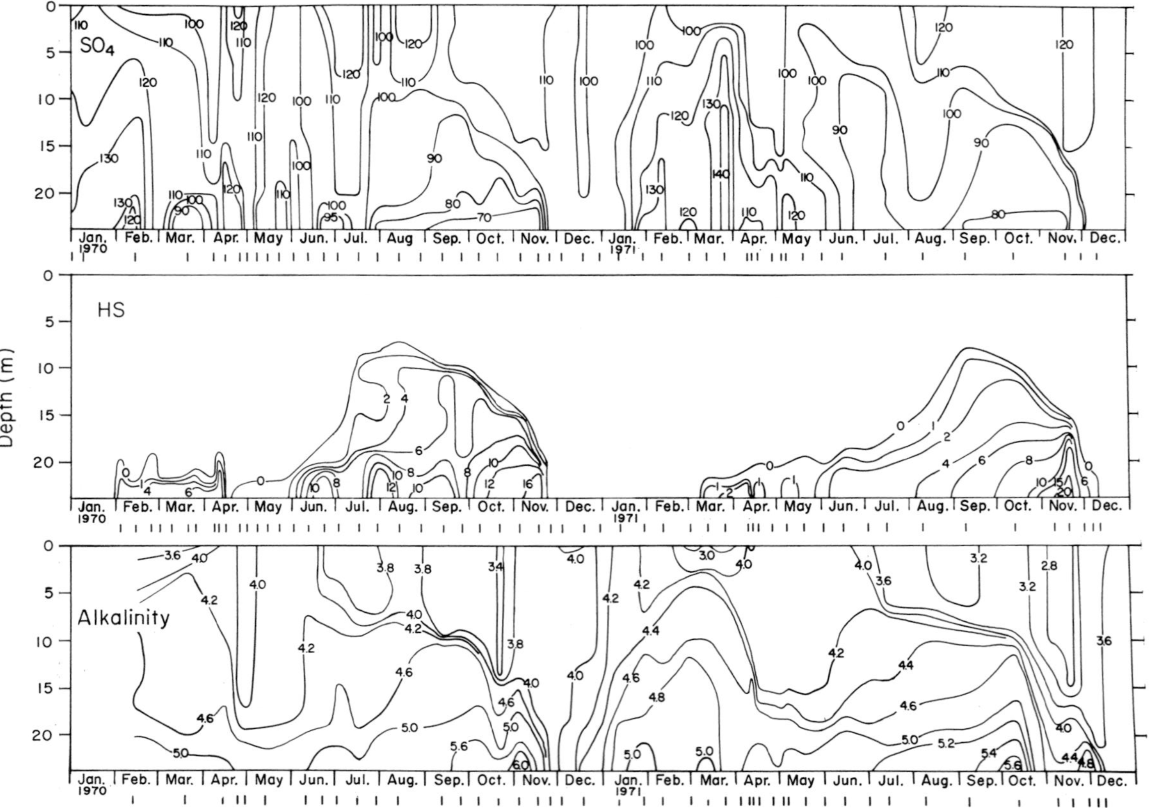

Fig. 15. Isopleths of sulfate and sulfide (mg/liter SO_4 and HS) and alkalinity (mEq/liter) at station 1 for 1970 and 1971. From data of Bubeck (1972).

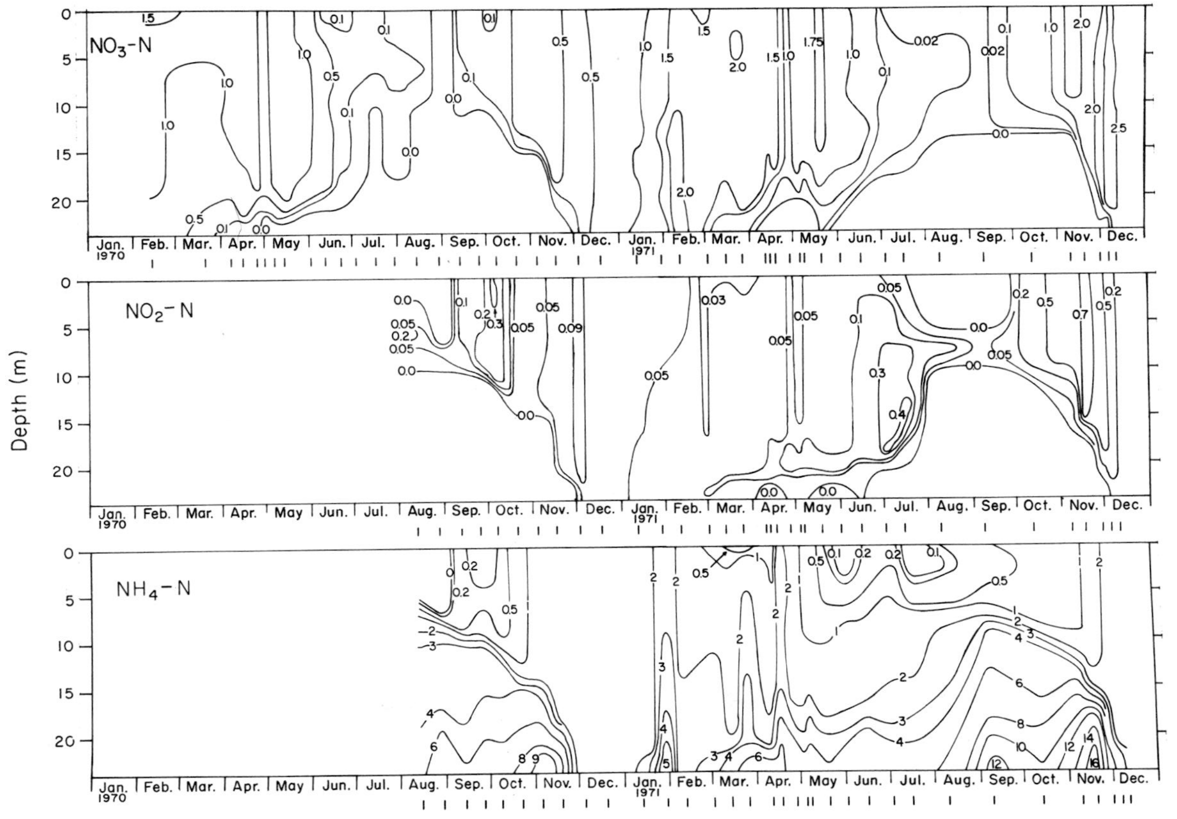

Fig. 16. Isopleths of NO_3–N, NO_2–N, and NH_4–N (mg/liter N) at station 1, for 1970 and 1971. From data of Bubeck (1972).

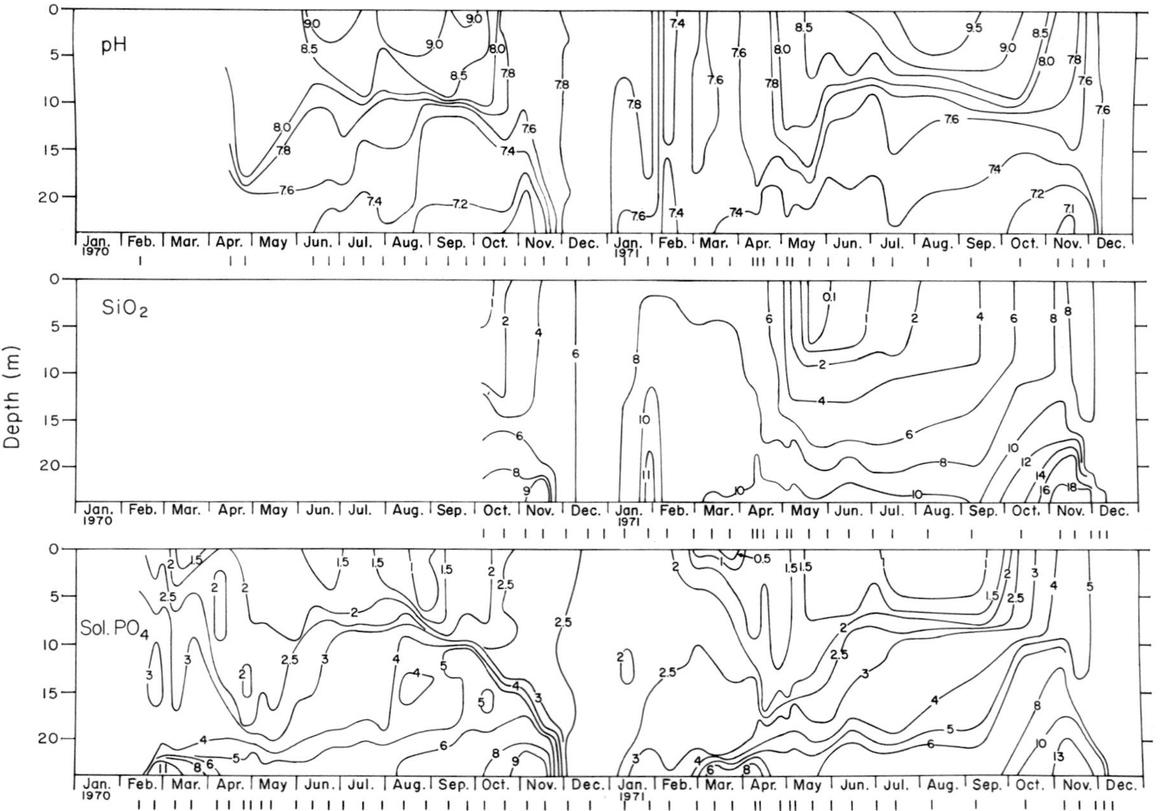

Fig. 17. Isopleths of pH and silica and soluble phosphate (mg/liter SiO_2 and PO_4) at station 1, for 1970 and 1971. From data of Bubeck (1972).

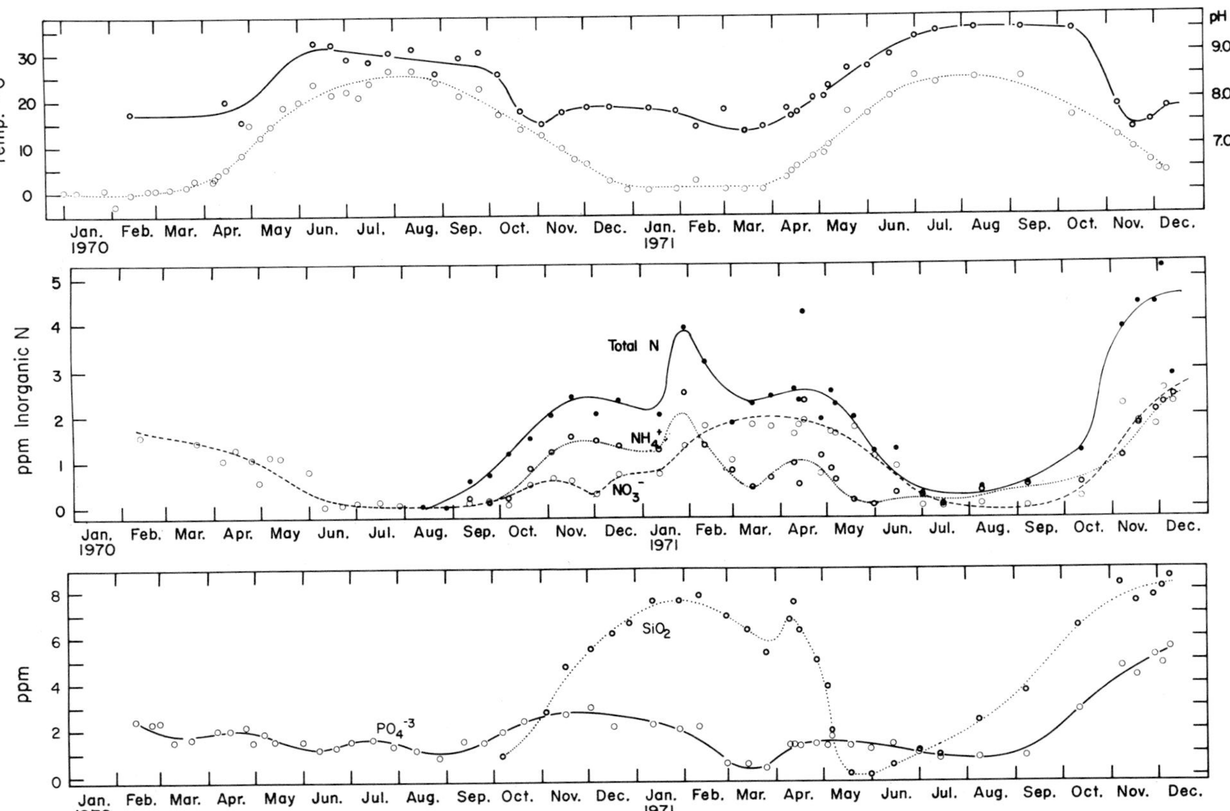

Fig. 18. Time course of pH, temperature, NH_4–N, NO_3–N, total inorganic N, SiO_2, and soluble PO_4 in surface water at station 1, for 1970 and 1971. From data of Bubeck (1972).

TABLE 9

Representative Values of Physical and Chemical Parameters at Station 1[a]

	1970–1971[b]				
	Jan.–Feb.	May 1 (end of spring mixis)	Jul.–Aug.	Dec. 1 (end of fall mixis)	1965[c] Summer surface water
Secchi depth (m)	4.5	1.5	<1.0	2	—
Temperature (°C)	0.5/1.5	7.2/2.5	25/5.5	7.2/7.5	24
pH	7.6/7.4	7.7/7.5	9.3/7.4	7.8/7.6	9.5
Alkalinity (mEq/liter)	3.8/5.1	4.1/5.0	3.5/5.2	3.9/4.6	3.1
Conductivity (μmho/cm)	1150/1800	1200/1900	950/1500	950/950	862
Dissolved oxygen	10/4	12/0.5	12/0	8/8	10.9
Cl^-	170/340	190/380	140/280	130/130	110
SO_4^{2-}	101/111	110/120	115/103	105/105	119
HS^-	0/0	0/0.2	0/5.5	0/0	—
Br^-	n[d]	0.4/0.1	1.2/0.2	n	—
F^-	n	n	0.7/0.7	n	—
PO_4^{3-}	2.2/3.1	1.7/5.0	1.2/6.0	2.9/2.2	0.73
SiO_2	7.5/10.5	4.8/9.8	1.0/9.0	5.5/5.7	1.79
NO_3–N	1.2/1.5	1.2/0.05	0.1/0	0.5/0.5	0.06
NO_2–N	0.05/0.05	0.05/0.05	0.04/0.00	0.28/0.28	—
NH_3–N	1.8/3.5	1.1/4.4	0.2/4.8	1.5/1.6	0.2
Ca^{2+}	70/100	100/200	90/80	n	86
Mg^{2+}	22/26	30/30	24/25	n	26
Na^+	n	135/245	120/150	n	68.4
K^+	n	3.6/5.0	2.3/n	n	3.9

[a] Value in surface water/value at 21 m. Units given in mg/liter except where otherwise noted.

[b] 1970–1971 data from Bubeck (1972).

[c] 1965 data from FWPCA.

[d] n, no data.

being as high as 14 m*M* (corresponding to 500 mg/liter Cl) in deep water in winter and spring. Alkalinity is also relatively constant throughout the year, but values at the surface (about 3.5 to 4.1 mEq/liter) are somewhat lower than those (4.6 to 5.2 mEq/liter) at 22 m.

Other parameters (pH, HS, soluble PO_4, NH_4–N, NO_2–N, NO_3–N, and SiO_2) show strong seasonal and depth dependences, which are governed by temperature, stratification, vertical mixing, and biological processes. At the end of the fall mixis in December, all parameters are approximately uniform with depth, and the water is well oxygenated at the bottom. The pH is about 7.7 and values of inorganic N (2.5 mg/liter in 1970 and 4.5 mg/liter in 1971), soluble PO_4 (about 2.5 mg/liter in 1970 and about 5 mg/liter in 1971), and SiO_2 (about 8 mg/liter) are all high. Chloride concentration (about 130 mg/liter) is at its annual minimum. Under ice cover from late December to about April 1, the water column stratifies and receives large inputs of cold, chloride-laden water which accumulate at the bottom. Oxygen depletion proceeds gradually and leads to anoxia at 22 m in February or March. High NH_4–N and soluble PO_4 and low NO_3–N values apply to the deep water at the time of ice breakup about April 1. At the end of April, after vernal mixis (which extends only to about 15 m), the upper mixed layer contains inorganic N, soluble PO_4, and SiO_2 at about the same levels that occurred in December and throughout the winter.

Stratification begins about the first of May. From this time to about July 1, massive blooms of diatoms, green algae, and blue-green algae occur in rapid succession in the warming epilimnion. Concurrently, SiO_2 falls precipitously to near zero in mid-May, and inorganic N declines steadily from 2.5 mg/liter on May 1 to about 0.2 mg/liter on July 1. In contrast, soluble PO_4 declines relatively little. By July 1, the pH reaches a maximum level (about 9.0 in 1970, about 9.5 in 1971). Also, by this time the hypolimnion and the lower part of the thermocline are anoxic and contain inorganic N, PO_4, SiO_2, and Cl at substantially higher concentrations than are found in the epilimnion.

From about July 1 to mid-September, chemical and algal parameters of the epilimnion remain relatively steady. The pH remains high (8.5 to 9.0 in 1970, about 9.5 in 1971), and soluble PO_4 remains at about 1.2 mg/liter. Inorganic N fluctuates between 0.05 and 0.5 mg/liter but shows no trend and was usually between 0.1 and 0.2 mg/liter. In 1970, but not 1971, there was a considerable replenishment of SiO_2 in the epilimnion. Throughout the summer, NH_4, PO_4, and HS concentrations increase steadily in the hypolimnion.

From mid-September to mid-November, the thermocline descends, and inorganic N, PO_4, and SiO_2 levels of the epilimnion increase as the hypolimnion waters are gradually incorporated. NO_2–N concentrations reached 0.6 mg/liter in late October. Thorough mixing to the bottom at 22 m begins in late November and continues until ice forms in late December.

During the mid-June to mid-September period, the depth of the well-mixed portion of the epilimnion varies considerably. On some occasions (July 16, 1970; August 28, 1970; June 15, 1971), temperature, DO, pH, PO_4 concentration and other properties were essentially constant to 6.1 m, and the sharp changes characteristic of anoxic water were first observed at 7.6 m. In these instances, the mixed layer was at least 6 m deep and included essentially all of the epilimnion. Infrequently in June, and more commonly in July and August, warm weather produced secondary stratification within the epilimnion. An extreme example was manifested by depth profiles on July 2, 1971. At depths of 0, 3, and 6 m, the temperatures were 24.7, 22.8, and 17.2; DO values were 14.8, 4.7, and 0.8; pH values were 9.3, 8.7, and 8.3; and inorganic N values were 0.30, 0.80, and 1.33 mg/liter, respectively. For July and August, the average situation (evidenced in profiles for July 2, July 29, and August 13, 1970 and for July 15, August 9, and September 8, 1971) was one in which pH, DO, and inorganic N were essentially identical at 0 and 3 m, nearly the same at 4.6 m, and appreciably different at 6.1 m. It appears, therefore, that the average depth of the well-mixed layer is about 4.0 to 4.5 m in summer. Within the well-mixed layer, the concentrations of nutrients in July and August ranged between the following limits: NO_3–N 0.00 to 0.10 mg/liter, NO_2–N 0.00 to 0.04 mg/liter, NH_4–N 0.00 to 0.38 mg/liter, soluble PO_4 0.78 to 1.63 mg/liter, and SiO_2 1.0 to 3.4 mg/liter. Within these ranges, higher levels of SiO_2 and lower levels of soluble PO_4 occurred in 1971 than in 1970. Mention has been made above of differences in summer pH levels and in the levels of inorganic N and soluble PO_4 reached in November in these 2 years. As will be detailed later (see Phytoplankton Growth in the Bay), the July–August algal crop also differed in these two years. Evidently, 2 years of study is inadequate to define normal levels of nutrients and algal components.

Harbison (1972) analyzed chemical components at 1.5-m depth intervals, at station 1, on five dates from June 28 to August 15, 1972. He found about the same pH and alkalinity and contents of NH_3–N and NO_2–N as Bubeck did in the previous year. Compared with 1971 values, those for 1972 for soluble PO_4 (0.60–0.75 mg/liter) and SO_4 (81–103 mg/liter) were lower, and NO_3–N (0.1–0.3 mg/liter) and SiO_2 (2.0–4.0 mg/liter) were somewhat higher. These differences are almost certainly attributable to the heavy runoff and dilution caused by Hurricane Agnes at the end of June 1972. Soluble PO_4–P averaged 89% of total P in both epilimnion and hypolimnion.

Harbison (1972) also reported heavy metal contents of water samples collected at various depths at station 1, on November 3, 1972 when the thermocline was at about 14 m. The results were 9–23 μg/liter zinc, <1 to 7 μg/liter chromium, 5–7 μg/liter copper, <1 μg/liter cadmium, and 17–18 μg/liter nickel; for these elements there was no clear dependence on depth. For manganese, the content was 4–8 μg/liter above and 1520 μg/liter below the thermocline. For iron, the content was 19 μg/liter above and 94 μg/liter below the thermocline.

Earlier Studies of Water Chemistry of the Bay

The earliest data are those of Whipple (1913) who reported temperature, chlorine, dissolved oxygen, and several other water quality parameters at several stations on August 8, 1912. His results are valuable in showing that the hypolimnion was anaerobic then, as it is today, and that chloride concentration was about 10 mg/liter, less than one-tenth that now occurring in summer at the surface.

Tressler *et al.* (1953) reported results obtained in monthly samplings at station 1, from August 1939 to June 1940. Temperature, conductivity, particulate organic matter, dissolved oxygen, pH, alkalinity, and total and soluble phosphorus (P) were reported for surface water and for water at a 22-m depth. Their data refer to a time when road deicing salt was not used and when nutrient loadings were undoubtedly smaller than in 1970–1971. In the absence of deicing salt runoff, the conductivity (about 500–700 micromhos/cm) in 1939–1940 was one-half to one-third that in 1970–1971. Moreover, temperature and dissolved oxygen data for 1939–1940, although showing mixing occurred at the same time (November and late April) as today, indicate that the spring mixing was then more nearly complete, the oxygen concentration at 22 m having reached 6 mg/liter in late April and having remained above zero until early June. The better mixing in 1939–1940 can be attributed to lower salt concentrations in deep water and a lower stability of the water column. Whether mixing to the bottom occurred somewhat earlier in the fall of 1939 is unclear; while temperatures were nearly equal at the surface and at 22 m in late October, dissolved oxygen concentration at 22 m was zero in late October but equal to the surface concentration in late November. Lower nutrient loads and the resulting smaller summertime algal crops in 1939–1940 can be inferred from the greater transparency (see Fig. 14), a lower maximum pH ($\leq$8.4) attained in spring and summer, and lower phosphorus concentrations. In 1939–1940, soluble phosphorus concentrations at the surface ranged from 0.4 to 0.2 mg/liter from late September to late April, half or less the corresponding values in 1970–1971. Furthermore, the soluble phosphorus concentration in May and June 1940 fell to well below 0.05 mg/liter; in constrast in 1970–1971 summer phosphorus levels remained at 0.3 mg/liter or higher.

For October 20 and November 9, 1955, and for September 13, 1962, the New York State Department of Health (1955, 1964) reported data for surface water at several bay stations. Temperature, dissolved oxygen, biological oxygen demand (BOD), chloride, alkalinity, suspended matter and, in 1962, NH_3, NO_3, and total PO_4 were determined. In 1955, the chloride concentration was 30–35 mg/liter. On September 13, 1962, the pH was 8.3, the chloride concentration about 70 mg/liter, NH_3 and NO_3 concentrations (presumably as N) were about 0.9 and 0.1 mg/liter, respectively, and total PO_4 was about 0.35 mg/liter. The pH, chloride, and PO_4 values are much lower, the NH_3 and NO_3 values about the same as for September 1971.

About May 20, July 25, and September 25, 1965, the Federal Water Pollution Control Administration (1965) sampled water at the surface and at one depth, at nine stations in the bay. Seven were offshore stations located between Stony Point and Glen Haven and, for these, the analyses were essentially the same. For the late July sampling date, Table 9 listed the average values of chemical parameters for the seven stations. The results are similar to those for 1970–1971, except that Na, Cl, conductivity, and alkalinity were somewhat lower. The FWPCA data are valuable in showing that soluble phosphorus was 80–90% of total phosphorus. In relation to the July data, those for May and September indicate the same seasonal trends evident in 1970–1971.

The Monroe County Department of Health (unpublished data, 1965) analyzed surface and near-bottom samples weekly between June and September 1965 at several bay stations. For surface samples taken in July–August, the pH ranged from 8.0 to 9.5, NO_3–N from 0.02 to 0.10 mg/liter, NH_4–N from 0.006 to 0.06 mg/liter, albumen N from 0.3 to 0.6 mg/liter, PO_4 from 0.5 to 1.25 mg/liter, and chloride from 100 to 140 mg/liter. Except for the low values for NO_3 and especially NH_4, the results agree with the FWPCA data.

Historical Trends in Water Chemistry of the Bay

The above studies clearly document the increases in chloride and conductivity resulting from applications of road deicing salt. Diment *et al.* (1974) showed that salt applications to roads began in the 1950's and that the quantity employed increased exponentially during the 1960's. The very rapid increase in chloride concentration in the bay, from 30 mg/liter in 1955 to about 150 mg/liter in 1970, reflects the accelerating use of deicing salt and contrasts sharply with the much slower rate of increase in the preceding 43 years (from 13 mg/liter in 1912 to 30 mg/liter in 1955).

The studies began too late to document the historical trends in nutrient concentrations which must have begun to increase by 1930 if not much earlier. The earliest values of summer concentrations of NO_3–N and NH_4–N, obtained in 1962 and 1965, are essentially the same as those obtained in 1970 and 1971. Figure 19, which assembles all the data for soluble PO_4 obtained between 1939 and 1971, indicates that the PO_4 concentration in surface water probably rose two- to fourfold between 1939 and the mid 1960's, but, as in the case of nitrogen, there is no clear evidence of any increase thereafter. Since the banning of phosphorus-based detergents in 1973, the discharge of phosphorus by Irondequoit Creek has declined about 50% (see next section). The concentration of phosphorus in the bay has probably also declined. The record of pH values in surface water in summer shows only that the value did not exceed 8.4 in 1939–1940 and that values of 9.0 to 9.5 occurred in 1965 and 1970–1971. Altogether, the pH, PO_4, and

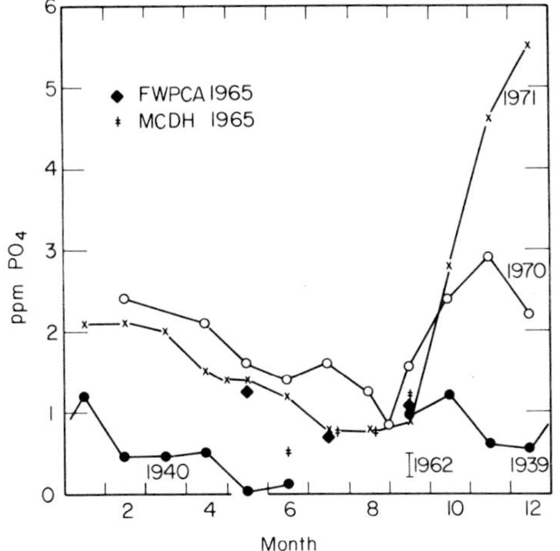

Fig. 19. Soluble PO_4 concentration (mg/liter PO_4) in surface water of Irondequoit Bay, 1940–1971. Data for 1939–1940 from Tressler *et al.* (1953), for 1962 from NYSDH (1963), for 1965 from FWPCA (1965) and MCDH (unpublished data 1965), and for 1970–1971 from Bubeck (1972).

nitrogen data are consistent with the view that nutrient concentrations and summer algal growth in the bay increased substantially between 1940 and the mid 1960's, but did not change further up to 1972.

Chemical Limnology of Irondequoit Creek

Water quality surveys of Irondequoit Creek and its tributaries were reported by Faigenbaum (1940) and the New York State Department of Health (1955, 1964). These surveys showed that sewage and industrial waste discharges led to anoxia, high biological oxygen demand, and sludge deposits in some reaches of the creek at least as early as 1939. Since 1970, chemical studies have been carried out by Bubeck (1972), Diment *et al.* (1974), Thorne (1973), and, routinely, by the Monroe County Department of Health. Table 10, based on Thorne's study in summer 1972, lists the chemical characteristics of Irondequoit Creek at Empire Boulevard. As expected, the values for alkalinity, calcium, sulfate, and chloride are similar to those found in bay water. Essentially similar values of these parameters have also been obtained by the Monroe County Department of Health (unpublished data for the years 1968–1975). Routine sampling by the Department of Health, from 1968 to 1974, at Empire Boulevard, shows that dissolved oxygen concentration is commonly between 2.0 and 4.0 mg/liter (extreme values were 1.5 and 8.0 mg/liter) and that biological oxygen

TABLE 10

Mean Values of Chemical Parameters in Irondequoit Creek at Empire Boulevard[a,b]

Parameter	Mean value
DO	4.3
pH	7.9
Suspended sediment	16.80
NH_4-N	0.36
NO_2-N	0.15
NO_3-N	0.92
Organic N	0.45
Kjeldahl N	0.80
Soluble PO_4-P	0.41
Hydrolyzable P	0.50
Ca (as $CaCO_3$)	279
Alkalinity (as $CaCO_3$)	202
SO_4	120
Cl	109
SiO_2	5.1

[a] Calculated from the data of Thorne (1973) for ten dates between early June and mid-August 1972.

[b] All values except pH in mg/liter.

demand is usually between 4.0 and 8.0 mg/liter (extreme values being 1.8 and 12.0 mg/liter).

Bubeck (1972) and Diment *et al.* (1974) made many measurements of chloride concentration and water flow at Browncroft Boulevard (see Fig. 4). During summer and fall, the chloride concentration was about 100 mg/liter, but was much higher (up to 600 mg/liter) during thaws in winter and early spring. Diment *et al.* calculated that Irondequoit Creek discharged 28,500 metric tons of NaCl in 1970–1971 and 39,100 tons in 1971–1972. After estimating the contributions of other tributaries and inputs, the total loading of NaCl on the bay was calculated to be 38,900 and 46,700 metric tons, respectively, in the 2 years. In each year about 70,000 metric tons of road deicing salt were applied in the bay drainage basin. Evidently 25,000 to 30,000 tons of applied salt were transferred to soils and to groundwater. Diment *et al.* (1974) discussed the rising salt concentration in area wells.

Since 1970, nitrogen and phosphorus concentrations in Irondequoit Creek and its tributaries have been investigated by R. C. Bubeck (unpublished data, 1975), Thorne (1973), and the Monroe County Department of Health (unpublished data). From July 1970 to August 1971, at fortnight intervals, Bubeck assayed nitrate and soluble phosphate in

Irondequoit Creek at Empire and Browncroft Boulevards. Thorne (1973) reported analyses of samples collected weekly between early June and mid-August 1972 at many stations in the drainage basin. Ammonia, nitrite, and total inorganic nitrogen were measured in filtered samples and nitrate nitrogen calculated by difference. Organic nitrogen was calculated as the difference between ammonia nitrogen and Kjeldahl nitrogen, the latter determined in unfiltered samples. Soluble phosphate phosphorus and hydrolyzable phosphorus (which includes soluble phosphate phosphorus and approximates total phosphorus) were also determined. The Monroe County Department of Health assayed samples collected at weekly intervals during the summers of 1972–1975 at several locations in the basin. Nitrate plus nitrite nitrogen, total phosphorus (after digestion of unfiltered samples), and soluble phosphate phosphorus (samples passed through 0.45 μm filters, no digestion) were determined.

Available data on nitrogen and phosphorus concentrations in Irondequoit Creek, at Empire and Browncroft Boulevards, are summarized in Tables 11 and 12. The tables provide evidence for the following conclusions: First, the data of Thorne (1973) indicate the ratio of NO_3–N to total N is about 0.50. The data of Thorne and of the Department of Health show that the ratio of soluble PO_4–P to total P was about 0.85 in 1972 and 0.56–0.74 in 1973–1975. The proportion of soluble PO_4–P is high, as expected in waters in which a large part of the phosphorus derives from municipal effluents.

Second, to a fair approximation, the concentrations of nitrogen and phosphorus are the same at Empire and Browncroft Boulevards. Thorne noted, however, that the concentrations of both elements were about 10% higher at the Browncroft station. He suggested that the difference might arise from nutrient uptake by vegetation of the wetlands.

Third, Bubeck's data show a marked seasonal dependence of nitrogen and phosphorus concentrations. Soluble PO_4–P (and presumably total P) concentration was nearly twice as high in the period from July through October (when creek flow was low) as in the period from November to May (when flow was high). The opposite behavior was exhibited by the concentrations of NO_3–N and total N, which were about 60% higher in the winter period of high flow.

Fourth, the concentration of phosphorus, characteristic of the summer period of low flow, has declined about 50% since 1970–1971, when the estimated concentration of total P was about 1.2 mg/liter. In 1972 the concentration (0.53 mg/liter) was sharply reduced, a consequence of the unusually high flow resulting from hurricane Agnes. The partial ban on phosphorus-containing detergents, which took effect in 1972, may also have contributed to the reduction. In 1973 and 1974, the concentration of total P (0.50–0.69 mg/liter) continued at about 50% of the 1970–1971 values. In

TABLE 11

Phosphorus Concentration (mg P/liter) in Irondequoit Creek at Empire (E) and Browncroft (B) Boulevards[a,b]

			Soluble PO_4–P			Total P				
Sampling period	Station	Number	Minimum	Mean	Maximum	Number	Minimum	Mean	Maximum	Source
Jul.–Oct. 1970	E	8	0.46	1.10	2.00	—	—	1.30[c]	—	Bubeck
Nov. 1970–May 1971	E	12	0.20	0.63	1.10	—	—	0.74[c]	—	Bubeck
Jun.–Aug. 1971	E	5	0.62	1.03	1.34	—	—	1.21[c]	—	Bubeck
Nov. 1970–May 1971	B	13	0.16	0.51	0.95	—	—	0.60[c]	—	Bubeck
Jun.–Aug. 1971	B	4	0.53	0.99	1.15	—	—	1.17[c]	—	Bubeck
Jun.–Sep. 1971	E	3	1.08	1.14	1.28	—	—	1.34[c]	—	MCDH
Jun.–Aug. 1972	E	10	0.18	0.41	0.57	8	0.34	0.50	0.72	Thorne
Jun.–Aug. 1972	B	10	0.29	0.49	0.70	8	0.22	0.56	1.07	Thorne
Jun.–Sep. 1973	E	13	0.35	0.51	0.79	13	0.22	0.69	1.12	MCDH
Jun.–Sep. 1974	E	12	0.12	0.28	0.59	13	0.23	0.50	0.90	MCDH
Jun.–Aug. 1975	E	8	0.12	0.48	0.62	7	0.31	0.84	1.20	MCDH
Jun.–Sep. 1975	B	16	0.12	0.52	0.86	16	0.23	0.77	1.60	MCDH

[a] Sources: Thorne (1973) and unpublished data of R. C. Bubeck and Monroe County Department of Health.
[b] For each data set, number of data and minimum, arithmetic mean, and maximum concentrations are listed.
[c] Calculated value based on ratio of soluble PO_4–P to total P of 0.85, as indicated by data of Thorne.

TABLE 12

Nitrogen Concentrations (mg N/liter) in Irondequoit Creek at Empire and Browncroft Boulevards[a,b]

Sampling period	Station	NH_4-N	NO_2-N	NO_3-N	TIN[c]	Organic N	Total N	Source
Jul.-Oct. 1970	Empire	—	—	0.83(9)	—	—	1.59[d]	Bubeck
Nov. 1970-May 1971	Empire	—	—	1.47(12)	—	—	2.97[d]	Bubeck
Jun.-Aug. 1971	Empire	—	—	0.95(5)	—	—	1.92[d]	Bubeck
Nov. 1970-May 1971	Browncroft	—	—	1.37(13)	—	—	2.77[d]	Bubeck
Jun.-Aug. 1971	Browncroft	—	—	1.06(3)	—	—	2.03[d]	Bubeck
Jun.-Aug. 1972	Empire	0.36(10)	0.16(10)	0.93(10)	1.45(10)	0.45(9)	1.90	Thorne
Jun.-Aug. 1972	Browncroft	0.45(10)	0.20(9)	1.04(9)	1.72(9)	0.35(8)	2.07	Thorne
Jun.-Sep. 1973	Empire	—	0.57(13)		—	—	0.97[d]	MCDH
Jun.-Sep. 1974	Empire	—	0.69(12)		—	—	1.18[d]	MCDH
Jun.-Aug. 1975	Empire	—	1.97(7)		—	—	3.39[d]	MCDH

[a] Sources: Thorne (1973) and unpublished data of R. C. Bubeck and Monroe County Department of Health.
[b] For each data set, arithmetic mean concentration and, in parentheses, number of data are listed.
[c] TIN = total inorganic nitrogen.
[d] Calculated value based on component proportions indicated by Thorne.

1975, the concentration was 0.77–0.84 mg/liter, about two-thirds the level in 1970–1971. The reduced concentrations of total phosphorus measured in 1973–1975 can be ascribed to the total ban on phosphorus-containing detergents which took effect in June 1973. Summertime concentrations of nitrogen have not exhibited any clear trend since 1970. The mean concentration of total nitrogen was 1.6–2.0 mg/liter from 1970 to 1972, was lower (about 1.0 mg/liter) in 1973 and 1974, but higher (3.4 mg/liter) in 1975.

Fifth, rates of discharge of nitrogen and phosphorus from Irondequoit Creek into the Bay can be estimated from the flow at Browncroft Boulevard and the concentrations at Browncroft and Empire Boulevards. The flow at Empire Boulevard into the Bay should equal the sum of (1) the flow at Browncroft Boulevard, (2) a negligible contribution (about 0.04 m^3/sec) by runoff from a small area (about 3 km^2) lying along the creek between Browncroft and Empire Boulevards, and (3) the contribution ($\leq$ 0.20 m^3/sec) from the Rochester interceptor overflow at Tryon Park. Since the interceptor discharge of water was accounted for separately (see Table 5) and the interceptor discharges of nitrogen and phosphorus will be estimated separately later, the creek discharges of nitrogen and phosphorus can be correctly calculated from the flow at Browncroft Boulevard. During the periods when mean concentrations of nutrients were determined (see Tables 11 and 12), the mean flow can be calculated from the monthly mean flows given in Table 4. Thus calculated, the mean flow in summer 1972 (the year of the hurricane) was 4.71 m^3/sec, higher than the value (4.1 m^3/sec) reported by Thorne (1973). We think our value, based on Bubeck's daily measurements of flow throughout the period, is more accurate. Mean flow in the June–October periods of 1973, 1974, and 1975 was assumed to be the same as the mean for 1971.

Table 13 lists the calculated mean discharge rates of nitrogen and phosphorus, in the periods when mean concentrations were determined. In 1970–1971 the phosphorus discharge rates were approximately the same in summer and winter. Apparently, phosphorus inputs (largely municipal effluents) in the basin were approximately constant, and the concentration of phosphorus at Browncroft Boulevard was determined mainly by dilution. Since the rate of discharge of phosphorus is approximately independent of season, the mean discharge rates listed in Table 13 are estimates of the mean annual discharge rate. Thus the mean annual discharge rate of total phosphorus was about 250 kg/day in 1970–1971 and about 140 kg/day in 1973–1975. The decline in phosphorus discharge (by 44% since 1971) is attributed to the ban of phosphorus-containing detergents.

In contrast to phosphorus, the nitrogen discharge rates were markedly higher during periods of high flow (winter 1970–1971 and summer 1972)

TABLE 13

Estimated Mean Discharges of Total Nitrogen and Total Phosphorus of Irondequoit Creek into Irondequoit Bay[a]

Sampling period	Flow[b] (m^3/sec)	Total N concentration (mg N/liter)	Discharge (kg/day)	Total P concentration (mg P/liter)	Discharge (kg/day)
Jul.-Oct. 1970	2.03	1.59	279	1.30	228
Nov. 1970-May 1971	4.60	2.87	1140	0.67	266
Jun.-Aug. 1971	2.77	1.98	474	1.19	285
Jun.-Aug. 1972	4.71	1.98	806	0.53	216
Jun.-Sep. 1973	2.36	0.97	198	0.69	141
Jun.-Sep. 1974	2.36	1.18	241	0.50	102
Jun.-Sep. 1975	2.36	3.39	692	0.80	163

[a] Concentrations from Tables 11 and 12; mean of concentrations at Empire and Browncroft Boulevards used when available.

[b] For the periods in 1970–1972, the mean flows were calculated from the mean monthly flows given in Table 4 and from unpublished data of R. C. Bubeck for 1970. For the periods in 1973, 1974, and 1975, the mean flow was assumed to be the same as in June–September 1971.

than during summers of low flow (1970, 1973–1975). Because of this behavior, the mean annual discharge rate of nitrogen should be estimated after summing up the total discharge throughout the year. Unfortunately, mean discharge in the November–May period is available only for 1970–1971, and in view of the fluctuations in summer mean discharges, there is little assurance that the single winter mean value (1140 kg N/day) is representative. A crude estimate is that the mean nitrogen discharge rate has been 450 $\pm$ 150 kg N/day during the June–September period and 1140 kg N/day in the November–May period. From these figures, a mean annual discharge rate of 910 kg N/day can be calculated. In the 6-year record, mean summer discharge of nitrogen shows no trend; clearly, year-to-year variations of twofold and, rarely, fourfold may occur.

Thorne (1973) also analyzed nitrogen and phosphorus concentrations in Irondequoit, Thomas, and Allen's Creeks in upstream areas where there are no municipal inputs. Thorne's data from stations in the sector of Irondequoit Creek between the village of Fishers and the Barge Canal are summarized in Tables 14 and 15. Also listed are data of the Monroe County Department of Health for the Thornell Road station in the same sector. Thorne's results indicate that the concentration of total phosphorus was about 0.05 mg/liter at the upstream stations, about 10% of the concentration at Browncroft Boulevard. The data of the Department of Health show much higher concentrations of total phosphorus at the upstream station,

TABLE 14

Phosphorus Concentration (mg P/liter) in the Powder Mill Sector of Irondequoit Creek[a,b]

		Soluble PO_4–P				Total P				
Sampling period	Station	Number	Minimum	Mean	Maximum	Number	Minimum	Mean	Maximum	Source
Jun.–Aug. 1971	Thornell Road	4	0.04	0.10	0.15					MCDH
Jun.–Aug. 1972	East Street	6	0.00	0.003	0.10	4	0.00	0.06	0.15	Thorne
	Powder Mill Park	6	0.00	0.01	0.02	4	0.02	0.05	0.13	Thorne
	Fishers Road	10	0.00	0.04	0.15	8	0.00	0.04	0.16	Thorne
	Fishers Village	9	0.00	0.012	0.04	7	0.02	0.04	0.15	Thorne
Jun.–Sep. 1972	Thornell Road	4	0.05	0.05	0.05	2	0.10	0.35	0.61	MCDH
Jun.–Sep. 1973	Thornell Road	13	<0.05	0.07	0.33	13	<0.05	0.15	0.43	MCDH
Jun.–Sep. 1974	Thornell Road	13	<0.05	0.05	0.14	13	<0.05	0.32	0.96	MCDH
Jun.–Aug. 1975	Thornell Road	6	0.05	0.05	0.05	6	0.15	0.40	0.75	MCDH

[a] Sources: Thorne (1973) and Monroe County Department of Health (unpublished data).

[b] For each data set, the number of data and the minimum, arithmetic mean, and maximum concentrations are listed.

TABLE 15

Nitrogen Concentration (mg N/liter) in the Powder Mill Sector of Irondequoit Creek[a,b]

Sampling period	Station	NH_4–N	NO_2–N	NO_3–N	TIN[c]	Organic N	Total N	Source
Jun.–Aug. 1972	East Street	0.095(6)	0.017(5)	0.45(6)	0.56(6)	—	0.62[d]	Thorne
Jun.–Aug. 1972	Powder Mill	0.047(6)	0.007(6)	0.79(6)	0.82(6)	—	1.03[d]	Thorne
Jun.–Aug. 1972	Fishers Road	0.032(10)	0.006(10)	1.18(10)	1.22(10)	—	1.61[d]	Thorne
Jun.–Aug. 1972	Fishers Village	0.024(8)	0.006(9)	1.15(8)	1.16(9)	0.40(8)	1.56	Thorne
Jun.–Sep. 1972	Thornell Road	—	1.24(4)		—	—	1.62[d]	MCDH
Jun.–Sep. 1973	Thornell Road	—	0.62(13)		—	—	0.81[d]	MCDH
Jun.–Sep. 1974	Thornell Road	—	0.53(13)		—	—	0.69[d]	MCDH
Jun.–Aug. 1975	Thornell Road	—	1.23(6)		—	—	1.60[d]	MCDH

[a] Sources: Thorne (1973) and Monroe County Department of Health (unpublished data).
[b] For each data set, the arithmetic mean concentration and, in parentheses, the number of data are listed.
[c] TIN, total inorganic nitrogen.
[d] Calculated value based on component proportions indicated by data of Thorne.

0.15–0.40 mg P/liter or 30–50% of the value at Browncroft. Nitrogen concentrations in the upstream sector are only a little lower than at Browncroft.

From his estimates of flow and of nutrient concentrations, Thorne (1973) calculated nitrogen and phosphorus discharges at a number of upstream and downstream locations in the Irondequoit Creek drainage basin. He then estimated the fractions of the discharges of nitrogen and phosphorus at Browncroft attributable to municipal inputs. Thorne calculated the discharge of total phosphorus at Browncroft to be 198 kg/day. He further calculated phosphorus discharges of 13.1 kg/day for Allen's Creek at Glen Road, 5.7 kg/day for Irondequoit Creek at Ketchum Road (just upstream from the Barge Canal overflow), and about 5 kg/day for Thomas Creek above the treatment plants. Thus, a total of 23.8 kg/day were ascribed to "headwaters sources." Thorne attributed the difference (174 kg/day) between the discharges at Browncroft and from headwaters sources to municipal inputs and concluded that diversion of municipal effluents would have reduced the phosphorus discharge at Browncroft in 1972 by at most 90%. Thorne was fully aware that his estimate was a crude one and that it rested on uncertain assumptions about Barge Canal overflow, septic tank leaching, agricultural practices, and behavior of suspended solids.

Thorne may have overestimated the contribution of municipal effluents to the phosphorus discharge at Browncroft Boulevard. First, Thorne's own value of upstream discharge (23.8 kg/day) was 12%, not 10%, of the discharge at Browncroft. Secondly, of the total area (391 km^2) of the Irondequoit Creek drainage basin contributing to the discharge at Browncroft, about 65 km^2 lies downstream from the stations at which Thorne evaluated headwaters discharges. Assuming that runoff contributes to phosphorus discharge in proportion to area, the total contribution of runoff to the discharge at Browncroft would be 28.6 kg/day. Thirdly, compared with Thorne's data, those of the Department of Health (Table 14) indicate sixfold higher total phosphorus concentrations in Irondequoit Creek above the Barge Canal. Since the flow was almost certainly not six times higher in 1972, it appears possible that Thorne underestimated the phosphorus discharge by Irondequoit Creek. Probably a better conclusion from Thorne's data would be that municipal effluents contributed at most 85% of the phosphorus discharged at Browncroft Boulevard in 1972. Therefore, of the 216 kg/day total phosphorus which we estimate was the mean discharge at Browncroft during summer 1972, a maximum of 184 kg/day originated in municipal inputs, and a minimum of 28.6 kg/day was due to runoff or other sources. Table 13 indicated that, in 1973–1975 (after the banning of detergent phosphorus), the total discharge of phosphorus at

Browncroft Boulevard was about 140 kg/day. Of this, 111.4 kg/day at most can be attributed to sewage-treatment plants. We conclude that diversion could reduce the present-day discharge of phosphorus at Browncroft by no more than 80%.

Thorne carried out similar calculations for nitrogen discharge. His values were 705 kg/day at Browncroft and 118 kg/day for the sum of the discharges at the three upstream stations. A discharge of 270 kg/day was obtained for Allen's Creek just below the treatment plant and 391 kg/day for Irondequoit Creek just downstream from the mouth of Thomas Creek. The sum of these last two discharges (761 kg/day) exceeds that at Browncroft suggesting that denitrification, sedimentation, or uptake by vegetation cause an appreciable loss during summer. Since there are additional inputs from runoff and municipal effluents downstream from Allen's and Thomas Creeks, the loss from lower Irondequoit Creek must have considerably exceeded 60 kg/day. Because of this loss, the contribution of municipal effluents to the discharge of nitrogen at Browncroft cannot be reliably estimated. Like Thorne, we suppose that roughly half of the summertime discharge at Browncroft may be eliminated by diversion. From our estimate of roughly 450 kg/day for the mean discharge at Browncroft in an average summer, we suppose that the postdiversion value may be about 225 kg/day. Possibly, the mean annual discharge, 910 kg/day by our estimate, will also be halved as well.

Other Nutrient Inputs to the Bay

In addition to the discharge from Irondequoit Creek, the bay receives nutrients from four other sources: (1) runoff from the East Irondequoit and Webster–Northwest Penfield subbasins, (2) effluents from two small treatment plants which discharge directly into the bay, (3) discharge from the Rochester interceptor overflows, and (4) precipitation onto the bay proper. At least in recent years, biological nitrogen fixation has probably not been an important input since, during the summer, the concentration (100–200 mg N/m^3) of dissolved inorganic nitrogen remained high and since the predominant blue-green genus *Microcystis* is not a nitrogen fixer.

The combined discharges of nitrogen and phosphorus associated with runoff from the East Irondequoit and Webster–Northwest Penfield subbasins may be roughly estimated as the product of the runoff component of the nutrient discharges (455 kg N/day, 28.6 kg P/day) from Irondequoit Creek and the ratio (0.26 m^3/sec :2.67 m^3/sec) of the runoff components of water discharges from the two subbasins and from Irondequoit Creek. Thus estimated, total discharges from the two subbasins are 45 kg N/day and 2.9 kg P/day.

The two small sewage-treatment plants (Penfield Northwest and Irondequoit Southeast) discharge a total of about 0.08 m^3/sec directly into the bay. A third very small plant (Webster, Bay Vista) is believed to have ceased operation. The combined discharge of the two operating plants is about one-seventh of the total discharge of municipal effluents into Irondequoit Creek and its tributaries. Therefore, the nutrient discharges of the two bay plants may be estimated as one-seventh of the effluent components of discharges from Irondequoit Creek. Total discharge by the two plants is found to be 65 kg N/day and 31.6 kg P/day in 1970–1971, and 15.9 kg P/day in 1973–1975 (after the detergent phosphorus ban).

In addition to measuring water discharge from Rochester interceptor overflows, O'Brien and Gere Engineers, Inc. (1975) analyzed water quality parameters in a very large number of discharge samples. For the samples collected between February and June 1975 the mean concentrations appear to have been about 2 mg/liter total Kjeldahl nitrogen and 0.3 mg/liter total inorganic phosphorus. With these values, together with the estimated mean annual discharge (0.35 m^3/sec) from the three overflows, nutrient discharges of 60 kg N/day and 9 kg P/day are obtained. It will be recalled that the water discharge is uncertain. In 1970–1971, prior to the ban of detergent phosphorus, the discharge of phosphorus was probably twice as large, about 18 kg P/day.

Mean annual precipitation on the bay is about 78.5 cm (see Table 6). The United States Geological Survey (1974) has reported nitrogen concentrations in precipitation at Mays Point in Wayne County (about 20 miles east of the Irondequoit Bay drainage basin). For the years 1970–1973, the total concentrations of NH_4–N and NO_3–N ranged from 0.42 to 1.10 mg N/liter and the mean was 0.79 mg N/liter. Of the total, NH_4–N comprised 10–50%. Phosphorus concentrations were also determined in 1970 and 1971; the mean concentration was about 0.01 mg P/liter. Thorne (1973) analyzed the total inorganic nitrogen in precipitation falling in the town of Brighton on eight dates in the summer of 1972. Concentrations ranged from 0.006 to 3.65 and averaged 0.84 mg N/liter. From the mean annual precipitation (0.785 m), a mean inorganic nitrogen concentration of 0.80 mg/liter, and the area (6.7 km^2) of the bay, a mean annual input of 11.5 kg N/day can be calculated. Similarly, the precipitational input of phosphorus is about 0.15 kg P/day. (It is interesting to note that, for a mean concentration of 0.8 mg/liter in nitrogen in precipitation, the calculated precipitational input to the 391 km^2 area which contributes to the discharge of Irondequoit Creek at Browncroft Boulevard is 673 kg N/day, 1.5 times the discharge previously estimated to originate in runoff.)

Nutrient Loadings on Irondequoit Bay

The mean annual discharges of water, nitrogen, and phosphorus into the bay, as estimated in preceding sections, are compiled in Table 16. The contributions of runoff, Barge Canal discharge, rainfall, and treatment plant and interceptor discharges for the several areas of the drainage basin are listed. Also included are the estimated discharges in 1970–1971 (before the ban of detergent phosphorus), in 1973–1975 (after the ban), and in future, postdiversion years (after sewage inputs will have been eliminated). It will be recalled that there are substantial uncertainties in the estimates. The runoff contribution to the phosphorus discharge by Irondequoit Creek may be underestimated. Discharges of the Rochester interceptor may be too large. Because of both annual and seasonal fluctuations, and because of apparent losses in the downstream sector, neither the absolute value of the nitrogen discharge from Irondequoit Creek nor the relative contributions of runoff and municipal sources is well established.

The principle characteristics of inputs to the bay are as follows. (1) Irondequoit Creek accounts for about 81% of the water input at present; it appears that the Creek contribution will rise to 88% after diversion. (2) Of the total water input, treatment plant effluents currently account for 13% and the Rochester interceptor discharges for 8%; apparently about two volumes in ten originate from sanitary or storm sewage sources. (3) The ban of detergent phosphorus appears to have reduced the phosphorus input from 303 kg/day in 1970–1971 to about 168 kg/day currently. Possibly, total diversion of municipal inputs will reduce the input to about 32 kg/day. Of the potential future reduction of 135 kg/day, elimination of the discharges from the treatment plants on the bay and from Rochester overflows would only account for 12% and 7%, respectively. Most of the potential reduction depends on eliminating the municipal inputs into Irondequoit Creek and its tributaries. (4) Probably, complete diversion will only about halve the nitrogen input into the bay; as for phosphorus, most of the reduction depends on eliminating upstream inputs. (5) For 1970–1975, the ratios of nitrogen to phosphorus loadings are low (≤ 6.5) compared with the N:P ratio in algae (about 16). Therefore, current algal growth, if limited by either P or N, would be limited by N. After diversion, the ratio of the loadings will rise and may reach, but is unlikely to exceed, 16. Thus, diversion may result in nitrogen and phosphorus becoming equally limiting for algal growth, but phosphorus is unlikely to become more limiting than nitrogen. At present, the absolute loadings of both nitrogen and phosphorus are so high that algal growth may not be limited by either element. Whether, after diversion, algal growth will become limited by either element will be considered later.

TABLE 16

Estimated Mean Annual Discharges of Water, Nitrogen, and Phosphorus into Irondequoit Bay in 1970–1971, in 1973–1975 (after the Detergent Phosphorus Ban), and in Postdiversion Years (PD) after Treatment Plant Effluents and Interceptor Overflows Have Been Eliminated

			Water (m^3/sec)		Nutrients (kg/day)					
					1970–1971		1973–1975		PD	
Subbasin(s)	Area (km^2)		1970–1975	PD	N	P	N	P	N	P
Upstream subbasins contributing to Irondequoit Creek at Browncroft Boulevard	391	Runoff	2.67	2.67	455	28.6	455	28.6	455	28.6
		Barge Canal	0.50	0.50						
		STP effluent	0.53	0.00	455	221.4	455	111.4	0	0
East Irondequoit and Webster–Northwest Penfield subbasins (including wetlands)	37.4	Runoff	0.26	0.26	45	2.9	45	2.9	45	2.9
		STP effluent	0.08	0.00	65	31.6	65	15.9	0	0
Rochester subbasin	26.1	Interceptor overflow	0.35	0.00	60	18.0	60	9.0	0	0
Irondequoit Bay	6.7	Precipitation	0.17	0.17	11.5	0.15	11.5	0.15	11.5	0.15
Total	461.2		4.56	3.60	1091.5	302.6	1091.5	168.0	511.5	31.6
Mean areal water loading (m/day)		—	0.059	0.046	—	—	—	—	—	—
Mean areal nutrient loading (mg m^{-2} day^{-1})		—	—	—	163	45.2	163	25.1	76.3	4.7
Ratio of N:P loadings		—	—	—	3.6		6.5		16.2	

PHYTOPLANKTON GROWTH IN THE BAY

Measurements in 1970–1971

From July to November 1970 and from May to November 1971, Bannister performed phytoplankton and chlorophyll analyses of surface samples taken two or three times weekly from the Rochester Canoe Club dock (where depth is about 4.5 m). During 1971, phytoplankton analyses were also carried out on samples taken fortnightly at 10-ft (3.05-m) depth intervals at station 1. Methods employed in the phytoplankton and chlorophyll analysis are described in the Appendix.

Table 17 lists the genera and species found in 1970–1971, along with times of occurrence and date and number per milliliter at the time of greatest abundance. Table 18 compares species, counts, and chlorophyll concentration in the surface water at the Canoe Club dock and at station 1 at various dates in 1971. While there are some discrepancies, especially among blue-greens which tend to accumulate at the surface, in general the same species in about the same numbers were found at the two stations on a given date. Both stations therefore appear to be representative of the central portion of the bay. Table 19 shows how algal volume varied with depth at station 1 during 1971. On May 19, substantial volumes (made up largely of circular diatoms, *Diatoma,* and *Chlamydomonas*) were found at all depths. In June the pattern was different: substantial biomass (consisting of mixed greens earlier in the month and of bluegreens later) was restricted to the epilimnion (0–6 m). In July and August, high biomass was sometimes restricted to an even shallower layer as a result of secondary stratification. (The very high biomass at the surface on August 9, due entirely to *Microcystis,* was unusual; on the same day, the biomass at the Canoe Club dock, also mainly *Microcystis,* was only one-third as large.) In November, mixing again distributed the biomass throughout the water column.

Figure 20 shows how chlorophyll and biomass concentrations varied during 1970 and 1971, in surface water at the Canoe Club dock. Although there is some short-term, trendless fluctuation, the general pattern is one of high, constant values from early-May to mid-October. In both years, the chlorophyll concentration was maintained at about 100 μg/liter and the total cell volume at about 10^7 μm^3/ml. Assuming carbon is about 10% of the fresh weight, these values imply a carbon:chlorophyll ratio of about 0.01 gm C/mg chlorophyll. About the same value of the ratio has been found for phytoplankton grown in the laboratory under conditions of light limitation and nutrient saturation (Myers and Graham, 1971; Bannister, 1974b).

Figure 20 also shows how the numbers of some of the more abundant phytoplankton changed with time in the surface water at the Canoe Club

TABLE 17

Phytoplankton Species Collected in 1970–1971[a]

Genus	References[a]	Unit	Unit volume[b] and (error factor)	Peak Date	Peak Number[c]	Months present
Chlorophyta						
Chlamydomonas sp. } *Carteria* sp.	—	1 cell	0.7(6.1)	May 21	4000	4–12
				Jul. 1	100	—
Eudorina elegans Ehrenberg	1	1 colony	13.3(2.2)	Jun. 1	10	5–6
Pandorina morum (Muell.) Bory.	2	1 colony	14.9(4.2)	Jun. 1	60	5–6
Pleodorina illinoisensis Kofoid	1	1 colony	13.3(2.2)	—	—	5–6
Pteromonas aculeata Lemmermann } *Pteromonas angulosa* Lemmermann	—	1 cell	1.1	—	—	5–9
Elaktothrix sp.	—	1 cell	0.1	Rare in summer 1970 Not seen summer 1971		—
Gloeocystis sp.	—	1 cell	4.2	Jul. 1	70	5–7
Sphaerocystis schroeteri Chodat.	1	—	—	Common in Jul. 1970 Absent in 1971		—
Actinastrum hantzschii Lagerh.	1	4–8 cells	0.89(3.5)	Jun. 1	600	5–6
Ankistrodesmus falcatus (Corda) Ralfs.	1	1 cell	0.02(2.1)	Jun. 11	350	5–8
Characium sp.	—	—	—	Rare in summer 1970 Absent in summer 1971		—
Chlorella sp.	—	—	—	—	—	—
Closteriopsis sp.	—	1 cell	0.12(3.7)	Jun. 1	70	5–8
Coelastrum sp.	—	1 colony	2.1	Jul. 1	150	5–11
Crucigenia quadrata Morren	3	4 cells	0.05(2.8)	Jun.11	200	5–7
Dictyosphaerium ehrenbergianum Naegeli	1	16 cells	0.07	Jun. 1	80	5–6
Micratinium pusillum Fres.	2	4 cells	0.45	Jun. 1	1000	5–6
Nephrocytium sp.	—	4 cells	0.32(5.0)	Common in summer 1970 Absent in summer 1971		—

Oocystis sp.	—	4 cells	1.7(4.9)	Jul. 1	500	5–11
Pediastrum duplex Meyen						
Pediastrum simplex (Meyen) Lemmermann	1	1 colony	2.2(5.8)	Jun. 1	45	4–11
Pediastrum tetras (Ehrenb.) Ralfs.						
Polyedriopsis spinulosa Schmidle	2	1 cell	0.52	Infrequent in summer		—
Scenedesmus dentriculatus Lagerh.	4	4 cells	0.38(6.0)	Jul. 1	35	6–7
Scenedesmus falcatus Chodat	5	4 cells	0.38(6.0)	Jun. 1	100	5–7
Scenedesmus opoliensis Richter	6	4 cells	0.38(6.0)	Jul. 1	800	4–11
Selenastrum bibraianum Reinsch	1	8 cells	1.5(1.2)	—	—	—
Selenastrum minutum (Naegeli) Collins	3	1 cell	0.05(1.5)	Jun. 14	800	5–10
Sorastrum sp.	—	—	—	Rare	—	—
Tetraedron caudatum (Corda) Hansgirg						
Tetraedron constrictum G. M. Smith	1	1 cell	0.52	Jun. 11	270	5–7
Tetraedron haustatum (Reinsch) Hansgirg						Rare in fall
Tetraedron regulare Kützing						
Treubaria triappendiculata Bernard	2	1 cell	0.69(2.6)	—	—	Rare
Closterium sp.	—	1 cell	1.3(9.2)	May 14	60	5–11
				Jun. 1	140	
Cosmarium sp.	—	1 cell	4.5(5.0)	Infrequent	—	5–7
Staurastrum paradoxum Meyen	2	2 cells	0.52	Infrequent	—	—
Euglenophyta						
Euglena sp.	—	1 cell	10.3	May 5	200	4–6, 11
Chrysophyta						
Synura sp.	—	1 colony	14.0	Rare	—	—
Coccinodiscus sp.						
Cyclotella meneghiniana Kützing	9					
Stephanodiscus astrae (Ehr.) Grunow	11	1 cell	1.8(6.3)	May 6	20,000	4–11
Stephanodiscus niagarae (Ehr.) Grunow	12			Jul. 1	1,000	
Melosira binderana Kützing	11	8 cells	7.2(3.2)	May 6	70	4–6, 11
Asterionella formosa Hassal	7	—	Rare in spring	—		

(Continued)

TABLE 17 *(Continued)*

Genus	References[a]	Unit	Unit volume[b] and (error factor)	Peak: Date	Peak: Number[c]	Months present
Diatoma tenue var. *elongatum* Lyngbya	7	1 cell	1.5	May 9	2000	4–6
Surirella sp.	—	1 cell	6.5	May 6	80	4–6
Amphora ovalis (Breb.) Kützing	9					
Cymatopleura solea (Breb.) Wm. Sm.	9					
Cymbella sp.	—					
Epithemia sp.	—					
Fragilaria crotonensis Kitton var. *crotonensis*	10	1 cell	0.9(13.8)	May 1	100	4–12
Opephora sp.	—			Jun. 1	250	
Navicula sp.	—					
Nitzschia acicularis (Kütz.) Wm. Sm.	9					
Synedra sp.	—					
Tabellaria sp.	—					
Pyrrophyta						
Peridinium sp.	—	1 cell	22.4	May 11	300	4–11
				Jun. 1	90	—
Ceratium hirundinella (O. F. Muell.) Schrank	1	1 cell	14.0	Sep. 10	22	9
Cryptophyta						
Cryptomonas sp.		1 cell	1.8(2.5)	May 15	170	—
				Jun. 8	1000	5–11
Cyanophyta						
Chroococcus sp.	—	1 cell	0.2	Infrequent in Aug. 1970		
				Absent in 1971		
Gomphosphaeria lacustris Chodat	1	100 cells	0.84(2.5)	Abundant Aug.–Sep. 1970		
				Infrequent in Aug.–Sep. 1971		

Microcystis aeruginosa Kützing	1	500 cells	22.3	Jul. 7	400	6–10
Anabaena flos-aquae (Lyngb.) Breb.	1	20 cells	0.5	Jun. 17	1000	6–10
				Sep. 7	75	
Anabaena spiroides Klebahn	1	20 cells	7.6	Jun. 20	1000	6–7
Aphanizomenon flos-aquae (I.) Ralfs	2	1 filament	0.5(3.2)	Jul. 1	200	6–9
Oscillatoria sp.	—	1 filament	20.0	Infrequent in spring		

[a] Taxonomic references: (1) G. W. Prescott, "Algae of the Western Great Lakes Area." Cranbrook Inst. Sci., Bloomfield Hills, Michigan, 1951. (2) G. M. Smith, "The Freshwater Algae of the United States," 2nd ed. McGraw-Hill, New York, 1950. (3) H. S. Forest, "Handbook of Algae." Univ. of Tennessee Press, Knoxville, 1954. (4) N. T. Hortobagyi, Algen aus den Fisch-teichen von Buzsak. I. Scenedesmus-Arten. *Nova Hedwigia* **1,** 41–64 and 345–381 (1959). (5) N. T. Hortobagyi, Algen aux den Fisch-teichen von Buzsak. III. Scenedesmus-Arten. *Nova Hedwigia* **2,** 173–190 (1960). (6) A. F. Hallet, Some observations on algae (excluding diatoms) of two sewage oxidation pond schemes. *Nova Hedwigia* **4,** 483–493 (1962). (7) L. A. Whitford and G. J. Schumacher, A manual of the freshwater algae in North Carolina. *N. C., Agric. Exp. Stn., Tech. Bull.* **188** (1969). (8) H. B. Ward and G. C. Whipple, in "Freshwater Biology" (W. T. Edmondson, ed.), 2nd ed., Wiley, New York, 1959. (9) C. S. Boyar, "The Diatomaceae of Philadelphia and Vicinity." Lippincott, Philadelphia, Pennsylvania, 1916. (10) R. Patrick and C. Reimer, "The diatoms of the United States exclusive of Alaska and Hawaii. Vol. I, Monographs of the Academy of Natural Sciences of Philadelphia, No. 13." (11) L. Rabenhorst, "Kryptogamen-Flora," Vol. VII, p. 1. Akad. Verlagsges. Leipzig, 1930. (12) F. Wolle, "Diatomaceae of North America." Comenius Press, Bethlehem, Pennsylvania, 1890.

[b] Estimated unit volumes (as 10^3 μm^3) are arithmetic or geometric means (see Appendix).

[c] Peak numbers are given as units/ml.

TABLE 18

Comparison of Phytoplankton Species and Numbers (units/ml) at Station 1 and the Rochester Canoe Club Dock[a]

Genus	Date						
	May 18–19	Jun. 1–2	Jun. 15	Jul. 15–16	Sep. 8–9	Oct. 12–13	Nov. 17–18
Chlamydomonas sp.	755–602	187–267	64–46	13–14	4–26	9	Trace–4
Eudorina elegans	—	7–11	Trace–4	—	—	—	—
Pandorina morum	Trace	4–36	—	—	—	—	Trace
Pteromonas sp.	11	41–147	10–9	17–14	4	Trace	—
Gloecystis sp.	—	-4	17	4–19	4	—	—
Actinastrum sp.	Trace	411–656	10–4	—	Trace	—	—
Ankistrodesmus falcatus	22	30	32–31	5	—	—	—
Closteriopsis sp.	13	15–48	5–4	—	—	—	—
Coelastrum sp.	—	76–91	19–118	9–2	13–9	4	—
Crucigenia quadrata	—	51	96–41	4	—	—	—
Dictyosphaerium sp.	—	62–64	—	—	—	—	—
Micractinium sp.	Trace–45	716–1108	—	—	—	—	—
Nephrocytium sp.	—	—	Trace	—	—	—	—
Oocystis sp.	11–1	30–47	86–52	42–41	8–16	15–Trace	7–4
Pediastrum sp.	4	39–45	0.5–4	17–5	0.5–Trace	4–Trace	4–Trace
Scenedesmus sp. (excluding *S. falcatus*)	166–56	381–464	115–271	206–131	30–39	22–9	30–17
S. falcatus	—	137–68	19–26	Trace	—	—	Trace

Selenastrum minutum	—	17	800–815	—	4	—	—
S. bibraianum	—	—	—	—	13	—	—
Tetraedron sp.	9	112	38–75	9	9	—	—
Treubaria triappendiculata	—	26	5	—	4	—	—
Closterium sp.	47	289–90	5–4	Trace–14	4–13	17–22	Trace–13
Cosmarium sp.	—	—	—	5	—	—	—
Euglena sp.	3	37	—	—	—	—	4–Trace
Circular diatoms	400–411	411–215	Trace–13	9–9	43–198	4–9	1080–830
Melosira sp.	8	42–23	4	—	—	—	9–22
Diatoma tenue var. *elongatum*	900–1689	440–306	—	—	—	—	Trace–17
Other pinnates	55–78	7–215	—	9–Trace	4	Trace–4	Trace–132
Ceratium hirundinella	—	4	Trace	—	Trace	4	—
Peridinium sp.	11–14	48	13	—	—	—	9
Cryptomonas sp.	18	14–350	19–75	2–37	Trace–39	Trace	4–26
Gomphosphaeria lacustris	—	—	—	—	9–3	Trace	—
Microcystis aeruginosa	—	0.4	7–307	216–211	63–400	48–82	4–13
Anabaena sp.	—	6	460–4899	—	21	—	Trace
Aphanizomenon flos-aquae	—	—	23–Trace	295	86	4	Trace–13
Unknowns	133–67	594–340	289–323	52–98	26–47	9–13	13–40
Chlorophyll (μg/liter)	67	110	79–112	56–88	12–59	34–60	12–17

[a] First figure refers to Station 1, the second figure to Canoe Club Dock.

TABLE 19

Total Algal Volume as a Function of Depth at Station 1 in 1971[a]

Depth (m)	May 19	Jun. 1	Jun. 15	Jul. 2	Jul. 15	Aug. 9	Sep. 8	Oct. 13	Nov. 8	Nov. 18
0	3141	4272	1828	2922	5154	14,781	1595	1216	793	2156
3.0	2344	3614	2002	246	1573	11,992	7281	1024	1205	1704
6.1	5265	2868	2228	117	432	4,511	1420	1290	540	1414
9.1	1712	404	206	42	302	548	338	410	672	986
12.2	2698	426	231	40	208	552	970	101	1786	1023
15.2	1852	175	186	7	127	204	688	57	1835	609
18.3	2441	267	72	7	95	279	331	178	184	258
21.3	1207	235	88	6	45	178	256	90	76	105

[a] Units given as 10^3 μm^3/ml.

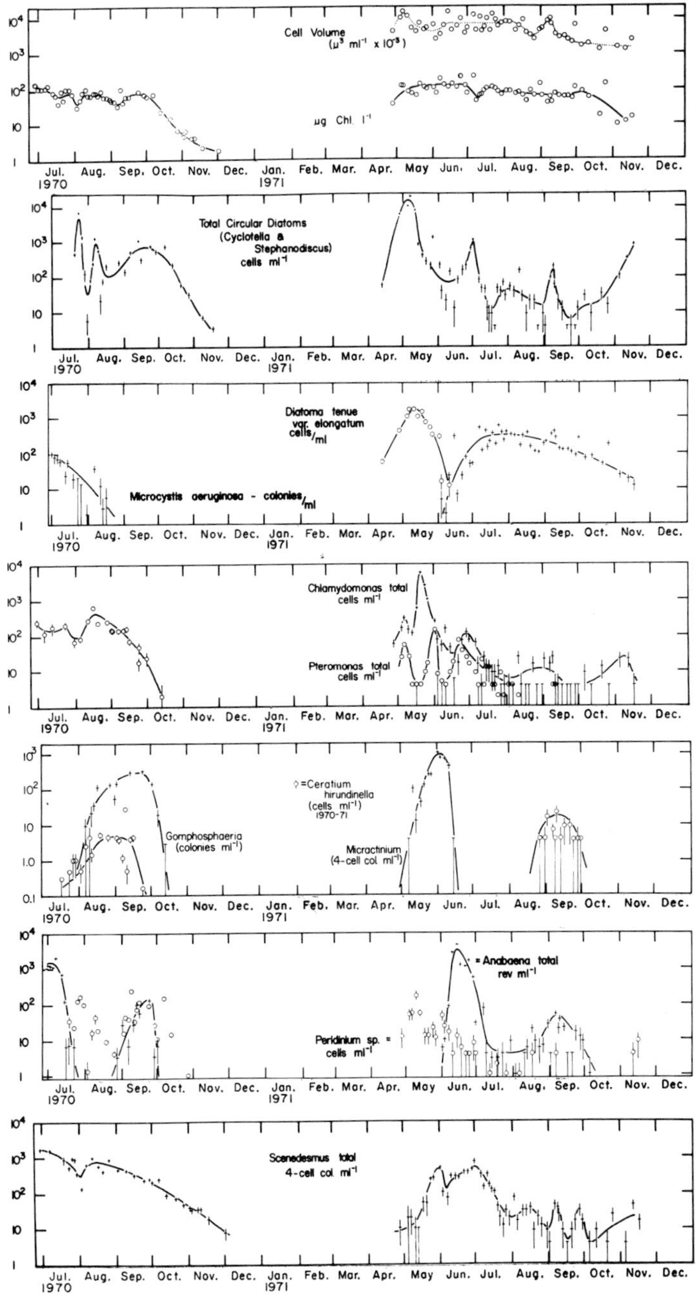

Fig. 20. Time courses of total chlorophyll (mg/m^3), phytoplankton cell volume (in thousands m^3/ml), and concentrations (counting units/ml) of some phytoplankton species, in 1970–1971.

dock. From these graphs and similar ones for other species, the successional pattern could be described in terms of nine phytoplankton groups, each group comprising species which reached maximum number at the same time, and were present either briefly for a short time or persistently over long periods. The groups and the maximum species concentrations (units/ml, units defined in Table 17) were as follows:

Early May, Brief

In 1971, *Diatoma* (1850), *Euglena* (99), *Surirella* (51), and *Melosira* (62) were detected under the ice in the first week of April, reached maxima between May 4–7, disappeared in mid-June, and reappeared in small numbers in November.

Early May, Persistent

Circular diatoms (*Stephanodiscus, Cyclotella,* and small numbers of *Coccinodiscus*) in numbers totalling 21,000, along with *Peridinium* (178), *Cryptomonas* (178), and *Closterium* (65), reached well-defined maxima between May 5 and 14. The *Cyclotella-Stephanodiscus* bloom was spectacularly large, but at its height most cells were small and poorly silicified. In the second half of May, grazing became intense (as judged by the abundance of zooplankton with guts filled with diatoms) and must have contributed to the rapid decline in number to 215 on June 2. These species occurred in fluctuating numbers throughout the summers of 1970 and 1971. Sometimes, well-defined secondary blooms took place; for example, circular diatoms (1120) on July 1, 1971, *Cryptomonas* (783) on June 8 and July 9, 1971, and *Peridinium* (100) in early August 1970.

Late May, Persistent

Chlamydomonas and *Carteria* (5908) reached a maximum on May 21, fell to a minimum (4) on June 29, and were present in fluctuating low numbers the remainder of the year.

Late May, Brief

Eudorina (17), *Pandorina* (39), *Pleodorina* (4), *Dictyosphaerium* (64), *Micractinium* (1108), and *Crucigenia* (111) appeared in early May and reached their maxima between May 21 and June 11. All then declined rapidly and were not observed from July 1970 to April 1971 or from July to December 1971.

Early June, Persistent

Actinastrum (656), *Scenedesmus falcatus* (68), *Ankistrodesmus falcatus* (345), *Selenastrum* (815), *Tetraedron* (293), *Coelastrum* (129), and *Pte-*

romonas (100) reached their maxima in the first half of June, and were seen occasionally throughout the summer of 1971 and regularly in higher numbers in the summer of 1970. *Coelastrum* was particularly numerous in August 1970.

Early July, Persistent

Oocystis (137), *Scenedesmus opolienis* (731), and *Gloeocystis* (65) occurred in small numbers in May, reached their maxima about July 1, and, except for *Gleocystis*, were found in small numbers throughout the year. In 1970, *Oocystis* and *S. opoliensis* were abundant from July to October.

Mid-June, Brief

Anabaena spiroides (784) and *A. flos-aquae* (4253) first appeared in early June, reached spectacular maxima around June 20, then rapidly declined. *Anabaena spiroides* was not seen after July 14, 1971, nor from July 1970 to May 1971. *Anabaena flos-aquae* was present in low numbers (50) throughout summer 1971 and disappeared in early October; in both September 1970 and 1971, it exhibited lesser secondary blooms.

July, Persistent

Microcystis (514 units of 500 cells each per milliliter, corresponding to 70 μl/liter) and *Aphanizomenon* (229) appeared in early June and reached their maxima in early July. In 1971, they remained at near-maximum levels through September and only disappeared in November. In contrast, in 1970, both species gradually declined during July and disappeared in mid-August.

Late Summer, Brief

Gomphosphaeria (340) in 1970 and *Ceratium* (28) occurred in well-defined maxima in late August and mid-September, respectively. In 1971 *Gomphosphaeria* was observed at about the same time but in much smaller numbers (18). Neither species was seen from October through June. *Chroococcus* in small numbers seems to accompany *Gomphosphaeria.*

Species other than those mentioned appeared in numbers too small to allow a definite conclusion about their temporal pattern. In Table 20, the biomasses of each group are listed for Canoe Club dock samples at selected dates.

Some information is available about the net rates of growth and decline. For 13 species, time courses of number (such as those in Fig. 20) were sufficiently precise to permit estimation of rates. In Table 21 are listed times, t_{10} in days, during which an approximately exponential, ten-fold rise or fall in number occurred. Also listed are corresponding values of the net growth

TABLE 20

Group Volumes in Surface Water at Canoe Club Dock in 1971[a,b]

Groups and principle species	May 7	May 21	Jun. 2	Jun. 15	Jun. 25	Jul. 1	Jul. 7	Sep. 10	Oct. 21
1. *Diatoma*	4,225	1,337	1,199	29	42	0	0	0	11
2. *Cyclotella*	10,717	824	2,194	451	585	2,032	209	203	153
3. *Chlamydomonas*	215	4,065	184	31	50	53	44	3	9
4. *Micractinium*	0	955	1,386	298	45	0	0	0	0
5. *Actinastrum*	62	21	1,025	349	80	376	50	9	0
6. *Scenedesmus*	46	68	417	275	650	832	524	17	31
7. *Anabaena*	0	0	3	7,058	6,422	29	58	19	0
8. *Microcystis*	0	0	9	6,849	1,809	1,152	11,565	6,306	1,137
9. *Ceratium*	0	0	0	0	0	0	0	319	0
Other	292	280	450	354	142	910	394	23	5
Total	15,567	7,550	6,867	15,694	9,825	5,384	12,844	6,899	1,346

[a] Units given as 10^3 μm/ml.

[b] Dates selected to show maximum volume of each group.

TABLE 21

Rates of Increase and Decrease during Blooms of Several Species

Species	During increase		During decline	
	t_{10}[a]	μ_n[b]	t_{10}	μ_n
Chlamydomonas	12	0.19	10	−0.23
Actinastrum	9	0.26	3	−0.77
Selenastrum	5	0.46	4–5	−0.51
Dictyosphaerium	8	0.29	3	−0.77
Crucigenia	24	0.10	9	−0.26
Micractinium	9	0.26	2	−1.15
Scenedesmus	18	0.13	—	—
Diatoma	15	0.15	4–5	−0.77
Circular diatoms	12	0.19	12	−0.19
Cryptomonas	8	0.29	2–3	−0.92
Microcystis	9	0.26	—	—
Aphanizomenon	12	0.19	—	—
Anabaena	8	0.29	2–3	−0.92

[a] t_{10}, time in days for a tenfold change in number.
[b] μ_n (per day) = net growth rate constant.

rate constant μ_n (as defined by the exponential growth equation $N_t = N_0 e^{\mu_n t}$, where N_0 and N_t are the concentrations of a species at times zero and t). The faster rise times were 5–9 days, the slower ones two or three times longer. In contrast, half the fall times, mainly of the ephemeral species, were much shorter (2–5 days) than the shortest rise times. The net growth rate constant equals the cellular growth rate constant less the sum of the rate constants of loss processes such as sedimentation and comsumption by grazers. During a period of exponential increase, the value of the net growth rate constant is a lower limit for the cellular growth rate constant. In Table 21, the values of μ_n during increase are relatively low compared with potential phytoplankton growth rates which can reach two doublings per day (μ = 1.39/day) or more. Whether the cellular growth rates were actually low or whether loss rates were large is unknown. During a period of decline, the magnitude of the net growth rate constant is a lower limit for the sum of the loss rate constants. The large magnitudes of μ_n during decline, for many of the species listed in Table 21, show that the loss processes can be very rapid and suggest that extremely active grazing or some form of pestilence must occur.

According to Bannister's observations in 1970–1971, the main features of phytoplankton growth in Irondequoit Bay are the following. From May through September a dense phytoplankton crop inhabits the upper 4 m (the well-mixed portion) of the epilimnion. Despite changes in species, total algal biomass is remarkably constant; the concentration of chlorophyll ($\sim 100\ \mu g$ chlorophyll *a*/liter) and cell volume ($\sim 10^7\ \mu m^3$/ml) are maintained throughout the 5-month period. Because of high algal concentration and substantial nonphytoplankton absorption, light transmission (about 10%/m) is low, and the mixed layer is essentially totally absorbing. In the meta- and hypolimnion, photosynthetic growth is not possible, and the low concentrations of phytoplankton presumably result from sedimentation from the mixed layer. In July and August, the concentration of inorganic nitrogen (0.1–0.2 mg N/liter) in the mixed layer remains relatively high, and algal growth is probably nearly nitrogen saturated. Since the concentration of PO_4–P (≥ 0.25 mg P/liter) is even higher, growth is certainly phosphorus saturated. In May and June, when higher concentrations occur, both phosphorus and nitrogen must be saturating for growth. In May and June, intense blooms of diatoms, greens, and blue-greens occur in rapid succession, and both algal growth rate and the rate of loss to sedimentation and especially to consumption must be high. Depletion of SiO_2, steadily rising temperature and pH, and, perhaps, rapid expansion of consumer populations may be the determining factors of succession. In July–August, temperature and pH, the concentrations of N and P, and algal species and numbers remain relatively constant; thus, a nutrient-saturated, or a slightly nitrogen-limited, steady state seems to occur. Rapid changes in algal species numbers not occurring, it is unknown whether algal turnover is less rapid than in May–June. In September small blooms of *Anabaena flos-aquae* and *Ceratium* are correlated with some increase of nutrient levels and deepening of the thermocline.

Earlier Studies

Whipple (1913) reported phytoplankton analyses of samples collected at several stations on August 8, 1912. Table 22 lists the counts at station 1 at the surface, at the thermocline (reported to be centered at 9.2 m), and at the bottom (24.6 m). Except for the abundance of *Tabellaria, Synedra,* and *Melosira,* and the early appearance of *Ceratium,* the results for surface water are similar to Bannister's in August 1971. (Presumably Whipple's *Coelosphaerium* is equivalent to *Gomphosphaeria.*) At other stations, small numbers of *Tetraspora, Anthophysa, Diatoma, Eudorina, Pandorina, Pediastrum,* and *Euglena* were reported by Whipple.

Tressler and Austin (1940) and Tressler *et al.* (1953) summarized their phytoplankton analyses of surface samples collected monthly between August 15, 1939 and June 13, 1940. Their list of genera is identical to Bannister's with the following exceptions. They reported, but Bannister did not, *Chlorococcum, Lagerheimia, Mougeotia, Aphanothece, Merismopedia,* and

TABLE 22

Phytoplankton Analysis at Station 1 for August 8, 1912[a,b]

Species	At surface	At thermocline	At bottom
Cyclotella	0	8–32	32
Melosira	96	16–24	0
Navicula	0	0	4
Stephanodiscus	0	16	0
Synedra	104	12–88	0
Tabellaria	304	8–112	40
Coelastrum	8	0	0
Closterium	0	4	0
Protococcus	36	48–60	48
Raphidium	8	4–12	0
Scenedesmus	8	4	0
Staurastrum	24	4–16	0
Anabaena	240	8–124	0
Aphanocapsa	0	40–100	0
Clathrocystis	40	48–160	0
Coelosphaerium	456	24–212	88
Microcystis	476	52–156	24
Oscillaria	24	32–160	0
Ceratium	160	40–60	0
Trachelomonas	0	0	120

[a] Source: Whipple (1913).
[b] Values are presumed to be number/ml.

Dinobryon. Of these *Chlorococcum* and *Aphanothece* probably correspond to Bannister's *Chlorella* and *Microcystis*. Bannister did not identify the rare filamentous green algae in his samples. On the other hand, Bannister found, but Tressler *et al.* did not, *Eudorina, Pandorina, Pleodorina, Pteromonas, Chlorella, Micractinium, Polyedriopsis, Sorastrum, Tetraedron, Treubaria, Closterium, Synura, Coccinodiscus, Diatoma, Surirella, Peridinium, Cryptomonas,* and *Microcystis*. Significant changes between 1939–1940 and 1970–1971 may be disappearance of *Dinobryon* and *Merismopedia* and the appearance in large numbers of *Diatoma, Micractinium, Peridinium,* and *Cryptomonas*.

Concerning numbers and seasonal succession, the observations of Tressler *et al.* can be summarized as follows. On May 16, 1940, diatoms numbered 21,000 cells/ml (of which *Cyclotella* numbered 8,700) and the green alga *Chlamydomonas* 717/ml. On June 13, the initial bloom had largely subsided, and persistent greens (*Scenedesmus, Sphaerocystis, Pediastrum,* etc.) totaled 1730 cells/ml, about the same as in mid-June 1971. However, no ephemeral greens and almost no blue-greens were reported. On August 15, 1939, mixed greens totaled 1345/ml and included, besides those just mentioned, *Crucigenia* and *Oocystis* in significant quantities. In addition, substantial numbers of *Aphanizomenon* (147 units/ml) and *Merismopedia* (130 units/ml) were counted. A similar distribution was found in September 1939 with the addition of *Ceratium* (242 cells/ml). Then as now, *Cyclotella, Melosira, Navicula, Stephanodiscus, Scenedesmus, Ankistrodesmus,* and *Pediastrum* were found throughout the year.

The data for 1939–1940 indicate a May diatom bloom as intense as present ones (through Diatoma was not observed); presumably exhaustion of silicon caused the diatom decline then as in 1971. While Tressler *et al.* (1953) recorded smaller numbers of *Chlamydomonas,* it can be doubted that they observed the peak of the bloom. Their data for mid-June are decidedly different from ours: they found neither ephemeral greens nor significant numbers of blue-greens, but rather only persistent greens and diatoms such as were characteristic of the steady state in July–August 1970. One can speculate that nitrogen and phosphorus levels in April 1940 were sufficient for intense diatom and *Chlamydomonas* blooms, but became limiting before an intense blue-green bloom could develop in June. As a result, a nutrient-limited steady state characterized by restricted numbers of mixed greens and blue-greens may have commenced in late May 1940, whereas in 1970–1971, a nearly nutrient-saturated steady state began only about July 1. Consistent with this interpretation is the decline of soluble phosphorus, from 180 mg P/m^3 in April to about 10 mg P/m^3 in mid-May 1940, and, also, the relatively low pH, about 8.4, recorded in June, August, and September. Since chlorophyll and nitrogen were not assayed by Tressler *et al.,* it is impossible to know whether the summer algal crop was smaller than at present and whether N or P was the more limiting.

Data for 1972

C. Pedersen, a student of Harbison (see Harbison, 1972) performed phytoplankton analyses of surface samples collected on 21 dates between May 28 and August 2, 1972 from station 1 and the Rochester Canoe Club dock. The principal species, their numbers, and their temporal patterns appear similar to those for 1971. Diatoms were not identified.

ZOOPLANKTON, AQUATIC VEGETATION, BENTHIC FAUNA, AND FISH

Zooplankton and Protozoa

These have been little studied. Whipple (1913) reported the presence of *Vorticella, Codonella, Enchelys, Anuraea, Polyarthra, Actinophrys, Cy-*

clops, and crustacea. Tressler and Austin (1940) and Tressler *et al.* (1953) recorded zooplankton analyses carried out in 1939–1940. Tressler *et al.* (1953) listed the following limnetic zooplankton: Copepoda: *Cyclops bicuspidatus* (Claus), *Cyclops leukarti* (Claus), and *Diaptomus siciloides* (Lilljeborg); Cladocera: *Alona* sp., *Bosmina longirostris* (Muller), *Daphnia longispina* (Muller) var. *hyalina* (Leydig) form *galeata, Daphnia pulex* (de Geer), *Daphnia retrocurva* Forbes, *Leptodora kindtii* (Focke), and *Sida crystallina* (Muller); Rotifera: *Asplanchna priodonta* Gosse, *Asplancha* sp., *Keratella cochlearis* (Gosse), *Keratella quadrata* (Muller) var. *divergens, Monostyla bulla* Gosse, *Notholca longispina* Kellicott, *Polyarthra trigla* Ehrenberg, *Rattulus* sp., and *Synchaeta stylata* Wierzejski.

According to Tressler *et al., Cyclops* was most abundant of the Copepoda and occurred throughout the year. Its numbers ranged from about 20,000/m^3 in December and April to 594,000/m^3 in May and 90,000/m^3 in September. *Diaptomus siciloides* occurred all year, reached a maximum of 40,000/m^3 in September and December, but was rare in April. A nauplii maximum of 600,000/m^3 occurred in May, a minimum of 7000/m^3 in January–March. Of the five genera of Cladocerans, only *Daphnia* appeared in any abundance; it was present all year and reached 87,000/m^3 in September. *Bosmina* and *Sida* reached maxima of 2000 and 5000 in August and October. *Leptodora,* which can elude traps, was recorded only once in September. Of the rotifers, four were abundant most of the year. *Keratella cochlearis* numbered 158,000/m^3 in October, *Asplancha* reached 131,000 in May. *Polyarthra* reached a maximum of 32,000 in September, and *Notholca* 52,000 in October. Other rotifers occurred mainly in the fall in small numbers. Protozoan genera listed by Tessler *et al.* (1953) included only *Difflugia, Mallomonas, Phacus,* and *Vorticella.*

Emergent and Submerged Aquatic Plants

Clausen (1940) listed species found in Irondequoit Bay in 1939. Abundant species included *Typha angustifolia* and *Spirodela polyrhiza.* Common species were *Typha latifolia, Heteranthera dubia, Ceratophyllum demersum,* and *Nymphaea tuberosa.* Species classified as frequent were *Potamogeton pectinatus, Sagittaria latifolia, Scirpus validus, Lemna minor, Wolffia punctata, Pontederia cordata, Polygonum coccineum,* and *Azolla caroliniana.* Rare species were *Sparganium eurycarpum, Potamogeton americanus, Potamogeton crispus, Najas flexilis, Sagittaria heterophylla, Vallisneria americana, Scirpus acutus, Scirpus americanus,* and *Utricularia vulgaris.* In an annotated list of aquatic plants, Clausen noted several species collected earlier but not recorded in 1939: *Sparganium minimum* collected in "clear, deep water" in 1866, *Najas marina, Zannichellia palustris,* and *Sagittaria cuneata* collected in 1915, and *Fimbristylis mucronulata* taken in 1922. Whereas 12 species of *Potamogeton* were found in nearby Sodus Bay, only three species (of which two were rare) occurred in Irondequoit Bay. From this, Clausen considered the ensemble of aquatic species in the bay in 1939 to be impoverished, and he believed that pollution and turbidity were the causes. Clausen stated that "no plants were found on the mucky bottom below a depth of two or three feet."

In 1969–1972, Forest and associates collected and identified aquatics found in Irondequoit and Sodus Bays and several of the western Finger Lakes. Biomass per unit area in the weed beds and the greatest depths of aquatic growth were also measured. Forest *et al.* (1973) noted the disappearance from the bay between 1939 and 1969 of *Vallisneria americana, Potamogeton americanus, Sagittaria heterophylla, Sagittaria latifolia, Najas flexilis,* and *Scirpus acutus*. Forest *et al.* also listed several *Potamogeton* species which were collected between 1866 and 1924, but not found in 1939 or 1969–1972; these were *P. epihydrus, P. friesii, P. natans, P. perfoliatus, P. pulcher,* and *P. richardsonii.* Two species, *Lemna trisulca* and *Phragmites communis* were recorded for the first time in 1969. A map showed the location of present weed beds; those of *Myriophyllum exalbescens* were the most extensive.

According to Forest *et al.* (1975) and Forest (1976), the maximum depths of rooted aquatics are about 1.6 m in Irondequoit Bay, about 2.6 m in Honeoye Lake, 3.9 m in Sodus Bay, and up to 7.2 in Conesus Lake. In this series the increase in depth correlates with increasing clarity of the water. In Conesus Lake, where aquatics are rooted in deep water and grow to considerable height, biomass was generally greater than 0.5 kg dry weight/m^2 and occasionally attained 1.8 kg dry weight/m^2. In contrast, in Irondequoit Bay, biomass in the shallow *Myriophyllum* and *Potamogeton* beds did not exceed about 30 gm dry weight/m^2. Plant forms also differed in the two lakes. In Conesus Lake, extensive branching occurred close to the bottom; in Irondequoit Bay a toadstool-like form was found in which extensive branching only occurred very close to the surface. Forest *et al.* (1975) and Forest (1976) concluded that light attenuation by the very dense phytoplankton crop is responsible for the shallowness of the weed beds in Irondequoit Bay. Apparently, pollution was already heavy enough in 1939 to restrict aquatic growth to the upper 1.6 m.

Following the methods developed by Gerloff and Krombholz (1966) and Gerloff (1969), Harbison (1972) analyzed the nitrogen content of *Myriophyllum* collected in beds at various locations in the bay. The nitrogen content (as % dry weight) was 4% in beds at the extreme south end, about 1.7% at Snider Island, 1.2–1.4% at Stony Point, and about 0.7% in the Northeast Arm. Applying Gerloff's conclusions, *Myriophyllum* appeared to be nitrogen saturated at the south end and strongly limited in the Northeast Arm.

Both the beds of cattails and of submerged aquatics were once much more extensive than at present. Local residents report that large areas of the cattail beds in the shallows at the north end of the bay disappeared during the 1950's, apparently dislodged by high water levels. Local boatsmen recall towing floating islands of cattails ashore during this time. Odell (1940) recorded reports that submerged weed beds were much more extensive prior to about 1933 than in 1939. he considered several possible causes—cold weather in 1933, pollution, and high turbidity resulting from waves, speedboats, and carp (but not absorption of light by phytoplankton)—but drew no conclusions. A local resident who grew up near the bay recalls watching sunfish building nests on the bottom in the 1920's. These reports agree in indicating that a dramatic diminution of the submerged weed beds and increases in turbidity and phytoplankton occurred in the decade or so between about 1925 and 1935.

Benthic Fauna

Little is known about this fauna. In a study of benthic fauna as food for fish in bays of Lake Ontario, Burdick (1940) reported an average of 7.6 gm of food/m^2 for Irondequoit Bay. He commented on the lack of weed beds, the occurrence of anoxia, the absence of fauna at depths greater than 7 m, and the low food supply compared with Sodus Bay. He remarked that shell debris showed that an extensive mollusc fauna had existed, but that no live specimen was found. An annotated list of snail and clam species of the Lake Ontario watershed was given, but there are no specific references to Irondequoit Bay.

The FWPCA (1965, 1968) analyzed benthic fauna at a number of stations, on June 10 and August 25, 1965 and on June 24, 1968. In a total of 20 samples (each enclosing an area of about 0.1 m^2), nine had no organisms, ten had 1–200 organisms, and one sample allegedly contained 61,334 snails (*Helisoma, Amnicola, Gyraulus, Physa, Goniobasis, Stagnicola, Promenters,* and *Valvata*). In these same samples, chironomids (*Polypedium* sp., *Procladius riparius, Procladius culciformis,* and unknown species) numbered 0–8, *Oligochaeta* 0–150, *Isopoda* 0–2, *Trichoptera* 0–2, and *Pelecypoda* (*Sphaerium*) 0–2. In Irondequoit Creek, some changes in benthic fauna associated with pollution were reported by Thorne (1973).

Fish

Fish of the bay were studied extensively in 1939 as part of the Biological Survey of the Lake Ontario Watershed (New York State Department of Conservation, 1940). Odell (1940) listed 35 species present in numbers ranging from abundant to rare. Greeley (1940) presented an annotated list of species of the Lake Ontario watershed (collection of a few species in the bay in the 1920's are cited) and he reported some age–length data for game fish species. Greeley also referred to the study of Stone (1937) on the growth, habits, and fecundity of Irondequoit Bay herring, a local race of *Coregonus artedi* now extinct. Tressler and Austin (1940) compared the stomach contents of fish in Irondequoit and Sodus Bays. Since 1939, collections have been made periodically by the New York State Department of Environmental Conservation. In addition, collections were made in 1967 by R. M. Roecker (State University College at Geneseo), in 1971 by Gittelman and Buchanan (1971), and by T. A. Haines (State University College at Brockport) in 1976. Some data were incorporated in the report of Marine Resources, Inc. (1970). Species lists have been reviewed by Forest *et al.* (1973), Gehris *et al.* (1974), and most expertly by Haines (see Ellis *et al.*, 1976). Table 23, based on the list of Haines, records the species found in the various collections.

Some passages in the 1940 report give glimpses of conditions at that time. Westman and Fahy (1940) wrote "In Long Pond and Irondequoit Bay, carp are present in such numbers as to discourage the anglers and this fact, coupled with the recent disappearance of large weed beds in these waters, has brought up the question of controlling the species," and, in another passage, "Long Pond, Irondequoit Bay, and Sodus Bay are among several bodies of water surveyed which are seined commercially for carp." Odell (1940) wrote that in view of the silt, turbidity, loss of vegetation, and anoxia below 15 to 20 feet, ". . . it is remarkable that the bay provides good fishing which prevails for bullheads, largemouth bass, small common sunfish, calico bass, and fair fishing for yellow perch. The catch of northern pike is reported to have fallen off markedly during the past few years. Irondequoit Bay has a large number of ciscoes or lake herring . . . It is reported that large runs of yellow perch and rockbass occur during the spring and fall."

As shown in Table 23, a number of species were collected in 1939 but not in 1967–1976, and a few species were collected recently but not in 1939. At first glance, these instances might seem to indicate changes in species since 1939. However, sampling since 1939 has mainly consisted of "spotchecks" to monitor selected species, not intensive collections to identify all species present (Holmes, 1975). Thus, some apparent changes, for example the decline in minnow species, may not be real. Also, the lack of a recent record of American eel (a large specimen of which was seen taken by an angler in 1972) and the existence of recent records of lamprey, channel cat, and burbot (long known to be present in Lake Ontario) are not indications of trends in the bay. Of changes which are probably real, most (the declines of cisco, yellow perch, white bass, walleye, and drum and the increase in white perch) are undoubtedly related to fish stocks in Lake Ontario which have shown the same trends. In our view, the collection records do not

TABLE 23

Fish Species Reported in Irondequoit Bay

Family, species	Year											
	1939[a,b]		1946[c]	1960[c]	1967[d]	1969[c]	1971[e]	1973[c]	1974[c]	1975[c]	1976[c,f]	
Petromyzonitidae												
Petromyzon marinus (sea lamprey)	+[g]	–[h]	–	–	–	R[i]	–	–	–	–	–	–
Lepisosteidae												
Lepisosteus osseus (longnose gar)	–	F[j]	–	–	–	R	–	–	+	+	+	–
Amiidae												
Amia calva (bowfin)	–	F	–	+	–	R	–	+	–	+	–	–
Anguillidae												
Anguilla rostrata (American eel)	29	F	–	–	–	–	–	–	+	–	–	–
Clupeidae												
Alosa pseudoharengus (alewife)	–	F	–	+	+	–	+	+	+	+	–	+
Dorosoma cepedianum (gizzard shad)	–	F	+	+	–	–	+	+	+	+	+	+
Salmonidae												
Coregonus artedii (freshwater herring, cisco)	27,30	F	–	–	–	–	–	–	–	–	–	–
Salmo trutta (brown trout)	–	–	–	–	–	–	–	–	–	+	–	–
Esocidae												
Esox lucius (northern pike)	–	F	–	+	–	C[k]	–	+	+	+	+	–
Cyprinidae												
Campostoma anomalum (stoneroller minnow)	–	R	–	–	–	–	–	–	–	–	–	–
Carassius auratus (goldfish)	–	–	–	–	–	–	–	+	–	+	+	+
Cyprinus carpio (carp)	–	A[l]	–	+	+	C	+	+	+	+	+	+
Cyprinus carpio × *Carassius auratus*	–	–	–	–	–	–	–	–	–	–	–	+
Hypognathus nuchalis (silvery minnow)	–	F	–	–	+	–	–	–	–	–	–	–
Notemigonus crysoleucas (golden shiner)	–	F	–	–	–	–	+	+	–	–	–	+
Notropis anogenus (pugnose shiner)	–	R	–	–	–	–	–	–	–	–	–	–
Notropis atherinoides (emerald shiner)	–	F	–	–	+	–	+	–	–	–	–	+
Notropis cornutus (common shiner)	–	F	–	–	–	–	–	–	–	–	–	–

(Continued)

TABLE 23 *(Continued)*

Family, species	Year											
	1939[a,b]		1946[c]	1960[c]	1967[d]	1969[c]	1971[e]	1973[c]	1974[c]	1975[c]	1976[c,f]	
Notropis hudsonius (spottail shiner)	–	F	–	–	–	–	+	–	–	–	–	+
Notropis spilopterus (spotfin shiner)	–	F	+	–	–	–	–	–	–	–	–	–
Notropis stramineus (sand shiner)	–	F	–	–	–	–	–	–	–	–	–	–
Notropis volucellus (mimic shiner)	–	R	+	–	–	–	–	–	–	–	–	–
Notropis whipplei (= *N. analostanus*) (steelcolor shiner)	–	+	–	–	–	–	–	–	–	–	–	–
Pimephales notatus (bluntnose minnow)	–	C	+	–	–	–	–	–	–	–	–	–
Catostomidae												
Carpiodes carpio (river carpsucker)	–	–	–	–	–	–	–	+	+	–	–	–
Catostomus commersoni (white sucker)	–	F	–	+	–	–	+	+	–	+	–	+
Moxostoma anisurum (silver redhorse)	–	–	–	–	–	–	–	+	–	–	+	–
Moxostoma macrolepidotum (= *M. aureolum*) (shorthead redhorse)	–	–	–	+	–	–	–	–	–	–	–	–
Moxostoma valenciennesi (= *M. rubreques*) (greater redhorse)	29	–	–	–	–	–	–	–	–	–	–	+
Ictaluridae												
Ictalurus melas (black bullhead)	–	–	–	–	–	–	+	–	–	–	–	+
Ictalurus nebulosus (brown bullhead)	–	C	–	+	–	C	–	+	+	+	+	+
Ictalurus punctatus (channel catfish)	–	–	–	–	–	–	+	–	–	–	–	+
Gadidae												
Lota lota (burbot)	–	–	–	–	–	R	–	–	–	–	–	–
Cyprinodontidae												
Fundulus diaphanus (eastern banded killifish)	–	R	–	–	–	–	+	–	–	–	–	–
Atherinidae												
Labidesthes sicculus (brook silversides)	–	C	+	–	–	–	–	–	–	–	–	–
Gasterosteidae												
Gasterosteus aculeatus (three-spine stickleback)	–	–	–	–	–	–	–	–	–	–	–	+

Percichthyidae												
Morone americana (white perch)	–	–	–	+	–	A	+	+	+	+	+	+
Morone chrysops (white bass)	–	F	–	–	–	–	–	–	+	+	+	+
Centrarchidae												
Ambloplites rupestris (rockbass)	–	F	–	–	–	C	–	+	+	+	–	–
Lepomis gibbosus (pumpkinseed sunfish)	–	A	–	–	+	C	+	+	+	+	–	+
Lepomis macrochirus (bluegill sunfish)	–	C	+	–	+	–	–	+	+	+	+	+
Micropterus dolomieui (smallmouth bass)	–	F	–	–	–	R	–	+	–	+	–	+
Micropterus salmoides (largemouth bass)	–	C	+	–	+	R	+	–	–	+	–	–
Pomoxis nigromaculatus (black crappie)	–	C	+	–	–	C	–	+	+	+	+	+
Percidae												
Etheostoma nigrum (johnny darter)[m]	–	C	+	–	–	–	+	–	–	–	–	+
Perca flavescens (yellow perch)	–	C	–	–	–	R	+	+	+	+	–	+
Percina caprodes (logperch)	–	F	–	–	–	–	–	–	–	–	–	–
Stizostedion v. *vitreum* (walleye)	–	F	–	+	–	–	–	–	+	+	–	–
Sciaenidae												
Aplodinotus grunniens (freshwater drum)	–	R	–	–	–	–	–	+	–	+	+	+

[a] Pre-1939 data (first column) from Greeley (1940); year of collection listed.

[b] 1939 Data (second column) from Odell (1940).

[c] 1946, 1960, 1969, 1973, 1974, 1975, 1976 (First column) data from the New York State Department of Environmental Conservation file reports for collections in these years.

[d] 1967 Data from R. M. Roecker, New York State University College at Geneseo (unpublished data).

[e] 1971 Data from Gittleman and Buchanan (1971).

[f] 1976 Data (second column) from T. A. Haines, New York State University College at Brockport (unpublished data).

[g] + Indicates species present.

[h] – Indicates species absent.

[i] R, Rare.

[j] F, Fairly common.

[k] C, Common.

[l] A, Abundant.

[m] In addition to *E. nigrum*, Gehris *et al.* (1974) reported *E. exile* (Iowa darter); Haines has questioned the identification of the latter.

identify any real changes in fish species which can be specifically attributed to altered conditions in the bay. In this conclusion, we disagree with Gehris *et al.* (1974).

Coliform Counts in the Bay and Its Watershed

Whipple (1913) recorded the presence of *Bacillus* (*Escherichia*) *coli* in both the bay and Irondequoit Creek. Reports of Metcalf and Eddy (1929) and Bonner (1938) included some coliform counts. The extent of sewage pollution in 1939 is not made clear in the Biological Survey of the Lake Ontario Watershed (New York State Department of Conservation, 1940). Although Faigenbaum (1940) noted anoxia in Irondequoit Creek near East Rochester, he attributed this to industrial rather than sewage pollution. Clausen (1940) described the Bay as ". . . turbid and badly polluted. . . . ," but Tressler *et al.* (1953) termed the pollution "slight." Tressler *et al.* did attribute the high productivity of the bay to enrichment by pollution.

Since 1929 and probably earlier, sewage discharges have led to high coliform counts in Densmore Creek and at various locations in Irondequoit Creek and its tributaries. The New York State Department of Health (1955) reported total coliform counts of 100,000/100 ml or more at the mouth of Irondequoit Creek, in Densmore Creek, and in the small Bay tributary P-113-3a which passes by the Tryon Park sewage pumping station and enters the bay at the southwest corner near Orchard Park. A count of over a million was recorded for the small tributary P-113-3-2 (Thompson Creek) which like Densmore Creek receives Rochester interceptor overflow. Elsewhere, in Allen Creek and in Irondequoit Creek upstream to Thomas Creek, counts of 20,000 to 46,000 were common. In Thomas Creek near its mouth, the count was 1.5 million. In Irondequoit Creek upstream from Thomas Creek to Powder Mill Park, counts were 4300 or less. Along Irondequoit Creek, from its mouth to Thomas Creek, counts of several tens of thousands up to 230,000 were again recorded in 1962 (NYSDH, 1964). In the mid-1960's, a number of the sewage treatment plants along Irondequoit Creek were overloaded, and their discharges, as well as the overflows from the Rochester interceptor, were not disinfected (Rochester Committee for Scientific Information, 1964, 1967a,b,c). In 1967 chlorination was being employed to sterilize the effluents, and coliform counts approached zero in Densmore Creek, Thomas Creek, and Irondequoit Creek at Browncroft and Empire Boulevards (Rochester Committee for Scientific Information, 1967d). During June to August 1972, Thorne (1973) analyzed total and fecal coliform and fecal streptococci at stations on Irondequoit Creek and a number of its tributaries. High counts in Irondequoit Creek below the Barge Canal showed that sewage inputs were continuing.

An extensive record of coliform concentration in surface water of the bay for the years 1965–1975 has been compiled by the Monroe County Department of Health (unpublished data). A few additional analyses are available for 1954 (NYSDH, 1955), 1962 (NYSDH, 1964), and 1972 (Harbison, 1972). In 1954 and 1965–1970, high total coliform counts (frequently more than 30,000 and occasionally more than 100,000 counts/100 ml) were registered at the extreme south end near Orchard Park and the mouth of Irondequoit Creek. Similarly, high counts were often obtained at the extreme north end, at or in the vicinity of the outlet; these high counts may have been due to effluents from the Irondequoit Northwest treatment plant, which discharged into the bay up to about 1970, or to mixing with contaminated Lake Ontario water. At other off-shore locations on the bay, from Stony Point to Snider Island, total coliform counts in the surface water have generally been in the range 200–2400/100 ml. On a few dates in 1968, 1969, and 1970, higher values (from 2400 to 92,000) were obtained at one or more, but not all bay stations.

Fecal coliform and fecal streptococcus counts were also made in 1973–1975 (MCDH, unpublished data; Harbison, 1972). In surface water at offshore locations, fecal coliform generally did not exceed 100/100 ml, but on a few dates in September 1975 reached several hundred, and on September 25 reached 7900/100 ml. Fecal streptococcus, analyzed a few times in 1974, ranged from 0–22 counts/100 ml. Some depth profiles of total and fecal coliform and fecal streptococcus were determined at station 1 in 1974–1975; there is an indication that counts are somewhat higher in the region of the thermocline and upper hypolimnion than at the surface.

At a near-shore station (Rochester Canoe Club docks), 20 assays in August–October 1975 gave 20 to 4900 total coliform and half the counts exceeded 2000/100 ml. Fecal coliform ranged from 20 to 3300 and 5 out of 20 assays gave more than 2000/100 ml.

In summary, the records show that at least since 1954, Irondequoit, Densmore, and Thompson Creeks have often been heavily polluted with sewage. Despite these inputs and the discharges of the treatment plants on the bay, coliform counts in surface water of the bay have generally been low except at the extreme north and south ends. Probably, high coliform counts occur transiently in the bay after storms, but quickly fall back to low levels as a result of natural processes. The Monroe County Pure Waters Agency (1969) will divert sewage from the Irondequoit Creek watershed, eliminate the two treatment plants which still discharge into the bay, and reduce Rochester interceptor overflows to a frequency of once in five years. These steps should eliminate the high counts at the south end of the bay and assure maintenance of low counts elsewhere. Also the improved and expanded Van Lare treatment plant can be expected to eliminate the present severe contamination of Lake Ontario water in the Rochester

Embayment; thus low counts in the vicinity of the bay outlet also seem assured. Possibly, some localized contamination of near-shore waters will persist where older cottages and houses are served by septic tanks (for example, in Ides and Massaug Coves).

CONCLUSIONS

State of Knowledge about the Bay

Available information provides fairly complete descriptions of bay morphology, stratification, vertical mixing, the loadings of water, nitrogen, and phosphorus, the chemistry of dissolved inorganic constituents, and the characteristics of phytoplankton growth. Based on what is now known, some interpretations are presented in the following sections concerning the effects of salt on mixing, regulation of the phytoplankton crop, and the probable effects of sewage diversion. Uninvestigated up to now are a number of areas important to understanding the dynamics of chemical and biological processes in the bay.

Aside from chlorophyll assays, available chemical data refer only to dissolved components; the concentrations of carbon, nitrogen, and phosphorus in algal and other particulates are unknown. Also, the rates of sedimentation of algal and other particulates have not been determined. As a result, the relative importance of losses of nutrients by dilution and sedimentation are uncertain. Consider, for example, nitrogen losses from the epilimnion during the summers of 1970 and 1971, losses which in sum must have equaled the external nitrogen loading (estimated at about 80 mg $\mathrm{Nm^{-2}}$ $\mathrm{day^{-1}}$, or one half the mean annual loading of about 160 mg N $\mathrm{m^{-2}}$ $\mathrm{day^{-1}}$). From the water loading (0.04 m/day) and the concentration of dissolved inorganic nitrogen (100–200 mg $\mathrm{N/m^3}$), the dilution loss of dissolved inorganic nitrogen is only 4–8 mg N $\mathrm{m^{-2}}$ $\mathrm{day^{-1}}$. Assuming algae contain about 1.7 mg N/mg chlorophyll *a*, the chlorophyll concentration (100 $\mathrm{mg/m^3}$) implies that the dilution loss of algal nitrogen is also small, about 7 mg N $\mathrm{m^{-2}}$ $\mathrm{day^{-1}}$. Evidently, most of the total loss, about 65 mg N $\mathrm{m^{-2}}$ $\mathrm{day^{-1}}$, must be attributed to (1) dilution of dissolved organic nitrogen, (2) sedimentation of algal nitrogen, (3) dilution and sedimentation of other particulate nitrogen, and (4) denitrification. The relative rates of these processes cannot be estimated from available data. Were the total concentration of epilimnetic nitrogen (dissolved plus particulate) known, the total dilution loss and, by difference, the sum of sedimentation and denitrification losses could be estimated.

Rates of diffusional transport of dissolved components between epilimnion and hypolimnion have been little studied; as a result the possibility of a significant loading of the epilimnion by hypolimnetic nutrient cannot be ruled out. As an illustration, the concentration gradient of inorganic nitrogen across the thermocline approached about 1000 mg N m^{-4} in the late summer of 1971. Assuming a molecular diffusion coefficient of 10^{-4} m^2 day^{-1} (equivalent to 10^{-5} cm^2 sec^{-1}), the rate of transport into the epilimnion would have been about 0.1 mg N m^{-2} day^{-1}, insignificant compared to the external loading of about 80 mg N m^{-2} day^{-1}. However, if eddy diffusion were 100-fold faster—which is conceivable—the hypolimnion loading component would have been significant.

Hypolimnion and sediment chemistry is poorly understood. For example, Fig. 16 showed that the concentration of NH_4–N in the hypolimnion increased slowly in June and July 1971, but very rapidly in August and September. Why was rapid generation of ammonia delayed for 2 months after deep water became anoxic, and about 3 months after a dense algal crop was established in the epilimnion? It is also unclear whether the high content of ammonia in the hypolimnion in late September arises from the organic particulate material which sedimented into the hypolimnion in the preceding 3 months, or whether from benthic deposits accumulated over many years. In the 120-day period from May to mid-September, inorganic nitrogen concentration of the hypolimnion (10–22 m) increased about 4 mg/l at all depths. If this were due solely to sedimentation of organic materials during the 4-month period, the average flux across the area (about 2.0×10^6 m^2) within the 10-m contour would have had to have been about 4.2×10^8 mg N/day. Since the total input to the bay was about 5.4×10^8 mg N/day, one would have to conclude that nearly 80% of the nitrogen entering the bay settled into the hypolimnion, which underlies only about one-third of the bay surface. Alternatively, if sedimentation of organic particles were actually slower, then a portion of the ammonia in the hypolinnion in September must have arisen from organic nitrogen accumulated in the bottom sediments over many years.

Of the biological properties, only concentrations of chlorophyll and limiting nutrient have been studied. Other experimentally accessible properties important to characterizing crop dynamics have not been measured, for example, algal production and nutrient uptake, grazing rate, and nutrient regeneration. Whether the phytoplankton growth rate constant is high in summer, as the grazing rate determinations of Haney (1973) indicated for Heart Lake, Ontario, or low, as would be concluded from observations that blue-green algae are only reluctantly grazed, is unknown. Our interpretation (see below) that the growth rate constant is high (about 0.8/day) is theoretical, and based on the observed concentrations of chlorophyll and limiting nutrient.

Effects of Road Deicing Salt Runoff

The principal direct effect of salt runoff has been to impede mixing (Bubeck, 1972; Diment *et al.,* 1974), and the hindrance of spring mixing is especially apparent. Under winter ice, the water column is stratified, and deep water becomes anoxic. In 1940, when ice-out occurred in mid-April, mixing was sufficiently complete to raise the dissolved oxygen concentration to about 7 mg/liter by late April (Tressler *et al.,* 1953). At that time, deicing salt was not used, and the low summer chloride concentration (<30 mg/liter) ruled out the possibility of significant salt gradients having existed under the winter ice. In contrast, prior to ice-out in the springs of 1970, 1971, and 1972, chloride concentrations ranged from 200 mg/liter Cl or less at the surface to about 400 to 500 mg/liter Cl at 22 m, and the gradient averaged 9 to 13 mg liter^{-1} Cl m^{-1} between 10 and 22 m. Within a few days after ice-out, mixing was complete to 11 to 13 m, but deeper mixing was prevented by the steep chloride gradient in the underlying water.

The duration of spring mixing extends from ice-out to the onset of thermal stratification in early May. In 1971 and 1972 (when ice-out took place about April 5 and 14, respectively), mixing never extended below 12–14 m. In 1970, mixing to 12 m occurred within a few days of ice-out (April 5) and subsequently extended gradually to about 18 m in late April. In 1973, ice-out occurred unusually early (about March 10); during the long period of mixing (from March 10 to mid-April), the mixed layer deepened from 12 to 20 m. In 1975, after a mild winter in which salt usage was reduced, ice-out again occurred in March, and according to Burton (1975, 1976) complete mixing to the bottom took place.

It is plausible that the steep chloride gradient (about 12 mg liter^{-1} Cl m^{-1}), which occurred between 10 and 22 m in 1970–1972, did limit the depth of mixing achieved shortly after ice-out. The density gradient which would be associated with this chloride gradient would be roughly 0.001 gm cm^{-3} m^{-1}. We know that by early May thermal stratification became stable enough to resist disruption by winds. At that time, the temperature gradient in the thermocline was about 2°C/m, and the associated density gradient would have been about 0.004 gm cm^{-3} m^{-1}. In view of the greater depth of the chloride gradient, and therefore the greater difficulty of coupling surface winds to mixing, it is reasonable to conclude that the chloride density gradient was sufficient to prevent initial mixing below 11 to 13 m. Gradual extension of mixing below this depth (as in 1973 and 1975) probably depends on mild winters (when salt usage is reduced and early thawing leads to longer than normal periods between ice-out and thermocline formation) and possibly also on stronger than normal winds during the mixing period.

From early May to late November, in 1970–1972, chloride gradients persisted in the hypolimnion and were particularly steep across the thermocline. Bubeck (1972) and Diment *et al.* (1974) concluded that the associated density gradients probably delayed the onset of fall mixing. They noted that mixing to the bottom occurred about November 13 in 1969, November 25 in 1970, December 10 in 1971, and about December 1 in 1972. Chloride concentrations at the bottom were higher, and temperatures at the onset of mixing were lower, in the years when mixing occurred at later dates. Diment *et al.* pointed out that, if deicing salt applications were to continue to increase, complete mixing to the bottom might be prevented in both spring and fall; thus the bay might become meromictic. Anoxia and the presence of hydrogen sulfide would then be continuous throughout the year in deep water.

It is unknown whether other changes in the bay have been caused by high salt concentration, either directly or as a result of incomplete mixing or lengthened periods of anoxia in deep water. Because only about 15% of the bay volume lies below 12 m, the consequences of incomplete mixing seem likely to be small. We are unaware of evidence that the salt concentration (150–200 mg/liter Cl) in the epilimnion in summer might be harmful to freshwater organisms. Much higher concentrations (1500 mg/liter Cl) commonly occur in Onondaga Lake, where phytoplankton species and numbers appear to be similar to those in the bay (Sze and Kingsbury, 1972). Such changes as are indicated by the historical record, both of phytoplankton and emergent and submerged aquatics, appear to have commenced long before 1955 when summer chloride concentration was still only 30 mg/liter (New York State Department of Health, 1955).

Control of Phytoplankton Growth in Summer, 1970 and 1971

During July and August in both years, constancy of algal and chemical properties of the mixed layer indicates a steady state in which carbon and nutrient incorporation into phytoplankton was balanced by algal losses to dilution into Lake Ontario, sedimentation into the hypolimnion, and consumption by bacteria, fungi, zooplankton, etc. Also, inputs of dissolved nutrients into the epilimnion, both from external sources and from regeneration, must have been balanced by losses to dilution and algal uptake. In general, steady-state algal properties might be determined by a number of environmental factors. Under conditions of very high nutrient loadings and either low solar illumination or strong light absorption by water and nonphytoplankton components of the water column, steady-state algal properties might be limited primarily by low subsurface illumination.

In this case, further increases in nutrient loadings would have no effect, and algal growth would be "nutrient saturated." On the other hand, low nutrient loadings could be the primary factor regulating growth; in this case the algal crop would be "nutrient limited." In most lake and sea epilimnia in summer, phytoplankton growth is nutrient limited. Compared with other bodies, however, Irondequoit Bay is subject to extremely high loadings. In 1970–1971, the estimated mean annual loading of nitrogen was about 163 mg N m^{-2} day^{-1} or 60 gm N m^{-2} $year^{-1}$ and that of phosphorus was about 45 mg P m^{-2} day^{-1} or 16 gm P m^{-2} $year^{-1}$ (see Table 16). Among 30 mainly eutrophic lakes analyzed by Vollenweider (1968, 1972), only Boden–Obersee, Lake Norrviken, and Lake Waubesa have loadings approaching (but in no case equaling) those of Irondequoit Bay. To decide whether the summer algal crop in Irondequoit Bay is nutrient saturated or nutrient limited, two questions must be answered: (1) what nutrient is most nearly limiting, and (2) does the most nearly limiting nutrient actually limit growth?

As candidates for the most nearly limiting nutrient, carbon, nitrogen, and phosphorus have been considered. Silicon, although probably limiting diatom growth in mid-May, cannot be limiting in July–August because the summertime crop is mainly composed of green and blue-green species, and because Si levels are higher in summer (and steadily increasing) than in late May. Of nitrogen and phosphorus, there are several reasons to believe that nitrogen is more limiting. Whereas the succession of blooms in May–June depletes inorganic nitrogen from about 2 to 0.2 mg/liter N, the concentration of soluble phosphorus declines proportionately less (from about 0.5 to 0.3 mg/liter P), and the summer concentration (0.3 mg/liter P) is higher than that of inorganic nitrogen. When samples of summer bay water are illuminated in the laboratory, algal growth occurs only when the nitrogen concentration is augmented. This is consistent with the fact that the N:P ratio in bay water in July–August is much lower than the ratio of N:P in algae. The ratio of the concentration of a dissolved nutrient to that concentration K_s which half-saturates algal growth rate is a measure of the degree of nutrient limitation. Of the two nutrients, that having the smaller ratio will be the more nearly limiting. For phosphorus, the concentration in the bay is about $10 \mu M$ and K_s values have been reported to be about 0.01 to $0.03 \mu M$ (Rhee, 1973; Soeder *et al.*, 1971); the ratio is therefore between 300 and 1000. For nitrogen, the concentration in the bay during July–August averages 7 to 14 μM, and reported K_s values range between about 0.02 and $1.0 \mu M$ (Eppley and Thomas, 1969; Caperon and Meyer, 1972b; Eppley and Renger, 1974); the ratio is therefore between 7 and 700. Thus nitrogen is probably more limiting than phosphorus. Droop (1974) showed that the more limiting of two nutrients will be that for which the ratio of the loading and the minimum content in algae is smaller. For nitrogen, the summer loading on the bay is about 80 mg N m^{-2} day^{-1} and the minimum content in algae is about 50 mg N gm C (Caperon and Meyer, 1972a); the ratio is then about 1.6. For phosphorus, the loading is about 45 mg P m^{-2} day^{-1}, and the minimum content in algae is about 3 mg P/gm C (Rhee, 1973); the ratio is then about 15. Nitrogen is therefore more limiting than phosphorus. Both Droop (1974) and Rhee (1974) have offered evidence that algal growth responds only to the more limiting of two nutrients; when two nutrients are equally limiting, growth is the same as if only one were limiting.

Whether nitrogen is more nearly limiting than carbon, in the July–August steady state, is uncertain. Application of the criterion of the ratio of dissolved nutrient concentration to the concentration K_s which half-saturates growth does not lead to a secure conclusion. For July–August, the alkalinity of the water was about 3.5 mEq/liter, and the pH was about 9.0 in 1970 and 9.5 in 1971. The corresponding concentrations of dissolved CO_2 (including H_2CO_3) would be about 8 and 2 μM. According to Goldman *et al.* (1974) and King and Novak (1974), the concentration of dissolved CO_2 which half-saturates growth in continuous cultures is about 2.5μM. For phytoplankton photosynthesis, the half-saturating concentration is lower, 1 μM or less (Whittingham, 1952; Emerson and Green, 1938; Steeman Nielsen, 1955) although Brown and Tregunna (1967) reported a somewhat higher value. If K_s is taken as 1 μM, then values of the ratio $[CO_2]/K_s$ are about 8 (at pH 9.0) and 2 (at pH 9.5), lower than the corresponding value for nitrogen. Thus CO_2 would seem to be more limiting than nitrogen. However, the growth and photosynthesis studies were carried out in strong, saturating illuminations. Since K_s for CO_2 is known to decline as illumination is reduced (Rabinowitch, 1951; Slovacek and Bannister, 1973) the value of K_s applicable in the strongly absorbing mixed layer of Irondequoit Bay could be considerably lower than 1 μM, and, conceivably, carbon is not more nearly limiting than nitrogen.

Another argument against carbon limitation is based on the study of Lake 227 by Schindler *et al.* (1973) and Schindler and Fee (1973). In this softwater lake, nitrogen and phosphorus fertilization produced dense algal crops despite the fact that dissolved CO_2 concentration was only about 0.01 μM during most of the day. Although this extremely low concentration of CO_2 demonstrably limited daily production and undoubtedly also reduced growth rate, high algal densities (60 to 160 μg chlorophyll *a*/liter) were nevertheless maintained throughout June and July. Since the concentration of dissolved CO_2 in Irondequoit Bay is more than 100-fold higher than in Lake 227, carbon limitation of phytoplankton in the bay seems unlikely.

Assuming nitrogen is more nearly limiting than carbon or phosphorus, the question of whether growth is actually nitrogen limited, or whether growth is nutrient saturated, can be considered. The question is important for predicting the future response to sewage diversion: if the present nitrogen loading were already much more than saturating, then a moderate reduction of, say, 50%, might be insufficient to bring any response. Alternatively, if growth is presently nutrient limited, or only barely saturated, then any future reduction in loading will assuredly diminish the algal crop. We think that the July–August steady-state crop, in 1970–1971, was almost, but not quite completely nutrient saturated. That the summer steady state was approximately nutrient saturated we infer from the fact that crop density (measured as chlorophyll and as algal cell volume) was the same in July–August (when nitrogen and carbon might have been limiting) and in May–June (when both nitrogen and carbon dioxide concentrations were much higher).

That the July–August steady state is not quite nutrient saturated is an inference based on the application of a new theory of phytoplankton growth in nutrient-limited and nutrient-saturated mixed layers (Bannister, 1976a,b). Although details of the theory cannot be presented here, a summary of some aspects of the theory is relevant. The theory treats a steady state in which algae are subject to losses to dilution, sedimentation, and consumption, and in which nutrient inputs from the environment and from regeneration are balanced by losses of dilution and algal uptake. Intrinsic phytoplankton properties, as characterized in recent physiological studies of continuous cultures, are incorporated in the form of several constant parameters and four adaptive parameter functions. Inputs include values of six environmental variables (day length, noon illumination at the surface, mixed-layer depth, nonphytoplankton extinction coefficient, and water and nutrient loadings) and an estimated value of the regeneration fraction (the fraction of algal nutrient taken up by consumers which is returned to the water in a form available to algae). Given these inputs, the theory predicts values of various crop functions (algal density as chlorophyll, carbon, and limiting nutrient, gross and net production, dissolved nutrient concentration, algal growth rate constant, algal uptake of nutrients, and regeneration) as functions of the first-order rate constant R_n for loss of algae to consumers. Because the crop functions exhibit different dependences on R_n, correct predictions of several crop functions only occur for a single value of R_n.

Approximately steady-state phytoplankton growth has been documented in a number of waters: in July–August in Irondequoit Bay, in November in the North Pacific Gyre (Eppley *et al.*, 1973), in July–August in Lake Tahoe (Goldman 1974, 1975), and in August–September in Lake Kinneret (Berman, 1973; Berman and Pollingher, 1974). For all four of these epilimnia, which differ greatly in values of environmental variables and also in the limiting nutrient (phosphorus in Lake Kinneret, nitrogen in the others), observed values of crop functions are correctly predicted provided (1) the value of the regeneration fraction lies between 0.90 and 1.0, and (2) the value of R_n is close to that which gives a maximum steady state harvest of algal carbon by consumers. In all four waters, daily production and the concentrations of chlorophyll and dissolved nutrient are correctly predicted; for the Gyre, experimental values of algal uptake of nitrogen and regeneration of nitrogen by consumers are also approximately correctly predicted.

In applying the theory to Irondequoit Bay, values of environmental variables were as follows: daylength, 0.63 days; noon illumination, 137 einsteins visible m^{-2} day^{-1}; nonphytoplankton extinction coefficient, 0.75/m; and mixed layer depth, 4 m (Bannister, 1974b). Water and nutrient loadings were 0.04 m/day and 81 mg N m^{-2} day^{-1} (see Tables 5 and 16). The regeneration fraction was taken as 0.94, close to the experimental value (0.93) for phosphorus regeneration determined by Peters and Rigler (1973). With these inputs, theory predicts a chlorophyll concentration of 107 mg/m^3 and a dissolved nitrogen concentration of 172 mg N/m^3 (in good agreement with the experimental values of 100 mg chlorophyll/m^3 and 100–200 mg N/m^3). These predicted values occur when R_n is 0.7/day. This value of R_n is that giving the maximum steady-state harvest of algal carbon by consumers, and it is close to the value (0.80/day) which Haney (1973) reported as the experimental mean for eutrophic Heart Lake in July–August. Gross daily production in Irondequoit Bay is predicted to be about 5.5 gm C/m^2 daily, in close agreement with the value found by Megard (1972) for the Tanager Lake basin of Lake Minnetonka; Tanager Lake and Irondequoit Bay appear to have similar properties including high sewage inputs. The good agreement between predicted and experimental values of crop functions, for Irondequoit Bay and the three other waters, gives some confidence that the theory itself is essentially correct.

The inference that Irondequoit Bay is not quite nitrogen saturated is the result of an analysis of the dependence of predicted crop functions on nitrogen loading. It was assumed that the regeneration fraction remains constant, and that the maximum harvest condition continues to apply. At a nitrogen loading of about 120 mg N m^{-2} day^{-1}, predicted crop density reaches a nutrient-saturated upper limit of about 150 mg chlorophyll *a*/m^3; further increase in nutrient loading does not result in higher crop density. Also, with a loading of 120 mg N m^{-2} day^{-1}, the predicted concentration (about 500 mg N/m^3) of dissolved inorganic N is much higher than that in the bay. Thus, theoretical analysis indicates that algal growth in the July–August steady state, in 1970–1971, was not quite nutrient saturated.

Probable Effects of Sewage Diversion

According to Table 16, complete diversion of treatment plant and interceptor discharges, together with the detergent phosphorus ban, will reduce the mean annual loading of nitrogen by about 50% (from 163 mg N m^{-2} day^{-1} in 1970–1975 to 76 mg N m^{-2} day^{-1} after diversion) and of phosphorus by possibly 90% (from 45 mg P m^{-2} day^{-1} in 1970–1971 to 4.7 mg P m^{-2} day^{-1} after diversion). The mean annual water loading will also decline about 25% (from 0.059 m/day to 0.046 m/day). Since the phytoplankton crops in the summers of 1970–1971 appear to have been slightly nitrogen limited (or only barely nutrient saturated), crop concentration should respond to any reduction in the nitrogen loading. Thus smaller summer crops should result from diversion. The concentrations of dissolved nitrogen and phosphorus in winter and early spring should also decline since the nutrient loadings will fall more than will the water loading.

These estimates of nutrient loadings provide a basis (albeit an imperfect one) for predicting the extent of changes in nutrient and algal concentrations which will result from diversion. Before presenting these predictions, it is well to recall that there are uncertainties not only about the estimated values of the mean annual loadings but also about the relationship between summer algal concentration and the mean annual loadings of total nitrogen and total phosphorus. With respect to nitrogen, Table 13 showed considerable year-to-year and seasonal variation in the nitrogen discharge from Irondequoit Creek. Thus the estimated mean annual loading for the 1970–1975 period is not accurate for each individual year. As previously noted, the mean annual loading probably overestimates by twofold the mean loading in summer months. Further, as discussed earlier, the projection that diversion will halve the nitrogen discharge from Irondequoit Creek is quite uncertain. For phosphorus, annual and seasonal variations have been smaller (see Table 13), and the mean loading for the 1970–1971 period is probably a reasonably accurate estimate of both the annual and summer loadings in both years. Uncertainties remain however.

As previously described, the estimated mean annual loading after diversion is a minimum one based on the low phosphorus concentrations reported by Thorne (1973) at upstream stations of Irondequoit Creek. After elimination of the large, relatively steady municipal inputs of phosphorus, seasonal variations in phosphorus discharge may become significant, and the mean summer loading may be smaller than the mean annual loading. The fraction of the total phosphorus, which will be available to phytoplankton after diversion, has not been determined; Table 14 suggested that soluble PO_4–P may comprise only 10–20% of the total. Finally, biological uptake of both nitrogen and phosphorus in the wetlands during summer, although apparently trapping only a small fraction of Irondequoit Creek nutrients at present, might withhold a larger fraction after diversion. The predictions which follow are based on the assumptions that summer algal crops have been and will continue to be determined by the mean nutrient loadings during summer, and that the mean summer loading is one-half the annual loading in the case of nitrogen, and equal to the annual loading in the case of phosphorus.

Summer algal growth will probably continue to be nitrogen limited after diversion. According to Table 16, the ratio of the mean annual loadings of nitrogen and phosphorus was about 3.6 in 1970–1971 and 6.5 in 1973–1975, and will reach 16.2 after diversion. Values of the ratio of the mean summer loadings are only half as large, and the value (8.1) after diversion will remain well below the threshold value (about 16) for phosphorus limitation. Only if a large fraction of the total phosphorus input proves to be unavailable to algae is phosphorus limitation likely. It is well known that sewage diversion will generally shift lakes toward phosphorus limitation. In the case of Lake Washington, sewage diversion resulted in a definite change from nitrogen limitation to phosphorus limitation (Edmondson, 1970, 1972). In the case of Irondequoit Bay, the shift will be in the same direction, but attainment of phosphorus limitation seems improbable.

The quantitative dependence of summer chlorophyll concentration on the loading of the limiting nutrient (or on the concentration of the limiting nutrient in winter or at spring mixing—this concentration being proportional to the loading) has been investigated extensively for phosphorus-limited, but not nitrogen-limited lakes. For a large number of phosphorus-limited lakes, a single empirical equation describes quite accurately the relation between summer chlorophyll and spring phosphorus concentrations (Sakamoto, 1966; Welch *et al.*, 1973; 1975; Dillon and Rigler, 1974). Interestingly, the equation is a power function, the chlorophyll concentration being proportional to the phosphorus concentration raised to a power of about 1.42 (Welch *et al.*, 1975), 1.6 (Sakamoto, 1966), or 1.45 (Dillon and Rigler, 1974). Apparently, annual production is an even higher power function of phosphorus loading (Schindler and Fee, 1974a). Paralleling these empirical findings, the theory of steady-state algal growth (Bannister, 1976b) predicts that both chlorophyll concentration and daily production will be proportional to the loading raised to the power 1.25 in nitrogen-limited waters. Altogether, these results suggest that the decline of the summer phytoplankton crop in Irondequoit Bay may be proportionately greater than the decline in the nitrogen loading. If, to the contrary, crop concentrations were supposed to be directly proportional to the loading, then diversion should reduce the crop concentration from 100 mg chlorophyll a/m^3 to about 47 mg chlorophyll a/m^3 (given summer loadings of 81 and 38 mg N^{-2} day^{-1}, before and after diversion). If crop concentration is proportional to loading raised to the power 1.25 or 1.45, then the corresponding postdiversion concentrations will be 38 or 30 mg chlorophyll a/m^3, respectively. Thus, summer chlorophyll concentrations should fall by at least 50% and possibly as much as 70% as a result of diversion.

The decline in summer phytoplankton concentrations can be expected to increase transparency considerably. From the estimated values of the algal extinction coefficient (0.016 m^2/mg chlorophyll *a*) and the nonphytoplankton extinction coefficient in Irondequoit Bay (about 0.75/m) (see Bannister, 1974b), the depth for 1% transmission can be calculated. For chlorophyll *a* concentrations of 100, 50, and 30 mg/m^3, the depths for 1% transmission are 2.0, 3.0, and 3.7 m, respectively. Diversion should therefore increase the depth from 2.0 m at present to at least 3.0 m, and possibly to between 3.0 and 4.0 m. Correspondingly, Secchi disk depths should increase from 0.8 m at present to at least 1.2 m, and possibly to about 1.5 m. As a result of increased transparency, rooted aquatic weeds may proliferate and compete with phytoplankton for the limiting nutrient; phytoplankton concentrations might therefore fall off still more.

Diversion can also be expected to alter phytoplankton growth in May and June. Presently, growth during this period is believed to be nutrient saturated, and the approximately constant chlorophyll concentration (about 100 mg chlorophyll/m^3) is determined by the balance between light-limited production and algal losses to dilution, sedimentation, and consumption. The concentration of dissolved nitrogen, 1.5 to 2.0 mg/liter on May 1, is determined by the balance between input and loss by dilution only. The development of a dense phytoplankton crop during the first half of May generates an additional, much faster loss process (sedimentation of algae and consumption) which causes the nutrient concentration in the epilimnion to fall. The 2-month period (May 1 to July 1) is the time presently required to reduce the concentration of dissolved nitrogen to the steady-state summer level. After diversion, the concentration of dissolved nitrogen about May 1 can be expected to be about 1.0 mg/liter, and the concentration of phosphorus about 0.07 mg/liter. These concentrations are high enough to support nearly nutrient-saturated growth initially, but the time required to reduce them to the levels of the nutrient-limited steady state will be shortened. Thus, it is probable that nutrient limitation may set in before the heavy green algal bloom (now occurring about June 1) and certainly before the heavy blue-green bloom (now occurring in mid-June) can develop. We anticipate, therefore, little change in the nature of the early diatom bloom, but some reduction in the green algal bloom, and a large reduction (perhaps complete disappearance) of the June blue-green bloom. The postdiversion summer steady state will probably commence about June 1 (rather than July 1 as at present), and will be characterized by lower pH and lower concentrations of dissolved nitrogen, phosphorus, and chlorophyll. These predictions appear to be supported by the observations of Tressler *et al.* (1953) in the spring of 1940; they observed a May diatom blood as intense as present ones but gave no indication of a blue-green bloom in June. Also, in 1940, the pH did not exceed about 8.5 and phosphorus concentration fell to about 0.03 mg/liter P by June 1.

Response to diversion can be expected to be prompt. The ratio of mean depth (6.8 m) and the mean annual water loading (0.059 m/day) is the time constant for dilution; the value is 115 days. From the oversimplified conception of the bay as a well-mixed volume subject only to dilution, the change of nitrogen and phosphorus concentration, from pre- to postdiversion steady-state levels should be 95% complete in about one year. More realistically, during summer stratification, about half the bay volume is isolated within and below the thermocline and only the epilimnetic volume undergoes dilution; the time constant for dilution of the whole bay is therefore longer than 115 days. A slow release of nutrients accumulated in the sediments during many years of nutrient enrichment would prolong attainment of the postdiversion steady state still more.

The response time of phytoplankton concentration in the epilimnion in the summer can be expected to be very much shorter, because the rate constant associated with loss of epilimnetic nutrient, due to sedimentation of algae and consumers, appears to be about ten times larger than the rate constant for loss of dissolved, algal, and consumer nutrient by dilution. Summer phytoplankton concentration should therefore respond with a time constant of about 10 days. It follows that, if complete diversion were to occur abruptly about July 15, the transition of phytoplankton concentration from pre- to postdiversion steady states should be 95% complete by about August 15. Responses of summer phytoplankton concentration, more rapid than can be attributed to dilution alone, have been recorded. For Lake Washington the dilution time constant is 3.2 years (Edmondson, 1961); nevertheless, the decline of summer chlorophyll concentrations during the period of gradual sewage diversion (1963–1968) did not lag behind by more than about 1 year the decline in sewage input (Edmondson, 1972). The mean summer chlorophyll concentration was about 40 mg chlorophyll a/m^3 in 1962 and 1963 and about 10 mg chlorophyll a/m^3 in 1968; relative sewage input was reduced to about 0.6 in March 1963, to about 0.5 in 1965, and to zero in mid-1967. Lake 304, in western Ontario, also has a dilution time constant of 3.2 years (Brunskill and Schindler, 1971). During 1971 and 1972, artificial fertilization with nitrogen, phosphorus, and carbon led to much increased chlorophyll concentrations in summer. In 1972, fertilization with phosphorus was stopped, and summer chlorophyll concentrations never exceeded prefertilization levels (Schindler and Fee, 1974b).

The possibility of increased growth of rooted aquatics over a period of several years following diversion could result in a slower, second phase of decline of summer phytoplankton concentration.

Might the Hypolimnion Become Aerobic?

Figure 10 showed that the mean depth of water below 6 m (the lower limit of the epilimnion) is about 4 m. If the water below the epilimnion contained 10 ppm oxygen at the time of stratification (shortly after May 1), its capacity for complete oxidation of carbohydrate carbon would be 3.7 gm C/m^3 or 14.8 gm C/m^2. To maintain an oxygen concentration of 5 ppm from May 1 to November 1, the average rate of sedimentation of particulate carbon from the epilimnion would have to be less than 0.04 gm C m^{-2} day^{-1}. Given an epilimnetic concentration of chlorophyll of 100 mg chlorophyll a/m^3, an approximate ratio of carbon to chlorophyll in algae of 0.015 gm C/mg chlorophyll *a*, and a typical velocity (0.3 m/day) (Smayda, 1970) for phytoplankton sedimentation, the rate of sedimentation of algal carbon is 0.4 g C m^{-2} day^{-1}. Sedimentation of fecal pellets and other forms of particulate carbon implies that the rate of sedimentation of total carbon is greater than 0.4 gm C m^{-1} day^{-1}. It follows that the summer algal concentration would have to be reduced more than tenfold to achieve an aerobic hypolimnion. Almost certainly, such a large reduction will not result from diversion; even if it did, oxidation of organic matter already accumulated in the sediments could require many years. It is worth recalling that the hypolimnion was entirely anaerobic on August 8, 1912 (Whipple, 1913), a time when nutrient loadings must have been much smaller than now.

Effects of "Opening" the Outlet

To permit small craft passage between Lake Ontario and Irondequoit Bay, the U.S. Army Corps of Engineers (1968) proposed that the highway and railroad bridges be removed, that the present outlet be widened to about 30 m and deepened to about 2.4 m, that an approximately 250-m-long channel of the same cross section be dredged into Lake Ontario, and that a 1.8-m-deep channel be dredged across the shallows at the north end of the bay. On the Lake Ontario side, 200-m-long groins would be required to prevent shoal formation. Gehris *et al.* (1974) have reviewed the undesirable effects that dredging would have but concluded that these would be short lived.

Proponents of opening the bay have suggested that enlargement of the outlet would enhance mixing of Lake Ontario and bay waters and would thereby improve water quality in the bay. Reverse flow from Lake Ontario into the bay is a transitory phenomenon occurring when strong northerly winds abruptly raise the Lake Ontario level. Reverse flow can only occur when the time rate of increase of the Lake Ontario level exceeds the water loading (about 0.059 m/day) due to runoff from the drainage basin. In order for dilution of bay water by Lake Ontario water to be significant, the total volume of Lake Ontario water entering the bay in the course of a year would have to be comparable to the volume of annual runoff. Whether a significant dilution by Lake Ontario will occur depends on two factors: (1)

the frequency and duration of the transient periods when Lake Ontario level increases rapidly, and (2) the area of the outlet channel. So far as we are aware, no estimate of the annual volume of reverse flow has been made based on the known character of level fluctuations in Lake Ontario.

Some perspective is provided by the following facts. (1) The proposed enlargement of the outlet channel would increase the cross-sectional area about tenfold; the enlarged channel would still be relatively small and Lake Ontario currents could not contribute to mixing. (2) No significant mixing occurs at present. Short-term increases in the surface level of the bay rarely attain 0.05 to 0.10 m in 1 day, corresponding to less than 1.5% increase in bay volume. Of this small volume increase, most is contributed by runoff rather than injected Lake Ontario water.

Future Studies

Public interest, long-range management, and improved scientific understanding will all require continued study of the bay and its watershed. The public will demand to learn if sewage diversion, undertaken at great expense, will improve the condition of the bay. To satisfy this interest, transparency, salt and nutrient concentrations, phytoplankton analyses, and bacteriological assays will have to be continued routinely for many years. The public will also have immediate interest in possible changes in fish stocks and any proliferation of aquatic weed beds.

Management directed to long-term preservation of the recreational and aesthetic qualities of the bay will require expanded monitoring and, in addition, some studies not yet begun. Water, nutrient, and particulate inputs need to be assayed routinely in the future. Establishment of a permanent, continuously recording gauging station on lower Irondequoit Creek (perhaps at Browncroft Boulevard) is an urgent requirement. Present Irondequoit Creek flows in dry summers approach 1 m^3/sec and may fall close to zero when treatment plant effluents (totaling about 0.51 m^3/sec) are diverted. Increased Barge Canal inputs may be required to maintain Creek flow. Postdiversion nutrient inputs to the bay, and nonsewage nutrient inputs into the drainage basin will have to be determined. Of great importance to long-term preservation is the rate of silting of the bay. Determination of the silt load of Irondequoit Creek and the rate of erosion of the silt bluffs and hillsides along the bay shore are urgent needs; measures to stabilize the bluffs and shoreline may be required. Enlightened management will also require a better understanding of the roles of the wetlands and of bottom sediment in the nutrient dynamics of the bay.

Sewage diversion from Irondequoit Bay will offer an important opportunity to advance limnological science. While many diversion projects are underway (Dunst *et al.*, 1974) in the United States, the results of diversion have been thoroughly investigated in relatively few cases. To our knowledge, none of these cases concerned a hard-water lake as heavily loaded with nutrients, nor as densely populated with phytoplankton, as the bay. In any case, the quantitative relationship between nutrient loading and crop functions such as density, production, growth rate, grazing rate, and regeneration are as yet poorly understood, and, because of this, reliable quantitative predictions of the effects of changes in nutrient loadings are not yet possible. Thorough study of postdiversion changes in the bay will help provide the basis for reliable prediction, useful in managing many other waters. A number of bay processes remain to be investigated; some are undoubtedly of critical importance in determining overall bay functioning. Vertical diffusion rates, relative losses by dilution and sedimentation, chemistry of the hypolimnion and bottom sediments, algal growth rate, grazing, and regeneration all remain to be characterized. How these processes will respond to altered nutrient loadings may be the key to understanding overall response.

APPENDIX

Phytoplankton counting was carried out with a Sedgewick–Rafter chamber or with permanent slides. The latter were prepared by filtering samples through Millipore filters (RAWP, 1.2 μm), fixing with 2% osmic acid, dehydrating with 20, 50, 70, and 95% ethanol and 100% isopropanol, clarifying with xylene and cedar wood oil, and mounting the filter and deposited algae in Canada balsam. In these slides, algae are uniformly distributed on the filter, pigmentation and cellular structure is preserved, and the filter is almost perfectly transparent so that high dry and oil immersion objectives can be employed. Counting units were defined in Table 17. Counting was performed over an area of the chamber or slide such that 100 or more units of the more numerous species were counted. The number of units per milliliter of bay water and the standard deviation (proportional to the square root of the number counted) were calculated. In counting, some genera and species were grouped: (1) *Chlamydomonas* and *Carteria*, (2) all *Pediastrum* species, (3) all *Tetraedron* species, (4) all *Scenedesmus* species other than *S. falcatus*, (5) *Cyclotella*, *Stephanodiscus*, and *Coccinodiscus*, but not *Melosira*, and (6) all pinnate diatoms other than *Diatoma* and *Surirella*. For each species, the cell volume (not including sheath) of a counting unit was estimated. For each species in which cell size did not vary much, an arithmetic mean volume was calculated. For species which varied greatly in size, volumes of large and small examples were determined and the geometric mean was calculated. Such a mean overestimates the volume of a small cell by the same factor f as it underestimates the volume of a large cell. For a large error factor, large errors may occur in computed biomass if most cells are either small or large. An instance of an overestimate was the computed volume ($4 \times 10^7\ \mu^3$/ml) of circular diatoms on May 7, 1971. Since the cells were almost all small, the true volume was actually smaller. The unit volume estimates agree fairly well with those of Nalewajko (1966).

The concentration of chlorophyll *a* was assayed by filtering usually 50 ml of water through a Millipore filter (RAWP, 1.2 μm). The filter and the algal pigments were dissolved in 10 ml of 90% pyridine maintained at 70°C for 10 minutes. After brief centrifugation, absorbance (A) was measured at 670 and 740 nm, and the chlorophyll *a* concentration calculated as

$$C\,(\mu\text{g chlorophyll } a/\text{liter}) = 10^3\, D\,(A_{670} - A_{740})/(\epsilon_{670})$$

where D is the dilution factor and ϵ_{670} ($= 8.0 \times 10^4$ liters mole^{-1} cm^{-1}) is the molar extinction coefficient of chlorophyll *a* in 90% pyridine at 670 nm. The method is less accurate than standard methods employing 80% acetone, in which chlorophylls *a* and *b* and pheophytins can each be determined. The advantages of pyridine are quick solution of the filter, more complete extraction of chlorophyll, and completion of the assay in about 20 minutes. Because the calculation ascribes all absorbance to chlorophyll *a*, the calculated concentration will be correct only if chlorophyll *b* and pheophytins are absent. The true concentration of chlorophyll *a* will be about 90% of the calculated concentration if chlorophylls *a* and *b* are present in a 3:1 mole ratio (as in green algae) and less if pheophytins are also present.

ACKNOWLEDGMENTS

The study of the bay in 1969–1972, in which the authors participated, was supported initially by the Rochester Committee for Scientific Information and the National Science Foundation (GH-1114), and later by the National Oceanic and Atmospheric Administration–National Sea Grant Program (GH-106). The investigation was supervised by W. H. Diment (formerly Professor, Department of Geology, University of Rochester, and now with the United States Geological Survey, Menlo Park, California) and assisted by B. L. Deck, A. L. Baldwin, S. I. Lipton, D. Chrapkiewicz, and N. Connors. The study was further supported by the Geological Survey and the Environmental Protection Agency, which lent equipment, and especially by R. G. Austin who made his boat, dock, and property on Irondequoit Bay available throughout the study.

In the preparation of this chapter, we acknowledge the generous help of many: H. S. Forest (Department of Biology, State University College at Geneseo) provided his data on rooted aquatic growth and supplied copies of many reports, manuscripts, and maps; K. G. Harbison (formerly Assistant Professor of Chemistry, University of Rochester, and now Analytical Science Division, Research Labs, Eastman Kodak Corp.) provided some of the data he obtained in the summer of 1972; C. B. Murphy and G. Welter (O'Brien and Gere Engineers, Inc., Syracuse) kindly made available data on overflows from the Rochester East Side Interceptor; R. S. Burton (Monroe County Department of Health) and K. Walker and R. Flint (EPA, Rochester Field Office) supplied data from their files; D. Day (Monroe County Department of Health) supplied descriptions of treatment plant discharges in 1969–1970; and D. Adler (Lamont–Doherty Geological Observatory), a participant in the study reported by Thorne (1973), discussed his conclusions with us. We appreciated also discussions with N. Holmes (New York State Department of Environmental Conservation, Avon, New York), U. B. Stone (formerly of the same agency), and T. A. Haines (Department of Biology, State University College at Brockport) on interpreting the fish trapping records. We are indebted to W. H. Diment, K. G. Harbison, H. S. Forest, D. H. Schindler (Freshwater Institute, Winnipeg), and K. M. Stewart (Division of Biology, State University at Buffalo) for reviewing the manuscript. We thank D. McCumber, R. Eaton, and R. Rasmussen for able assistance in drafting, photography, and typing.

REFERENCES

Bannister, T. T. (1974a). Production equations in terms of chlorophyll concentration, quantum yield, and upper limit to production. *Limnol. Oceanogr.* **19,** 1–12.

Bannister, T. T. (1974b). A general theory of steady state phytoplankton growth in a nutrient saturated mixed layer. *Limnol. Oceanogr.* **19,** 13–30.

Bannister, T. T. (1976a). Algal parameters as functions of the growth rate constants μ_s and μ, for nutrient saturated and nutrient limited growth. (Unpublished manuscript.)

Bannister, T. T. (1976b). A general theory of steady state algal growth limited by a single nutrient. (Unpublished manuscript.)

Berman, T. (1973). "Lake Kinneret Data Record," p. 74.

Berman, T., and Pollingher, U. (1974). Annual and seasonal variations of phytoplankton, chlorophyll, and photosynthesis in Lake Kinneret. *Limnol. Oceanogr.* **19,** 31–54.

Black and Veatch, Consulting Engineers. (1969). "Report on Comprehensive Sewerage Study for City of Rochester, New York," p. 284. Report to New York State Department of Health and to City of Rochester.

Bonner, J. F. (1938). "A Survey of the Pollution of the Surface Waters of Monroe County, New York." Monroe County Division of Regional Planning, Rochester, New York. pp. 28.

Brown, D. L., and Tregunna, E. B. (1967). Inhibition of respiration during photosynthesis by some algae. *Can. J. Bot.* **45,** 1135–1143.

Brunskill, G. J., and Schindler, D. W. (1971). Geography and bathymetry of selected lake basins, Experimental Lakes Area, northwestern Ontario. *J. Fish. Res. Board Can.* **28,** 139–155.

Bubeck, R. C. (1972). Some factors influencing the physical and chemical limnology of Irondequoit Bay, Rochester, New York. Ph.D. Thesis, Part I, p. 290. University of Rochester, Rochester, New York. [Data are included in Part II of this thesis, p. ~250.]

Bubeck, R. C., Diment, W. H., Deck, B. L., Baldwin, A. L., and Lipton, S. D. (1971). Runoff of de-icing salt: Effect on Irondequoit Bay, Rochester, New York. *Science* **172,** 1128–1132.

Burdick, G. E. (1940). Studies on the invertebrate fish food in certain lakes, bays, streams and ponds of the Lake Ontario watershed (inclusive of ecological data on the mollusca). *In* A biological survey of the Lake Ontario watershed. Supplement to the 29th Annual Report, pp. 147–166. N.Y.S. Dept. Conservation, Albany.

Burton, R. 1976. Improvement in Irondequoit Bay following decrease of road salting in the watershed. Rochester Committee for Scientific Information Bulletin No. 198.

Burton, R. 1975. Monroe County Dept. of Health (personal communication).

Caperon, J., and Meyer, J. (1972a). Nitrogen-limited growth of marine phytoplankton. I. *Deep-Sea Res.* **19,** 601–618.

Caperon, J., and Meyer, J. (1972b). Nitrogen-limited growth of marine phytoplankton. II. *Deep-Sea Res.* **19,** 619–632.

Chadwick, G. H. (1917). The lake deposits and evolution of the lower Irondequoit valley. *Proc. Rochester Acad. Sci.,* **5,** 123–160.

Clausen, R. T. (1940). Aquatic vegetation of the Lake Ontario watershed. *In* "A Biological Survey of the Lake Ontario Watershed," New York State. Suppl. to 29th Ann. Rep., pp. 167–187. Dept. of Conservation, Albany.

Dardenne, B. 1974. "Lost Map from Paris Shows New Denonville Trail," Monday, 26 August. pp. 1C, 16C. Rochester Times-Union.

Dillon, P. J., and Rigler, F. H. (1974). The phosphorus-chlorophyll relationship in lakes. *Limnol. Oceanogr.* **19,** 767–773.

Diment, W. H., Bubeck, R. C., and Deck, B. L. (1973). Some effects of de-icing salts on Irondequoit Bay and its drainage basin. *Highw. Res. Rec.* **425,** pp. 23–35.

Diment, W. H., Bubeck, R. C., and Deck, B. L. (1974). Effects of de-icing salts on the waters of the Irondequoit Bay drainage basin, Monroe County, New York. *In* "Fourth Symposium on Salt" (A. H. Coogan ed.), Volume I, pp. 391–405. Northern Ohio Geol. Soc., Inc., Cleveland, Ohio.

Droop, M. R. (1974). The nutrient status of algal cells in continuous culture. *J. Mar. Biol. Assoc. U.K.* **54,** 825–855.

Dunn, B. (1962). "Hydrology of the Irondequoit Creek Basin," Rochester, New York. Open File Rep., p. 40. U.S. Geol. Survey and New York State Dept. of Health.

Dunst, R. C., Born, S. M., Uttermark, P. D., Smith, S. A., Nichols, S. A., Peterson, J. O., Knauer, D. R., Serns, S. L., Winter, D. R., and Wirth, T. L. (1974). "Survey of Lake Rehabilitation Techniques and Experiences," Tech. Bull. No. 75, p. 179. Wisconsin Department of Natural Resources, Madison.

Edmondson, W. T. (1961). Changes in Lake Washington following an increase in the nutrient income. *Verh. Int. Ver. Theor. Angew. Limnol.* **14,** 167–175.

Edmondson, W. T. (1970). Phosphorus, nitrogen, and algae in Lake Washington after diversion of sewage. *Science* **169,** 690–691.

Edmondson, W. T. (1972). Nutrients and phytoplankton in Lake Washington. *Am. Soc. Limnol. Oceanogr., Spec. Symp.* **1,** 172–188.

Ellis, R. H., Haines, T. A., and Makarewicz, J. C., (1976). Aquatic biological survey, Irondequoit Bay, Monroe County, New York. Report to the U.S. Army Engineers, District Buffalo.

Emerson, R. E., and Green, L. (1938). Effect of hydrogen ion concentration on Chlorella photosynthesis. *Plant Physiol.* **13,** 157–168.

Eppley, R. W., and Renger, R. H. (1974). Nitrogen assimilation of an oceanic diatom in nitrogen-limited continuous culture. *J. Phycol.* **10,** 15–23.

Eppley, R. W. and Thomas, W. H. (1969). Comparison of half-saturation constants for growth and nitrate uptake of marine phytoplankton. *J. Phycol.* **5,** 375–379.

Eppley, R. W., Renger, E. H., Venrick, E. L., and Mullin, M. M. (1973). A study of plankton dynamics and nutrient cycling in the central gyre of the North Pacific Ocean. *Limnol. Oceanogr.* **18,** 534–551.

Faigenbaum, H. M. (1940). Chemical investigation of the Lake Ontario watershed. *In* "A Biological Survey of the Lake Ontario Watershed," Suppl. to 29th Annu. Rep., pp. 117–146. New York State Dept. of Conservation, Albany.

Fairchild, H. L. (1894). The geological history of Rochester, New York. *Proc. Rochester Acad. Sci.* **2,** 215–223.

Fairchild, H. L. (1895). Glacial lakes of western New York. *Geol. Soc. Am. Bull.* **6,** 353–374.

Fairchild, H. L. (1896a). Kame area in western New York south of Irondequoit and Sodus Bays. *J. Geol.* **4**(2).

Fairchild, H. L. (1896b). Glacial Genesee lakes. *Geol. Soc. Am. Bull.* **8,** 423–452.

Fairchild, H. L. (1906). The geology of Irondequoit Bay. *Proc. Rochester Acad. Sci.* **3,** 236–239 (abstr.).

Fairchild, H. L. (1909). Glacial waters in central New York. *N.Y. State Mus., Bull.* **127.**

Fairchild, H. L. (1913). Pleistocene geology of New York State. *Science* **37,** 237–249 and 290–299. Also *Geol. Soc. Am. Bull.* **24,** 133–162.

Fairchild, H. L. (1919). The Rochester canyon and the Genesee River base-levels. *Proc. Rochester Acad. Sci.* **6,** 1–55.

Fairchild, H. L. (1928). "Geologic Story of the Genesee Valley and Western New York." Rochester, New York.

Fairchild, H. L. (1930). Artesian water in the Genesee valley. *The Rochester Engineer.* **8,** 236–243.

Faichild, H. L. (1932). Closing stage of New York glacial history. *Geol. Soc. Am. Bull.* **43,** 603–626.

Fairchild, H. L. (1930). Artesian water in the Genesee valley. *The Rochester Engineer.* **8,** 236–243.

Federal Water Pollution Control Administration. (1955). Data on file at Rochester Field Office, EPA.

Federal Water Pollution Control Administration. (1965). Data on file at Rochester Field Office, EPA.

Federal Water Pollution Control Administration. (1968). Data on file at Rochester Field Office, EPA.

Forest, H. S., Maxwell, T. F., and Doby, D. C. (1973). "Environmental Studies of Irondequoit Bay, Monroe County, New York," Contrib. No. 28. Environ. Resour. Cent., New York State University College, Geneseo. (A report for the Monroe County Environmental Management Council.)

Forest, H. S. (1978). Submerged aquatic vascular plants in northern glacial lakes. *Folia Geobot. Phytotaxon.* (in press).

Forest, H. S., Mills, E. L., and Maxwell, T. F. (1975). The phytoplankton and macrophyte communities of the Finger Lakes. (Unpublished manuscript.)

Gehris, C., Stoss, F. W., and Robb, A. E., Jr. (1974). Possible biological impacts of dredging the existing channel from Irondequoit Bay to Lake Ontario in Rochester. *N.Y. Sea Grant Rep.* **NYSSGP-RS-74-017,** pp. 31.

Genesee Finger Lakes Regional Planning Board. (1969). "Economic Analysis Regional Summary and Monroe County Profile, Rep. No. 2, p. 124.

Genesee Finger Lakes Regional Planning Board. (1970). "Land Use Regional Inventory and Analysis," Rep. No. 9, p. 57.

Gerloff, G. C. (1969). Evaluating nutrient supplies for the growth of aquatic plants in natural waters. *In* "Eutrophication: Causes, Consequences, Correctives," pp. 537–555. Natl. Acad. Sci., Washington, D.C.

Gerloff, G. C., and Krombholz, P. H. (1966). Tissue analysis as a measure of nutrient availability for the growth of angiosperm aquatic plants. *Limnol. Oceanogr.* **11,** 529–537.

Gittelman, S., and Buchanan, C. (1971). A Survey of the Fish of Irondequoit Bay, Bull. No. 130, p. 6. Rochester Comm. Sci. Inf., Rochester, New York.

Goldman, C. R. (1974). Eutrophication of Lake Tahoe emphasizing water quality. *Ecol. Res. Ser.* **EPA-660/3-74-034,** pp. 408.

Goldman, C. R. (1975). "Trophic Status and Nutrient Loading for Lake Tahoe, California-Nevada." Report prepared by Tahoe Research Group for NSF-RANN GI-22.

Goldman, J. C., Oswald, W. J., and Jenkins, D. (1974). The kinetics of inorganic carbon limited algal growth. *J. Water Pollut. Control Fed.* **46,** 554–574.

Greeley, J. R. (1940). Fishes of the area with annotated list. *In* "A Biological Survey of the Lake Ontario Watershed," Suppl. to 29th Annu. Rep., pp. 42–81. Albany, New York State Dept. of Conservation.

Grossman, I. G., and Yarger, L. B. (1953). Water resources of the Rochester area, New York. *U.S. Geol. Surv. Circ.* **246,** p. 30.

Haney, J. F. (1973). An in situ examination of the grazing activities of natural zooplankton communities. *Arch. Hydrobiol.* **72,** 87–132.

Harbison, K. G. (1972). Data and records in his possession.

Harbison, K. G. (1973). "Environmental Investigations of Irondequoit Bay," p. 14. Report to Monroe County Environmental Management Council.

Healey, J., Nelson, R., and Stewart, R. (1975). "Solar Energy for New York State," ASRC-SUNY Publ. No. 365. *Atmos. Sci. Res. Cent.,* SUNY, Albany. (A report for the New York State Legislature.)

Hennigan, R. D. (1970). "Reconnaissance Report Environmental Study Irondequoit Bay Pure Waters District," p. 117. Report to Division of Pure Waters, Monroe County.

Holmes, E. N. (1975). Fisheries biologist, New York State Dept. of Environmental Conservation, Avon, New York (personal communication).

Hubbard, J. E. (1975). Dept. of Earth Sciences, State University College at Brockport (personal communication). [Hubbard's solar irradiance data are included in Healey *et al.* (1975).]

King, D. L., and Novak, J. T. (1974). The kinetics of inorganic carbon-limited algal growth. *J. Water Pollut. Control Fed.* **46,** 1812–1816.

Kuichling, E. (1889). "Report on the proposed Trunk Sewer for the East Side of the City of Rochester, N.Y., made by Emil Kuichling, Civil Engineer, to the Common Council." April 29, 1889. Rochester, New York.

Kuichling, E. (1907). "Report on the disposal of the sewage of the city of Rochester, N.Y. by E. Kuichling, consulting engineer. Presented to the Mayor, February 1907." The Morrison Press, Rochester, New York. pp. 43.

Leggette, R. M., Gould, L. O., and Dollen, B. H. (1935). "Ground Water Resources of Monroe County, New York," p. 141. Monroe County Regional Planning Board, Rochester, New York (in cooperation with U.S. Geological Survey).

Lozier Engineers, Inc. (1967). "Appendix Report of the Comprehensive Sewerage Study—East Area Monroe County, New York," WPC-CS-42, p. 617. Rochester, New York.

Marine Resources, Inc. (1970). "Irondequoit Bay Study for the Monroe County Department of Public Works Pure Waters Agency." Mar. Resour., Inc., New York.

Megard, R. O. (1972). Phytoplankton, photosynthesis, and phosphorus in Lake Minnetonka, Minnesota. *Limnol. Oceanogr.* **17,** 68–87.

Metcalf and Eddy, Engineers, Boston, Mass. (1929). "Report to Harold W. Baker, Commissioner of Public Works upon Sewage Disposal Problem, Rochester, New York." December 20, 1929. p. 371.

Monroe County Department of Planning. (1973). "Irondequoit Bay Plan," p. 100. Prepared in cooperation with the Irondequoit Bay Policy Committee.

Monroe County Department of Planning. (1975). "Proposed Monroe County Comprehensive Plan," p. 102. Rochester, New York.

Monroe County Pure Waters Agency. (1969). "Pure Waters Master Plan Report," p. 146. County of Monroe, New York.

Myers, J. and Graham, J. (1971). The photosynthetic unit in Chlorella measured by repetitive short flashes. *Plant Physiol.* **48,** 282–286.

Nalewajko, C. (1966). Dry weight, ash, and volume data for some freshwater phytoplanktonic algae. *J. Fish. Res. Board Can.* **23,** 1285–1288.

New York State Department of Conservation. (1940). "A Biological Survey of the Lake Ontario Watershed." Suppl. to 29th Annu. Rep., p. 261. NYSDC, Albany.

New York State Department of Health. (1955). "Irondequoit Bay Drainage Basin," Lake Ontario Drainage Basin Surv. Ser. Rep. No. 2, p. 73. NYSDH, Albany.

New York State Department of Health. (1964). "Water Pollution Study Portion of Irondequoit Creek Drainage Basin, 1962. Lake Ontario Drainage Basin," Spec. Surv. Rep., p. 157. NYSDH, Albany.

New York State Museum and Science Service. (1970). "Geological Map of New York. Finger Lakes Sheet Map and Chart Ser.," No. 15.

O'Brien and Gere Engineers Inc. (1975). 1304 Buckley Road, Syracuse, New York. (Unpublished data provided by C. B. Murphy and G. Welter.)

O'Brien and Gere Engineers, Inc. (1976a). Genesee River Water Quality Investigations. [A report to Joint Venture, c/o Lozier Engineers, Rochester, submitted on April 15, 1976. Appendix IV-C of the report (pp. 114–122) presents estimates of Rochester interceptor discharges to Irondequoit Bay tributaries.]

O'Brien and Gere Engineers, Inc. (1976b). Memorandum from D. J. Carleo and G. J. Welter to C. B. Murphy, dated 20 Dec. 1976.

Odell, T. T. (1940). Bays and ponds of the shore area. *In* "A Biological Survey of the Lake Ontario Watershed," New York State Department of Conservation, Supplement to 29th Annual Report, pp. 82–97. Albany.

Peters, R. H. and Rigler, F. H. (1973). Phosphorus release by Daphnia. *Limnol. Oceanogr.* **18,** 821–839.

Rabinowitch, E. (1951). "Photosynthesis and Related Processes," Vol. II. Wiley (Interscience), New York.

Rhee, G-Y. (1973). A continuous culture study of phosphate uptake, growth rate and polyphosphate in Scenedesmus sp. *J. Phycol.* **9,** 495–506.

Rhee, G-Y. (1974). Phosphate uptake under nitrate limitation by Scenedesmus sp. and its ecological implications. *J. Phycol.* **10,** 470–475.

Rochester Committee for Scientific Information. (1964). "Report on Water Pollution," Rep. No. 1. RCSI, Rochester, New York.

Rochester Committee for Scientific Information. (1965). "Third Report on Water Pollution," Rep. No. 3. RCSI, Rochester, New York.

Rochester Committee for Scientific Information. (1965). "Hearings on Irondequoit Creek," Rep. No. 4. RCSI, Rochester, New York.

Rochester Committee for Scientific Information. (1966). "Water Pollution in Monroe County," Rep. No. 12. RCSI, Rochester, New York.

Rochester Committee for Scientific Information. (1967a). "Sewage Pollution of Oatka Creek and Lower Irondequoit Creek," Rep. No. 22. RCSI, Rochester, New York.

Rochester Committee for Scientific Information. (1967b). "Sewage Pollution of Densmore and Thompson Creeks: A Follow-Up Report," Rep. No. 25. RCSI, Rochester, New York.

Rochester Committee for Scientific Information. (1967c). "Pollution of Densmore Creek," Rep. No. 26. RCSI, Rochester, New York.

Rochester Committee for Scientific Information. (1967d). "Improved Conditions of the Irondequoit Creek Watershed," Rep. No. 30. RCSI, Rochester, New York.

Rochester Committee for Scientific Information. (1967e). "Continued Pollution of Thompson Creek, Densmore Creek and Slater Creek with Undisinfected Sewage," Rep. No. 33. RCSI, Rochester, New York.

Rochester Committee for Scientific Information. (1967f). Dissolved Oxygen in Monroe County Waters. Lower Irondequoit Creek," Rep. No. 34. RCSI, Rochester, New York.

Rochester Committee for Scientific Information. (1967g). "Sewage Pollution of Irondequoit Creek in East Rochester," Rep. No. 36. RCSI, Rochester, New York.

Rochester Committee for Scientific Information. (1967h). "Pollution of Irondequoit Bay with Diesel Fuel or Fuel Oil," Rep. No. 38. RCSI, Rochester, New York.

Rochester Committee for Scientific Information. (1968a). "Fecal Pollution of Densmore Creek and Streams in the Town of Greece," Rep. No. 41. RCSI, Rochester, New York.

Rochester Committee for Scientific Information. (1968b). "Dissolved Oxygen Levels in Irondequoit Bay," Rep. No. 45. RCSI, Rochester, New York.

Rochester Committee for Scientific Information. (1968c). "Landfills Threaten Irondequoit Bay," Rep. No. 84. RCSI, Rochester, New York.

Rochester Committee for Scientific Information. (1969). "Phosphate in Irondequoit Creek and Tributaries, 1968–1969, "Rep. No. 61. RCSI, Rochester, New York.

Rochester Committee for Scientific Information. (1970a). "Water Temperature and Dissolved Oxygen in Irondequoit Bay," Rep. No. 67. RCSI, Rochester, New York.

Rochester Committee for Scientific Information. (1970b). "Use of Irondequoit Creek by William Stappenback, Inc., "Rep. No. 101. RCSI, Rochester, New York.

Rochester Committee for Scientific Information. (1971a). "Nutrient Studies of Irondequoit Creek," Rep. No. 123. RCSI, Rochester, New York.

Rochester Committee for Scientific Information. (1971b). "Run-off of Deicing Salt: Effect on Irondequoit Bay, Rochester, New York," Rep. No. 125. RCSI, Rochester, New York.

Rochester Committee for Scientific Information. (1971c). "A Survey of the Fish of Irondequoit Creek," Rep. No. 130. RCSI, Rochester, New York.

Rochester Committee for Scientific Information. (1972). "Environmental Research in the Irondequoit Bay Watershed," Rep. No. 137. RCSI, Rochester, New York.

Sakamoto, M. (1966). Primary production by phytoplankton community in some Japanese lakes and its dependence on lake depth. *Arch. Hydrobiol.* **62,** 1–28.

Sassaman, W. (1970). "How Deep was the Outlet to Irondequoit Bay." Monroe County Planning Council. Rochester, New York.

Schindler, D. W., and Fee, E. J. (1973). Diurnal variation of dissolved inorganic carbon and its use in estimating primary production and CO_2 invasion in Lake 227. *J. Fish. Res. Board Can.* **30,** 1501–1510.

Schindler, D. W., and Fee, E. J. (1974a). Primary production in freshwater, *In* "Structure, Functioning, and Management of Ecosystems," pp. 155–157. Cent. Agric. Publ. Doc., Wageningen, Netherlands.

Schindler, D. W., and Fee, E. J. (1974b). Experimental Lakes Area: Whole lake experiments in eutrophication. *J. Fish. Res. Board Can.* **31,** 937–953.

Schindler, D. W., Kling, H., Schmidt, R. V., Prokopowich, J., Frost, V. E., Reid, R. A., and Capel, M. (1973). Eutrophication of Lake 227 by addition of phosphate and nitrate: the second, third, and fourth years of enrichment, 1970, 1971, and 1972. *J. Fish. Res. Board Can.* **30,** 1415–1440.

Slovacek, R. E., and Bannister, T. T. (1973). Effects of carbon dioxide concentration on oxygen evolution and fluorescence transients in synchronous cultures of Chlorella pyrenoidosa. *Biochim. Biophys. Acta* **292,** 729–740.

Smayda, T. J. (1970). The suspension and sinking of phytoplankton in the sea. *Oceanogr. Mar. Biol.* **8,** 353–414.

Soeder, C. J., Muller, H., Payer, H. D., and Schulle, H. (1971). Mineral nutrition of planktonic algae: Some considerations, some experiments. *Mitt. Int. Ver. Limnol.* **19,** 39–58.

Soil Survey Monroe County, New York. (1973). U.S. Department of Agriculture, Soil Conservation Service, in cooperation with Cornell University Agricultural Experiment. Station, p. 172.

Steeman Nielsen, E. (1955). CO_2 as carbon-source and narcotic in photosynthesis and growth of Chlorella pyrenoidosa. *Physiol. Plant.* **8,** 317–335.

Stone, V. B. (1937). Growth, habits, and fecundity of the ciscoes of Irondequoit Bay, New York. *Trans. Am. Fish. Soc.* **67,** 234–245.

Sze, P., and Kingsbury, J. M. (1972). Distribution of phytoplankton in a polluted saline lake, Onondaga Lake, New York. *J. Phycol.* **8,** 25–37.

Thompson, K. W. (1956). The story of Irondequoit Bay. Pamphlet, p. 8. Publisher and place of publication not given.

Thorne, J. (1973). "The Irondequoit Creek System: A Drainage Basin Before Sewage Diversion." University of Rochester, Rochester, New York. [Final report of a project of the Student Originated Studies Division, National Science Foundation, carried out under supervision of Prof. Kenneth Harbison.]

Tressler, W. L., and Austin, T. S. (1940). A limnological study of some bays and lakes of the Lake Ontario watershed. *In* "A Biological Survey of the Lake Ontario Watershed," Suppl. to 29th Ann. Rep., pp. 188–210. New York State Dept. of Conservation, Albany.

Tressler, W. L., Austin, T. S., and Orban, E. (1953). Seasonal variations of some limnological factors in Irondequoit Bay, New York. *Am. Mid. Nat.* **49,** 878–903.

U.S. Army Corps of Engineers. (1957). "Report of the Chief of Engineers, Dept. of the Army, on Irondequoit Bay, New York. February 9, 1956." US Govt. Printing Office, Washington, D.C.

U.S. Army Corps of Engineers. (1968). "Chart of Proposed Irondequoit Bay Outlet Modification. 30 June 1968." Buffalo, New York.

U.S. Army Corps of Engineers. (1969). "Dredging and Water Quality in the Great Lakes," Vols. 1–12. Buffalo District.

U.S. Army Corps of Engineers. (1971). Amendment: Conditions of improvement of Irondequoit Bay.

U.S. Department of the Interior, Geological Survey. (1968). "Water Resources Data for New York State, Part 1: Surface Water Records." [See also volumes for 1961–1967 under same title.]

U.S. Department of the Interior, Geological Survey. (1974). "Water Resources Data for New York State, Part 2: Water Quality Records." [Additional data on nitrogen and phosphorus content of precipitation at Mays Point, N.Y., were obtained in the 1971, 1972, and 1973 volumes of the same title.]

Vollenweider, R. A. (1968). "Scientific Fundamentals of the Eutrophication of Lakes and Flowing Waters," Tech. Rep. DAS/CSI 68.27, p. 193. Organ. Econ. Coop. Dev., Paris.

Vollenweider, R. A. (1975). "Input-Output Models," *Schweiz. Z. Hydrologic* **37,** 53–84.

Welch, E. B., Rock, C. A., and Kroll, J. D. (1973). Long-term lake recovery related to available phosporus. *In* Modeling the Eutrophication Process" (E. J. Middlebrooks, D. H. Falkenborg, and T. E. Maloney, eds.), pp. 5–14. Ann Arbor Sci. Publ., Ann Arbor, Michigan.

Welch, E. B., Hendrey, G. R., and Stoll, R. K. (1975). Nutrient supply and the production and biomass of algae in four Washington lakes. *Oikos* **26,** 47–54.

Westman, J. R. and Fahy, W. E. (1940). The carp problem of the area. *In* "A Biological Survey of the Lake Ontario Watershed," New York State. Suppl. to 29th Ann. Rep., Dept. of Conservation, pp. 226–231. Albany.

Whipple, G. C. (1913). Irondequoit Bay. *In* "Report on the Sewage Disposal System of

Rochester, New York" (E. A. Fisher, City Engineer) Appendix No. 5, p. 198–200, 216–217, 231–236.

Whittingham, C. P. (1952). Rate of photosynthesis and concentration of carbon dioxide in Chlorella. *Nature* (*London*) **170,** 1017–1018.

Witten, A. (1971). "Seiching in Irondequoit Bay." Dept. of Geology, Univ. of Rochester, (Manuscript in possession of R. C. Bubeck.)

Onondaga Lake

Cornelius B. Murphy, Jr.

ISBN 0-12-107302-5

INTRODUCTION

Location

Onondaga Lake (20.5 m maximum depth) is a highly eutrophic water body situated at the northern edge of the city of Syracuse in central New York State. Although the lake is only 11.7 km² (4.5 square miles) in area, the drainage basin is 600 km² (240 square miles), contains approximately 325,000 people, and essentially all of Onondaga County's 140 industries. The lake flows from southeast to northwest and ultimately discharges to the Seneca River. The confluence of the Seneca and Oneida Rivers forms the Oswego River which discharges at the southeastern shore of Lake Ontario, 64 km (40 miles) north of Onondaga Lake. A map locating Onondaga Lake relative to significant defining features and political subdivisions is shown in Fig. 1.

History

The earliest recording of Onondaga Lake and its salt springs was by a Jesuit Missionary, Father Simon LeMoyne. Father LeMoyne's account of the lake, which at that time was called Ganentaka by the Onondaga Indians, was:

> We arrive at the entrance of a small lake in a large half dried basin; we taste the water of a spring that they (the Indians) durst not drink, saying that there is a demon in it, which render it fetid. Having tasted it, I found it a fountain of salt water; and, in fact, we made salt from it as natural as that from the sea, of which we carried a sample to Quebec. (Geddes, 1860)

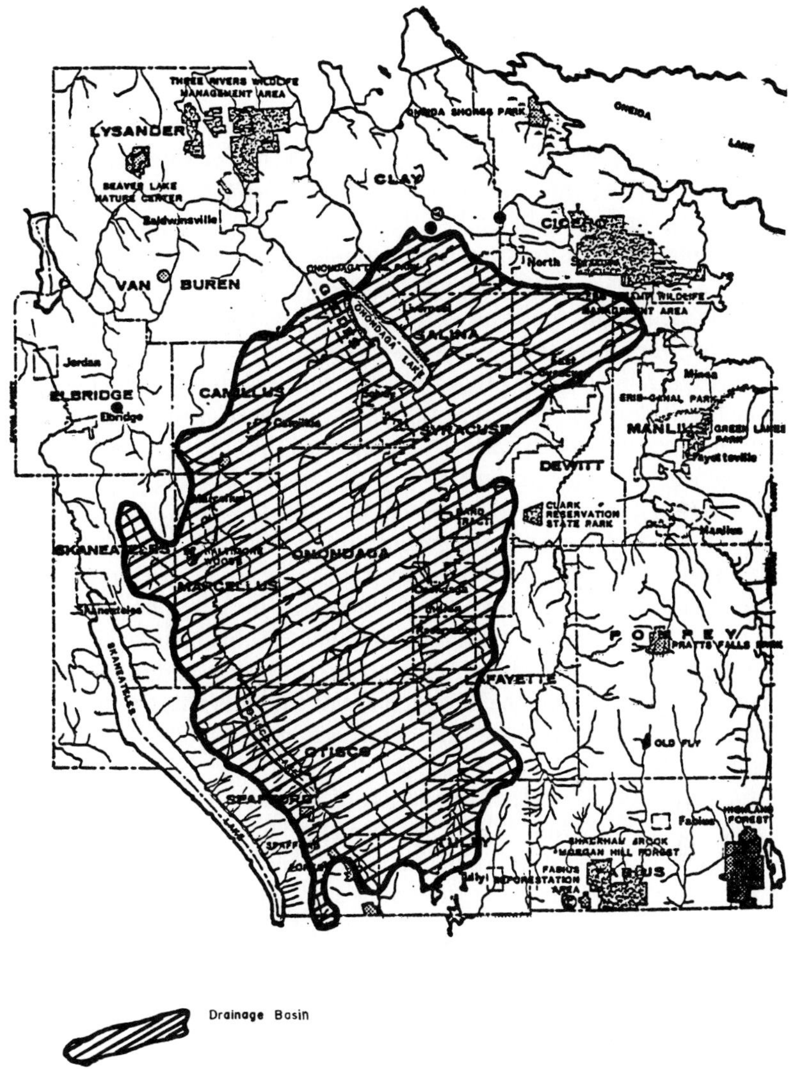

Fig. 1. Onondaga Lake drainage basin.

Geddes (1860) and Clark (1849) provided detailed histories concerning the gradual utilization of salt by the Indians, the white man's influence in the development of a salt industry on the shores of the lake and the eventual takeover of the lands around the lake to the exclusion of the Indians. The latter author also relates some interesting features of the shoreline of Onondaga Lake:

> The shores of the Onondaga Lake, at an early period of the settlement of the County, were composed of soft, spongy bog, into which a pole could be thrust to an almost interminable depth.

Schultz (1810) described the local belief that the lake was bottomless and that the lower waters were extremely saline. Schultz attempted to find an area where the bottom could not be found. His efforts resulted in a measured maximum depth of 19.5 m (63 ft). Using a cork and bottle, he also attempted to get a representative sample of the bottom waters to check the local hypothesis as to its highly saline nature. Schultz "found the water a little cooler, but not otherwise different from that on the surface."

The most important event in the lake's early history involved the lowering of the lake level in 1822. At that time, the canal commissioners cut a channel which was designed to permit the waters of the lake to reach the evaluation of the Seneca River. Clark (1849) reported that the lowering of the lake waters laid bare a wide surface of the salt marsh. By studying the map of Sweet (1874), it has been determined that the lowering of the lake level resulted in a 20% decrease in the surface area of the lake.

The initial industrial development within the vicinity of Onondaga Lake involved the utilization of salt and brine deposits located along the southern shore of the lake. After the initial development of this resource in 1812, the area quickly became the major salt supplier in the nation during the 1800's. It has been proposed that the development of the salt industry on the shores of Onondaga Lake prompted the construction of the Erie Canal system (Wright, 1969).

In the late 1800's a wider distribution of industry began to locate within the confines of the Onondaga Lake drainage basin, manufacturing such products as soda ash, steel, vehicular accessories, and pottery. During the first four decades of the 1900's the area surrounding the lake witnessed a development of major pharmaceutical, air conditioning, general appliance, and electrical manufacturing facilities. Through this period of very significant industrial growth, the metropolitan area of Syracuse increased in population from approximately 110,000 in 1900 to the present population of greater than 500,000.

Previous Studies

A significant number of previous studies have been conducted on Onondaga Lake and its tributaries. The first significant work is contained in a report prepared by Metcalf and Eddy (1920) to the Syracuse Intercepting Sewer Board. The report recommended that the then existing discharge of sewage from the city of Syracuse to the Barge Canal at the mouth of

Onondaga Creek be discharged instead directly into Onondaga Lake. The report also recommended the installation of facilities to remove solids by screening and settling.

In addition to the presentation of dissolved oxygen measurements, estimates of the rate of oxygen depletion in the lake waters were presented. The first relatively detailed hydrographic map of Onondaga Lake was presented by Metcalf and Eddy (1920). The maximum soundings found in the northwest basin ranged from 10.2 to 20.1 m (33.6 to 66 ft) and a corresponding range from 12.2 to 22.8 m (40 to 75 ft) in the southeastern basin.

A minimal amount of data accumulated on the lake and the discharges to the lake is contained within the annual reports of the Syracuse Intercepting Sewer Board. The sixteenth report (Holmes, 1922) contains results of dissolved oxygen measurements made at five to six locations along the lake's longitudinal axis following implementation of the recommendations presented by Metcalf and Eddy (1920).

Onondaga Lake and its basin received little attention after the 1920 work until the reports of Agar and Sanderson (1947), Pitts (1949), and Burdick and Lipschuetz (1947). The New York State Health Department and New York State Conservation Department reports, Agar and Sanderson (1947), and Burdick and Lipschuetz (1947) present data on the fish populations as well as on the water chemistry and coliform levels. The dissolved oxygen data contained in the above-mentioned reports again indicate significant oxygen depletion, a condition which is still observed.

In 1951 a survey was conducted by the New York State Department of Health for the purpose of obtaining data necessary for the reclassification of waters in the Onondaga Lake drainage basin. The survey was carried out between August 8, 1950 and February 7, 1951 and summarized in the New York State Department of Health report (1951). The 1951 survey and drainage basin report represented the most complete description of the lake and its tributaries developed up to that point in time. In addition to defining conditions of tributaries and sources of pollutants in the Onondaga Lake drainage basin, the report also presents primary data collected on the tributaries as well as a synopsis of data drawn from the Agar and Sanderson (1947) and Burdick and Lipschuetz (1947) reports.

Subsequent to the New York State Department of Health report, a large volume of data has been collected and a number of reports written regarding Onondaga Lake. O'Brien & Gere Engineers, Inc. (1952) developed a number of alternatives for adequate sewage treatment for the Syracuse area. Topics considered were: population growth, definition of the city of Syracuse municipal sewerage system, the dissolved oxygen, biochemical oxygen demand (BOD), and coliform counts measured in Onondaga Lake, a dissolved oxygen balance for the lake, as well as general lake development. The data on dissolved oxygen and sulfide concentrations indicate that anaerobic conditions of the lower waters existed from 1946 through 1952.

The Onondaga County Department of Public Works has been responsible for the collection of a wide range of chemical and microbiological information since 1959. The information collected by Onondaga County from 1959–1970 represents data collected at ten stations on Onondaga Lake at the surface and at a depth of 6.1 m (20 ft). The data show that most parameters measured remained reasonably constant from 1959 to 1970.

A number of engineering studies have been conducted relative to the condition and recommended remedial measures for Onondaga Lake following the original O'Brien & Gere report. In an engineering study, Weber (1958) analyzed those factors affecting the concentration of dissolved oxygen in Onondaga Lake. Weber determined the theoretical detention time for the lake to be 180 days for the 1946 water year. Additionally, Weber concluded that reaeration through the surface of the lake was the most important process supplying dissolved oxygen to the lake waters.

The most widely acknowledged publication concerning Onondaga lake prior to 1970 was that published by Berg (1963). Berg highlighted the very significant distinction between Onondaga Lake and other lakes in New York State, its unique ionic composition, and its attendant high level of total dissolved solids.

The Syracuse University Research Corporation (SURC) presented a number of conclusions and new data relative to Onondaga Lake in 1966 following its investigation conducted during the previous two years. The SURC investigation involved a characterization of the chemistry and phytoplankton composition of the lake. It was concluded that the chemical characteristics of the lake in the time frame of 1964–1965 was almost identical to that presented by O'Brien & Gere (1952). SURC also stressed that the algal population of the lake had not changed appreciably since 1962 and that the sodium–calcium ratio of the lake was approaching that characteristic of Ninemile Creek prior to confluence with Onondaga Lake. Of particular importance, the SURC study also presents bioassay experiments which indicate the inhibition of algal growth with increasing salinity.

In an unpublished report, Brennan *et al.* (1968) dealt largely with the possibilities associated with the restoration of Onondaga Lake. Consideration was given to the treatment of saline wastes discharged to the lake as well as the impact of existing benthic deposits on the quality of the lake waters. Brennan *et al.* (1968) also presented the observation of the existence of *Enteromorpha,* a salt-water form of algae, in Onondaga Lake on 1965 and 1966. A retention time of the lake was calculated to be 270 days with an attendant volume of 1.51×10^8 m^3 (40×10^9 gallons).

The Allied Chemical Corporation (Anonymous, 1968) has collected some of the more interesting data concerning Onondaga lake. Available data include information on water chemistry, thermal profiles, and biological measurements. The most important limnological data are the concentrations of dissolved oxygen and hydrogen sulfide measured in the lower waters of Onondaga Lake. A 1945 memorandum from the Allied Chemical Corporation reports that hydrogen sulfide was present in the mill water as far back as August 22, 1910. The Allied Chemical Corporation's data suggest the depletion of dissolved oxygen in the lower waters in the mid to late summer extending as far back as the beginning of the century.

Jackson (1968) presented a brief geographical, morphometric, and historical review of Onondaga lake and the results of the bioassay stimulation tests conducted utilizing 25 algal cultures. He concluded on the basis of the results of the bioassay stimulation tests that the lake could be more eutrophic (and unaesthetic) if it was not so saline and polluted.

In a second publication, Jackson (1969) described attempts to measure the primary productivity of phytoplankton in Onondaga lake. The presented photosynthetic rates (expressed as microliters of oxygen per hour per mg/liter ash-free dry weight) always exceeded the respiration rate. He reported an algae succession which included the green algae, *Chlamydomonas* and *Scenedesmus,* dominant in May, June, and July with *Chlorella,* dominant in August. The diatoms, *Cyclotella* and *Stephanodiscus,* were reported as dominant in September.

The most complete limnological investigation of Onondaga Lake was conducted by Onondaga County (1971) with the assistance of O'Brien & Gere Engineers, Inc., under a grant provided by the Water Quality Office of the Environmental Protection Agency. The baseline study was conducted under the technical direction of an advisory committee composed of the following individuals: Dr. John Forney (Ichthyology), Dr. John M. Kingsbury (Phycology), Dr. Richard Noble (Ichthyology), Dr. Philip Sze (Phycology), Dr. George Waterman (Zooplankton), Dr. Myrton C. Rand (Chemistry), Dr. Kenton M. Stewart (Limnology and Coordination of Advisors), and Dr. Jeffrey C. Sutherland (Geochemistry).

The baseline study, Onondaga County (1971), involved a broad range of physical, chemical, and biological investigations which were conducted to determine the trophic status of Onondaga Lake. It is largely from the materials developed in the process of this comprehensive baseline evaluation that an understanding of the limnology of Onondaga Lake evolved.

Following the completion of the baseline investigations, a monitoring program was instituted and funded by Onondaga County. A yearly analysis of compiled data has been published in the form of annual monitoring reports (O'Brien & Gere Engineers, Inc., 1972a,b, 1973a,b, 1975). The monitoring program involves the bimonthly analysis of up to 34 parameters on the lake water column at the deepest location in the southern basin and the major tributaries to the lake. The baseline and monitoring programs have fostered the preparation of a number of technical papers dealing with the program development and the findings of the baseline and monitoring programs.

DESCRIPTION OF THE DRAINAGE BASIN

Geology, Morphology, and Soils

The Onondaga Lake drainage basin lies partially within the Ontario Lowlands province and partially within the Appalachian Plateau. The lake drainage basin is situated on beds of Silurian and Upper Devonian sedimentary rock, including shale, siltstone, limestone, and gypsum. In general, the rock units dip gently southward beneath sequentially younger strata.

A unique characteristic of the Ontario Lowlands is the presence of a drumlin belt between Rochester and Syracuse; the Onondaga Lake drainage basin lies in the southeastern portion of this drumlin belt (Cressy, 1966). Drumlins have been described as being half-egg-shaped and having steeply sloped glacial features. Figures 2 and 3 show the significant geological and topographical features of the drainage basin.

Due to the steep grades characteristic of and in the upstream areas of drumlins, large amounts of soil are carried from the upland areas during intense rainstorms. The eroded material consists of the unconsolidated sediments and other materials washed into the streams by overland runoff.

Much of the soil lying on the upland areas is composed of unconsolidated sediment mixed with glacial till. In the flatter areas of the drainage basin, the soil is composed largely of sand, with a small percentage of clay and silt.

The land bordering the southern shore of Onondaga Lake consists of three distinct soil layers. The uppermost layer is largely fill material consisting of sand, silt, brick, ashes, and cinders. Below the fill material is a layer of chemical residues that dates back to a time when the Allied Chemical Corporation used the area for disposal of process wastes. The chemical residue is principally calcium carbonate, calcium chloride, and calcium oxide. The third and lowest layer consists of organic material, gray sand, and brown clay silt.

Climate

The climate of the Onondaga lake drainage basin is continental in character and comparatively humid (National Oceanic and Atmospheric

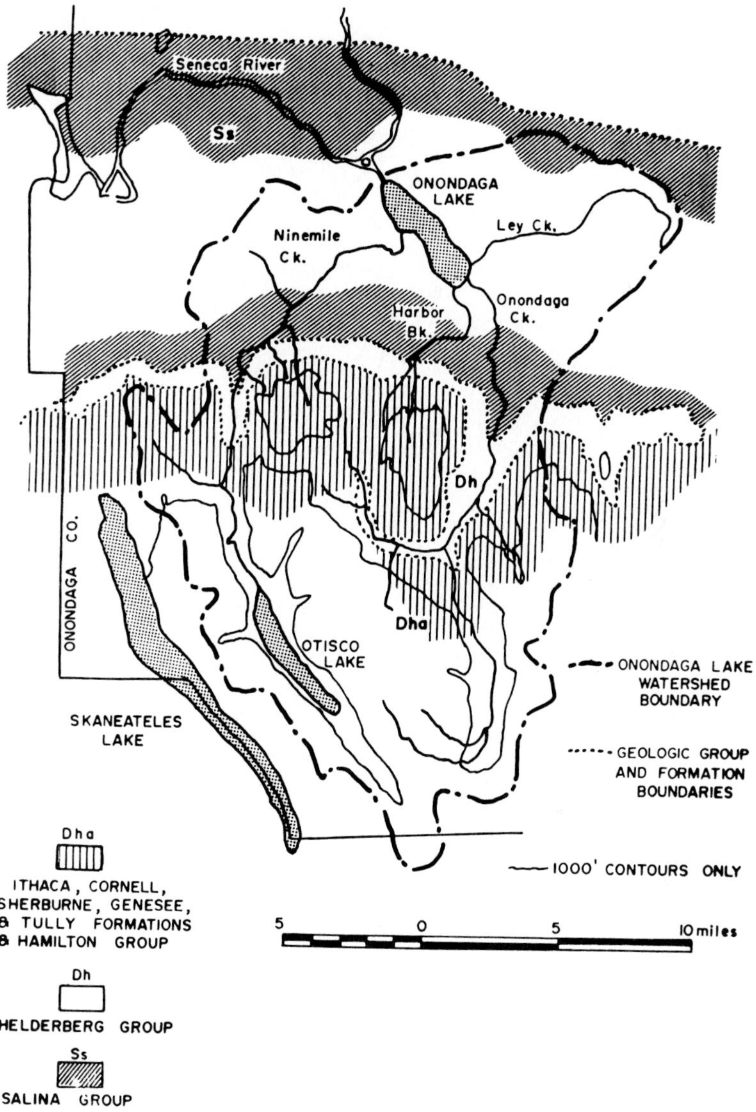

Fig. 2. Geologic map—Onondaga Lake basin. From Onondaga Lake Study (1971).

Administration, 1974). Fairly rapid changes in weather occur throughout the year since cyclonic systems moving through the interior of the country often follow the St. Lawrence Valley, and thereby affect the Onondaga Lake drainage basin.

Precipitation in the Syracuse area generally averages about 7.6 cm (3 in.) per month throughout the year with an annual average of 92.4 cm (36.33

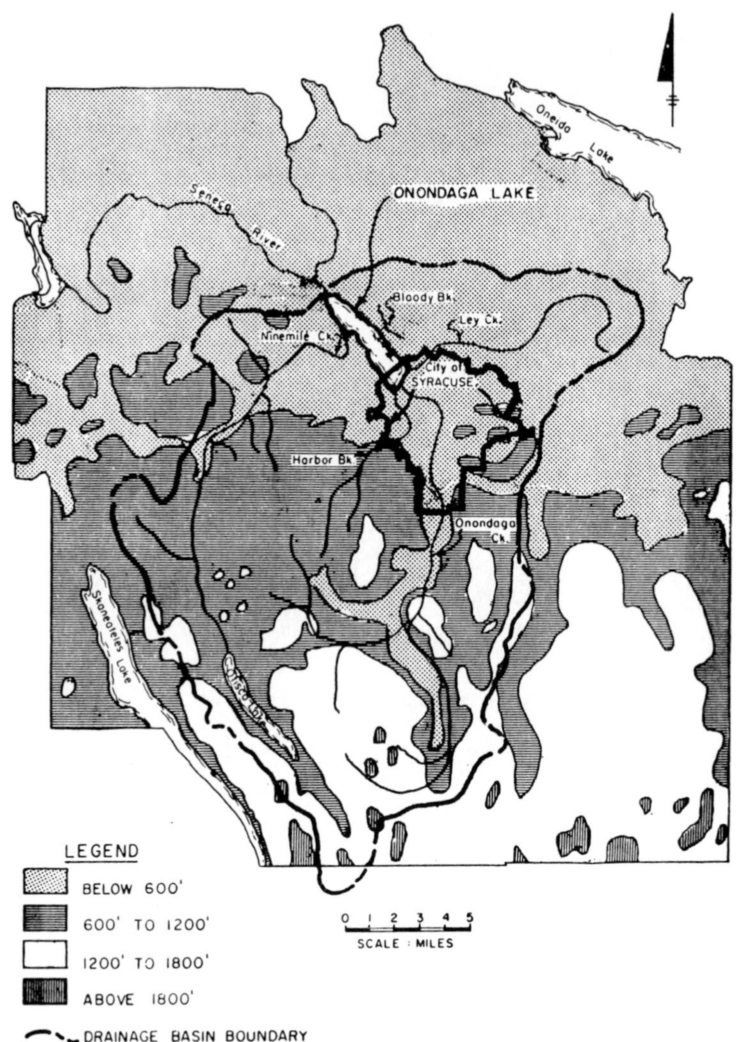

Fig. 3. Topographic map—Onondaga Lake basin. From Onondaga Lake Study (1971).

in.). Snowfall is moderately heavy with an annual average of just over 250 cm (100 in.). Monthly average precipitation values as recorded at the Syracuse, New York, Hancock International Airport for the years 1935 to 1974 are shown in Table 1.

The precipitation received in the Onondaga lake drainage basin is derived principally from cyclonic storms which travel the continent and pass through the St. Lawrence Valley. A significant source of winter precipitation is Lake Ontario. Snow squalls and attendant cloudiness, characteristic of winter, are frequently derived from saturated cold air sweeping off Lake Ontario.

The annual mean temperature for the years 1935 to 1974 is 8.7°C (47.7°F). As expected, the winters are cold and severe, with daytime winter temperatures generally averaging about 2.3°C (35°F) and nighttime lows of about −7.8°C (18°F). The lowest temperature ever recorded was 32.2°C (26°F). Table 2 shows the monthly average temperatures recorded for 1935 to 1974.

Other miscellaneous climatic data encountered in the Onondaga Lake drainage basin are outlined in Table 3. The wind over the lake is primarily out of the west-northwest at an average velocity of 15.8 km/hr (9.8 mph). The relative humidity averages approximately 80% and the average percentage of possible sunshine being approximately 40–50%.

Population Distribution

The major population centers in the drainage basin are the villages of Camillus, Marcellus, Otisco, LaFayette, Liverpool, and East Syracuse as well as the city of Syracuse. All or portions of the following areas are included in the Onondaga Lake drainage basin: the townships of Marcellus, Camillus, Otisco, Tully, LaFayette, Onondaga, Geddes, and Salina, the village of East Syracuse, the Onondaga Indian Reservation, and the city of Syracuse. The 1970 drainage basin population was approximately 318,000. The population breakdown by political subdivision is shown on Table 4.

Population projections for a number of municipalities have been made by the Syracuse–Onondaga County Planning Agency (1972). These sources indicate that the populations of each of the towns in the Onondaga Lake drainage basin are projected to increase slightly. In contrast, the projected population of Syracuse is expected to stabilize. This has been borne out in the recent past by comparing populations determined in 1960 and 1970. The Onondaga Lake drainage basin showed an increase in population of approximately 0.52% while Syracuse showed a 8.7% decrease in population over the same period.

Land Usage

The present population of the Onondaga Lake drainage basin is scattered throughout the basin, centered primarily on the fringes of industrial and commercial development. The most intensive industrially and commercially developed areas are located in the proximity of Onondaga Lake, in the city

TABLE 1

Monthly Precipitation (Inches)[a,b] Syracuse, New York, Hancock Airport (1935–1974)

Year	Jan.	Feb.	Mar.	Apr.	May	Jun.	Jul.	Aug.	Sep.	Oct.	Nov.	Dec.	Annual
1935	3.46	2.25	2.63	3.51	2.48	4.38	6.14	2.59	2.86	2.73	2.24	3.79	39.06
1936	3.18	1.90	5.97	2.85	1.63	0.57	0.67	2.44	3.01	3.44	3.80	1.89	31.35
1937	3.68	3.14	2.68	3.51	3.23	6.59	2.85	5.83	1.39	6.83	2.66	2.74	45.13
1938	2.01	2.56	2.02	3.09	4.36	2.56	4.02	5.05	4.90	0.56	2.74	2.72	36.59
1939	2.74	2.86	2.87	2.59	0.55	2.21	3.08	1.40	1.58	2.79	1.16	3.15	26.98
1940	1.72	5.22	5.42	3.98	4.08	3.27	2.32	1.23	1.91	1.89	3.32	3.78	38.14
1941	2.25	1.98	2.85	1.97	2.90	2.96	5.41	2.81	2.28	3.63	1.93	2.96	33.93
1942	1.59	3.55	4.52	2.43	3.71	2.09	2.72	2.40	3.32	2.99	3.44	6.55	39.31
1943	2.09	2.59	2.70	4.63	3.88	4.63	1.91	7.26	0.51	5.44	2.93	1.25	39.84
1944	1.75	2.55	2.90	3.65	3.19	5.89	1.43	2.65	2.18	1.92	4.23	5.00	37.34
1945	4.26	2.56	3.06	3.50	3.47	3.52	2.55	2.02	6.74	6.52	4.83	0.85	43.88
1946	2.58	2.72	1.52	2.14	2.43	3.85	3.24	2.61	2.83	4.78	2.20	3.14	34.04
1947	4.23	2.12	4.18	3.20	5.54	3.94	5.41	3.70	2.14	0.82	3.15	2.32	40.75
1948	1.93	2.19	2.63	4.10	3.87	3.89	2.05	2.53	1.02	1.87	4.34	3.28	33.70
1949	3.25	2.04	2.34	3.60	2.37	3.08	3.10	3.92	4.38	1.36	2.32	2.32	34.08
1950	4.20	4.12	5.15	1.57	2.43	2.48	2.05	2.52	4.22	3.41	3.91	3.59	39.65
1951	2.73	5.38	4.65	3.79	1.97	4.21	5.22	3.33	3.17	1.86	3.76	4.34	44.41
1952	2.45	2.00	2.39	2.57	3.88	1.18	4.36	2.05	1.52	2.88	2.36	3.75	31.39
1953	2.91	2.71	3.75	2.07	4.35	1.51	1.93	2.39	2.11	1.64	1.74	2.61	29.72
1954	2.43	3.74	3.48	4.75	3.17	3.66	1.12	7.19	3.06	1.99	4.84	4.03	43.46

1955	1.61	3.32	6.84	1.81	1.90	2.43	2.85	3.80	1.87	8.29	2.69	2.53	39.94
1956	2.28	3.90	4.63	3.35	3.03	1.71	3.75	8.41	4.27	1.28	2.38	3.02	42.01
1957	2.19	1.76	2.01	2.60	2.88	4.18	6.13	3.45	2.28	0.93	1.86	2.90	33.17
1958	4.46	5.28	1.31	3.36	3.70	5.24	3.63	2.06	4.89	3.37	3.76	1.73	42.79
1959	4.59	2.22	2.93	2.51	1.97	2.53	2.54	4.48	0.93	7.15	4.34	5.01	41.20
1960	3.11	4.90	2.48	2.94	3.96	1.86	1.03	2.69	2.93	2.77	1.68	1.86	32.21
1961	2.30	4.14	4.22	3.74	2.40	3.68	5.08	1.78	1.21	3.59	2.99	2.45	37.58
1962	2.87	2.96	1.96	3.57	1.05	1.10	2.74	4.63	1.99	3.30	2.22	2.25	30.64
1963	1.85	2.05	2.79	2.22	2.84	2.49	1.21	3.59	0.85	0.21	5.65	2.06	27.81
1964	2.18	1.13	3.83	3.66	2.31	1.41	2.15	3.09	0.75	1.52	2.20	2.87	27.10
1965	2.28	2.82	1.63	3.53	1.61	2.04	1.34	1.95	3.60	2.70	2.97	1.92	28.39
1966	3.98	2.96	2.27	3.05	1.79	2.73	2.09	2.64	4.75	0.90	2.05	3.93	33.14
1967	1.47	1.49	1.34	2.11	3.33	1.56	6.33	5.00	2.73	3.52	4.48	2.66	36.02
1968	2.08	1.10	3.13	2.40	3.46	6.14	3.77	4.17	3.43	5.81	4.07	4.67	44.23
1969	3.37	1.49	1.08	3.95	4.34	3.74	0.90	1.77	1.13	2.30	4.56	3.42	32.05
1970	1.02	1.84	2.45	3.68	2.79	2.93	4.42	4.07	4.33	3.84	3.53	3.33	38.23
1971	1.90	4.07	2.90	2.19	3.40	3.26	6.49	4.01	2.56	1.62	3.52	3.26	39.18
1972	1.10	2.87	2.49	4.03	6.19	12.30	3.45	3.76	4.12	4.36	6.79	3.95	55.41
1973	1.85	1.71	3.45	6.91	5.58	7.07	3.62	2.97	4.57	3.81	6.73	4.38	52.65
1974	2.08	1.70	4.34	3.09	5.78	4.67	9.52	4.60	4.45	1.58	4.95	3.47	50.23
Record mean	2.69	2.60	3.11	3.05	3.00	3.57	3.39	3.27	2.84	2.97	2.93	2.91	36.33

[a] Source: U.S. Department of Commerce (1974).
[b] 2.54 cm = 1 in.

TABLE 2

Monthly Temperatures (°F)[a,b] Syracuse, New York, Hancock Airport (1935–1974)

Year	Jan.	Feb.	Mar.	Apr.	May	Jun.	Jul.	Aug.	Sep.	Oct.	Nov.	Dec.	Annual
1935	21.5	24.5	37.6	44.4	52.6	66.2	75.4	71.0	60.2	52.6	42.2	23.3	47.6
1936	23.3	19.2	39.0	43.5	61.0	68.2	73.0	71.4	64.2	53.2	36.2	33.2	48.8
1937	33.8	29.8	28.6	45.5	59.4	67.2	73.4	75.4	61.4	50.4	41.5	28.9	49.6
1938	25.0	29.6	40.2	48.8	57.3	68.2	74.0	74.2	58.9	54.2	43.0	31.4	50.4
1939	25.6	28.2	31.2	43.1	60.8	68.4	72.6	74.2	64.6	52.0	36.8	30.4	49.0
1940	17.2	24.4	27.2	42.0	58.4	66.1	71.3	69.4	60.0	47.3	39.2	30.3	46.1
1941	22.6	21.7	26.3	50.6	57.8	68.7	73.0	67.0	63.6	52.7	43.3	31.0	48.2
1942	24.2	20.4	37.3	50.1	60.0	67.4	70.8	68.6	62.0	52.1	39.8	23.6	48.0
1943	19.2	25.6	32.0	38.5	56.8	70.6	72.3	68.6	60.0	48.6	37.2	24.8	46.2
1944	26.0	24.5	28.8	41.2	63.2	66.6	72.2	72.2	62.4	49.6	40.6	23.8	47.6
1945	14.4	26.3	44.4	50.8	52.6	64.8	70.6	69.4	64.7	49.9	40.5	22.6	47.6
1946	25.4	23.1	43.9	44.4	55.9	65.0	69.6	65.7	63.4	54.9	44.5	30.6	48.9
1947	27.9	20.4	30.8	44.7	55.5	65.6	71.2	74.2	63.4	57.8	37.2	25.8	47.9
1948	17.6	21.3	33.0	48.6	54.2	65.8	71.9	70.0	63.0	48.8	46.7	31.2	47.7
1949	29.4	30.6	35.4	46.8	58.4	73.2	74.6	72.8	60.0	56.2	38.0	31.6	50.6
1950	33.2	22.4	27.1	42.5	57.9	65.7	70.3	69.2	58.8	53.3	40.9	24.8	47.2
1951	26.7	25.4	35.1	46.5	58.0	65.5	71.1	67.6	61.6	53.1	35.2	30.2	48.0
1952	26.8	27.5	33.2	49.5	54.2	67.3	74.7	69.9	63.8	48.2	42.5	31.8	49.1
1953	30.4	29.5	36.8	45.7	58.4	67.7	72.2	69.7	63.6	53.0	43.9	34.9	50.5
1954	20.4	33.0	33.3	48.1	56.2	68.4	70.9	67.7	61.9	55.4	40.9	27.6	48.7
1955	21.2	25.5	33.5	51.4	60.4	68.3	76.7	74.6	61.7	54.0	38.4	21.7	48.9

1956	22.6	27.0	28.5	42.9	53.4	67.0	68.2	69.2	58.8	52.7	41.2	33.3	47.1
1957	18.8	30.4	35.1	48.4	56.0	70.4	70.4	67.1	62.6	50.1	42.0	34.2	48.8
1958	21.4	19.5	34.3	48.3	54.5	62.5	70.5	69.1	61.9	50.4	41.9	19.4	46.1
1959	21.0	21.8	31.1	47.6	60.0	67.6	73.5	74.1	67.3	52.3	38.6	31.1	48.8
1960	24.7	27.0	24.4	49.5	59.8	66.6	69.4	68.7	63.8	50.0	43.8	23.7	47.6
1961	18.8	25.9	33.2	42.9	55.5	66.6	71.8	70.4	69.5	55.8	40.7	29.1	48.3
1962	24.1	21.5	34.6	47.3	62.3	68.3	69.2	69.1	59.7	51.2	35.5	24.5	47.3
1963	20.8	18.4	34.2	45.6	54.7	66.6	71.7	66.1	57.2	56.3	44.9	20.8	46.4
1964	25.9	23.7	35.1	46.5	61.6	66.0	73.2	67.5	61.6	49.5	43.6	29.6	48.6
1965	20.5	24.6	30.3	42.3	59.4	63.7	67.5	69.1	62.8	47.9	38.7	32.1	46.6
1966	19.0	23.1	34.4	42.6	51.2	65.7	71.1	70.4	59.5	49.5	42.9	29.2	46.6
1967	30.6	19.2	31.6	44.9	50.2	69.5	67.7	66.6	60.7	51.8	37.6	32.7	46.9
1968	18.3	21.1	33.2	48.1	53.6	64.9	69.9	68.8	64.8	52.5	40.0	27.2	46.9
1969	24.3	23.6	30.4	46.9	55.7	64.5	69.9	71.3	63.8	51.1	40.4	23.8	47.1
1970	16.1	24.1	31.7	47.2	57.5	63.5	69.7	68.0	61.4	52.3	41.7	25.4	46.5
1971	18.5	26.5	31.2	42.8	55.8	67.9	69.0	67.1	65.6	56.6	36.9	33.2	47.6
1972	26.4	22.9	29.4	40.5	58.5	64.5	72.9	69.2	63.5	46.5	37.0	30.8	46.9
1973	28.4	21.4	42.6	46.8	54.3	69.6	72.7	73.5	62.0	53.7	40.7	29.5	49.6
1974	26.0	21.6	32.3	48.8	54.1	65.6	69.1	68.9	59.1	46.5	40.6	30.4	46.9
Record mean	24.2	23.9	33.3	45.4	56.8	66.3	71.2	69.2	62.4	51.6	40.1	28.2	47.7
Maxi-mum	31.9	31.8	41.2	54.5	66.7	76.2	81.0	78.9	71.9	60.6	47.0	34.9	56.4
Minimum	16.4	16.0	25.4	36.2	46.8	56.3	61.3	59.5	52.8	42.5	33.1	21.5	39.0

[a] Source: U.S. Department of Commerce (1974).
[b] °C = 0.556 (°F − 32).

TABLE 3

Miscellaneous Weather Data Normals, Means, and Extremes for Syracuse,

	Relative humidity (%)				Wind						
							Fastest mile				Mean sky cover, tenths, sunrise to sunset
Month	1 AM[d]	7 AM	1 PM	7 PM	Mean speed (mph	Prevailing direction	Speed (mph)	Direction	Year	% Of possible sunshine	
(a)	11	11	11	11	25	14	25	25		25	25
Jan.	75	75	68	73	10.8	WSW	60	W	1974	35	7.9
Feb.	75	77	66	72	11.3	WNW	62	W	1967	40	7.8
Mar.	77	78	62	69	11.1	WNW	56	SE	1956	46	7.5
Apr.	75	76	52	59	10.9	WNW	52	NW	1957	51	6.8
May	78	75	56	60	9.6	WNW	50	NW	1964	56	6.6
Jun.	82	78	56	63	8.6	WNW	49	NW	1961	62	6.1
Jul.	83	79	56	63	8.4	WNW	47	NW	1951	66	5.7
Aug.	85	84	57	68	8.2	WSW	43	NW	1958	63	5.9
Sep.	84	86	61	75	8.6	S	52	W	1962	57	6.0
Oct.	82	84	60	75	9.2	WSW	63	SE	1954	46	6.5
Nov.	81	82	69	77	10.4	WSW	59	E	1950	25	8.3
Dec.	81	82	74	79	10.5	WSW	52	W	1962	25	8.4
Year	80	80	61	70	9.8	WNW	63	SE	Oct. 1954	50	7.0

[a] Source: U.S. Department of Commerce (1974). Data taken at Hancock Airport.
[b] 1.609 km = 1 mile.
[c] °C = 0.556 (°F − 32).
[d] Local time.

of Syracuse along Erie Boulevard and Genesee Street, along interstates 690 and 81, and around the village of East Syacuse as shown on Fig. 4. The southern area of the Onondaga Lake drainage basin is the most sparsely settled and is primarily agricultural, with fairly intensive truck farming that support nearby population centers, particularly Syracuse. Figure 5 shows the viable farm areas as provided by the Onondaga County Environmental Management Council. The most intense residential population centers lie in the areas in the northern part of the basin, particularly concentrated in the towns of Camillus, Geddes, Solvay, Salina, and Dewitt. In all the areas, residential zoning of a quarter of an acre (0.405 hectare) or less is common.

In the Onondaga lake drainage basin there are two significant water bodies, Onondaga and Otisco Lakes, which presently support a limited amount of water related recreation. Onondaga Lake certainly has the potential to be a significant asset to the population in the drainage basin. Faro and Nemerow (1969) reported that almost 46% of the shoreline around

New York[a,b,c]

Mean number of days											
Sunrise to sunset							Temperature (°F)				
							Max.		Min.		
Clear	Partly cloudy	Cloudy	Precipitation (0.01 in. or more)	Snow, ice pellets (1.0 in. or more)	Thunderstorms	Heavy fog (visibility ¼ mile or less)	(b) 90° and above	32° and below	32° and below	0° and below	Average station pressure mb. (elevation 408 ft. m.s.l.)
25	25	25	25	25	25	25	11	11	11	11	2
3	7	21	19	8	*	1	0	16	28	5	1003.2
3	6	19	16	8	*	1	0	15	26	3	1003.1
5	6	20	17	5	1	1	0	7	24	*	1000.5
6	7	17	14	1	2	1	0	*	13	0	998.7
6	10	15	13	*	3	1	*	0	1	0	997.8
8	10	12	10	0	6	1	1	0	0	0	999.4
8	13	10	11	0	7	1	3	0	0	0	1000.4
8	11	12	10	0	6	1	2	0	0	0	1002.4
8	10	12	10	0	3	1	1	0	*	0	1002.9
7	8	16	11	*	1	1	0	0	5	0	1004.9
2	6	22	16	3	1	1	0	2	14	0	1000.7
2	5	24	19	8	*	1	0	12	26	1	1001.9
66	99	200	167	32	29	9	7	51	138	10	1001.3

TABLE 4

Past and Present Population of Selected Political Subdivisions in the Onondaga Lake Basin[a]

	1960	1970
Syracuse (city)	216,038	197,208
Salina (town)	33,076	38,281
Geddes (town)	19,679	21,032
East Syracuse (village)	4,708	4,333
Marcellus (town)	4,527	5,744
Camillus (town)	18,328	26,841
Otisco (town)	1,188	1,470
Onondaga (town)	13,429	16,555
LaFayette (town)	3,379	4,401
Tully (town)	1,633	1,901
Onondaga Indian Reservation	941	785
	316,886	318,551

[a] Source: U.S. Bureau of the Census (1962, 1972).

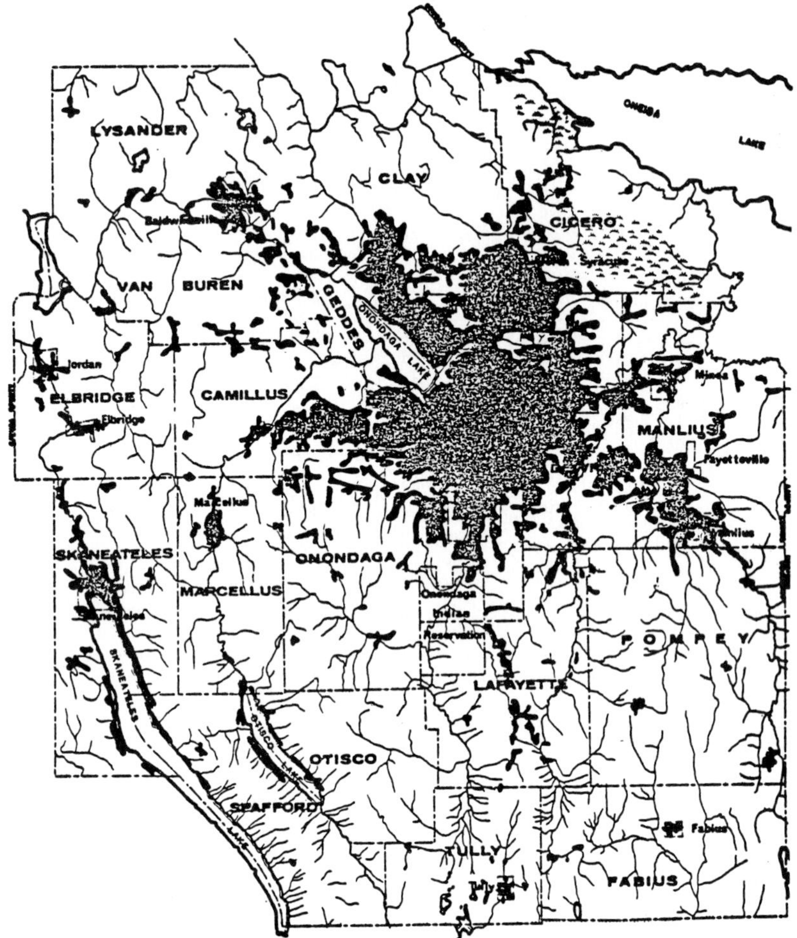

Fig. 4. Existing development. Map courtesy of SOCPA.

the lake is public parkland. Shattuck (1968) estimated that water-related activities drew 384,166 people to the Onondaga Lake County Park during fiscal 1967, making the park the principal recreational asset on the shores of Onondaga Lake. Figures 6 and 7 show the significant natural areas and major parks and recreation areas in the Onondaga Lake Basin.

Tributary Loadings and Point Sources of Pollutants

There are eight significant natural streams and wastewater discharges tributary to Onondaga Lake. Figure 8 shows the watersheds of each of the

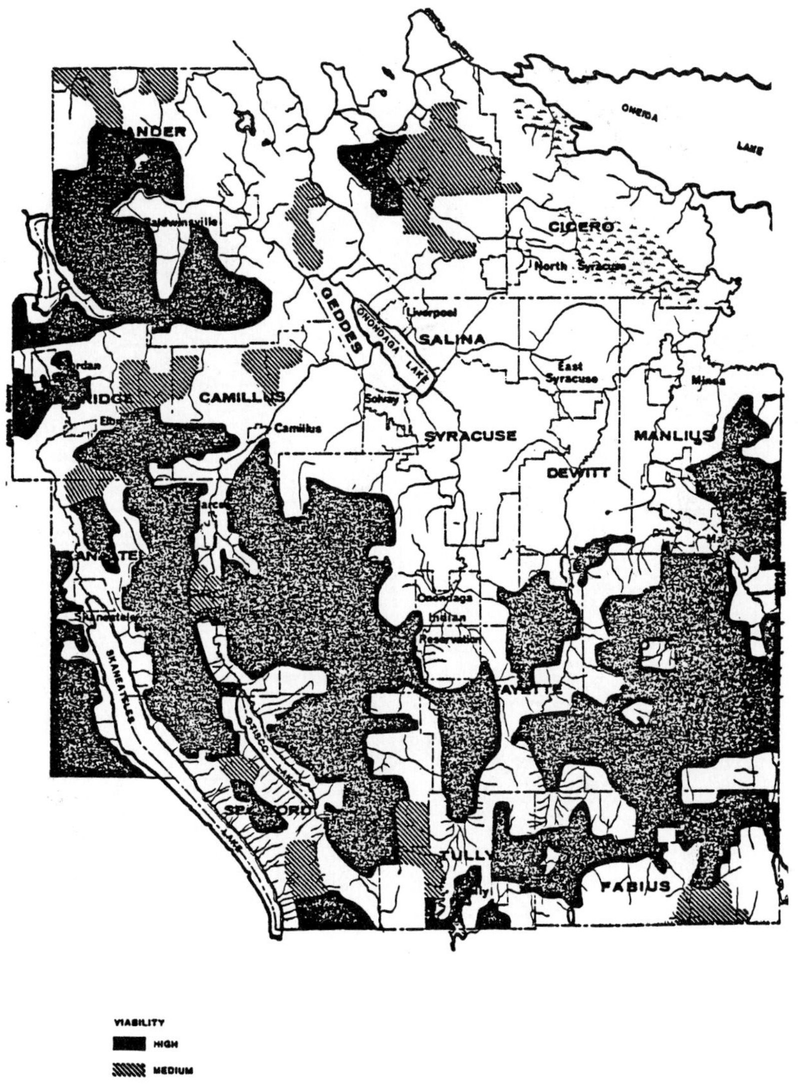

Fig. 5. Viable farm areas. Map courtesy of SOCPA.

tributaries to Onondaga Lake. Most of these tributaries discharge to the southern portion of the lake. The major domestic and industrial wastewaters discharged from the municipal ideas are discharged at the southwestern corner of the lake. A description of the tributaries and their loadings are described as follows.

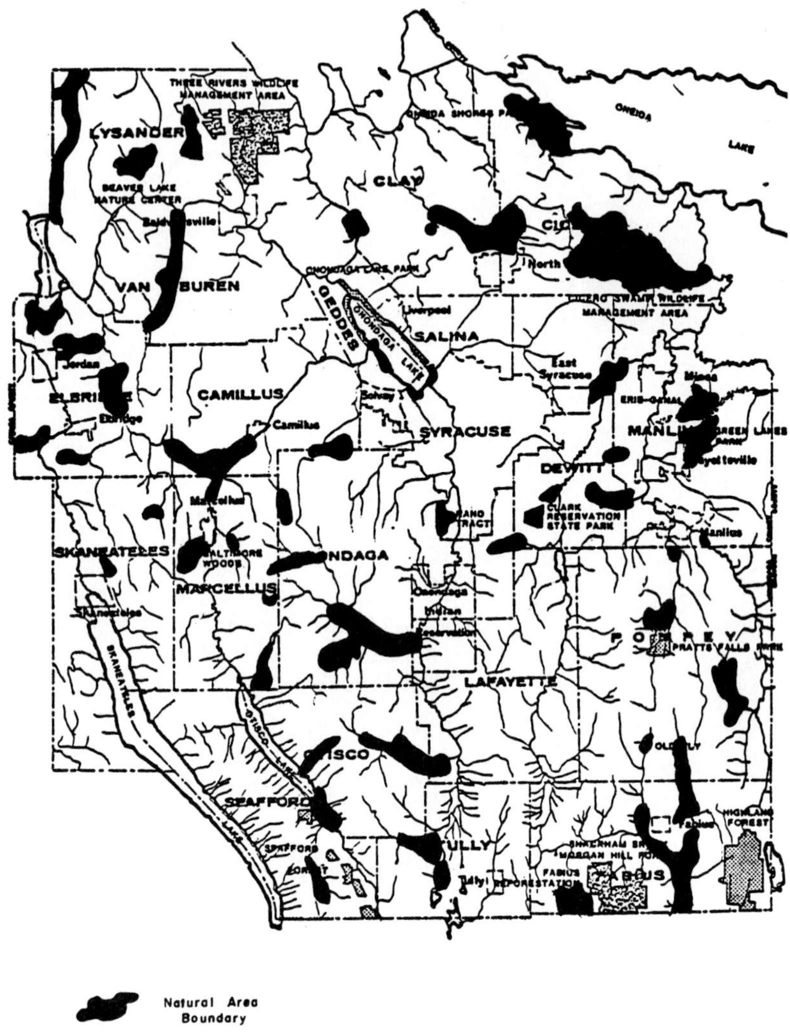

Fig. 6. Significant natural areas. Map courtesy of SOCPA.

Ninemile Creek

The watershed for Ninemile Creek extends to the west and south from the western shore of Onondaga Lake. The watershed area of Ninemile Creek is 323.0 km (124.8 miles) and the total length of the main stream is 44.2 km (27.5 miles). Ninemile Creek originates at the Otisco Lake outlet. For the most part, the watershed is characterized by steep valley walls and high stream gradients. Relatively flat stream gradients are experienced both in

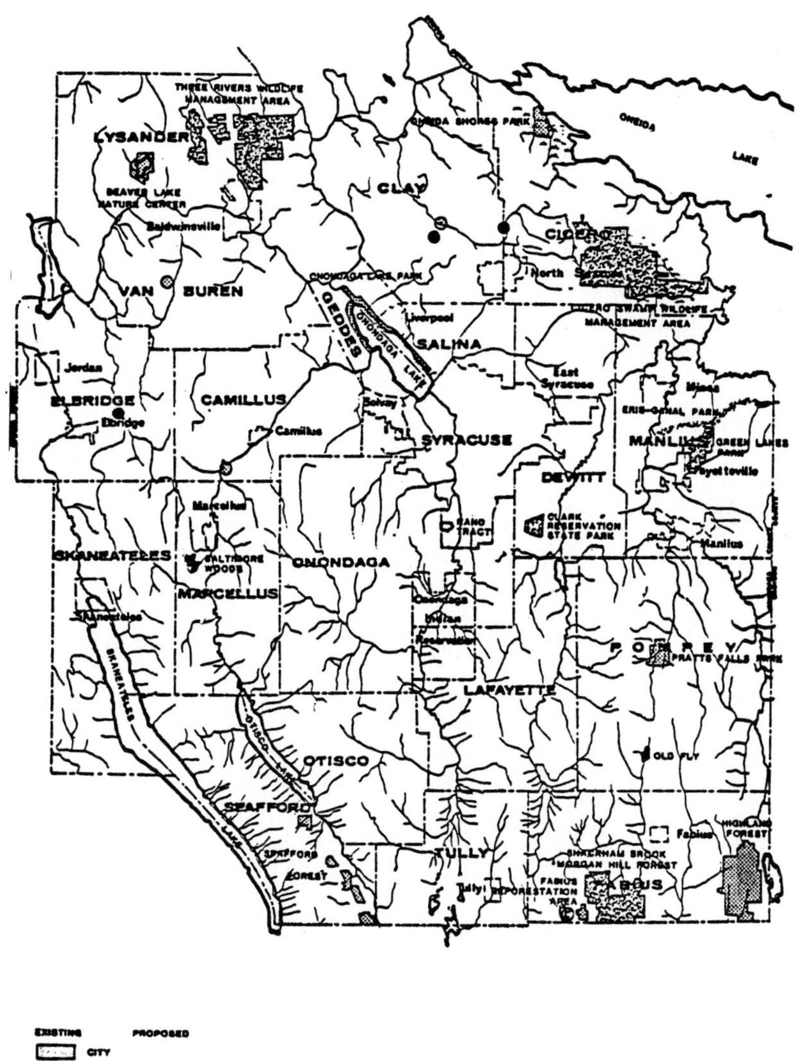

Fig. 7. Major parks and conservation areas. Map courtesy of SOCPA.

the lake plain and the upper plateau regions. The region between the lake plain and the highland plateau is characterized by rapid stream velocities.

The average flow measured on Ninemile Creek at Lakeland is 2.5 m/sec (88.4 cfs) as shown in Table 5. The analytical data illustrating the average constituent concentrations and loadings obtained in the course of the

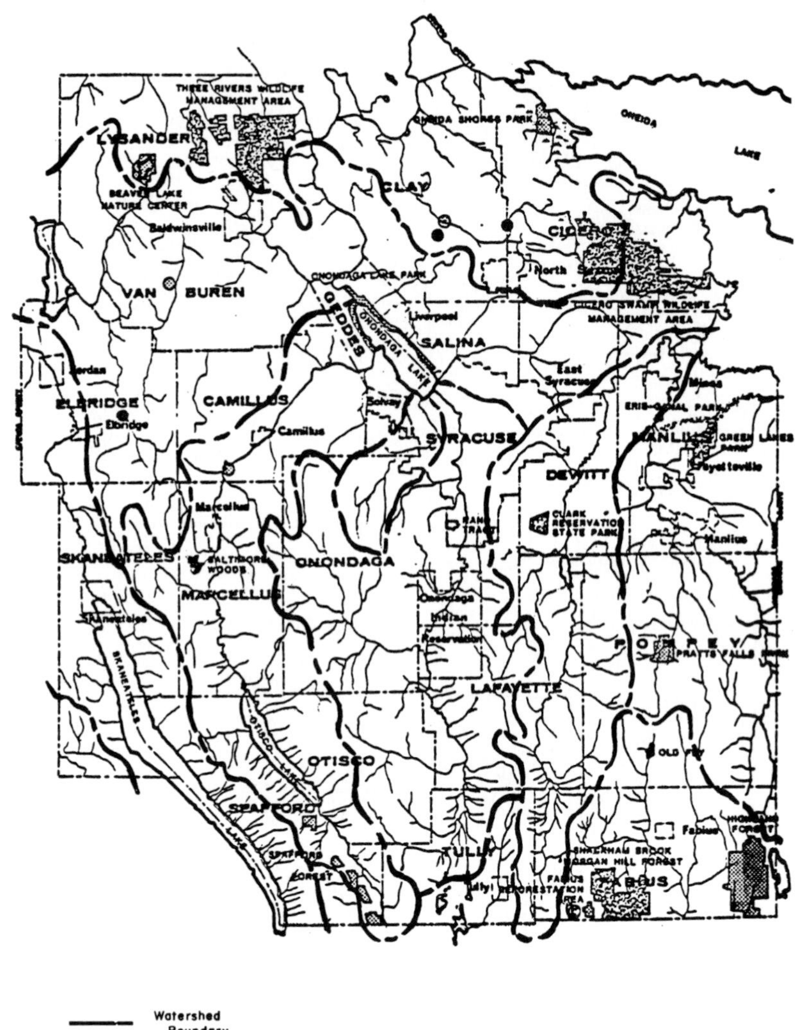

Fig. 8. Major watersheds. Map courtesy of SOCPA.

Onondaga Lake Study and the monitoring programs are shown in Appendix A. Ninemile Creek is the major contributor of most of the inorganic chemical species in Onondaga Lake. This stream discharges the major portion of calcium, chloride, sodium, iron, and potassium. The source of the dissolved inorganic constituents is primarily the supernatant from the Allied Chemical Corporation waste beds.

The principle sources of pollution to Ninemile Creek occur between its confluence with Onondaga Lake and just south of the village of Marcellus.

TABLE 5

Major Tributaries of Onondaga Lake

Tributary	Watershed area[a] (square miles)	Watershed area[a] (km²)	Mainstream length[b] (miles)	Mainstream length[b] (km)	Average flow[a] (cfs)	Average flow[a] (m³/sec)
Ninemile Creek, (including Otisco Lake drainage area)	115.0	298.0	34.3	55.2	88.4[a]	2.5
Onondaga Creek	109.0	282.0	27.5	44.2	213.0	6.03
Ley Creek	29.9	77.4	9.5	15.3	62.7[a]	1.8
Harbor Brook	11.3	29.3	7.5	12.1	18.1	0.5
Bloody Brook	4.5	11.7	2.2	3.5	Negligible	—
Seneca River	—	—	Outlet	—	No gauging station	

[a] From USGS Surface Water Records for New York State (1975).

[b] From USGS, Annual Reports to Onondaga County, Department of Public Works (1965, 1967, 1968).

The principal sources of pollution in addition to the Allied Chemical Corporation include the villages of Camillus and Marcellus and Camillus Cutlery. Pollution abatement programs have been established for most of the above-mentioned sources.

Onondaga Creek

Onondaga Creek has a wastershed area of approximately 298.0 km (115 miles) and a mainstream length of 44.2 km. The Onondaga Creek watershed extends generally to the south from the southern end of Onondaga Lake. The gradient of the main stream is comparatively flat for the length of the stream. The lower reaches of Onondaga Creek drain the lake plain region, including a significant portion of the city of Syracuse. The plateau region of the drainage basin is characterized by short, high-velocity feeder streams and a relatively narrow, steeply sloped valley.

The average measured flow on Onondaga Creek is 6.03 m/sec (213 cfs) as shown in Table 5. The analytical data obtained on Onondaga Creek during the course of the Onondaga Lake study and the monitoring programs are shown in Appendix A. A review of the data shows that Onondaga Creek is a major contributor of magnesium, total phosphorus, organic nitrogen, and orthophosphate. The relatively high levels of phosphate and organic nitrogen can be attributed to the combined sewer overflows from Syracuse, many of which are discharged to Onondaga Creek within the city limits.

Ley Creek

The area of the Ley Creek watershed is approximately 77.4 km^2 (29.9 square miles). This tributary has a main stream length of 15.3 km (9.5 miles). The Ley Creek watershed extends east from Onondaga Lake over the lake plain region. The main stem of the stream and the tributaries to it are characterized by low stream gradients with an attendant sluggish flow. The watershed is primarily residential and industrial in character although the upper reaches of the tributary drain agricultural land.

The average flow measured on Ley Creek upstream from its confluence with Onondaga Lake is approximately 1.8 m^3/sec (62.7 cfs). The tributary is a significant source of both ammonia and organic nitrogen. Ley Creek is also a major contributor of organic materials as assayed by biochemical oxygen demand.

The principal sources of wastes to Ley Creek are combined sewer overflows and treated wastewaters from the Carrier Corporation's Thompson Road plant as well as that from the Crouse–Hinds Company. The presence of additional sources in the form of leachate from sanitary landfills, street runoff, and runoff from the New York Central Railroad also contributes to the load carried by Ley Creek.

Harbor Brook

The Harbor Brook watershed outside of the Syracuse city limits is primarily agricultural land in which small farms and pastureland predominate. Within the city limits the stream is the recipient of wet weather flow from a large number of combined sewer overflows.

The tributary watershed extends to the southwest from the southern end of Onondaga Lake. The watershed has an area of 29.3 km^2 (11.3 square miles), the upper portion of which originates in the plateau or highland area. The watershed is relatively long, 12.1 km (7.5 miles), and narrow. The majority of the watershed lies in the lake plain with approximately the last 6.2 km (3 miles) within the city limits.

Bloody Brook

The Bloody Brook watershed extends to the northeast from approximately the midsection of the east shore of Onondaga Lake. The watershed lies completely within the lake plain and has a relatively small rectangular drainage area of about 11.7 km^2 (4.52 square miles). The total length of the main stream is only 3.5 km (2.2 miles), making it the smallest of the significant natural tributaries discharging to Onondaga Lake. The average annual flow measured for Bloody Brook is 0.5 m^3/sec (17.4 cfs).

The tributary receives no significant pollutant point source with the exception of some treated coolant and process wastewaters from the General Electric Corporation's Electronics Park complex. Bloody Brook does, however, receive a considerable amount of stormwater and nonpoint source loading characteristic of an area having a high population density. The pollutant load to Onondaga Lake contributed by Bloody Brook is relatively insignificant compared to the other tributaries. The only significant load discharged to Onondaga Lake by Bloody Brook during the baseline 1968–1969 Onondaga Lake Study is zinc, contributing 10.0% of the total load to the lake.

Point Sources Tributary to Onondaga Lake. Metropolitan Syracuse Sewage Treatment Plant

The metropolitan Syracuse sewage treatment plant, following the diversion of the Ley Creek treatment plant, has been the most significant contributor of BOD_5, ammonia nitrogen (NH_3–N), organic nitrogen, and total inorganic phosphate to Onondaga Lake. The metropolitan Syracuse sewage treatment plant provides primary treatment to a service district population of 260,854 (U.S. Bureau of the Census, 1972) as well as to the associated industry at an average flow of approximately 29.2×10^3 m^3/day—77.2 million gallons per day (MGD)—(1974). It is anticipated that the 1974 loading figures shown in Appendix A of BOD_5, NH_3–N, organic nitrogen, and total inorganic phosphorus will be significantly reduced by the added secondary and tertiary facilities presently being constructed.

East Flume, Allied Chemical Corporation

The Allied Chemical Corporation's "east flume" is primarily a cooling water discharge. Outside of the obvious thermal load carried by the thermal discharge, which results in a 15.6°C (60°F) average annual temperature differential between the east flume and the south center basin epilimnion temperature, the east flume discharge is a significant contributor of NH_3–N and inorganic salts.

Crucible Steel Discharge

The Crucible Steel wastewater discharge has characteristically been the significant source of iron, chromium, copper and grease and oil. Since wastewater treatment facilities were installed at the end of 1974, the loading of iron to the epilimnion has been reduced by 35%, chromium 47%, and copper 51% (O'Brien & Gere Engineers, Inc., 1975).

Nonpoint Sources of Pollutants

The majority of the tributary drainage areas feeding the Onondaga Lake drainage basin pass through primarily urban or fairly densely populated suburban areas. The contribution of urban runoff in the form of stormwater runoff represents a significant load to Onondaga Lake of fecal coliform, fecal streptococci, lead, cadmium, zinc, oils and grease, and suspended solids. In Syracuse there exists a combined sewer system (a system for conveying both stormwater and sanitary wastewater in a common conveyance system) which, under wet weather conditions, is insufficient to convey the combined wastewater flow. At a large number of points within the city, hydraulic relief is provided to the collection system, allowing excess flow to discharge to three of the tributaries passing through the city: Harbor Brook, Onondaga Creek, and Ley Creek. During overflow conditions the discharge carries contaminants that have been washed from roofs, streets, sidewalks, and other impervious surfaces, along with some sanitary sewage. These overflows seriously affect the water quality of the receiving streams.

The combined sewer overflow has characteristics which range from almost raw sanitary wastewater to urban stormwater. In the first flush from the combined sewer system, the contributions of NH_3 nitrogen, organic nitrogen, total inorganic phosphate, suspended solids, bacteriological indicators, heavy metals, chlorinated organics, and BOD_5 are significant. The city of Syracuse has approximately 100 overflow devices in its sewer system, as shown on Fig. 9. This number includes both direct overflows to the tributaries noted above and intrasystem overflows.

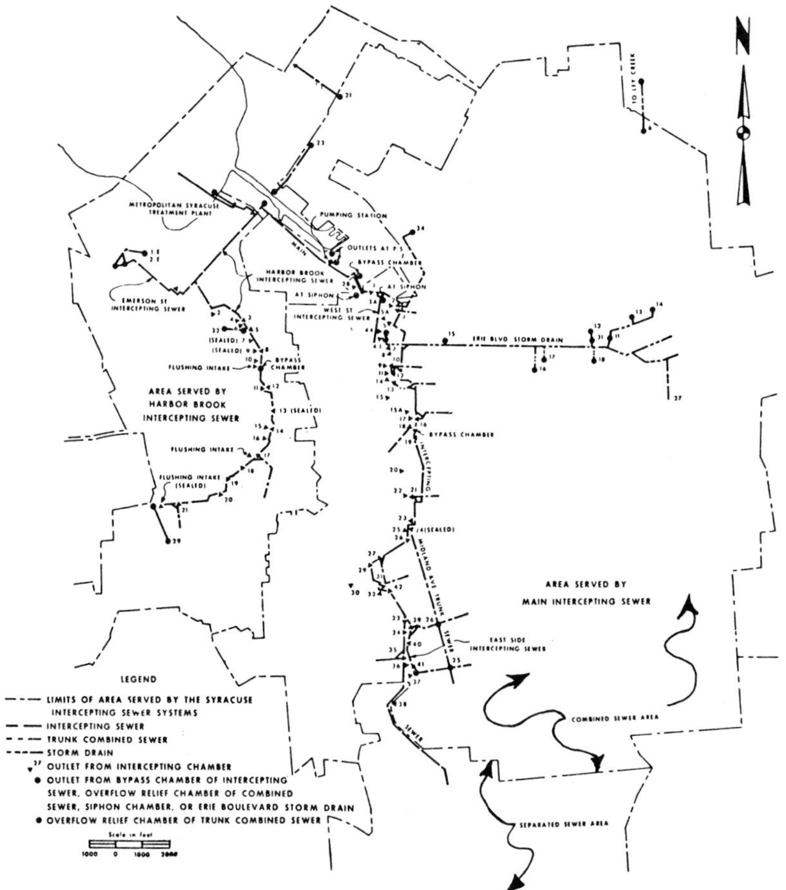

Fig. 9. Relief points on combined Syracuse sewer system. From Camp *et al.* (1968).

O'Brien & Gere Engineers, Inc. (1973a) has estimated that wet weather overflows occur on the average of nine times per month for a total duration of approximately 24 hours per month. Based on data obtained in the course of conducting an EPA combined sewer overflow demonstration program (EPA Project No. 11020 HFR) on a 46.5 hectare (115 acre) drainage area (see Figs. 10, 11, and 12) the order of magnitude of the pollutant load represented by the combined sewer overflow system has been estimated.

Making the rough approximations that the test drainage area is characteristic of the rest of the metropolitan Syracuse sewage treatment plant service area and that the storm event measured on September 23, 1974 is characteristic of those predicted, the following tabulation shows the areal

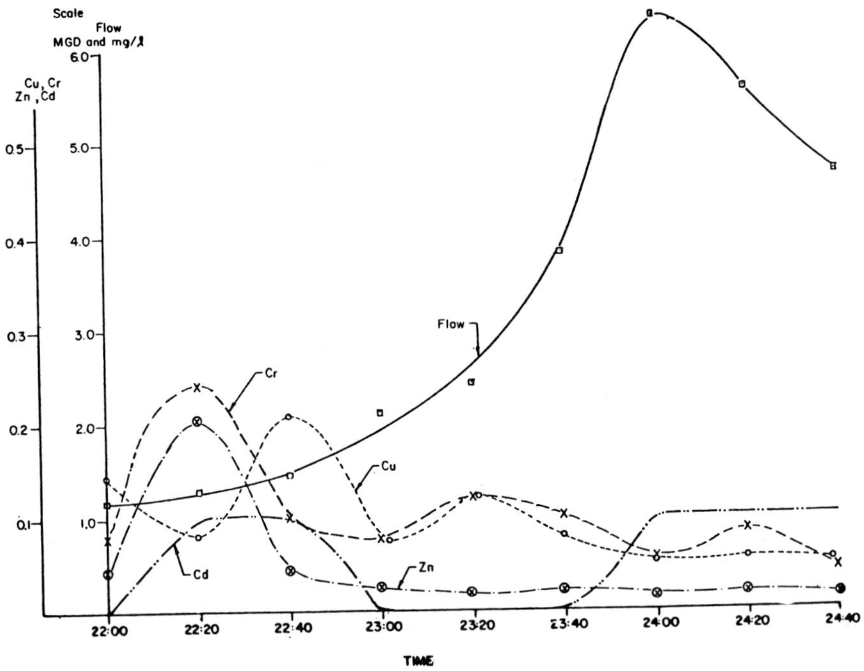

Fig. 10. Heavy metal contributions from combined sewer overflows (Cu, Cr, Zn, and Cd); storm of July 23, 1974.

wet weather stormwater and combined sewer loadings that have been developed:

	Parameter	Loading	
		lb/acre/year	kg/hectare/year
Total organic carbon	(TOC)	762	4150
Total kjeldahl nitrogen	(TKN)	13	71
	Fe	10	55
	Ni	15	82
	Pb	10	55
	Cu	6	33
	Cr	7	38
	Cd	2.5	14
	Zn	2	11

Socioeconomic Status of the Basin

The occupational distribution in the Onondaga Lake drainage basin is dominated by services and education, followed by manufacturing and commerce. Table 6 shows the occupations of employed residents within Onondaga County, which represents approximately 95% of the occupational activity within the Onondaga Lake drainage basin. The median income of residents of Onondaga County in 1960 was \$6691 per family unit while in 1970 the median income had increased to \$10,836 (U.S. Bureau of the census, 1962, 1972).

The Onondaga Lake drainage basin is characterized by a fairly stable and nontransient population. Approximately 70% of the residents living in Onondaga County have lived in the county for five or more years. The 1970 census also showed that at least 88% of the employed Syracuse residents worked in Onondaga County, supporting the fact that most of the employed residents are located in the vicinity of their place of work.

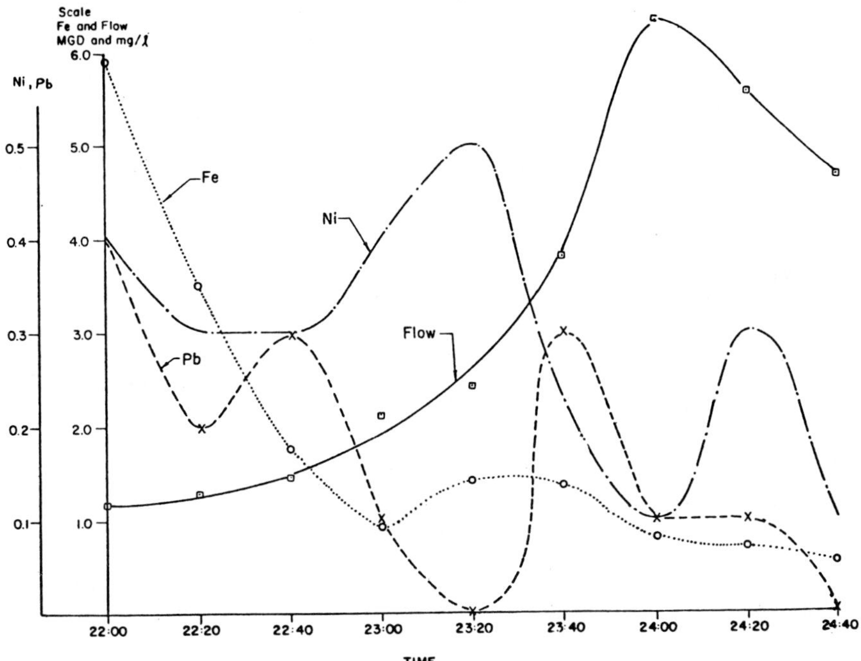

Fig. 11. Heavy metal (Fe, Ni, and Pb) contributions from combined sewer overflows; storm of July 23, 1974.

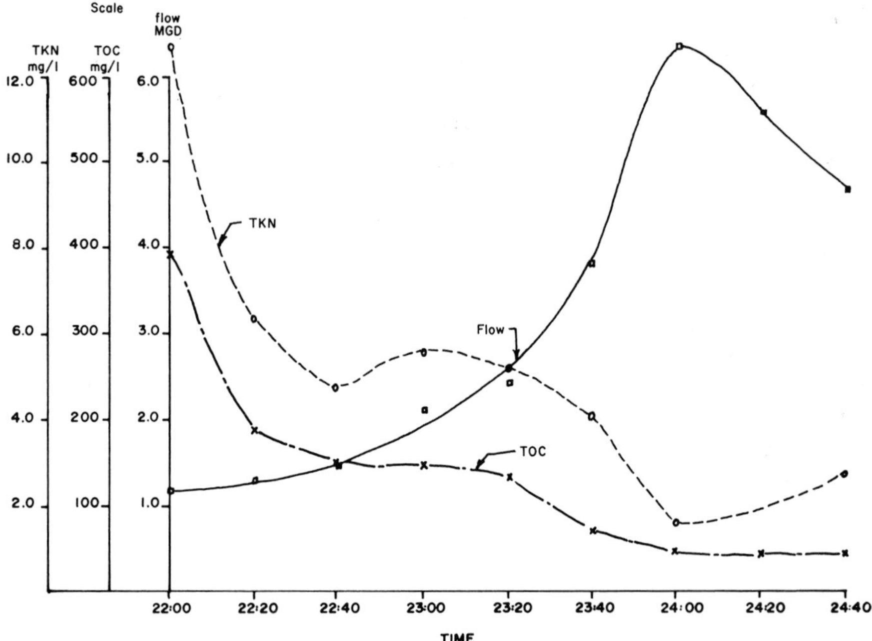

Fig. 12. TOC and TKN contributions from combined sewer overflows; storm of July 23, 1974.

TABLE 6

Occupations of Employed Residents of Onondaga County[a]

Occupation	1960	1970	Percent change, 1960–1970
Agriculture, forestry, fisheries, mining	2,925	2,231	−23.7
Contract construction	8,612	8,823	+2.5
Manufacturing	57,590	49,101	−14.7
Transportation, communication, public utilities	12,024	12,474	+3.7
Wholesale trade	6,662	10,706	+60.7
Retail trade	24,372	30,387	+24.6
Finance, insurance, and real estate	8,287	11,534	+39.1
Services and education	35,259	51,754	+46.7
Public administration	6,662	7,523	+12.9
Total	162,393	184,533	+13.6

[a] Source: U.S. Bureau of the Census (1962, 1972).

Onondaga County ranks tenth in New York State in both the number of available jobs and the size of the business payroll. A trend presently exists toward a decrease in blue-collar jobs in the manufacturing sector and a corresponding increase in white-collar jobs in the services and educational sector (U.S. Bureau of the Census, 1962, 1972). Within the manufacturing sector, the electrical equipment and supplies industry employs the greatest number of workers. The second and third largest employers are the nonelectrical machinery and chemical industries.

HYDROLOGY

Tributary Flows

Onondaga County is drained by two drainage basins. The majority of the county is drained through a series of subbasins by the Oswego River which ultimately discharges to Lake Ontario. A small portion of the southeastern corner of the county is drained through the Tioughnioga Creek subbasin to the Susquehanna River. The Onondaga Lake subbasin is a part of the Oswego River Basin and drains approximately 620 km^2 (240 square miles) of the central portion of Onondaga County.

Major tributaries to Onondaga Lake include Onondaga Creek, Harbor Brook, Ninemile Creek, Bloody Brook, and Ley Creek. The Lake also receives a direct discharge of wastewater from the metropolitan Syracuse sewage treatment plant and a number of industrial discharges. These discharges were discussed in more detail in the preceding section.

Based on the data given in Table 5, the annual inflow to Onondaga Lake from the major tributaries is estimated to be 28×10^7 m^3/year (74×10^9 gal/year). In addition to this flow, the metropolitan Syracuse sewage treatment plant annually discharges approximately 11×10^7 m^3/year (29×10^9 gal/year). The water supply of the area is primarily obtained from three lakes—Ontario, Otisco, and Skaneateles—and the discharge from the plant does represent a net increase to the inflow to the Lake. While the cooling water discharge of the Allied Chemical Corporation has a daily flow nearly equal to that of the metropolitan Syracuse sewage treatment plant, the water is withdrawn from Onondaga Lake and returned to it. Thus, there is no net increase in the inflow from this source. Other tributaries contribute relatively minor amounts to the inflow of the lake.

Precipitation in the Syracuse area averages 92.4 cm (36.33 in.) per year as was discussed in the previous section. It has been reported (Onondaga County, 1971) that the annual evaporation and precipitation were equal and opposite in their effects on Onondaga Lake.

Lake Residence Time

The hydraulic lake residence time can be calculated based on the volume of the lake and the amount of inflow to the lake. It has been determined by means of continuous echo soundings that the volume of Onondaga Lake is 1.405×10^8 m^3 (37.078×10^9 gal) (Onondaga County, 1971). Assuming the total inflow to the lake noted above is mixed throughout the depth of the lake, the residence time would be 185 days. Data reported in 1971 (Onondaga County, 1971) indicated that lake residence times for the years 1963 through 1969 (except 1966 when sufficient inflow data were not available) ranged from 99 to 250 days, assuming complete mixing.

Investigations conducted during 1968 to 1969 (Onondaga County, 1971) indicated that there is justification for utilizing only that volume of the lake contained in the upper 9 m (29.6 ft) when calculating the lake residence time. Variations of major ionic species with time, irrespective of depth, indicate that epilimnetic waters (0–9 m) of the lake respond to inflow variations. The hypolimnetic waters (>9 m) are not affected significantly. Lake residence times for the period of 1963 through 1969 (except 1966) ranged from 64 to 161 days for the epilimnion (Onondaga County, 1971).

Beginning in 1972 an attempt was made to measure the amount of bidirectional flow which was known to exist at the lake outlet. It was reported that in 1972 inflow to the lake at the outlet measured over a period of ten days was 76.4 MGD while the outflow was 409.9 MGD (O'Brien & Gere Engineers, Inc., 1973b). In 1973 inflow and outflow measured at the outlet were 13.0 MGD and 317.8 MGD, respectively (O'Brien & Gere Engineers, Inc., 1974). It would appear that the inflow measured at the lake outlet would have some influence on the lake residence time. However, at the present time no allowance has been made for the ill-defined bidirectional flow occurring at the lake outlet.

Groundwater Flow

Onondaga Lake has high chloride values, averaging nearly 1420 mg/liter in the epilimnion and 1960 mg/liter in the hypolimnion. (U.S. Environmental Protection Agency, 1974). These values are far in excess of those characteristic of a "normal" freshwater lake and also well above the U.S. Public Health Service recommended limit of 250 mg/liter for public drinking water supplies.

The early development of the Syracuse area was at least partially due to utilization of the salt and brine deposits as was discussed in the Introduction. Chloride concentrations of the groundwater in wells drilled near the southern end of Onondaga Lake were reported to be about 100,000 ppm

(U.S. Environmental Protection Agency, 1974). Analysis of the water found in the ponds located on the eastern shore of the lake show that chloride levels in the ponds ranged from 300 to 2480 mg/liter.

It would appear that the high chloride levels found in both Onondaga Lake and ponds adjacent to the eastern shore are influenced by the extremely high concentrations present in the groundwater. However, there have been no known estimates made of the groundwater contribution to the inflow of Onondaga Lake. The lake residence time calculations have made no allowance for any groundwater contribution.

PHYSICAL LIMNOLOGY

Morphometry

Onondaga Lake has a maximum depth of 20.5 m (67.5 ft) and a surface area of 11.7 km^2 (4.52 square miles). Bottom contour plots were developed in the course of the baseline study and are presented in Fig. 13. The contour information shows the maximum depth recorded in the northern basin was found to be 18 m. Between these two basins exists a saddle having a maximum depth of approximately 17 m.

The maximum slope of the lake bottom is exhibited along the northeast and northwest shore of Onondaga Lake with measured slopes of approximately 12%. The slope of the lake bottom is a minimum along the southern shore of the lake with slopes in the order of 2%.

Based on the seismic survey conducted in the source of the baseline study (Onondaga County, 1971), the volume of the lake was calculated based on a water surface elevation of 363.0 USGS datum. The lake volume was estimated to be $1.405 \times 10^{8}/m^{3}$($37.078 \times 10^{9}$ gal).

The lake residence time has been calculated assuming that (1) the inflows are completely mixed throughout the lake depth and (2) the inflows are confined to the epilimnion (0–9 m depth). The lake residence time has been calculated to vary from 64–250 days for the two cases noted above. Due to strong, chemical-density-induced statification as well as to two periods of vertical mixing, the true annual residence times are likely to lie between the two extremes.

Temperature Regime

Temperature data were collected on Onondaga Lake during the baseline study (Onondaga County, 1971) and for each of the subsequent monitoring years to the present. Figure 14 shows the isotherm profiles developed as a

Fig. 13. Onondaga Lake. Contour interval: 1 m. From Onondaga Lake Study (1971).

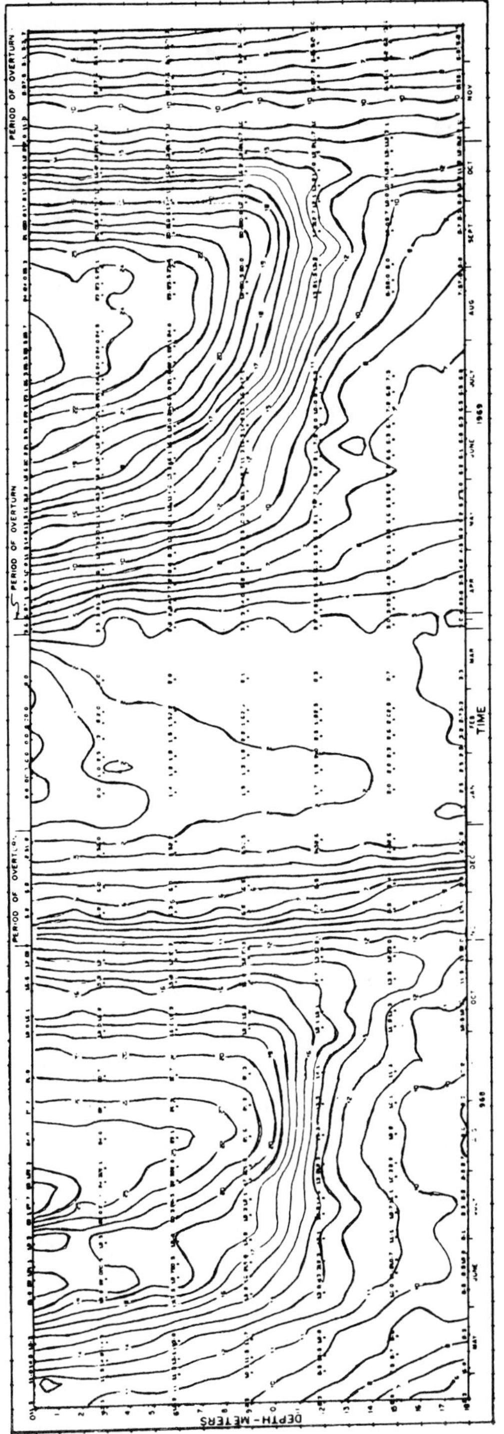

Fig. 14. Water column temperatures, April 1968–December 1969, Station 1.

function of depth and time for the period of April, 1968 to December, 1969. The data show the presence of two major periods of circulation for each year representing the vernal and autumnal circulation. For this reason Onondaga Lake has been classified as being dimictic (Hutchinson, 1957).

The lake exhibits a rapid increase in temperature at most levels following the ice breakup in the spring. This is followed by a continued increase in temperature in the upper waters while the lower waters show little change in temperature.

After achieving the summer maximum temperature and the resulting density stratification, there is a cooling of the surface waters with the subsequent loss in stratification. In the winter the entire lake waters cool to a temperature below 4°C where an inverse stratification is initiated. This inverse stratification holds through the winter with the cycle repeated following the loss of ice in the spring.

Thermal transects conducted during 1969 (Onondaga County, 1971) supports the growth and decay of the thermal structure of the lake and the external contributions of water of different temperature and their disposition within the lake. The subsurface discharge from the metropolitan syracuse sewage treatment plant appears at times to flow down the bottom slope of the lake into the southern basin. At other times the discharge appears to diffuse within a plane at the approximate depth of the discharge.

Mixing and Circulation

The lake waters seem to undergo little vertical mixing throughout the greatest portion of the year, as supported by the synoptic thermal data. In contrast, all chemical data indicate good horizontal mixing throughout the year. The difference between all the chemical constituents measured in the northern basin and southern basin in the period from 1968 to 1971 shows the majority of constituents to vary less than 10%.

This consistency of horizontal concentrations is significant in that the chlorides, calcium, and sodium discharged to the lake are contributed by Ninemile Creek at a point approximately three miles from the southern end.

The circulation of the lake has not been well defined. However, the analysis of concentration gradients measured within the lake indicate a general northern flow. Within this overall trend a countercurrent circulation pattern is exhibited, with the east shore surface waters exhibiting a northern flow and a corresponding southerly flow of west shore surface waters. There is also some intrabasin countercurrent circulation which is largely established by the influence of the Ninemile Creek discharge to the lake, as shown in Fig. 15.

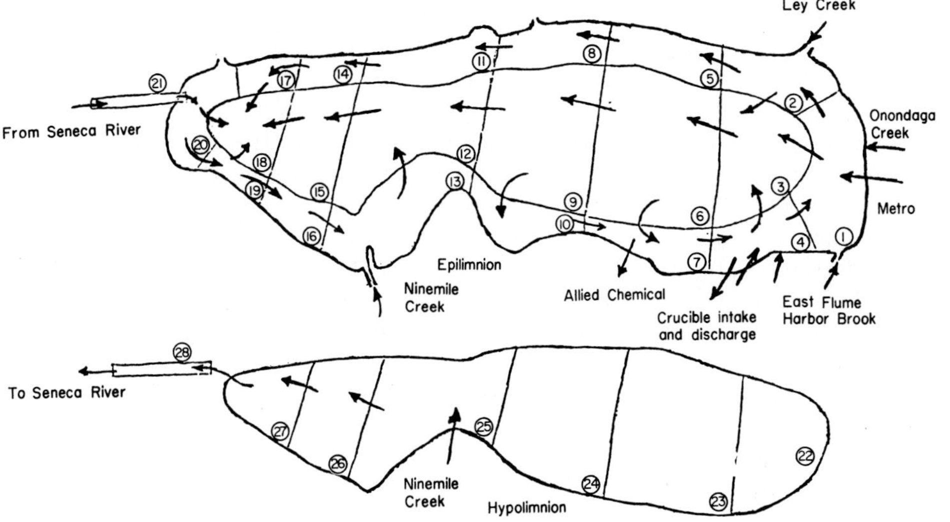

Fig. 15. Proposed circulation model of Onondaga Lake.

Due to the density differences of the Seneca River and the Onondaga Lake waters, the northeast/southwest alignment of the lake outlet with a predominant westerly wind direction, and the varying Seneca River/ Onondaga Lake water level elevations, considerable surface water flow reversal is exhibited by the lake outlet. Frequently there is bidirectional flow with the Seneca River acting as a source of lake surface waters and as a sink for the more dense hypolimnion lake waters. The extent to which this occurs has been documented to range between complete outflow and complete inflow (O'Brien & Gere Engineers, Inc., 1975).

Physical Characteristics

The annual average Secchi disk reading measured in Onondaga Lake is approximately 1 m. Minimum Secchi disk readings in the range of 0.2–0.4 m are commonly observed during the summer months at the time of intense algal blooms. Upon algal die-off and clearing of the lake waters, Secchi disk readings in the range of 2.5–3.0 m are observed. Extremes in clarity are commonly observed during the growth season (see Appendix C, Fig. C-13).

Total solids measured in the epilimnion of Onondaga Lake are approximately 3500 mg/liter with slightly higher values measured in the southern basin. Correspondingly, the hypolimnion measured total solids are approximately 4650 mg/liter with slightly higher values again measured in the southern basin (Onondaga County, 1971).

As discussed earlier, the water temperatures in Onondaga Lake go through a complete cycle once each year. The lake surface usually freezes over, except for the southern end, in late December or early January. With the coming of spring, the ice breaks up usually in early to mid-March. The temperature of the lake waters continues to rise throughout the summer months. The lake waters gradually cool during the fall months until freezing temperatures are again reached in December or January.

Maximum surface temperatures during the summer months are about 25°C (77°F) while the temperatures near the bottom reach about 10°C (50°F). As the surface and bottom waters approach an equilibrium temperature in the spring and again in the fall, the lake overturns or mixes. This phenomenon causes a mixing of the chemical constituents of the lake waters.

CHEMICAL LIMNOLOGY

Lake Water Column Composition

From 1969 to the present a continuing monitoring program on Onondaga Lake has resulted in the development of a data base of 34 chemical and physical parameters obtained at a frequency of every two weeks. The data set has been acquired for two stations located in the south deep and north deep basins at 3-m intervals from the surface to the 18-m depth. The 9-m depth readings are not included in either, as this is the approximate depth of the thermocline and therefore not representative of either region.

The mean values for the measured parameters in the hypolimnion and epilimnion are presented in Appendix A, Tables A-1, A-5, A-8, A-11, A-14, and A-18 for the monitoring years of 1969–1974. Chronological line plots of the parameter measurements at each depth are presented in Appendix C, Figs. C-1 through C-34.

Dissolved Oxygen

Dissolved oxygen is introduced to the lake waters chiefly by diffussion from the air–water interface and by photosynthetic production. The content of dissolved oxygen at equilibrium is primarily a function of the temperature and salinity of the water as well as of the surface tension. Sinks for the water column dissolved oxygen involve the presence of carbonaceous and nitrogenous constituents, as well as benthic oxygen demand.

The New York State Classification System has set standards of not less than 5 mg/liter for trout waters, and not less than 4 mg/liter for nontrout waters for all classes of waters with the exception of "D" waters which are specified as not having less than 3 mg/liter. Presently, Onondaga Lake falls under two classifications, namely "B" and "C." The southern fourth of the lake is classified C where as the northern three-fourths of the lake are classified B.

The geometric mean of dissolved oxygen as measured at Station 1 for the epilimnion in 1969 was 3.8 mg/liter with frequent measured values down to 3.5 mg/liter. Values greater than 10 mg/liter measured in the epilimnion occurred during the summer months due to the influence of photosynthetic oxygenation.

In 1969 dissolved oxygen was entirely absent most of the time in considerable parts of the hypolimnion except during the winter months. The hypolimnetic dissolved oxygen geometric mean was determined as 1.1 mg/liter in 1969. Even when dissolved oxygen was measured in the hypolimnion, values in excess of 1 mg/liter were rare.

The extremes in dissolved oxygen concentrations measured in 1969 were observed in the subsequent monitoring years of 1970 to the present. However, a general improvement in dissolved oxygen conditions has been observed for both the epilimnion and hypolimnion since 1969. The epilimnion-dissolved oxygen concentrations of 6.3 mg/liter measured in 1974 are 32.4% above the baseline concentrations. The average 1974 hypolimnion-dissolved oxygen concentrations of 2.2 mg/liter are 46.4% above the baseline measured concentrations.

In 1974 the period of zero dissolved oxygen is much reduced from the baseline study. Anoxic conditions existed in the deeper lake waters for a period of less than three months in 1974 as compared to the original span of nine months observed in 1969.

The presence of anoxic conditions in the deeper lake waters is correlated by significant increases in sulfide levels produced by anaerobic bacteria respiring on sulfates under conditions of minimal dissolved oxygen. The concentrations of BOD, inorganic phosphate, and ammonia were also found to peak during periods of zero hypolimnion-dissolved oxygen. These concurrent trends might be caused by the die-off of a preceding algal bloom. As the algal bloom terminated, dead cells and residual constituents settle from the surface to the hypolimnion and are subject to bacterial decomposition, resulting in the observed BOD peaks and depressed dissolved oxygen.

Biochemical Oxygen Demand

Onondaga Lake, in suffering from the cultural influence of its drainage basin, receives approximately 45,454 kg/day (100,000 lbs/day) of biochemical oxygen demand. This loading is significantly reflected in the biochemical oxygen demand measured in the epilimnion and hypolimnion water column. As measured in 1969, the geometric mean of the epolimnion biochemical oxygen demand concentration was 4.8 mg/liter while in the hypolimnion a value of 9.0 mg/liter was recorded.

Since the baseline investigations, considerable fluctuation in lake measured biochemical oxygen concentrations have been observed. As measured in 1974, the epilimnion biochemical oxygen demand concentrations had been measured at a level of 22.4% above the baseline level. The 8.5 mg/liter hypolimnion biochemical oxygen demand concentration was below the 1969 baseline data by 31.6%. The fluctuations observed on a year-to-year basis have been correlated with the alteration of biochemical oxygen demand loadings characteristic of the tributary discharges to the lake.

The most significant tributary source of biochemical oxygen demand is from the metropolitan Syracuse sewage treatment plant. As of 1974, the plant effluent was responsible for approximately 85% of the biochemical oxygen demand load. Upon completion of construction of the secondary facilities for the municipal treatment plant in 1979, the daily load from the plant will be approximately 9090 kg/day (20,000 lb/day) resulting in epilimnion biochemical oxygen demand concentrations below 30 mg/liter.

Chlorides

The chloride concentration of Onondaga Lake characterizes the lake as unique in New York State. The epilimnion chloride concentration measured during the baseline study was determined to be 1458.4 mg/liter with a corresponding hypolimnion concentration of 1930.2 mg/liter. A concentration gradient of approximately 50 mg/liter per meter of depth is exhibited during both summer and winter months. This chloride-established density stratification tends to inhibit vertical mixing and lake overturn, thereby enhancing the establishment of anoxic conditions in the bottom waters.

The presence of chlorides at concentrations greater than 400 mg/liter generally imparts a salty taste while levels of 4000 mg/liter can prove to inhibit the proliferation of bass, pike, and perch (McKee and Wolf, 1963). The chloride concentrations characteristic of the lake therefore would preclude its use as drinking water and likely inhibit the establishment of a population of higher order game fish.

Comparison of recent data with that collected by Metcalf and Eddy (1920) show a significant increase in the concentrations of chlorides at both the surface and near bottom waters over the last 50 years. Metcalf and Eddy had reported values ranging from 818 mg/liter at the surface to 1009 mg/liter at the bottom in the south deep portion of the lake. Comparable chloride concentrations for approximately the same period and same temperatures as observed by Metcalf and Eddy ranged from 1100 to 2200 mg/liter.

Onondaga Lake has often been referred to as a saltwater lake. Historical records dating back as far as the late 1700's strongly indicate that the lake was fresh water although salt springs existed along the southern and southeastern shores of the lake. This apparently led to the questionable conclusion that the lake itself was salty. Although the definition of the terms "fresh" or "salty" are difficult to quantitatively define, the lake appears to have increased its "saline" nature and decreased its "fresh" nature as judged by data of this century.

Two squid were reported caught alive in Onondaga Lake, each by nets and each by a different person in 1902 (Clarke, 1902). They were apparently caught where the first salt springs were discovered. Clarke had proposed that the squid could have been remnants from some bygone era when the lake was connected to Lake Champlain and the St. Lawrence Valley. Clarke also speculated on whether they developed from eggs off marine oyster and clam shells that were thrown into the lake from a hotel on the shore. Whatever the reason, the fact that they survived to be caught attests to the relative salinity of the lake in the proximity of the salt springs.

The discharges from Allied Chemical, reflected in the loadings presented by Ninemile Creek, account for approximately 60% of the chlorides that enter Onondaga Lake via surface discharges. It has been estimated that if the Allied Chemical contribution of chlorides were to cease, the lake's chloride level would be cut approximately in half (U.S. Environmental Protection Agency, 1974).

Phosphorus

Average total inorganic phosphorus levels measured in the epilimnion and hypolimnion of Onondaga Lake during the baseline investigations were 2.35 mg/liter and 3.17 mg/liter, respectively. The corresponding measured orthophosphate concentrations were 0.94 mg/liter and 1.57 mg/liter. These nutrient levels obviously indicate the existence of a very high level of cultural eutrophication. In the time frame following the 1968–1969 baseline investigations, phosphorus levels measured in the lake waters have declined to levels of 0.20 mg/liter and 0.47 mg/liter for the epilimnion and hypolimnion (O'Brien & Gere Engineers, Inc., 1975). The noted reductions are attributable to the New York State ban on high phosphate detergents which went into effect on July 1, 1973.

One trend observed for both total inorganic and orthophosphate is strong vertical stratification during the summer seasons with high levels recorded in the hypolimnion and the epilimnion concentrations dipping toward zero. This phenomenon probably results from breakdown of phosphate-bearing particulates settling after algal die-off during the growing season. The phosphate concentrations are then mixed well by the autumnal overturn and remain homogeneous until the next growing season.

The primary source of phosphorus in Onondaga Lake is from the metropolitan Syracuse sewage treatment plant which contributes approximately 545 kg/day (1200 lb/day) of total inorganic phosphorus (O'Brien & Gere Engineers, Inc., 1975). The present upgrading of the metropolitan Syracuse sewage treatment plant will include the addition of both secondary and tertiary facilities which will likely reduce the daily average loading to approximately 136 kg/day (300 lb/day).

Nitrogen

The total concentration of nitrogen present in the water column is highly important when considering the trophic aspects of the lake. The specific forms of nitrogen are of importance when considering specific physiological requirements of the various primary producers. Organic nitrogen in the form of amino acids and urea may inhibit biological growth whereas ammonia and nitrates often stimulate phytoplankton production. Sawyer (1957) has reported that the critical concentration of inorganic nitrogen below which algal growths were not troublesome was 0.30 mg/liter provided that inorganic phosphorus was kept below 0.015 mg/liter.

The baseline study average measured total Kjeldahl nitrogen concentrations were measured to be 4.10 mg/liter for the epilimnion region and 5.79 mg/liter for the hypolimnion. The ammonia nitrogen concentrations averaged 2.14 and 4.31 mg/liter for the epilimnion and hypolimnion during the baseline period. The concentration of ammonia nitrogen is particularly important in light of pH levels which approach 9.0 during periods of intense algal activity. The toxicity of ammonia and ammonium salts to aquatic animals is directly related to the amount of undissociated ammonium hydroxide present in the aquatic system. When the pH is above 8.0 and ammonia nitrogen concentrations greater than 2.0 mg/liter, the aquatic life is exposed to very toxic conditions.

Since 1969, the ammonia nitrogen concentration has been observed to decline steadily through 1972. However, in 1973 a sharp increase in hypolimnion ammonia concentrations was observed particularly at the 18-m depth. During this same period, the peak summer pH values have declined from 9.5 to 8.4. The decreasing ammonia nitrogen and pH levels has resulted in reduced ammonia toxicity.

It should be noted that the New York State water quality standards for Class B and Class C waters, both of which apply to Onondaga Lake, require that the ammonia concentration not exceed 2.0 mg/liter at a pH of 8.0 or above. The results of the Onondaga Lake monitoring program (O'Brien & Gere Engineers, Inc., 1972a,b, 1973a,b, 1974, 1975, 1976) indicate that these conditions have occurred simultaneously, resulting in a contravention of water quality standards.

The measured nitrate nitrogen concentrations of the lake are generally quite low with baseline average concentrations of 0.06 mg/liter and 0.08 mg/liter for the epilimnion and hypolimnion, respectively. The higher epilimnion concentrations likely reflect the higher level of bacterial nitrification occurring in the more highly oxygenated upper lake waters. During nitrification, *Nitrosomonas* bacteria oxidize ammonia to nitrite with *Nitrobacter* bacteria oxidizing nitrite to nitrate.

The average nitrate nitrogen concentrations measured during the baseline investigations ranged from 0.39 mg/liter for the epilimnion to 0.13 mg/liter for the hypolimnion. The lake average nitrate concentrations have been on the decline since 1969 except for the late spring and summer of 1974. The lower hypolimnion concentrations are primarily due to the presence of biological denitrification in the more anoxic bottom waters. This is supported by the fact that the lake waters at the 18-m depth are generally characterized by the lowest nitrate concentrations.

The presence of reduced forms of nitrogen generally present three important water quality problems: toxicity to aquatic life, reduction of dissolved oxygen concentrations, and a nutrient source. The primary source of reduced nitrogen being discharged into the lake are from the Syracuse metropolitan sewage treatment plant and the East Flume discharge of Allied Chemical. These two sources account for the discharge of nearly 6363 kg/day (14,000 lb/day) of total Kjeldahl nitrogen with the Syracuse metropolitan sewage treatment plant contributing approximately 57% of the total load to the lake.

Sulfides and Sulfates

The sulfides measured in Onondaga Lake range from near-zero concentrations measured in the surface waters to hypolimnion average concentrations of 10 mg/liter as measured during the baseline investigations. The high hypolimnion concentrations are due to the greater activity of sulfate-reducing bacteria under anaerobic conditions. The obligate anaerobic, autotrophic, *Desulfodibrio* bacteria are probably responsible for a very large proportion of the sulfides measured in the bottom waters. The breakdown of sulfur-containing organic compounds by sulfhydryl splitting bacteria also probably contribute at least in part to the observed sulfide concentrations.

It is interesting to note that as the period of anoxic conditions in the bottom waters of Onondaga Lake shorten so follows the reduction of sulfide concentrations. A review of the 18-m sulfide and sulfate depth synoptic data illustrates the inverse relationship between sulfates and sulfide concentrations. Bottom water sulfate concentrations tend to decline during the summer months and increase to a late autumn, early winter peak.

Sulfate concentrations have been recently found to range from 120 to 210 mg/liter with approximately equal epilimnion and hypolimnion average concentrations of 164.7 mg/liter and 169.5 mg/liter, respectively.

An estimate of the sulfate contribution from atmospheric sources to Onondaga Lake has been conducted based on the analysis of the 1966–1967 Onondaga County air pollution study data. From the stations monitored it was estimated that the monthly deposition of sulfate could be as high as 0.24 mg/cm^2 (Onondaga County, 1971). An estimate of the atmospheric contribution of sulfates has been calculated based on a lake residence time of 150 days. Only 1.3 mg/liter of sulfates can be attributed to this source.

The primary source of sulfates to the lake are the surface tributaries which were estimated to contribute 220,400 kg/day (485,077 lb/day) (Onondaga County, 1971). The most significant sulfate contributors are Ley Creek, Onondaga Creek, and Ninemile Creek.

Heavy Metals

Since the Onondaga Lake drainage basin is extremely urbanized, the lake finds itself the recipient of a significant load of a wide range of heavy metals. As measured during the baseline investigations, approximately 171 kg/day (376 lb/day) of copper, 163 kg/day (359 lb/day) of chromium, 437 kg/day (963 lb/day) of zinc, and 1589 kg/day (3498 lb/day) of iron were discharged from the eight significant tributaries to Onondaga Lake. These loadings resulted in the establishment of average epilimnion and hypolimnion copper concentrations of 0.05 mg/liter. The measured chromium concentrations were 0.02 mg/liter, while the zinc and iron concentrations averaged 0.08 and 0.23 mg/liter, respectively. The concentrations of each of the noted metals are approximately the same for both the hypolimnion and epilimnion.

Since the 1968–1969 baseline period, a very significant reduction of water column concentrations have been observed for each of the discussed metals. Copper concentrations have approached values of less than 0.01 mg/liter while chromium concentrations are consistently below 0.02 mg/liter and frequently approach zero. Iron concentrations measured in the water column have also declined most significantly from the 1973 peak, with values averaging approximately 0.10 mg/liter for the latter half of 1975. The reduction in water column metal concentrations has been attributed to the startup of the Crucible Steel wastewater treatment facility which occurred in the fall of 1974, and the incorporation of a procedure to reuse the wastewater and reduce discharge flow (O'Brien & Gere Engineers, Inc.,

1976). The occurrence of the above is extremely important since water column concentrations had begun to approach potential inhibitory levels.

Water column concentrations of mercury as measured in 1975 were 4.5 μg/liter and 6.1 μg/liter for the epilimnion and hypolimnion, respectively. The hypolimnion values are slightly higher due to coprecipitation and entrainment mechanisms and to the subsequent settling of particulate material into the hypolimnion. Methylation of the mercury contained within sediments under anaerobic conditions may also contribute to the higher hypolimnion values.

Sodium, Calcium, and Magnesium

The level of total dissolved solids measured in the Onondaga Lake water column during the baseline investigations averaged 3200 mg/liter for the epilimnion and 3896 mg/liter for the hypolimnion. As previously presented, chlorides and sulfates accounted for nearly 50% of the level of total disolved solids. The other primary contributors to the total dissolved solids are the contributions made by the cation component of the ion pairs, namely, sodium, calcium, and magnesium.

During the baseline investigations the epilimnion- and hypolimnion-measured average sodium concentrations were 554.5 and 669.7 mg/liter, respectively, which is equivalent to a lake average sodium chloride concentration of 1640 mg/liter. This is in light of the fact that NaCl has been found to kill or immobilize fresh-water fish at concentrations ranging from 1270 to 50,000 mg/liter.

The lake-measured sodium concentrations have been observed to follow a depth synoptic concentration profile which closely follows that exhibited by chlorides. Factors influencing the observed sodium concentrations include lake residence time, the production capacity of Allied Chemical, and the amount of total precipitation.

Calcium concentrations within the lake averaged 639.4 mg/liter for the epilimnion and 849.0 mg/liter during the baseline investigations. This compares with the lower values of 345 mg/liter and 426 mg/liter measured during 1975. The results generally indicate an improvement in calcium concentrations from the baseline investigations to the present.

The presence of calcium in the water column is very important to the aquatic system. The formation of hydroxyapatite and fluorapatite serves as a means of precipitating phosphorus from the overlying waters, thereby exercising a degree of *in situ* nutrient control. Calcium in water also reduces the toxicity of many chemical compounds to fish and other aquatic fauna. A concentration of 50 mg/liter of calcium has canceled the toxic effect upon some fish of 2 mg/liter of zinc and 0.7 mg/liter of lead (Jones, 1938).

As to the toxicity of calcium itself, fish have reported to survive calcium chloride concentrations of 2500 to 4000 mg/liter (Doudoroff and Katz, 1953).

The presence of high-water column calcium concentrations within the lake is one of the most determining characteristics of the water column chemistry. This importance is due to the continuous formation and deposition of $CaCO_3$ therefore providing a sediment self-sealing mechanism. It is also responsible in part for the high turbidity levels exhibited in the lake and its associated self-cleansing ability.

Average magnesium concentrations measured during the baseline period are 30.4 and 31.0 mg/liter for the epilimnion and hypolimnion, respectively. The nearly identical epilimnion and hypolimnion values indicate little relationship to the chloride distribution. The magnesium present in the water column probably contributes to a continual source of dolomite mineral formation. The presence of magnesium chloride may reduce the toxicity of calcium and potassium chlorides toward freshwater fish (Garrey, 1916).

Geochemical Interactions

The most significant active mineral equilibrium interactions present within the lake are those which primarily involve calcium combined with the carbonate, sulfate, phosphate, and silicate anions. Their nature and relative importance in determining the water column quality is presented.

Onondaga Lake is oversaturated with hydroxyapatite and fluorapatite, and is undersaturated with other orthophosphate compounds. The tributary loading of large quantities of hydroxyl alkalinity, orthophosphate, and fluorides readily contribute to the continual precipitation of these minerals according to the following reactions:

$$10\ Ca^{2+} + 6\ PO_4{}^{3-} + 2\ PH \rightarrow \text{hydroxyapatite}$$
$$10\ Ca^{2+} + 6\ PO_4{}^{3-} + 2\ F \rightarrow \text{fluorapatite}$$

These interactions result in minimizing orthophosphate fluoride and hydroxyl alkalinity levels in the water column. Relative to the formation of fluorapatite minerals, it is observed that the concentration of fluorides diminishes with depth in the epilimnion. The deeper waters have a longer residence time than the surface waters, resulting in longer exposure to conditions of oversaturation with respect to the fluoride minerals, thereby enhancing conditions for fluoride mineral formation.

Onondaga Lake is, in most months, oversaturated with regard to calcite and dolomite. Since the degree of saturation is a function of pH, only when the pH falls below 7 is Onondaga Lake undersaturated with regard to the following mineral formation:

$$Ca^{2+} + CO_3^- \rightarrow CaCO_3 \text{ (calcite)}$$
$$CaMg^{2+} + 2CO_3^- \rightarrow CaMg\,(CO_3)2 \text{ (dolomite)}$$

The formation of calcite and dolomite minerals in the epilimnion results in a reduction of inorganic carbon concentrations contributed from tributary loadings and the bacterial decomposition of organic matter present in the water column. This process tends to reduce the carbon source available for phytoplankton growth and may become increasingly important in future years. The extensive precipitation of calcium carbonate is reflected in the composition of the sediments and assists in the sealing of the sediment-entrained macronutrients.

During the baseline study (Onondaga County, 1971) the lake was shown to be slightly undersaturated with regard to fluorite, with the exception of occasional periods of near equilibrium. A stability diagram prepared for fluorite suggests that at times the formation of fluorite

$$Ca^{2+} + 2F^- \rightarrow CaF_2$$

may limit the concentrations of fluoride to values less than 2 mg/liter at the present levels of calcium.

With respect to sulfate minerals, Onondaga Lake is undersaturated with regard to gypsum ($CaSO_4$). Of the oxides, hematite and occasionally cuprite and pyrolurite have been determined to be stable in the hypolimnion while other metal sulfides have been determined to be unstable. Of the silicates, kaolinite was found to be stable in nearly all samples of water analyzed during the baseline study (Onondaga County, 1971). Additionally, in samples with concentrations of dissolved silica at high levels, equilibrium between kaolinite and calcium montmorillonite is approached.

Lake Sediments

The sediments from Onondaga Lake are generally very fine grained and rich in carbonate minerals. Samples taken from Onondaga Lake are divided into three categories by color and texture (Onondaga County, 1971). The black sediments are characterized by a considerable amount of the greases and clays of the southern basin, the color being largely determined by the presence of considerable quantities of iron sulfide. The black sediments are generally the first layer of sediments observed in cores obtained in the southern basin and reflect the intense cultural influence of the southern tributaries (see Fig. 16). This is shown by the fact that these sediments have been found to be 10% organic matter by weight.

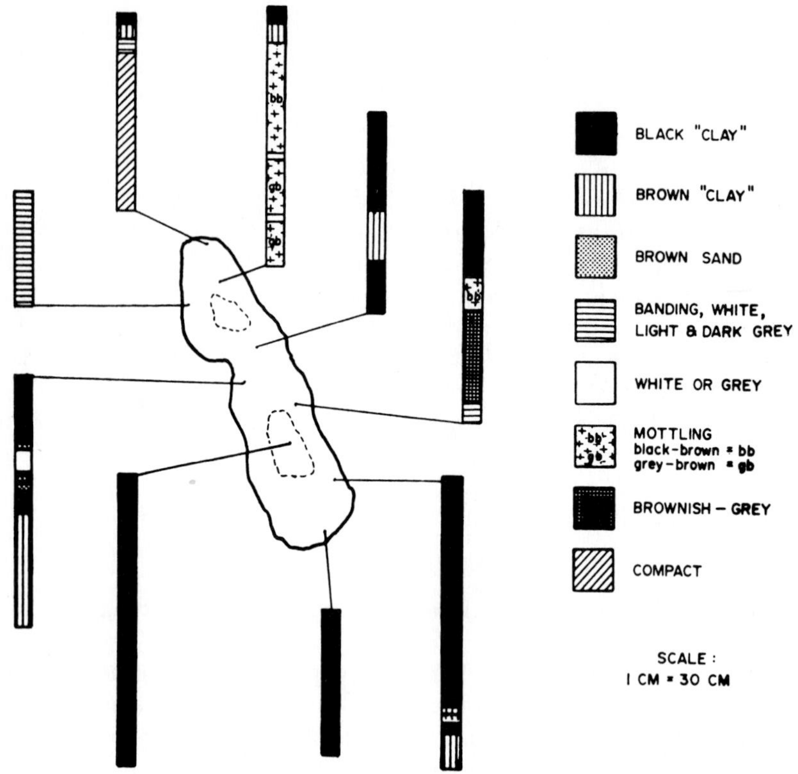

Fig. 16. Sediment layering–Onondaga Lake. From Onondaga Lake Study (1971).

The benthic cores acquired during the baseline study (Onondaga County, 1971) within 305 m (1000 ft) of shore in the southern basin exhibit a brown silty layer. The sediment characteristics of the brown layer are very similar in texture and minerology to the black sediments. The brown sediments have been found to be an oxidized form of the above-mentioned black sediments determined by the presence of oxidized forms of iron. The oxidation occurs when oxygen-rich waters contact with the sediments at times of vertical mixing.

The third major type of sediment involves the grayish white sediment frequently encountered in the shallower, near-shore regions and in the north basin. This sediment is composed primarily of calcite and other carbonate minerals. Calcareous material is also frequently encountered beneath a layer of black silt in layers of less than 1 cm to 1 m in thickness.

It is interesting to note that there is very little coarse-grained or sandy material found in the sediments of Onondaga Lake. This has been attributed to the significant lack of sand sources in the lower reaches of the tributaries to Onondaga Lake. During most months of the year, the waters of Onondaga Lake are oversaturated in calcium carbonate and very high in concentrations of calcium.

The sediments are prevalently inorganic or crystalline carbonates and silicates. The capacity of such sediments to exchange or adsorb ions is small. One would not expect an ion exchange capacity greater than about 10–100 milliequivalents (meq) of cations/100 gm of sediment.

During the baseline study (Onondaga County, 1971) an attempt was made to date the sediments using pollen and radiocarbon analysis. In the pollen-dating evaluations, the pollen of sour gum or black gum (*Nyssa sylvatica*) was present at depths below 180 cm (71 in.) in the saddle of the lake and absent at depths less than 180 cm. The historical background of the use and presence of sour gum indicates that sour gum pollen was contributed to the sediments before 1875 or 1880. It has therefore been deduced that the upper 180 cm of black sediments have been deposited after that date.

Entrained Sediment Waters

The entrained waters of the sediments exhibit concentrations of all the major cations that are significantly higher than those found in the lake waters. Concentrations of sodium and chloride ions found in the entrained sediment waters indicated that they exceed 10,000 mg/liter (Onondaga County, 1971). There appears to be no consistent concentration gradient in the enclosed sediments.

Analysis of the enclosed sediment waters indicated that sodium and chloride were present in a nearly exact simple proportion. Magnesium was also found to be in a close proportion to the chloride concentrations. The calcium concentrations, however, showed only a very weak correlation with the chlorides. These correlations are consistent with those noted for commercial brine wells in the vicinity of Onondaga Lake. These observations indicate that there may be an upwelling through the sediments of saline groundwater derived from natural deposits of rock salt.

During the baseline study (Onondaga County, 1971) the entrained sediment waters were also analyzed for the presence of heavy metals. It was concluded that the concentration of heavy metals in the enclosed waters is several times larger than that measured in the open waters, with no relationship established between the concentration of the heavy metals and the depth of sediments. For the metals—manganese, zinc, chromium, copper, and iron—the highest observed values were 5.6 mg/liter, 0.6 mg/liter, 0.17 mg/liter, 0.35 mg/liter, and 139 mg/liter, respectively.

BIOLOGICAL LIMNOLOGY

Phytoplankton

Prior to 1972, the annual succession of phytoplankton within Onondaga Lake consisted of a spring flora of diatoms and flagellates and a summer bloom of green algae replaced by a bloom of blue-green algae in August. The transition between the green and blue-green blooms was usually characterized by a general clearing of the lake. The fall flora consisted mainly of blue-greens, a growth of diatoms, or a second growth of greens. Shown in Appendix D are the annual succession of phytoplankton for the years of 1969 through 1975. Sze (1975; Sze and Kingsbury, 1972) presents extensive data on the distribution of phytoplankton in Onondaga Lake.

In 1972 the annual succession of phytoplankton was altered coincident with a 44.4% reduction in lake-measured total inorganic phosphate concentrations attributed to the implementation of phosphate concentrations attributed to the implementation of phosphate detergent legislation (Murphy, 1973a, 1973b). Since 1972 there has been a nearly complete disappearance of blue-green algae from Onondaga Lake with a dominance of green algae generally through the summer and into the fall.

The principal flagellate observed during the spring is generally *Chlamydomonas* sp. Diatoms of principal importance include *Cyclotella glomerata, Amphiprora alata,* and *Diatoma tenue.* A bloom of *Diatoma tenue* is characteristic of Onondaga Lake every year, occurring in the middle or end of May. The early summer period is generally dominated by green algae of the order Chlorococcales, *Chlorella vulgaris* being the most important species of the early summer greens. Also abundant in the summer are *Scenedesmus obliquus* and *Scenedesmus quadricauda,* which have generally paralleled *Chlorella* in annual occurrence and abundance.

The blue-green algae dominant in the late summer and early fall prior to 1972 generally consisted of *Aphanizomenon flosaquae. Polycystis aeruginosa* ("*Microcystis*") showed significant populations during the baseline study but has diminished in importance in subsequent years and has been absent since 1971. This observed alteration in the composition of the blue-green population and its subsequent absence from the seasonal succession may indicate a slight improvement in the trophic state of Onondaga Lake.

Green algae characteristic of the fall following 1972 consist primarily of *S. obliquus* and *Cryptomonas* sp. *Cryptomonas* sp. is generally found in the lake for most of the productive portion of the year (March to November).

Nutrient depletion is presently observed in the epilimnion of Onondaga Lake at various times during the growth season. In 1974, silica was depleted in both early June and September. Concentrations of both orthophosphate and total inorganic phosphate are depleted well below levels of 0.1 mg/liter during the summer. In recent years nitrate and ammonia nitrogen have shown summer declines, indicating that nutrients are no longer in excess.

Zooplankton

Appendix E contains the zooplankton counts exhibited within Onondaga Lake from 1969 through 1975. The seasonal succession of zooplankton is largely as characteristic of Onondaga Lake as the previously described phytoplankton succession.

Rotifers are found throughout the year with maxima generally observed in early summer, late summer, and mid fall. In addition to rotifer populations ranging up to 17,167 organisms/100 ml (1972), significant populations of both copepods and cladocerans are observed during the summer months. Copepod populations up to 8,842 organisms/100 ml (1972) and cladoceran populations up to 5,944 organisms/100 ml (1972) have also been observed. Both the copepod and cladoceran populations tend to peak much earlier in the season than the rotifers.

The zooplankton counts presented represent 12-m vertical net hauls at the southern basin sampling station. Increases in zooplankton populations, particularly the herbivorous cladocerans, are not associated with declines in phytoplankton, which indicates that grazing may not be a controlling factor in phytoplankton dynamics. It is assumed that bacteria and organic materials are the primary food sources for the zooplankton community.

Bacteria

The measurement of fecal coliform, fecal *Streptococcus,* and total coliform were conducted coincidently in 1974 (O'Brien & Gere Engineers, Inc., 1975) on both the lake and the influent tributaries. Prior to 1974, bacterial data on the lake were limited to total coliform which have been deleted in preference for the fecal coliform and fecal *Streptococcus* measurements. Total coliform values as high as 19,079 colonies/100 ml were recorded for the epilimnion in the southern basin (O'Brien & Gere Engineers, Inc., 1975).

The average tributary bacterial concentrations (see Appendix A, Table A-18) were found to be the highest for Harbor Brook, which had fecal coliform levels averaging 190,000 colonies/100 ml, followed by Onondaga Creek which had fecal coliform levels averaging 44,000 colonies/100 ml. Third on the list of significant contributors of bacterial input to Onondaga Lake is Ley Creek, which had fecal coliform levels of 12,200 colonies/100 ml. All three of the above-mentioned tributaries flow through heavily urbanized and/or industrialized areas and are subject to the influence of combined sewer overflows and possible unrecorded direct discharges.

The average fecal coliform discharge recorded for the metropolitan Syracuse sewage treatment plant is listed in Table A-18 of Appendix A as 2557 colonies/100 ml. This value is heavily skewed by an atypical concentration of 43,500 colonies/100 ml observed in the sample collected on May 2, 1974. Excluding this value, the average value is calculated to be 148 colonies/100 ml, indicating effective disinfection.

In the course of conducting the 1974 Monitoring Program, the ratio of the fecal coliform (FC) concentrations to fecal *Streptococcus* (FS) concentrations and calculated to indicate the type of contamination found within the tributaries to Onondaga Lake. These ratios are shown in Table 7. In human feces, fecal coliform are more numerous than fecal streptococci with an accepted ratio of 4.0. In the wastes of warm-blooded animals, the coliform bacteria are less numerous, with an accepted ratio of less than 0.7. A general interpretation of the ratio is given as: $FC/FS > 4$, strong evidence of human waste contamination; $2 < FC/FS < 4$, predominance of human wastes in mixed contamination; $1 < FC/FS < 2$, uncertain interpretation; $0.7 < FC/FS < 1$, predominance of animal wastes in mixed contamination; $FC/FS < 0.7$, strong evidence of animal waste contamination.

The FC/FS ratios for Harbor Brook are very high, indicating largely human contamination. This is probably due to existence of combined sewer overflows and exfiltration from the combined sewers in that area. The FC/FS ratios for Onondaga Creek are initially high but tend to decline after the summer. A possible explanation might be that the combined sewer overflows, which occur extensively in the spring, are less frequent during low flow periods later in the year, thereby reducing the volume of human contamination. In contrast to Harbor Brook and Onondaga Creek, the Ley Creek FC/FS ratios tend to lie in an intermediate range where interpretation is extremely difficult.

The Ninemile Creek tributary contains a very large load of inorganic materials of industrial origin. The very low FC/FS ratios observed in Ninemile Creek are probably due to the greater survival rate for fecal streptococci under stress conditions. The metropolitan Syracuse sewage treatment plant effluent exhibited very low FC/FS ratios for similar reasons.

Within the lake itself the fecal coliform levels generally range from approximately 10–1,000 colonies/100 ml at the surface or at 3-m depths. At increasing depths a wider range in fecal coliform concentrations exists. The average fecal coliform concentrations determined in the epilimnion during 1974 is approximately 341 colonies/100 ml. This contrasts with the average measured in the hypolimnion of 249 colonies/100 ml. The lake-measured

TABLE 7

Fecal Coliform/Fecal *Streptococcus* Ratios in Creek Discharges (1974)[a]

Location	Date													
	May 16	May 30	Jun. 13	Jun. 27	Aug. 1	Aug. 15	Aug. 29	Sep. 12	Sep. 26	Oct. 10	Oct. 24	Nov. 7	Dec. 11	Dec. 26
Harbor Brook	2.36	4.3	2.1	11.6	1.9	4.2	13	6.1	0.7	3.5	2.8	7.6	6.3	9.5
Onondaga Creek	5.02	3.9	0.9	2.3	1.6	—	3.4	0.4	0.6	0.6	7.5	1.6	0.6	4.3
Ley Creek	1.88	1.4	0.8	1.9	1.8	16	2.3	3.5	1.7	2	2.6	2.9	2.4	3.2
Ninemile Creek	0/0	0/2	0.2	0/0	0/0	0/12	0/1000	0/2	0/1	0/0	1	0/0	0.3	1.3
Metro	0/0	0/10	0.10	0.014	0/400	0.8	0/0	0.1	0.5	0.14	0.003	0.0007	0.23	—
Crucible	4.84	15	2.8	2.2	6.7	2.2	0/0	7.7	4.3	2	0.3	1.3	0.04	0.33
East Flume	10/0	0/0	0.5	28/0	0.4	3.2	4/0	2	4.7	0.06	0/0	0/0	0/0	2

[a] Source: Onondaga Lake Monitoring Program (January 1974–December 1974).

fecal *Streptococcus* concentrations are 270 and 357 colonies/100 ml for the epilimnion and hypolimnion regions of the lake, respectively.

Fish

The primary data on the fish within Onondaga Lake were acquired by Noble and Forney (1971) in the course of conducting the baseline study of Onondaga Lake (Onondaga County, 1971). Noble and Forney reported that Onondaga Lake has ". . . a fairly diverse fish fauna, typical of many warm water lakes in Central New York State." Table 8 shows a listing of fish species found in the lake during surveys conducted in 1927, 1946, 1969, and

TABLE 8

Fish Species Found in Onondaga Lake (1927, 1946, 1969, and 1972)

Scientific name	Common name
	1927[a]
Cyprinus carpio	Carp
Notemigonus crysoleucas	Golden shiner
Pimephales notatus	Bluntnose minnow
Esox americanus vermiculatus	Grass pickerel[b]
Fundulus diaphanus	Killifish
Perca flavescens	Yellow perch
Micropterus salmoides	Largemouth bass
Lepomis gibbosus	Common sunfish
Catostomus commersoni	Common sucker
Moxostoma sp.	White-nosed, red-fin sucker
	1946[c]
Stizostedion v. vitreum	Pike-perch (walleye)
Perca flavescens	Yellow perch
Esox lucius	Northern pike
Lepibema chrysops	Silver bass
Lepomis gibbosus	Common sunfish (mainly young)
Ictalurus l. lacustris	Catfish
Moxostoma aureolum	Redfin sucker
Moxostoma sp.	Redfin sucker
Cyprinus carpio	Carp
Notemigonus c. crysoleucas	Golden shiner
Pomolobus pseudoharengus	Alewife
Percina caprodes semifasciate	Logperch
Fundulus diaphanus	Killifish
Notropis atherinoides	Buckeye shiner
	1969[d]
Cyprinus carpio	Carp
Notropis atherinoides	Emerald shiner

(Continued)

TABLE 8 *(Continued)*

Scientific name	Common name
Catostomus commersoni	White sucker
Moxostoma macrolepidotum	Northern redhorse
Moxostoma sp.	Redhorse sucker
Ictalurus punctatus	Channel catfish
Ictalurus nebulosus	Brown bullhead
Culaea inconstans	Brook stickleback
Roccus americanus	White perch
Micropterus dolomieui	Smallmouth bass
Lepomis macrochirus	Bluegill
Lepomis gibbosus	Pumpkinseed
Lepomis sp.	Bluegill or Pumpkinseed
Perca flavescens	Yellow perch
Stizostedion v. vitreum	Walleye
Aplodinotus grunniens	Freshwater drum
1972[e]	
Ictalurus natalis	Yellow bullhead
Micropterus salmoides	Largemouth bass
Pomoxis annularis	White crappie
Stizostedion v. vitreum	Walleye
Morone chrysops	White bass
Catostomus commersoni	White sucker
Esox lucius	Northern pike
Cyprinus carpio	Carp
Notemigonus crysoleucas	Golden shiner
Roccus americanus	White perch

[a] Source: Greely (1928).
[b] Called little pickerel in original text.
[c] Source: Stone and Pasko (1946).
[d] Source: Nobel and Forney (1971).
[e] Source: U.S. Environmental Protection Agency (1973b).

1972. The data presented in the table support the contention that the species composition has remained essentially the same over the period of record. The only readily noted change in species observed in 1927 compared with that reported in 1969 was the presence of significant numbers of white perch in the 1969 survey.

In the course of conducting the 1969 study members of the Department of Conservation of Cornell University identified 16 species of fish in adult or juvenile stages of development. Noble and Forney (1971 have found that, despite the chemical and physical characteristics of Onondaga Lake, growth of most game and panfish compared favorably with the published growth rates for fish in other waters of the Northeast. It was reported that reproduction was limited in 1969 although the few undeveloped organisms taken were of good size and condition.

It is interesting to note that the distribution of fish is not uniform throughout the lake. Conditions along the northwest shoreline are less affected by the industrial and municipal wastewaters which discharge into the southern basin. As a result the waters in this area support a large and varied fish fauna. For the same reason adult fish are scarce in the southernmost part of the lake. It is hoped, however, that the present pollution abatement programs, having been instituted on point sources discharging to the southern basin, will result in improved water quality conditions capable of supporting populations of fish similar to those exhibited in the northern basin.

The level of dissolved oxygen in the hypolimnion during the growth season is considerably lower than that necessary to support fish populations. As a result, fish populations are limited to the epilimnion even though dissolved oxygen conditions can be marginal during the summer months when surface reaeration is limited. Variable and marginal dissolved oxygen conditions can have a most significant limiting effect on the reproductive activity of the species present. Also, most important to the development of a game and panfish population is the quality of substrate available in the littoral zone.

Aquatic Vegetation

Little published information is available on Onondaga Lake in the areas of both vascular plants and littoral and benthic invertebrates. Generally little growth of emergent vegetation is observed in Onondaga Lake. The areas of limited emergent growth are primarily limited to the points of tributary confluence and to embayments along the shore of the northern basin.

The general composition of the bottom sediments are very fine grained and rich in carbonate minerals which are diffuse in nature. The calcareous component of the sediments suggests that the most significant component is calcite, which is precipitated directly from the overlying waters. A very diffuse and continually building sediment mass is not supportive of either a littoral or benthic invertebrate population. Additionally, high turbidity is created in the lake waters by biomass formation and calcite precipitation. The resulting low light transmittance also inhibits extensive growth of vascular plants.

DISCUSSION OF THE ECOSYSTEM

Structure

Fish populations comprise the highest life form normally found within a lake's ecosystem. As such, fish are the major predators of the lower life forms present. Zooplankton, and to a lesser extent phytoplankton, are the major food sources of most fish species although other organisms, such as insects and smaller fish species, also contribute to the diet. The predation of zooplankton by fish tends to control their population in the ecosystem. In turn, the population of phytoplankton is controlled to some extent by the zooplankton which feed on the various phytoplankton species present.

It has been reported that other herbivores besides crustaceans and rotifers are present in the lake. At times ciliated protozoans are abundant and possibly may be preventing proportional phytoplankton growths.

It is also not uncommon at times in Onondaga Lake to find phytoplankters parasitized by fungi (probably chytrids). The significance of the fungi is unknown. However, they may bring about a decline of the phytoplankton population or they may be taking advantage of unhealthy cells in an already declining population.

Primary Productivity

Primary production of organic matter is governed by the equation of photosynthetic processes:

$$6\,CO_2 + 6\,H_2O \xrightarrow{\text{light}} C_6H_{12}O_6 + 6O_2$$

Carbon dioxide and water are assimilated in the presence of light and transformed into algal biomass with oxygen as a by-product. In an aquatic environment the rate of biochemical photosynthesis is dependent on such factors as ambient nutrient and inorganic carbon cell concentrations, chlorophyll concentration, incident solar radiation and its extinction through the water column, water temperature, and turbulence.

Primary production profiles were determined at a station in the south basin of Onondaga Lake four times in 1975 and on three occasions in 1974, using the dissolved oxygen light and dark bottle technique (Vollenweider, 1974; American Public Health Association, 1971). Samples were taken at 1-m intervals from 0.25 m to 6.25 m and were then returned to the same depths for incubation in 300-ml BOD bottles. For purposes of replication, two light bottles and two dark bottles were suspended at each depth.

Dissolved oxygen determinations were made using the Winkler technique with azide modification (American Public Health Association, 1971). Incubation times for the 1974 experiments were approximately 3½ hr (from 9:30 A.M. to 1:30 P.M.), while incubation times for the 1975 experiments were approximately 5 to 6 hr. (The exact durations in 1975 were set as the second and third fifths of the photic period, as suggested by Vollenweider, 1974). The production results are expressed as mg C/M^3 produced during the experimental period, assuming a photosynthetic quotient of unity (Murphy and Welter, 1976).

A summary of the primary productivity data and the corresponding chlorophyll *a*, biomass, solar radiation, Secchi depth, temperature, adenosine triphosphate, nutrient and standard plate count measurements are presented in Table 9. Plots of the gross production profiles as a function of depth for the dates of June 7, 1974; July 25, 1974; September 19, 1974; and May 8, 1975 are presented as Figs. 17, 18, 19, and 20.

Transparency was measured with a 20-cm diameter, all-white Secchi disk. In all the calculations discussed in this chapter the Secchi depth is converted to extinction coefficient (base e) using the relationship (Beeton, 1958)

$$K_e = 1.9/\text{Secchi depth}$$

which corresponds to the assumption that Secchi depth is the 15% light penetration depth. This assumption is supported by measurements made on a date when a submarine photometer was available and comparisons were made to the Secchi depth measured.

Incident solar radiation measurements during the 1975 experiments were obtained from an ongoing solar energy research program being conducted by W. Howard Card and associates at the electrical engineering department of Syracuse University. These measurements were made with an Eppley black and white pyranometer sensitive to wavelengths in the range 0.28 to 2.8 μm, with the signal routed through a time integrator and recorded on a strip chart. In 1974 pyranometer measurements were not available and estimates of incident solar radiation were calculated from the formula (Baker and Haines, 1969)

$$R = R^* (0.22 + 0.535S)$$

where R equals the estimated incident solar radiation; R^*, theoretical extraterrestrial radiation; S, percent of possible sunshine (as measured at Syracuse's Hancock International Airport by the National Weather Service). It should be pointed out that this formulation was derived on the basis of full-day measurements in the north central United States; consequently, it probably underestimates midday insolation for part of a day.

TABLE 9

Onondaga Lake Primary Productivity Data[a]

	Date						
Parameter	6/7/74	7/25/74	9/19/74	5/8/74	6/26/75	7/31/75	9/5/75
Dominant alga	*Chlamydomonas* sp.	*Chlorella vulgaris*	*Chlamydomonas* sp.	*Cryptomonos ovata*	*Cyclotella glomerata*	*Chlorella vulgaris*	*Cyclotella glomerata*
Areal gross production (mg C/m^2)	1950	2450	2475	642	1481	3523	2793
Chlorophyll *a* (μg/liter)	86.5	15.5	41.4	11.6	102.8	22.9	107.6
Biomass (μ^3/ml)	8,780,000	1,940,000	4,070,000	530,000	—	2,500,000	7,200,000
Solar radiation (Langleys)	286	125	202	411	248	378	258
Secchi depth (m)	0.5	0.7	0.7	1.0	0.3	0.7	0.5
Temperature (°C)	18.5	20.0	17.0	9.5	21.0	26.0	18.5
ATP (μg/liter)	—	—	—	1.21	3.9	3.04	1.76
PO_4 phosphorus (mg/liter)	0.04	0.02	0.15	0.26	0.01	0.02	0.02
NH_3 nitrogen (mg/liter)	1.43	1.41	3.08	2.93	1.20	1.15	0.41
NO_3 nitrogen (mg/liter)	0.71	0.25	0.03	0.26	0.04	0.08	0.05
Standard plate count (count/ml)	—	—	—	1,000	—	2,200	34,000
Silica (mg/liter)	0.4	2.1	0.9	4.8	0.6	1.7	0.2

[a] Source: Murphy and Welter (1976).

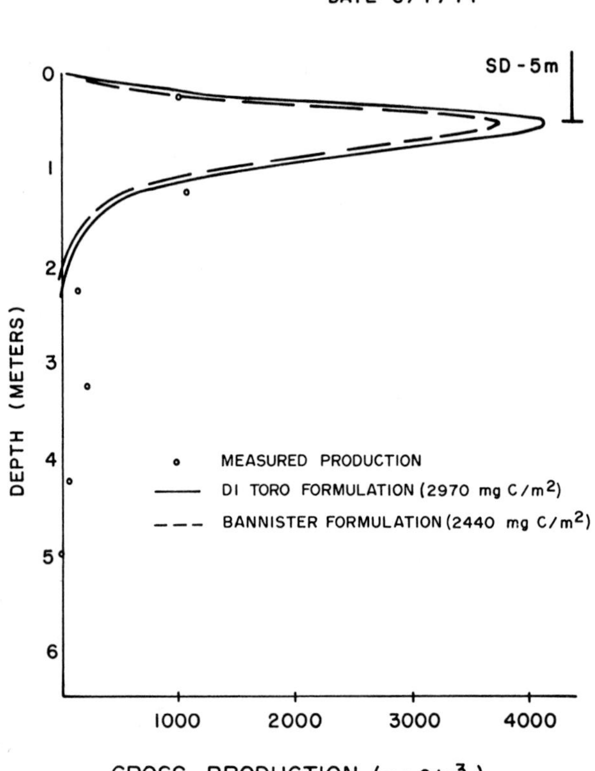

Fig. 17. Gross production (6/7/74). From Murphy and Welter (1976).

The measured and estimated total incident solar radiation values were multiplied by 48% to obtain the fraction in the photosynthetically active range of 400 to 700 nm (Vollenweider, 1974).

Chlorophyll *a* measurements were performed under the direction of John Barlow of Cornell University using spectrophotometric methods (Lorenzen, 1967). Chlorophyll samples were filtered through a Whatman GF/C filter with a small quantity of magnesium carbonate suspension. The filters were then dessicated in a dark refrigerator until dry (2 or 3 days), wrapped in aluminum foil, and transmitted to the Cornell laboratory for subsequent extraction and measurement. The determinations used are corrected for measured phaeophytin content.

Samples for algal counts and identification were collected at the surface and at depths of 3, 6, and 12 m, and a vertical net haul was made from the 12-m depth to the surface using a 9-in. diameter, #20 mesh net.

Unconcentrated samples were stained and filtered (De Noyelles, 1968), and filters were cleared and mounted on microscope slides using Karo corn syrup (Dawson, 1966). Samples from the net hauls were preserved with 2 parts commercial formaldehyde and 2 parts "Lugol's" solution. Identification and enumeration of phytoplankton and zooplankton was performed by Philip Sze of the State University of New York at Buffalo. Algal cell counts were converted to volumetric biomass using species specific cell volume dimensions as measured from Onondaga Lake samples.

An additional measure of biomass undertaken during 1975 was the determination of adenosine triphosphate during the regular monitoring program and at each depth of the primary productivity studies. These determinations were made in accordance with the boiling TRIS buffer extraction methodology (Chappelle and Piccolo, 1975).

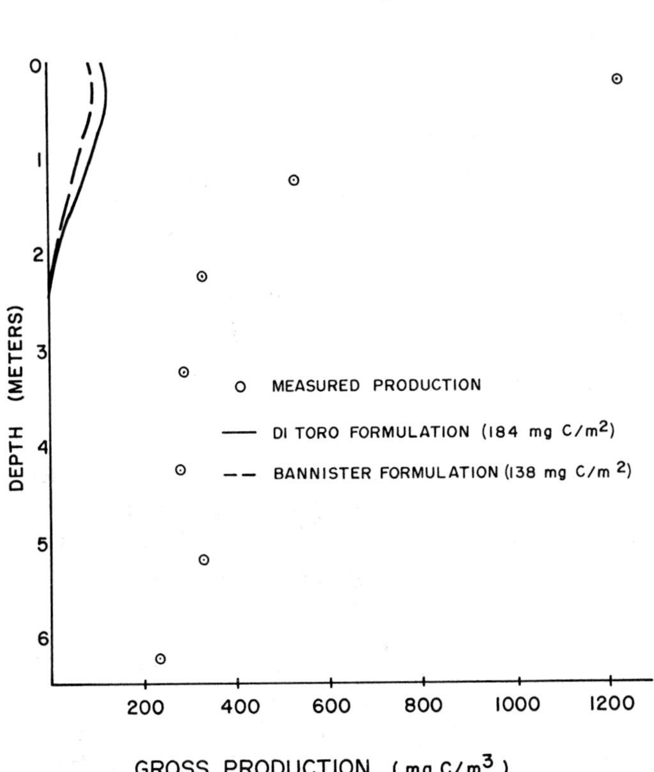

Fig. 18. Gross production (7/25/74). From Murphy and Welter (1976).

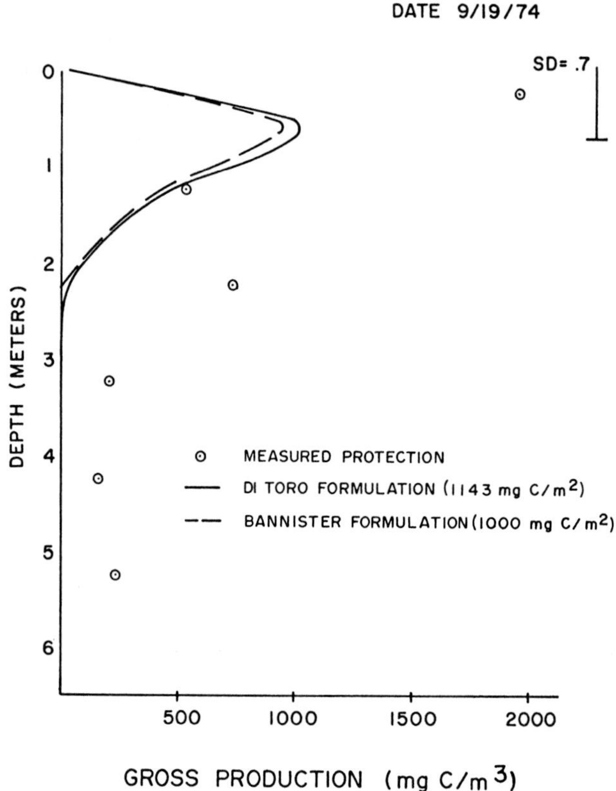

Fig. 19. Gross production (9/19/74). From Murphy and Welter (1976).

Dynamics and Trophic Status

General

Onondaga Lake can be considered a eutrophic, saline water body. Published and unpublished information on Onondaga Lake has generally supported an impression of Onondaga Lake as a eutrophic lake, rich in biomass, with a poor diversity of species. The most extensive investigation of Onondaga Lake and its trophic status is documented in the "Onondaga Lake Study" (Onondaga County, 1971).

The trophic status of any lake is a rough measure of the productivity or actual capacity of a lake to produce. Onondaga Lake, being very rich in nutrients and having a subsequently high level of supported biomass, is considered eutrophic. This is consistent with the fact that Onondaga Lake has served as the receptacle for major portions of domestic and industrial waste discharges from the Syracuse metropolitan area.

An average annual chlorophyll *a* concentration of 30.5 μg/liter was measured in 1974 with an associated total pigment concentration of 37.5 μg/liter. (See Fig. C-34 in Appendix C.) Total pigment concentrations in the surface waters have been measured as high as 167 μg/liter during the 1974 monitoring program. Associated full-day, gross primary production levels were found to range from 1950 to 2475 mg C/m² for the three determinations made during the 1974 growth season. This indicates a level and activity of supported biomass consistent with a eutrophic classification.

Brylinsky and Mann (1973) have investigated the relationships among phytoplankton standing crop, production rates, and photosynthetic efficiency using data collected under the International Biological Program, covering a number of lakes with wide ranges of geographic locations and limnological characteristics. In order to compare the data from Onondaga Lake to the compiled IBP data, the former were converted to other units according to transformations suggested by Brylinsky and Mann. These

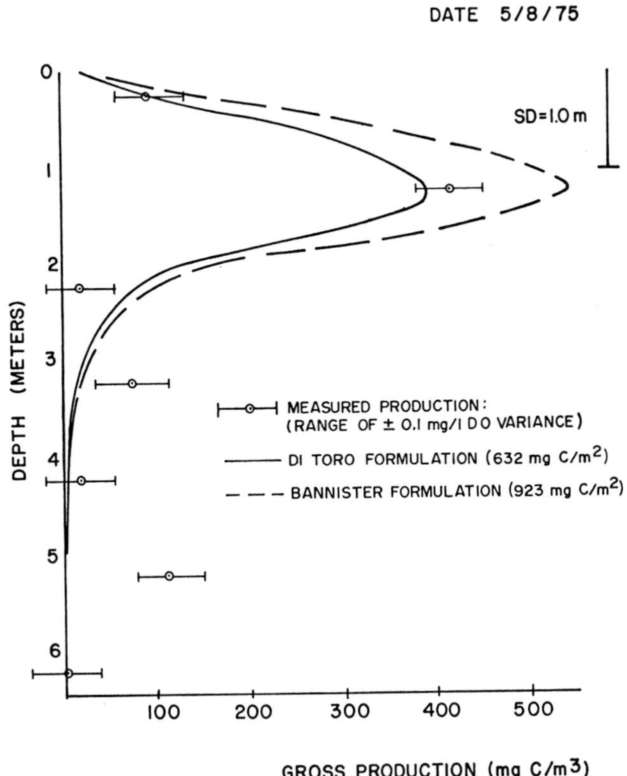

Fig. 20. Gross production (5/8/75). From Murphy and Welter (1976).

include the assumption that the average of the production rates measured in Onondaga Lake is typical of what occurs throughout the growing season (from May through September), and that gross primary production throughout the growing season constituted 85% of the annual production. In converting the production data to energy terms, it was assumed to 1 gm of carbon is equivalent to 9.4 kcal. The annual photosynthetic efficiency was calculated using a value of 3200 kcal/m^2 day for the average visible solar radiation intensity (Odum, 1971).

These conversions for Onondaga Lake data yielded growing season and annual production rates of 5300 and 6300 kcal/m^2, and a photosynthetic efficiency of 0.5%. The average epilimnion chlorophyll content for the 1974 and 1975 growing seasons was calculated to be 63 mg/m^3. These parameters representing Onondaga Lake are compared with the statistical regression equation developed for the IBP data (Murphy and Welter, 1976). These comparisons indicate that, while Onondaga Lake is quite highly productive, its internal production relationships among these parameters are in general agreement with those found typical of a wide range of other lakes.

Mathematical Relationships

Murphy and Welter (1976) have compared several measures of phytoplankton standing crop as applied to Onondaga Lake. The parameters evaluated include chlorophyll *a*, phytoplankton cell-count biomass, and adenosine triphosphate.

The most significant correlation, a correlation coefficient of 0.94, was found in comparing the response of chlorophyll *a* and cell-count biomass. The comparison of the limited amount of available paired ATP and biomass data from the epilimnion yielded a correlation of 0.53.

The ATP:chlorophyll pairing produced a correlation of only 0.17 for all 1975 epilimnion data (from 0- to 6-m depths). The correlation between the two parameters improves markedly when attention is confined to data from the surface samples. Using only surface water data, a correlation coefficient of 0.73 was obtained. The difference between the surface and epilimnion correlation coefficients has been attributed to the development of a substantial heterotroph community during the growing season at the lower depths of the epilimnion while algae totally dominate the surface water.

Two mathematical models developed by other researchers were formulated and calibrated in order to facilitate a greater understanding of the measured gross production data.

The first model formulated for use with these data is based on the phytoplankton growth model first applied to the Sacramento–San Joaquin Delta (DiToro *et al.*, 1971) and later to Lake Ontario (Thomann *et al.*, 1975) and will be referred to as the DiToro model (see Fig. 21).

DI TORO FORMULATION INTEGRATED OVER DEPTH

$$\frac{dC}{dt} = 2.1\ \mathrm{Ch}\ R\ \frac{e}{k_e (Z_l - Z_u)} \left[\exp\left(\frac{I_o}{I_s} e^{-k_e z_l}\right) - \exp\left(\frac{I_o}{I_s} e^{-k_e z_a}\right) \right] \left[\mathrm{TEMP.}\right] \left[\mathrm{NUT}\right]$$

BANNISTER FORMULATION INTEGRATED OVER DEPTH

$$\frac{dC}{dt} = 12.0\ \emptyset\ K_c\ \mathrm{Ch}\ \frac{I_s e}{k_e (Z_l - Z_u)} \left[\exp\left(\frac{I_o}{I_s} e^{-k_e z_l}\right) - \exp\left(\frac{I_o}{I_s} e^{-k_e z_u}\right) \right] \left[\mathrm{NUT}\right]$$

Fig. 21. Gross primary production models. Notation for production models: *dC/dt*, photosynthetic carbon assimilation rate; Ch, chlorophyll *a* concentration; *R*, phytoplankton carbon : chlorophyll; *T*, ambient water temperature; I_o, incident surface solar radiation (PAR); I_s, optimal growth solar radiation intensity (PAR); K_e, extinction coefficient (base *e*); *Z*, depth; P, N, Si, chemical nutrient concentrations; K_p, K_n, K_{si}, Michaelis–Menten half-sautration constants; ϕ, maximum photosynthetic quantum yield in weak light; K_c, radiation extinction due to absorption by a unit concentration of chlorophyll *a*; Z_u, Z_l, upper and lower depth boundaries of a homogeneous vertical layer in the water column.

In the DiToro model, at conditions of optimal incident solar radiation and nonlimiting nutrient concentrations, phytoplankton growth is assumed to progress proportionally to the standing algal crop at a rate determined by the ambient temperature. The utilized temperature effect is of an exponential form first postulated by Eppley (1972) based on a synthesis of maximum growth rates observed in the field by a number of investigators. The solar radiation effects in the DiToro model are based on a formulation proposed by Steele (1962) and account for both limitation at suboptimal light levels and inhibition above the optimal light level. Chemical nutrient effects are modeled as products of Monod growth kinetic terms. In cases where a diatom was the dominant alga, a silica nutrient term was used in addition to those indicated for phosphorus and nitrogen.

An alternate model formulation has been proposed by Bannister (1974) based on concepts of the quantum yield of photosynthesis in weak light. As outlined by Bannister, the photosynthetic rate of carbon assimilation is proportional to the amount of radiation absorbed by chlorophyll in the water column. This chlorophyll absorption of light energy makes up a part of the total extinction coefficient of the water column and can be coupled to a photosynthesis–light relationship, such as that formulated by Steele. The photosynthetic absorption of light energy is related to the rate of carbon assimilation by a quantum yield factor, ϕ, which has been demonstrated not to be a function of temperature. To complete the equation in a format comparable to the DiToro formulation, multiplicative Monod substrate kinetic terms were introduced and the Steele formulation for light utilization adopted (see Figure 21).

Comparison of the results from the two gross production models and the measured data were presented in Fig. 17 to 20. Correspondence between the measured data and model responses is quite variable, but on the whole encouraging. In general good correspondence is found between the model profiles and field data. In relation to the variance from the data to which they are applied, both model formulations appear to be equally applicable to the description of gross production in Onondaga Lake.

Nutrient Sources

Walker (1976) utilized the water quality data collected on Onondaga Lake and its tributaries from 1970 through 1974 to establish mass balances for a number of parameters. Those parameters for which mass balances were established include: chloride, nitrate and nitrite nitrogen, total Kjeldahl nitrogen, total nitrogen, silica alkalinity (as $CaCO_3$), total inorganic phosphorus, ortho phosphorus, condensed inorganic phosphorus, and 5-day biochemical oxygen demand (BOD_5). Walker's analyses are summarized in Table 10.

In order to establish a mass balance for total inorganic phosphorus attributable to various load sources, an analysis was made of data collected during 1975 on each of the major tributaries to Onondaga Lake. Four broad categories of possible loadings were determined as follows: (1) residential runoff and combined sewer overflows, (2) woodland and agricultural with minor residential runoff, (3) direct industrial discharge, and (4) sanitary sewage effluent.

The total inorganic phosphorus loading from Ninemile Creek was used as a baseline to represent the background levels in the other tributaries due to woodland and agricultural area runoff. Of the four major tributaries to Onondaga Lake, Ninemile Creek contains the most rural (woodland and agricultural) areas throughout its length. Based on the 1975 data, it was determined that a background level of 0.28 lb $acre^{-1}yr^{-1}$ was appropriate for the Onondaga Lake drainage basin. This value compares favorably with values reported by others for similar areas (Dornbush *et al.*, 1974).

It was determined that the sanitary sewage effluent discharged to the Lake from the metropolitan Syracuse sewage treatment plant contributed most of the total inorganic phosphorus loading. The loadings were as follows:

1. Residential runoff and combined sewer overflows (Harbor Brook, Onondaga Creek, and Ley Creek) 21.9%

TABLE 10

Onondaga Lake Mass Balances[a]

Component	Water years	Total input	Total output	Change in storage	Accumulation	Retention[b] coefficient	Fraction of input due to Metro STP
Chloride	1970–1974	72814.0	70753.0	322.0	1739.0	0.024	0.055
Nitrate and nitrite nitrogen	1970–1972	47.4	25.7[b]	0.2	21.5	0.453	0.120
			(28.0)		(19.2)	(0.405)	
	1973–1974	24.5	20.8	−0.3	4.0	0.163	0.150
			(22.7)		(2.1)	(0.086)	
Total Kjeldahl nitrogen	1970–1971	273.8	267.3	−0.2	6.7	0.025	0.434
			(241.8)		(32.2)	(0.118)	
	1972–1974	262.3	189.1	+0.8	72.4	0.227	0.516
			(171.1)		(90.4)	(0.345)	
Total nitrogen	1970–1971	321.5	288.7	1.3	31.5	0.098	0.387
			(263.8)		(56.4)	(0.175)	
	1972–1974	294.9	214.9	0.0	80.0	0.271	0.471
			(196.4)		(98.5)	(0.334)	
Silica (SiO_2)	1970–1974	394.2	235.8	−6.3	164.7	0.417	0.247
Alkalinity (as $CaCO_3$)	1970–1974	10500.0	9038.0	37.0	1425.0	0.136	0.162
			(8799.0)		(1659.0)	(0.158)	
Total inorganic P	1970	156.7	79.4	−20.8	98.1	0.626	0.431
	1971	68.2	70.3	10.1	−12.2	−0.179	0.545
	1972–1974	36.3	20.9	−7.6	23.0	0.634	0.743
Orthophosphate	1970	48.0	39.5	−14.3	22.8	0.477	0.613
	1971	32.5	40.0	11.7	−19.2	−0.592	0.592
	1972–1974	23.7	15.7	−5.2	13.2	0.560	0.697
Condensed inorganic P	1970	108.7	39.9	−6.5	75.3	0.693	0.351
	1971	35.7	30.3	−1.6	7.0	0.196	0.502
	1972–1974	12.6	5.2	−2.4	9.8	0.778	0.831
Five day BOD	1970–1974	1389.0	289.0	−5.0	1105.0	0.796	0.792
			(217.0)		(1177.0)	(0.847)	

[a] Data given as gm m^{-2} $year^{-1}$.
[b] From Walker, 1976.

2. Woodland and agricultural areas (Ninemile Creek and upstream areas of Harbor Brook, Onondaga Creek, and Ley Creek)	9.0%
3. Direct industrial discharge (East Flume, Crucible)	0.6%
4. Sanitary sewage effluents	68.5%
	100.0%

A mass balance analysis was also done on the data collected during 1975 to determine the various load sources of total nitrogen to Onondaga Lake. The same loading categories were considered as for the total inorganic phosphorus analysis.

Ninemile Creek was again utilized as the baseline to represent the background levels present in the other tributaries. Based on the 1975 data, the background level of total nitrogen discharged to Onondaga Lake is 3.43 lb acre-1 year-1. This value falls within the range reported by others (Dornbush *et al.*, 1974).

It was determined that the sanitary sewage effluent discharged to the lake from the metropolitan Syracuse sewage treatment plant constitutes the largest single loading of total nitrogen. Most of the loading was in the form of ammonia nitrogen (NH_3 nitrogen). The loadings attributable to each of the categories are as follows:

1. Residential runoff and combined sewer overflows (Harbor Brook, Onondaga Creek, and Ley Creek)	8.3%
2. Woodland and agricultural areas (Ninemile Creek and upstream areas of Harbor Brook, Onondaga Creek, and Ley Creek)	12.1%
3. Direct industrial discharge (East Flume, Crucible)	23.0%
4. Sanitary sewage effluent	56.5%
	100.0%

As can be seen from this analysis and the work reported by Walker (1976) the major source of both total inorganic phosphorus and total nitrogen is the metropolitan Syracuse sewage treatment plant. The effluent from the treatment plant contributes more than enough phosphorus and nitrogen, the primary nutrients for algal growth, to the lake to promote excessive algal growth. Current upgrading of the treatment provided at the plant will help to reduce nutrient loadings in the future.

Phytoplankton Enrichment Studies

The concept of a limiting nutrient was an outgrowth of the Law of Minimum proposed by Liebig in 1840. It states that the nutrient present in least amount relative to requirements of a population determined the level of its growth. Much ecological work has been based on this theory despite the general realization that it is an oversimplification to assume that a single nutrient acts alone to determine the level of growth. It has been well established that interactions occur between different elements and that algae can store nutrients in excess of their requirements for use when nutrients become depleted in the environment. It is also well known that different algal populations vary greatly in their requirements for specific elements.

Philip Sze of the State University of New York at Buffalo has conducted phytoplankton enrichment studies on populations characteristic of Onondaga Lake (O'Brien & Gere Engineers, Inc., 1976). The enrichment studies so described evaluated the response of naturally occurring populations in Onondaga Lake after a week's incubation following enrichment. The results of the enrichment studies are presented in Table 11 and discussed below.

Silica. Each spring and some years also in the fall, there has been a correlation between diatom growth and silica concentrations in the epilimnion of the lake (Sze, 1975). The enrichment studies support earlier evidence for the limitation of diatom growth by availability of silica. *Cyclotella glomerata* clearly was stimulated in lake water enriched with silica on June 20 and September 18.

Phosphorus and Nitrogen. During the summer period, the green algae *Chlorella vulgaris* and *Scenedesmus obliquus* were the dominant species. As the summer progressed, stimulation of their growth by added phosphorus was increasingly apparent. The consistency of these results strongly suggest that phosphorus is at times in critical amounts in the lake. *Stimulation by addition of nitrate may not mean that total concentrations of nitrogen in preference to other forms of nitrogen.* Thus, it appears that the green algae deplete major nutrients as the summer progresses and that neither nitrogen nor phosphorus is available in great surplus in the lake.

Trace Elements. Generally, trace elements appeared to inhibit algal growth and this effect was partially reversed by addition of a chelating agent. This supports earlier evidence (Keenan, 1970; Sze and Kingsbury, 1972) suggesting a likelihood of inhibition when metal concentrations in the lake are high.

Effect of Phosphate Detergent Legislation

Legislation was enacted by the Common Council of the City of Syracuse on July 1, 1971 affecting the urban watershed influent to Onondaga Lake by limiting the phosphate composition of detergents to 8.7%. The state of New

TABLE 11

Results of Enrichment Experiment Treatments

	N	½N	P	½P	(N + P)	½(N + P)	Si	½Si	Trace	Trace + EDNA
June 20, 1975										
Chlamydomonas	1.25	1.38	0.88	0.94	no	no	0.63	1.15	4.19*	1.35
Chlorella vulgaris	0.84	0.82	1.05	1.55*	no	no	0.83	0.56*	0.23*	0.66*
Scenedesmus obliquus	0.99	1.07	1.02	1.43	no	no	1.03	0.53*	0.27*	1.08
Cyclotella glomerata	0.58	0.89	0.58	0.07*	no	no	3.18*	2.66*	0.38*	0.64*
July 9, 1975										
Chlorella vulgaris	0.97	0.92	1.21*	1.62*	1.14	1.45*	1.25	1.04	2.05*	0.96
Scenedesmus obliquus	0.96	0.95	0.89	0.57*	1.79*	2.74*	1.04	0.70	0.39*	0.61*
August 21, 1975										
Chlorella vulgaris	1.05	0.96	2.18*	1.40*	1.92*	1.06	0.86	0.94	0.59*	0.79*
Scenedesmus obliquus	0.96	1.40	1.98*	1.85*	2.43*	2.31*	0.91	1.22	0.41*	0.79
Cyclotella glomerata	0.48*	1.14	0.04*	0.78*	0.03*	0.29*	0.81	0.76	0.14*	0.61*
September 18, 1975										
Chlorella vulgaris	1.01	0.91	1.87*	1.44*	1.72*	1.49*	0.69*	0.62*	0.21*	0.77
Scenedesmus obliquus	1.12	1.59	1.88*	1.97*	4.93*	4.21*	1.69	1.19	0.40*	1.59*
Scenedesmus quadricauda	0.72	1.25	1.19	1.07	1.29	1.61*	1.27	1.04	0.25*	1.00
Cyclotella glomerata	0.91	0.88	0.13*	0.36*	0.20*	0.44*	1.86*	2.16*	0.19*	0.77
October 11, 1975										
Chlorella vulgaris	1.27*	1.28*	2.15*	1.46*	3.98*	3.85*	1.00	1.22	0.24*	1.45*
Scenedesmus obliquus	1.53	1.16	1.04	1.11	4.61*	6.64*	2.03*	1.59	0.55	1.76
Cyclotella glomerata	0.93	1.12	0.14*	0.35*	0.06*	0.41*	1.24	1.11	0.14*	0.45*
November 2, 1975										
Chlorella vulgaris	0.47*	0.54*	0.80	0.74	0.66	1.82	0.70*	0.78	0.04*	0.58*
Scenedesmus obliquus	0.60*	0.83	0.49*	0.76*	0.54*	1.02	0.40*	0.76	0.03*	0.66*
Cyclotella glomerata	0.80	1.00	0.07*	0.18*	0.05*	0.08*	0.28*	0.79	0.00*	0.43*
December 5, 1975										
Chlamydomonas	1.17	0.99	0.91	1.38	1.61*	1.03	0.60	0.41*	0.03*	0.42*
Scenedesmus obliquus	1.19	1.22	0.44*	0.83	0.52	0.59	0.87	0.75	0.01*	0.83
Cyclotella glomerata	0.45*	0.59*	0.04*	0.06*	0.08*	0.24*	1.03	1.01	0.01*	1.04

York, following the examples of Erie County, Syracuse, Suffolk County, and Bayville, enacted legislation eliminating the use of inorganic phosphates as sequestering agents in detergents. The statewide legislation became effective January 1, 1972. Reductions in condensed inorganic phosphorus have been subsequently observed in the tributaries discharging to Onondaga Lake as well as in the water column of Onondaga Lake.

Following the implementation of the city- and statewide legislation, a 78.1% reduction in the level of condensed inorganic phosphorus was measured in epilimnion waters (Murphy, 1973a, 1973b). A corresponding 56.8% decline in the level of inorganic phosphorus was measured in hypolimnion waters. Corresponding decreases in the measured orthophosphate concentrations observed during this same timeframe were 26% and 20% for the epilimnion and hypolimnion, respectively. The orthophosphate reductions have been related to the phosphate detergent legislation by means of a zero-order kinetic relationship for the hydrolysis of the condensed forms.

The phosphate detergent legislation appears to have altered the seasonal succession of plankton. The nuisance blue-green algae normally encountered in the late summer and early fall have not been observed in significant numbers since 1971. Instead, the green algae has remained generally dominant all summer and into the fall (Sze, 1975). The shift of the blue-greens out of the normal seasonal succession, coincident with a decrease in total phosphate within the lake, is consistent with the recent findings of Shapiro (1973) who has determined that blue-green algae will dominate in a mixed culture with green algae if the level of nutrients is increased. Shapiro feels that phosphate uptake kinetics favor the dominence of blue-green algae over green algae. Thus, where excess nutrients are present, blue-green algae would be expected to dominate.

The absence of blue-green algae generally indicates an improvement in the trophic status of Onondaga Lake. The green algae are generally considered more desirable in that they do not form objectionable floating scums which have been characteristic of Onondaga Lake for many years. Additionally, the green algae are generally more ecologically compatible because of their more intimate and direct linkage within the food chain.

Attendant with the decrease in phosphorus levels in 1972 was the increase in algal diversity index over that measured in the previous year (O'Brien & Gere Engineers, Inc. 1972b). The 1971 average algal diversity index was calculated to be 0.695 compared with the 0.801 measured in 1972. This is not by itself significant, but taken together with a decrease in phosphorus levels and an improvement in the character of the algal succession, it supports the contention that the phosphate detergent legislation has resulted in the general improvement of the trophic status of Onondaga Lake.

Effect of Interceptor Maintenance Program

In 1971 a program was introduced whereby the combined sewer relief valves or weirs were maintained on a regular basis. Routine maintenance has resulted in a reduction in the volume of municipal wastewater being discharged to the receiving stream under dry weather conditions due to blockages in the interceptor system. This corrective program has resulted in rerouting nearly 5.7×10^3 m^3/day (15 MGD) of sanitary wastes, constituting dry weather overflow, to the metropolitan Syracuse sewage treatment plant.

Following the initiation of the above-mentioned corrective program, Onondaga Lake showed marked increases in dissolved oxygen concentrations, particularly in the bottom waters. In 1972, at the 9-m depth, dissolved oxygen was detectable above 1 mg/liter at all times of the year. This compares with a 3-month period of no dissolved oxygen measured in 1970. Similarly, the anaerobic period measured in 1970 at the 12-m depth has been reduced from 4 months to 1 month. As measured in the southern basin of Onondaga Lake, the hypolimnion and epilimnion dissolved oxygen concentrations have increased 21.2% and 23.2%, respectively, as a result of the implementation of the maintenance program.

Improvement in dissolved oxygen levels in Onondaga Lake is not the only result of the interceptor maintenance program. Onondaga Creek, one of the tributaries receiving the combined sewer overflow, has undergone a 76% reduction in organic nitrogen during the 1970–1972 time frame. This is reflected in a reduction of organic nitrogen in the epilimnion and hypolimnion of Onondaga Lake by 6.1% and 10.1%, respectively, when compared to earlier data.

Impact of the Construction of Industrial Wastewater Treatment Facilities

Among the significant changes observed in the chemistry of Onondaga Lake during the 1975 monitoring program are reductions in the concentration of a number of toxic metals, particularly copper and chromium (O'Brien & Gere Engineers, Inc., 1975). Epilimnion copper concentrations were observed to decline by 51% over those measured in 1974, followed by reductions in chromium and iron of 47% and 35%, respectively. The reductions in the levels of heavy metals is significant in the establishment of a less toxic aquatic environment within Onondaga Lake.

The reduction of the levels of heavy metals is attributed to the start-up of the Crucible Steel wastewater treatment facilities which was placed on line in the fall of 1974. The treatment system involves the significant reuse of process wastewaters and the chemical precipitation of the recycle bleed-off line.

Special Problems

Combined Sewer Overflow Discharges

One of the major water quality problems which interferes with the best usage of Onondaga Lake involves the excessive measured levels of bacterial indicator organisms within the lake waters. The average epilimnion total coliform level measured during 1974 was 19,079 colonies/100 ml with a corresponding average fecal coliform level of 341 colonies/100 ml (see Appendix A, Table A-18). Fecal *Streptococcus* average levels were found to be 270 colonies/100 ml. Hypolimnion levels were found to be lower than those measured in the epilimnion except for the fecal *Streptococcus* concentrations, the average value of which was determined to be 357 colonies/100 ml showing the greater resistance of the fecal *Streptococcus* indicator group to total dissolved solids induced die-away.

The most significant source of bacterial contamination of Onondaga Lake are the approximately 70 combined sewer overflow devices which discharge directly to Onondaga Creek, Harbor Brook, and Ley Creek. During wet weather events, wastewater flows in excess of the collection system or treatment plant's capacity are discharged to the three tributaries by means of the relief devices. It has been estimated that wet-weather overflows occur on an average of nine times per month with a total monthly overflow duration of approximately 24 hr. The combined sewer overflows convey wastewaters having five to eight log levels of indicator organisms.

A study is presently underway to evaluate the impact and subsequent treatment of the combined sewer overflow discharges under an EPA and Onondaga County sponsored program (EPA Project No. 11020 HFR). The relative impact of the combined sewer overflow discharges can be seen by evaluating the preliminary output of the coliform mathematical model of Onondaga Lake described in the following section and shown in Figs. 24 and 25.

A review of the mathematical model output shows that under dry weather conditions, all segments of Onondaga Lake should meet water quality standards for contact recreation except for a very small area in the vicinity of the outfall of the metropolitan Syracuse sewage treatment plant. Under wet-weather conditions, however, fecal coliform water quality standards are exceeded over an extensive area of Onondaga Lake.

A minimal treatment of combined sewer overflow wastewaters by primary treatment and high-rate disinfection would result in a significant improvement in the water quality of Onondaga Lake. The implementation of such an abatement program would render Onondaga Lake waters suitable for contact recreation.

Mercury Concentrations Found in the Lake's Fish Population

Discharges of mercury to Onondaga Lake have been greatly reduced in recent years. A soda ash manufacturing plant, located on the western shore of Onondaga Lake, was found to be discharging approximately 9.55 kg/day (21 lb/day) of mercury to the Lake (Onondaga County, 1971). The New York State Department of Environmental Conservation indicated that this discharge had been reduced to less than .454 kg/day (1 lb/day) (Onondaga County, 1971). Data for the period July 1971 through June 1974 show that daily mercury discharges averaged .21 kg/day (.470 lb/day) (New York State Department of Environmental Conservation, personal communication, 1977). Other possible sources of mercury to Onondaga Lake, including the levels in freshly fallen snow and background levels in the atmosphere were investigated by the NYSDEC. It was found that in 1970 the average mercury levels of freshly fallen snow collected from several locations in the Onondaga Lake drainage basin ranged from $.08 \times 10^{-3}$ to 1.16×10^{-3} ppm. Background levels in the atmosphere averaged 1.1×10^{-3} ppm (New York State Department of Environmental Conservation personal communication, 1977).

Significant quantities of mercury persist in the bottom sediments. Core samples of the bottom sediments were obtained during 1969 and 1970 as a part of the Onondaga Lake Study. Analysis of the bottom sediments revealed that average concentrations of mercury ranged from 0.15–1.15 ppm depending on core depth (New York State Department of Environmental Conservation, personal communication, 1977). The mercury contained in the bottom sediments is a continuing source of contamination which is exhibited by the accumulation of mercury compounds in fish flesh.

In 1970, fishing in Onondaga Lake was prohibited by New York State due to the high concentrations of mercury found in the Lake's fish population and in other organisms forming the lower food web. Analyses for mercury levels in fish from Onondaga Lake in 1970 showed that average concentrations for three species of fish ranged from 1.6–5.8 ppm, well above the .5-ppm actionable mercury level (New York State Department of Environmental Conservation, personal communication, 1977). Organisms belonging to the lower food web in Onondaga Lake, including zooplankton, algae, and weed growths, were found to have mercury concentrations ranging from .49–5.83 ppm (New York State Department of Environmental Conservation, personal communication, 1977).

Fish specimens analyzed since 1970 have shown a decidedly downward trend in mercury concentrations shown in Fig. 22. The ban on fishing continues at present and will likely continue for the immediate future. Restrictions on fishing will probably remain in effect until testing of fish from Onondaga Lake indicates that mercury concentrations have fallen below the .5-ppm level and are likely to remain below that level.

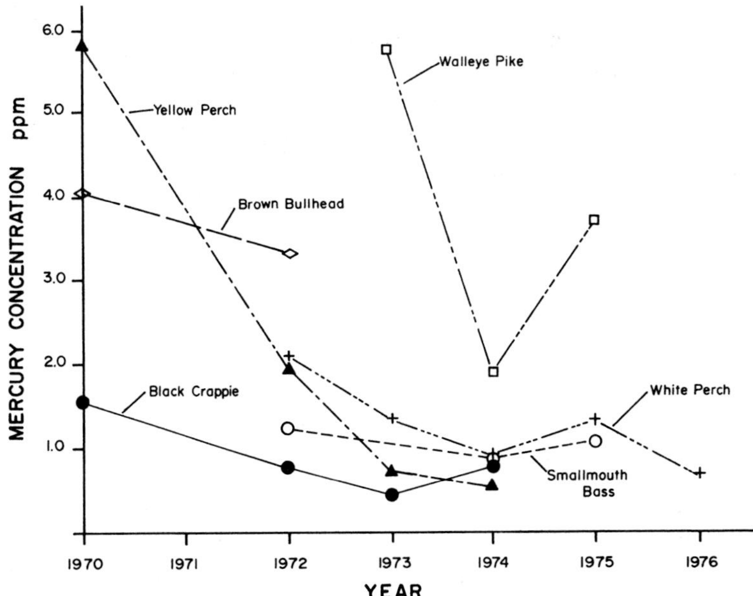

Fig. 22. Mercury concentrations in Onondaga Lake fish populations. [Source: New York State Bureau of Environmental Protection, Rome Analytical Laboratory, Rome, New York, (December 1977, personal communication).]

MATHEMATICAL MODELING

During the summer of 1973 (O'Brien & Gere Engineers, Inc., 1974) a study was initiated with the assistance of Professor Raymond Canale of the University of Michigan to develop mathematical models for water quality in Onondaga Lake. The first step was the development of a water-balance model for the lake and subsequent computation of the lake residence time. A 28-segment steady-state model for chloride, sodium, and potassium was then developed based on measured loadings and an assumed circulation pattern for the lake. This model was verified using observed field data. A non-steady-state or dynamic, four-segment model was then developed for chloride, sodium, and potassium using the simulation program. This model included the effects of lake turnover, time-variable flow and pollutant loading. The comparison between the model calculations and the observed field data was considered encouraging. The chloride model can be used to predict the effect of chloride loading control alternatives. The models were also used to obtain a preliminary estimate of the effect of combined sewer overflow management on coliform levels in the lake.

Because the combined and stormwater management project and the analysis of the effect of treatment on lake coliform levels are most important to optimum usage of Onondaga Lake, the dynamic fecal coliform model was developed (O'Brien & Gere Engineers, Inc., 1975). The fecal coliform model is a system of 28 simultaneous differential equations, each of which represents a mathematical statement of fecal coliform continuity for each model segment.

Factors in the model include transport due to fluid velocity and dispersion, reaction due to bacterial die-away, and point source loading. The relationship between these factors is described by the following mathematical relationship:

$$\frac{V_k dc_k}{dt} = \sum [-Q_{kj}(\alpha_{kj}c_k + \beta_{kj}c_j) + E'_{kj}(c_j - c_k)] - V_k K_k c_k + W_k$$

where c_k equals the concentration of fecal coliform in segment k; V_k, volume of segment k; Q_{kj}, net flow from segment k to segment j; α_{kj}, finite difference weighting factor, β_{kj} $(1 - d_{kj})$; E'_{kj}, dispersive mixing flow between segment k and j; K_k, bacterial die-away rate coefficient; W_k, loading of fecal coliform to segment k; c_j, concentration of fecal coliform in adjacent segments.

The model assumes that the rate of coliform die-away is proportional to the coliform concentration and that the proportionality factor is variable as a function time. The coliform die-away rates determined in Onondaga Lake are much greater than that determined in other fresh water bodies. This difference in derived die-away rate may be due to the high salinity of Onondaga Lake since it is well known that coliform die-away is more rapid in the ocean than in fresh water.

To date, the coliform loadings due to a combined sewer overflow event have not been developed. However, based on wet-weather tributary fecal coliform concentrations determined on 5/21/74, 6/27/74, and 8/1/74, the preliminary simulation of the wet-weather combined sewer overflow induced tributary loading has been developed. The preliminary background mass loading figures for the tributaries are as follows (O'Brien & Gere Engineers, Inc., 1975):

Tributary	Mass loading (colonies/day)
Harbor Brook	6.0×10^{13}
Onondaga Creek	6.7×10^{13}
Ley Creek	7.1×10^{12}
Ninemile Creek	—
Metropolitan discharge	2.0×10^{10}
Crucible discharge	1.5×10^{11}
East Flume	7.8×10^{10}

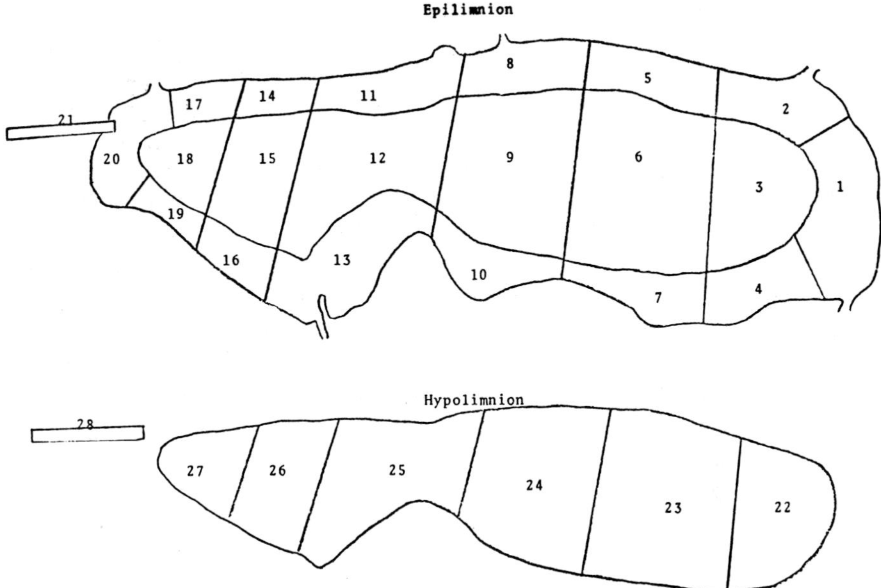

Fig. 23. Segmentation of Onondaga Lake, fecal coliform model.

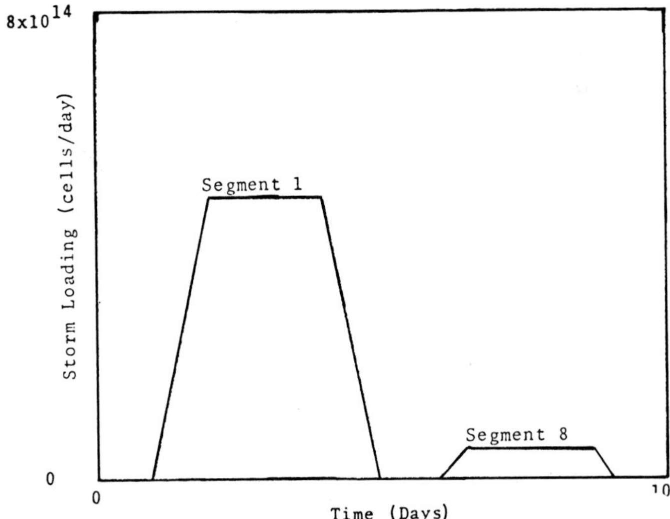

Fig. 24. Hypothetical fecal coliform loading.

For the initial fecal coliform simulation, it was assumed that a typical large overflow would have a discharge of about $11.3 \times 10^3 m^3$/day (30 MGD) and a fecal coliform concentration of 60,000 colonies/100 ml. This results in a load of about 6.7×10^4 colonies/day which was input into segment 1 of the model. The segmentation for the fecal coliform model is shown in Fig. 23 while the simulated stormwater fecal coliform loading is shown in Fig. 24.

Figure 25 shows the fecal coliform level in the lake assuming K = 0.5/day under background loading while Fig. 25 shows the impact of a simulated combined sewer overflow discharge into segment 1. It can be seen from Fig. 26 that the peak lake concentration is predicted to peak to 36,000 colonies/100 ml following a wet-weather event. The model also predicts little influence on the hypolimnion waters and the northernmost region of the lake, the area having the most potential contact recreational use.

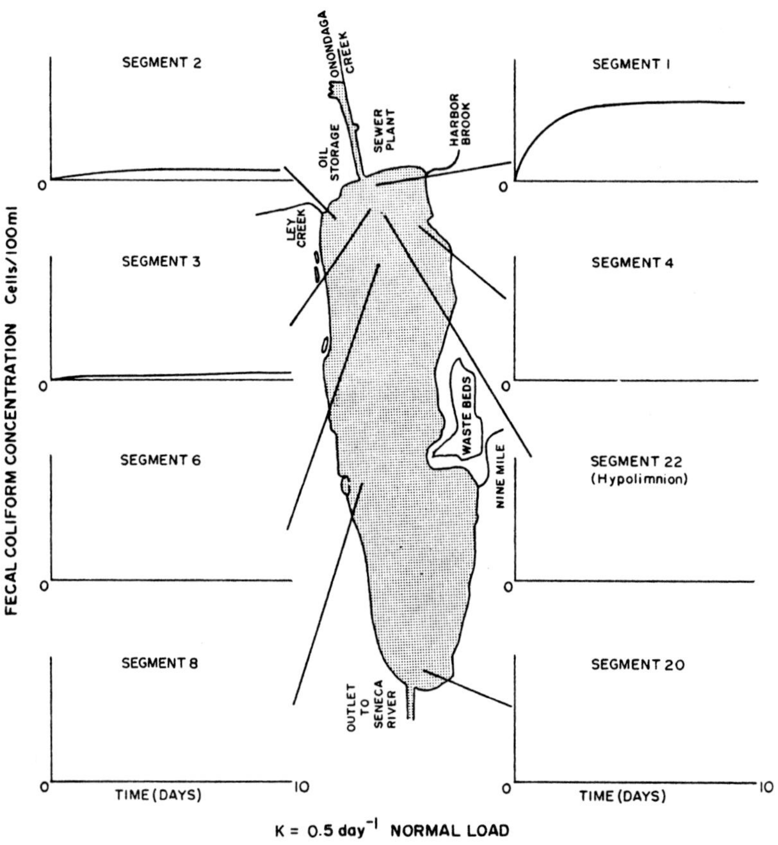

Fig. 25. Normal fecal coliform loading, Onondaga Lake.

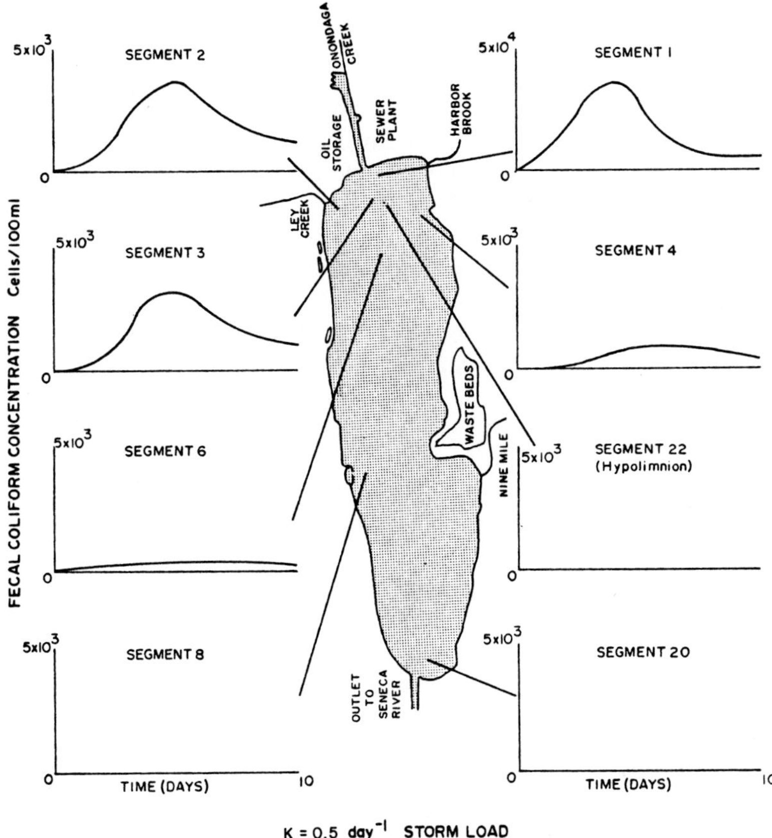

Fig. 26. Simulated fecal coliform loading during a stormwater overflow, Onondaga Lake.

The fecal coliform model for Onondaga Lake will be used to determine the level of disinfection required for the combined sewer overflow in order to protect the best usage of the receiving waters. As part of this analysis, an extensive verification of the fecal coliform model has been initiated (O'Brien & Gere Engineers, Inc., 1976).

CONCLUSIONS

Onondaga Lake is a eutrophic water body, the trophic status of which is principally determined by the intensive cultural influence on the drainage basin. The trophic status has shown some improvement since 1970 due to the influence of the interceptor maintenance system improvements, phosphate detergent legislation, and improvements in industrial wastewater treatment facilities tributary to the lake. The reduction in nutrient levels in Onondaga Lake has resulted in conditions which has altered the algal species present within the water column. During the summer months phosphorus reaches values likely to limit the level of algal productivity and the intensity and frequency of algal blooms.

The application of tertiary treatment to the metropolitan Syracuse sewage treatment plant effluent will probably reduce the epilimnion orthophosphate levels by approximately 50%. The algal standing crop will probably be significantly reduced following operation of the facilities. The completion of the modifications to the metropolitan Syracuse sewage treatment plant, involving biological contact stabilization, secondary clarification, and tertiary treatment, will significantly improve the present epilimnetic and hypolimnetic water quality of Onondaga Lake.

Primary treatment and adequate disinfection of the combined sewer overflow wastewaters which are presently discharged via overflows to Onondaga Creek, Harbor Brook, and Ley Creek should result in an improved environment in the lake to support contact recreation. This is supported by the preliminary fecal coliform model output.

The Onondaga Lake Monitoring Program, supported by Onondaga County, is one of the most unique limnological programs presently conducted in the United States. The monitoring program data base may well serve as one of the best records of a community's effort to reverse the cultural degradation of one of its most unique resources. The vision of Onondaga County should serve as an example to regulatory agencies and other communities.

RECOMMENDATIONS

The Onondaga Lake Monitoring Program and associated substudies should be continued and Onondaga County should continue to pursue an effective means of reducing the impact of combined sewer overflows on the lake.

Efforts should be continued to develop a predictive phytoplankton model of the lake. A requisite preliminary objective in this regard is the development of a model for algal nutrient concentrations, especially phosphorus. Algal enrichment studies should be continued to better define phytoplankton–nutrient interactions.

Special efforts should be devoted to the field and laboratory activities necessary to obtain undisturbed core samples from the lake sediments. Such study is required to better understand the transport, transformation reactions, and availability to the ecosystem of such parameters as phosphorus, heavy metals, PCB's, and sulfides. An understanding of these mechanisms is essential to the prediction of lake recovery time for pollution abatement strategies aimed at these constituents.

APPENDIX A: ONONDAGA LAKE TRIBUTARY DATA

TABLE A-1

Average Concentrations, 1969 (mg/liter)[a]

Parameter	Ley Creek	Bloody Brook	Onondaga Creek	Harbor Brook	East Flume	Steel Mill	Ninemile Creek	Metro plant
Alkalinity	274.0	156.7	217.7	227.7	34.8	16.3	147.0	214.2
BOD_5	198.7	0.0	5.7	19.3	0.0	13.4	2.7	90.1
Chloride	252.0	655.7	446.7	1916.2	805.9	1158.1	3282.3	838.7
Ammonia nitrogen	7.5	0.3	0.5	3.2	0.9	0.0	0.0	8.8
Organic nitrogen	4.0	0.1	1.6	3.5	1.5	1.0	0.0	1.4
Nitrate	0.5	0.5	0.9	1.8	0.4	10.9	0.2	0.4
Nitrite	36.0	0.0	34.0	52.8	0.0	0.0	53.1	64.0
Total phosphate	5.7	1.8	3.7	4.7	1.1	1.0	1.0	11.4
OPO_4	3.2	0.6	0.6	1.9	0.3	0.1	0.1	3.5
Sulfate	312.1	163.5	189.2	358.1	40.1	107.7	221.4	182.5
Dissolved oxygen	2.5	6.0	9.5	6.2	15.1	4.4	5.3	1.1
Calcium	217.8	178.7	282.5	389.3	252.3	623.3	2227.2	183.2
Sodium	379.8	53.2	380.7	1306.7	355.8	830.9	1490.0	367.8
Magnesium	37.0	25.6	65.2	67.1	5.2	38.6	28.2	32.0
Potassium	20.5	13.6	7.3	13.4	18.5	48.4	21.1	23.7
Copper	0.1	0.1	0.1	0.1	0.2	0.3	0.1	0.1
Chromium	0.1	0.1	0.0	0.0	0.1	2.6	0.0	0.1
Iron	1.4	1.0	0.6	0.7	0.8	14.0	1.8	0.9
Manganese	0.4	0.4	0.8	1.1	0.0	0.0	0.0	0.1
Zinc	0.5	2.4	0.2	0.2	0.5	0.5	0.1	0.1
Fluoride	2.2	4.2	2.1	2.3	1.0	3.1	1.5	1.1
Silicate	12.4	6.7	8.2	10.3	1.6	2.4	5.2	14.0

[a] Source: Onondaga Lake Study (April, 1971).

TABLE A-2

Average, 1969 (lb/day)[a]

Parameter	Ley Creek	Bloody Brook	Onondaga Creek	Harbor Brook	East Flume	Steel Mill	Ninemile Creek	Metro plant	Total
Alkalinity	82,266	4,770	138,725	13,636	23,219	882	90,801	86,716	441,015
BOD_5	53,576	0	3,614	1,300	0	726	1,506	38,503	99,230
Chloride	67,869	25,997	271,905	171,611	537,687	62,779	1,744,830	350,058	3,232,736
Ammonia nitrogen	1,286	5	210	110	603	0	0	3,097	5,311
Organic nitrogen	686	2	561	184	994	54	0	537	3,018
Nitrate	93	10	305	118	298	590	104	160	1,678
Nitrite	26,041	1	52,538	11,685	22	1	71,779	32,090	194,157
Total phosphate	1,760	73	2,434	308	703	55	732	4,883	10,948
OPO_4	961	24	405	119	200	6	116	1,374	3,205
Sulfate	101,619	5,422	112,524	20,604	26,768	5,839	138,734	73,567	485,077
Dissolved oxygen	974	212	6,361	422	10,084	236	2,938	420	21,647
Calcium	73,189	5,844	217,596	25,248	168,321	33,787	1,493,598	76,640	2,094,223
Sodium	127,891	1,787	284,165	115,551	237,371	45,045	990,170	151,892	1,953,872
Magnesium	12,338	859	46,851	3,774	3,457	2,090	21,169	13,267	103,805
Potassium	5,697	281	5,897	976	12,356	2,623	14,134	10,015	51,979
Copper	42	4	90	9	105	18	73	35	376
Chromium	28	2	23	4	80	142	28	52	359
Iron	427	34	409	43	533	756	942	354	3,498
Manganese	140	15	472	70	0	0	0	41	738
Zinc	174	96	145	16	329	25	122	56	963
Fluoride	385	117	1,151	163	661	166	492	399	3,534
Silicate	3,993	235	5,383	586	1,074	132	3,183	5,542	20,128

[a] Source: Onondaga Lake Study (April, 1971).

TABLE A-3

Percent of Total Pounds/Day, 1969[a]

Parameter	Ley Creek	Bloody Brook	Onondaga Creek	Harbor Brook	East Flume	Steel Mill	Ninemile Creek	Metro plant
Alkalinity	18.7	1.1	31.5	3.1	5.3	0.2	20.6	19.7
BOD_5	54.8	0.0	3.7	1.3	0.0	0.7	1.5	39.1
Chloride	2.1	0.8	8.4	5.3	16.6	1.9	54.0	10.8
Ammonia nitrogen	24.2	0.1	4.0	2.1	11.3	0.0	0.0	58.3
Organic nitrogen	22.7	0.1	18.6	6.1	32.9	1.8	0.0	17.8
Nitrate	5.5	0.6	18.1	7.1	17.8	35.2	6.2	9.5
Nitrite	13.4	0.0	27.1	6.0	0.0	0.0	37.0	16.5
Total phosphate	16.1	0.7	22.2	2.8	6.4	0.5	6.7	44.6
OPO_4	30.0	0.7	12.6	3.7	6.2	0.2	3.6	42.9
Sulfate	20.9	1.1	23.2	4.2	5.5	1.2	28.6	15.2
Dissolved oxygen	4.5	1.0	29.4	2.0	46.6	1.1	13.6	1.9
Calcium	3.5	0.3	10.4	1.2	8.0	1.6	71.3	3.7
Sodium	6.5	0.1	14.5	5.9	12.1	2.3	50.7	7.8
Magnesium	11.9	0.8	45.1	3.6	3.3	2.0	20.4	12.8
Potassium	11.0	0.5	11.3	1.9	23.8	5.0	27.2	19.3
Manganese	19.0	2.1	63.9	9.5	0.0	0.0	0.0	5.6
Zinc	18.1	10.0	15.1	1.7	34.2	2.6	12.6	5.8
Fluoride	10.9	3.3	32.6	4.6	18.7	4.7	13.9	11.3
Copper	11.2	1.0	24.0	2.4	27.9	4.8	19.4	9.3
Chromium	7.7	0.6	6.5	1.0	22.4	39.5	7.8	14.6
Silicate	19.8	1.2	26.7	2.9	5.3	0.7	15.8	27.5
Iron	12.2	1.0	11.7	1.2	15.2	21.6	26.9	10.1

[a] Source: Onondaga Lake Study (April, 1971).

TABLE A-4

Lake Residence Equivalents, 1969 (mg/liter)[a]

Parameter	Ley Creek	Bloody Brook	Onondaga Creek	Harbor Brook	East Flume	Steel Mill	Ninemile Creek	Metro plant
Alkalinity	39.9	2.3	67.3	6.6	11.3	0.4	44.0	42.1
BOD_5	26.0	0.0	1.8	0.6	0.0	0.2	0.7	18.7
Chloride	32.9	12.6	131.9	83.2	260.8	30.4	846.3	169.8
Ammonia nitrogen	0.6	0.0	0.1	0.1	0.3	0.0	0.0	1.5
Organic nitrogen	0.3	0.0	0.3	0.1	0.5	0.0	0.0	0.3
Nitrate	0.0	0.0	0.1	0.1	0.1	0.3	0.1	0.1
Nitrite	12.6	0.0	25.5	5.7	0.0	0.0	34.8	15.6
Total phosphate	0.9	0.0	1.2	0.1	0.3	0.0	0.4	2.4
OPO_4	0.5	0.0	0.2	0.1	0.1	0.0	0.1	0.7
Sulfate	49.3	2.6	54.6	10.0	13.0	2.8	67.3	35.7
Dissolved oxygen	0.5	0.1	3.1	0.2	4.9	0.1	1.4	0.2
Calcium	35.5	2.8	105.5	12.3	81.6	16.4	724.4	37.2
Sodium	62.0	0.9	137.8	56.0	115.1	21.8	480.2	73.7
Magnesium	6.0	0.4	22.7	1.8	1.7	1.0	10.3	6.4
Potassium	2.8	0.1	2.9	0.5	6.0	1.3	6.9	4.9
Copper	0.02	0.00	0.04	0.00	0.15	0.01	0.04	0.02
Chromium	0.01	0.00	0.01	0.00	0.04	0.07	0.01	0.02
Iron	0.2	0.0	0.2	0.0	0.3	0.4	0.5	0.2
Manganese	0.1	0.0	0.2	0.0	0.0	0.0	0.0	0.0
Zinc	0.1	0.0	0.1	0.0	0.2	0.0	0.1	0.0
Fluoride	0.2	0.1	0.6	0.1	0.3	0.1	0.2	0.2
Silicate	1.9	0.1	2.6	0.3	0.5	0.1	1.5	2.7

[a] Source: Onondaga Lake Study (April, 1971).

TABLE A-5

Average Concentrations, 1970 (mg/liter)[a]

Parameter	Ley	Bloody	Onon-daga	Harbor	East Flume	Crucible	Ninemile Creek	Metro plant	Landfill	Sawmill Creek	Outlet
BOD	24.759	6.149	5.604	19.977	14.106	10.532	10.797	127.102	21.753	5.519	3.433
Dissolved oxygen	5.049	6.866	9.004	7.697	4.376	4.942	9.376	3.640	3.024	7.333	5.189
pH	7.547	7.666	7.926	7.753	0.378	1.566	6.832	7.497	7.843	7.671	7.845
Alkalinity	208.958	172.086	218.625	279.136	25.554	18.317	192.507	214.913	818.687	234.277	176.200
Conductivity	1493.409	1516.904	1490.681	1644.523	322.640	655.937	6267.232	1770.047	5264.286	1870.625	3757.777
Ca	253.952	269.085	444.285	365.736	5.780	7.400	1728.693	588.619	488.346	471.785	261.111
Mg	51.628	31.313	52.847	64.637	10.331	13.321	70.802	42.316	108.319	58.933	33.935
Na	216.514	192.044	268.485	85.510	120.806	61.796	1465.315	646.428	944.146	411.953	231.994
K	7.943	6.920	4.469	4.931	7.319	7.982	23.780	25.129	51.771	7.333	7.511
Cl	1039.583	732.608	616.666	240.909	1917.142	1406.031	4686.883	660.869	2331.562	643.055	1623.750
SiO_2	7.768	7.335	6.161	9.169	2.779	2.711	10.664	16.632	15.399	6.131	5.944
SO_4	192.833	185.954	138.541	333.142	35.189	81.911	207.623	216.478	433.538	240.117	186.700
Total phosphate	1.545	2.530	1.720	3.168	0.799	1.061	2.012	5.460	4.099	2.411	1.384
Ortho-P	0.579	0.325	0.627	0.897	0.245	0.227	0.259	2.632	2.493	0.764	0.074
F	0.170	14.671	0.551	0.744	0.497	5.683	0.150	0.886	0.461	0.461	0.279
NH_3 nitrogen	1.815	0.912	2.612	2.874	2.004	0.573	0.757	4.951	23.106	0.750	1.952
Organic nitrogen	1.977	1.334	1.811	2.928	1.586	0.624	0.960	9.542	7.273	1.304	1.536
NO_3	0.552	0.490	0.864	1.493	1.026	2.970	1.050	0.457	1.456	0.614	0.306
NO_2	0.052	0.015	0.028	0.089	0.054	0.591	0.013	0.105	0.074	0.176	0.010
Cr	0.113	0.069	0.060	0.532	0.217	1.144	0.138	0.154	0.062	0.052	0.037
Cu	0.262	0.121	0,136	0.100	0.441	0.141	0.096	0.146	0.109	0.134	0.108
Fe	0.713	0.372	0.245	0.181	0.963	1.253	0.661	0.586	0.381	0.352	0.143
Mn	0.501	0.248	0.206	0.111	0.229	0.459	0.276	0.089	0.514	0.668	0.659
Zn	0.214	0.172	0.181	0.097	0.185	0.115	0.153	0.291	0.316	0.159	0.139

[a] Source: Onondaga Lake Study (1970).

TABLE A-6

Average, 1970 (lb/day)[a]

Parameter	Ley	Bloody	Onondaga	Harbor	East Flume	Crucible	Ninemile Creek	Metro Plant
BOD	8,846.230	724.225	11,431.302	3,245.972	9,411.726	570.949	24,582.687	63,080.421
Dissolved oxygen	3,307.674	832.172	18,653.656	1,280.655	2,920.301	267.919	27,167.554	1,886.091
pH	4,296.955	868.801	16,436.753	1,246.223	252.701	84.929	19,240.769	3,747.496
Alkalinity	119,839.390	20,142.972	437,913.813	47,649.679	17,049.726	992.968	502,282.313	106,603.359
Conductivity	772,775.751	187.641.250	2,388,820.505	251,909.531	215,265.750	35,558.390		872,388.626
Ca	157,090.312	36,861.671	1,094,290.503	59,234.320	3,856.418	401.163	3,815,105.005	297,354.250
Mg	27,515.566	3,001.222	167,563.562	9,706.406	6,893.126	722.143	148,550.719	21,848.816
Na	138,092.937	33,519.289	753,103.126	13,606.126	80,602.203	2,350.011	3,238,357.505	330,997.563
K	3,989.105	1,164.493	9,541.039	794.043	4,883.581	432.755	51,082.875	12,703.792
Cl	520,948.625	111,903.578	2,837,961.505	35,332.656	1,279,116.752	76,220.953		339,863.813
SiO_2	4,039.449	779.279	11,133.310	1,523.921	1,854.510	146.996	31,379.761	8,265.101
SO	110,123.484	19,540.316	291,508.813	51,554.851	23,478.476	4,440.428	556,977.126	107,783.093
Total phosphate	869.666	248.899	2,285.362	487.515	533.203	57.548	5,630.959	2,687.502
Ortho-P	392.106	45.485	844.646	140.530	164.019	12.358	953.765	1,307.218
F	69.094	1,236.389	1,418.244	120.866	331.709	308.098	340.897	477.052
NH_3 nitrogen	940.842	123.342	7,202.110	434.946	1,337.364	31.108	2,591.598	2,275.165
Organic nitrogen	1,519.024	212.889	5,422.439	497.410	1,058.345	33.847	3,061.074	4,404.936
NO_3	251.346	52.477	1,116.383	191.574	684.832	161.044	2,435.909	230.818
NO_2	11.239	0.765	32.953	12.880	36.695	32.070	21.475	51.578
Cr	71.017	7.248	279.783	68.275	145.153	62.046	194.116	79.474
Cu	114.443	12.124	469.642	16.640	294.679	7.683	151.862	72.763
Fe	283.984	28.348	529.047	26.470	642.735	67.963	909.196	295.174
Mn	201.863	25.180	643.546	18.645	153.455	24.884	406.079	43.549
Zn	103.903	22.173	668.741	13.567	123.506	6.254	233.723	141.950

[a] Source: Onondaga Lake Monitoring Report (1970).

TABLE A-7

Lake Residence Equivalents, 1970 (mg/liter)[a]

Parameter	Ley	Bloody	Onondaga	Harbor	East Flume	Crucible	Ninemile Creek	Metro plant
BOD	1.515	0.124	1.958	0.556	1.612	0.097	4.212	10.810
Dissolved oxygen	0.566	0.142	3.196	0.219	0.500	0.045	4.655	0.323
pH	0.736	0.148	2.816	0.213	0.043	0.014	3.297	0.642
Alkalinity	20.536	3.451	75.045	8.169	2.921	0.170	86.076	18.268
Conductivity	132.430	32.156	409.372	43.169	36.890	6.093	3075.635	149.501
Ca	26.920	6.316	187.528	10.150	0.660	0.068	653.794	50.957
Mg	4.715	0.514	28.715	1.663	1.181	0.123	25.457	3.744
Na	23.665	5.744	129.059	2.331	13.812	0.574	554.957	56.723
K	0.683	0.199	1.635	0.136	0.836	0.074	8.754	2.177
Cl	89.275	19.176	486.341	6.054	219.202	13.061	1797.841	58.242
SiO_2	0.692	0.133	1.907	0.261	0.317	0.025	5.377	1.416
SO_4	18.871	3.348	49.955	8.834	4.023	0.760	95.449	18.470
Total phosphate	0.149	0.042	0.391	0.083	0.091	0.009	0.964	0.460
Ortho-P	0.067	0.007	0.144	0.024	0.028	0.002	0.163	0.224
F	0.011	0.211	0.243	0.020	0.056	0.052	0.058	0.081
NH_3 nitrogen	0.161	0.021	1.234	0.074	0.229	0.005	0.444	0.389
Organic nitrogen	0.260	0.036	0.929	0.085	0.181	0.005	0.524	0.754
NO_3	0.043	0.008	0.191	0.032	0.117	0.027	0.417	0.039
NO_2	0.001	0.000	0.005	0.002	0.006	0.005	0.003	0.008
Cr	0.012	0.001	0.047	0.011	0.024	0.010	0.033	0.013
Cu	0.019	0.002	0.080	0.002	0.050	0.001	0.026	0.012
Fe	0.048	0.004	0.090	0.004	0.110	0.011	0.155	0.050
Mn	0.034	0.004	0.110	0.003	0.026	0.004	0.069	0.007
Zn	0.017	0.003	0.114	0.002	0.021	0.001	0.040	0.024

[a] Source: Onondaga Lake Monitoring Report (1970).

TABLE A-8

Average Concentrations, 1971 (mg/liter)[a]

Parameter	Harbor	Onondaga	Ley Creek	Ninemile Creek	Metro plant	Steel mill discharge	East Flume	Outlet
BOD	34.35	7.43	8.24	2.60	203.36	7.79	4.72	4.57
Dissolved oxygen	7.64	9.41	5.95	7.86	4.03	4.06	3.22	7.99
Alkalinity	213.94	225.13	223.30	199.98	230.48	37.31	31.53	164.10
Conductivity	1,305.85	3,845.54	1,106.31	11,660.35	1,920.85	548.66	1,141.38	2,000.10
Ca	1,251.36	422.04	423.40	1,859.14	730.09	1,677.56	2,060.12	705.42
Mg	264.61	196.05	137.54	70.75	294.08	134.39	262.03	51.94
Na	337.34	640.29	409.29	1,684.80	1,388.50	1,132.45	2,463.19	437.43
K	4.00	4.20	7.63	31.64	19.35	6.39	5.04	10.82
Cl	161.83	390.38	177.08	4,989.66	413.42	222.81	460.09	810.18
SiO_2	7.23	6.23	7.50	6.31	15.30	2.60	1.76	2.35
SO_4	336.00	159.00	171.67	163.67	217.00	62.36	41.37	194.00
Total phosphates	1.34	0.28	0.42	0.08	3.67	0.13	0.23	0.38
Ortho-P	0.56	0.16	0.31	0.09	2.30	0.14	0.17	0.22
NH_3 nitrogen	2.04	1.13	1.75	0.47	8.92	1.11	3.57	1.35
Organic nitrogen	5.16	2.84	2.43	1.05	13.79	0.93	1.86	3.13
NO_3	1.01	0.75	0.56	1.20	0.81	1.41	0.63	0.41
NO_2	0.16	0.04	0.05	0.07	0.18	0.60	0.11	0.03
Cr	0.06	0.06	0.06	0.08	0.10	1.12	0.03	0.05
Cu	0.09	0.05	0.07	0.09	0.17	0.15	0.03	0.05
Fe	0.94	0.94	0.90	0.86	0.77	0.74	1.46	0.18
pH	7.71	7.72	7.50	5.25	7.37	1.66	0.29	7.89
Temp.	52.28	50.94	53.12	58.15	59.32	64.04	76.46	56.45
Flow	8.57	91.84	104.65	185.93	60.83	6.50	80.00	—

[a] Source: Onondaga Lake Monitoring Report (1971).

TABLE A-9

Average, 1971 (lb/day)[a]

Parameter	Harbor	Onondaga	Ley Creek	Ninemile Creek	Metro plant	Steel mill discharge	East Flume
BOD	2,399.12	5,847.70	6,996.26	5,506.87	102,611.09	422.24	3,147.44
Dissolved oxygen	577.16	7,451.13	7,166.45	15,553.99	2,081.64	220.41	2,147.27
Alkalinity	16,292.44	150,883.84	186,184.81	325,181.06	117,419.59	2,022.73	21,039.70
Conductivity	77,994.56	1,366,890.00	289,248.56	7,595,911.01	944,890.63	29,743.16	761,529.25
Ca	93,721.81	422,219.56	480,799.44	3,100,669.51	371,702.44	90,940.33	1,374,509.50
Mg	22,262.18	201,963.69	166,392.78	171,301.59	150,505.03	7,285.12	174,829.66
Na	27,810.35	558,584.38	498,767.06	2,868,433.00	710,146.00	61,389.94	1,643,440.25
K	295.30	3,768.56	6,659.23	45,829.20	9,685.66	346.66	3,360.26
Cl	10,670.20	193,022.22	130,033.53	5,084,641.01	208,430.06	12,078.73	306,974.56
SiO_2	496.66	3,764.70	4,585.24	7,410.20	7,655.28	140.85	1,173.16
SO_4	23,776.10	121,180.00	65,895.16	221,374.84	108,441.64	3,380.56	27,599.84
Total phosphate	97.34	177.22	284.92	121.57	1,855.28	7.12	154.84
Ortho-P	33.71	53.18	80.69	96.09	1,123.81	7.84	111.99
NH_3 nitrogen	129.68	1,035.94	1,130.37	939.07	4,487.23	60.35	2,380.16
Organic nitrogen	370.13	2,865.43	2,534.53	2,466.86	7,075.56	50.23	1,240.08
NO_3	78.98	883.66	434.16	1,670.66	413.20	76.38	418.82
NO_2	11.62	22.46	13.43	62.99	96.45	32.25	76.73
Cr	4.49	50.37	55.53	113.17	49.99	60.47	18.35
Cu	6.61	29.04	18.40	53.24	83.68	8.00	22.26
Fe	80.46	371.57	186.69	534.41	380.37	79.31	110.63

[a] Source: Onondaga Lake Monitoring Report (1971).

TABLE A-10

Lake Residence Equivalents, 1971 (mg/liter)[a]

Parameter	Harbor	Onondaga	Ley Creek	Ninemile Creek	Metro plant	Steel mill discharge	East Flume
BOD	0.53	1.30	1.56	1.23	22.86	0.09	0.70
Dissolved oxygen	0.13	1.66	1.60	3.46	0.46	0.05	0.48
Alkalinity	3.63	33.61	41.48	72.44	26.16	0.45	4.69
Conductivity	17.38	304.52	64.44	1692.22	210.50	6.63	169.65
Ca	20.88	94.06	107.11	690.77	82.81	20.26	306.21
Mg	4.96	44.99	37.07	38.16	33.53	1.62	38.95
Na	6.20	124.44	111.12	639.03	158.21	13.68	366.13
K	0.06	0.84	1.48	10.21	2.16	0.07	0.75
Cl	2.36	43.00	18.97	1132.76	46.43	2.69	68.39
SiO_2	0.11	0.84	1.02	1.65	1.71	0.04	0.26
SO_4	5.30	0.00	14.68	49.32	24.16	0.75	6.15
Total phosphate	0.02	0.04	0.06	0.03	0.41	0.00	0.03
Ortho-P	0.01	0.01	0.02	0.02	0.25	0.00	0.02
NH_3 nitrogen	0.03	0.23	0.25	0.21	1.00	0.01	0.53
Organic nitrogen	0.98	0.64	0.56	0.55	1.58	0.01	0.28
NO_3	0.02	0.20	0.10	0.37	0.09	0.02	0.09
NO_2	0.00	0.00	0.00	0.01	0.02	0.01	0.02
Cr	0.00	0.01	0.01	0.02	0.01	0.01	0.00
Cu	0.00	0.01	0.00	0.01	0.02	0.00	0.00
Fe	0.02	0.08	0.04	0.12	0.08	0.02	0.03

[a] Source: Onondaga Lake Monitoring Report (1971).

TABLE A-11

Average Concentrations, 1972 (mg/liter)[a]

Parameter	Harbor	Onondaga	Ley Creek	Ninemile Creek	Metro plant	Crucible	East Flume	Outlet (In)
BOD	26.309	3.536	4.838	0.702	163.863	3.691	3.981	3.811
Dissolved oxygen	7.469	10.330	6.727	7.451	4.991	3.589	2.771	9.117
pH	7.634	7.891	7.595	7.765	7.395	7.695	7.708	7.858
Alkalinity	226.782	217.217	200.000	135.639	215.000	35.552	39.039	147.117
Conductivity	1096.136	1135.454	942.619	8105.245	2016.363	405.224	1109.181	2982.500
Ca	173.895	117.513	114.034	1278.559	136.700	31.139	50.148	327.164
Mg	37.795	27.654	26.361	24.580	27.254	6.476	3.618	25.449
Na	124.936	184.431	142.523	1061.671	345.500	81.258	164.608	349.777
K	3.433	3.792	5.120	15.526	15.618	6.232	4.970	8.221
Cl	110.434	213.913	152.954	3073.133	530.652	109.369	393.819	1014.705
SiO_2	6.447	5.960	7.136	4.855	10.047	3.053	0.894	3.558
SO_4	0.000	0.000	0.000	0.000	0.000	0.000	0.000	0.000
Total phosphate	0.941	0.185	0.346	0.082	2.583	0.124	0.380	0.304
Ortho-P	0.543	0.107	0.257	0.059	1.547	0.093	0.356	0.206
F	0.000	0.000	0.000	0.000	0.000	0.000	0.000	0.000
NH_3 nitrogen	2.729	0.747	1.627	0.218	8.356	0.616	5.108	1.324
Organic nitrogen	2.751	0.844	0.877	0.374	9.594	0.693	0.824	1.492
NO_3	1.048	0.712	0.417	0.337	0.686	2.163	0.259	0.318
NO_2	0.069	0.032	0.072	0.013	0.128	0.318	0.207	0.065
Cr	0.23	0.16	0.051	0.026	0.145	1.813	0.006	0.026
Cu	0.057	0.054	0.048	0.062	0.249	0.196	0.035	0.050
Fe	0.613	1.802	1.508	0.582	0.778	6.349	0.118	0.193
Mn	0.000	0.000	0.000	0.000	0.000	0.000	0.000	0.000
Temp. (°C)	9.652	9.409	10.650	11.931	14.022	17.200	24.886	13.266

[a] Source: Onondaga Lake Monitoring Report (1972).

TABLE A-12

Average, 1972 (lb/day)[a]

Parameter	Harbor	Onondaga	Ley Creek	Ninemile Creek	Metro plant	Crucible	East Flume	Outlet (In)
BOD	6,049.988	4,960.450	5,889.333	1,870.958	95,713.609	210.623	2,750.529	2,596.258
Dissolved oxygen	1,568.718	11,611.251	7,945.277	18,431.933	2,506.615	204.288	1,840.862	8,082.410
Alkalinity	42,508.976	229,938.281	221,338.781	318,179.063	109,413.937	2,023.650	18,361.335	85,843.359
Conductivity	198,670.281	1,035,811.876	867,902.251		1,134,714.753	21,967.214	740,045.751	1,481,115.503
Ca	33,189.679	122,625.406	113,137.515	1,925,925.752	80,044.093	1,688.078	33,459.171	149,587.594
Mg	7,030.524	27,690.953	26,858.726	53,414.601	14,970.361	351.092	2,414.047	14,755.615
Na	22,836.472	170,272.156	131,565.469	1,563,410.253	186.936.187	4,405.011	109,827.093	179,338.281
K	684.291	5,737.519	4,979.747	26,531.277	8,630.177	337.882	3,316.586	3,939.300
Cl	19,442.812	191,184.687	124,921.781	4,960,332.013	277,018.875	6,225.374	272,461.125	466,382.063
SiO_2	1,220.378	6,228.289	7,090.293	9,015.837	5,588.811	173.812	561.053	1,667.450
SO_4	0.000	0.00	0.000	0.000	0.000	0.000	0.000	0.000
Total phosphate	175.362	285.992	341.581	219.975	1,514.240	7.095	258.994	109.216
Ortho-P	102.106	166.925	252.499	151.923	895.355	5.081	238.068	55.773
F	0.000	0.000	0.000	0.000	0.000	0.000	0.000	0.000
NH_3 nitrogen	448.313	1,161.778	1,502.804	629.517	4,759.525	33.401	3,408.174	314.328
Organic nitrogen	487.215	665.396	999.178	773.546	4,659.389	37.584	550.106	780.135
NO_3	219.406	705.052	420.527	701.455	246.229	117.285	172.865	60.374
NO_2	12.498	65.488	58.430	24.015	70.558	17.289	138.261	33.691
Cr	3.408	37.644	58.135	60.764	80.654	98.287	4.459	0.800
Cu	10.910	64.034	57.161	114.926	81.633	10.639	23.655	30.523
Fe	146.158	3,694.331	1,625.245	1,079.605	424.992	344.222	78.881	136.914
Mn	0.000	0.000	0.000	0.000	0.000	0.000	0.000	0.000

[a] Source: Onondaga Lake Monitoring Report (1972).

TABLE A-13

Lake Residence Equivalents, 1972 (mg/liter)[a]

Parameter	Harbor	Onondaga	Ley Creek	Ninemile Creek	Metro plant	Crucible	East Flume	Outlet (In)
BOD	1.173	0.962	1.142	0.362	18.568	0.040	0.533	0.503
Dissolved oxygen	0.304	2.252	1.541	3.575	0.486	0.039	0.357	1.568
Alkalinity	8.246	44.608	41.000	61.727	21.226	0.392	3.562	16.653
Conductivity	38.542	200.951	168.376	2485.184	220.139	4.261	143.571	287.342
Ca	6.438	23.789	21.949	373.637	15.528	0.327	6.491	29.020
Mg	1.363	5.372	5.210	10.362	2.904	0.068	0.468	2.862
Na	4.430	33.033	25.524	303.307	36.266	0.854	21.306	34.792
K	0.132	1.113	0.066	5.147	1.674	0.065	0.643	0.764
Cl	3.771	37.090	24.235	962.323	53.742	1.207	52.858	90.479
SiO_2	0.236	1.208	1.375	1.749	1.084	0.033	0.108	0.323
SO_4	0.000	0.000	0.000	0.000	0.000	0.000	0.000	0.000
Total phosphate	0.034	0.055	0.066	0.042	0.293	0.001	0.050	0.021
Ortho-P	0.019	0.032	0.048	0.029	0.173	0.000	0.046	0.010
F	0.000	0.000	0.000	0.000	0.000	0.000	0.000	0.000
NH_3 nitrogen	0.086	0.225	0.291	0.122	0.923	0.006	0.661	0.060
Organic nitrogen	0.094	0.129	0.193	0.150	0.003	0.007	0.106	0.151
NO_3	0.042	0.136	0.081	0.136	0.047	0.022	0.033	0.011
NO_2	0.002	0.012	0.011	0.004	0.015	0.003	0.026	0.006
Cr	0.000	0.007	0.011	0.011	0.015	0.019	0.000	0.000
Cu	0.002	0.012	0.011	0.022	0.015	0.002	0.004	0.005
Fe	0.028	0.716	0.315	0.209	0.082	0.066	0.015	0.026
Mn	0.000	0.000	0.000	0.000	0.000	0.000	0.000	0.000

[a] Source: Onondaga Lake Monitoring Report (1972).

TABLE A-14

Average Concentrations, 1973 (mg/liter)[a,b]

Parameter	Harbor	Onondaga	Ley Creek	Ninemile Creek	Metro plant	Crucible	East Flume	Outlet (In)	Outlet (O)
BOD	16.16	3.54	6.60	0.21	102.54	3.41	1.06	4.75	5.65
Dissolved oxygen	6.75	9.81	6.23	3.62	4.71	4.02	2.82	7.37	6.03
pH	7.63	7.81	7.56	2.28	7.23	1.59	0.18	7.72	7.59
Alkalinity	232.08	212.67	183.17	52.79	193.83	53.63	68.37	147.96	149.00
Conductivity	996.69	1401.25	926.00	2588.40	2076.25	333.32	708.01	2810.94	3016.88
Ca	186.28	196.28	138.71	1952.86	162.43	43.42	1.26	256.14	370.00
Mg	41.28	33.00	22.14	35.14	21.57	13.59	4.24	22.57	25.57
Na	60.28	305.00	113.14	1929.29	292.14	21.31	89.16	300.71	382.86
K	3.04	5.27	5.71	32.00	14.62	1.65	0.24	7.93	9.74
Cl	218.33	318.12	279.13	970.37	523.75	52.40	398.42	879.17	1153.54
SiO_2	7.19	6.60	7.62	1.89	11.51	4.09	1.42	3.43	3.52
Total phosphorus	0.72	0.12	0.48	0.03	2.11	0.08	0.20	0.24	0.29
Ortho-P	0.44	0.07	0.38	0.02	1.25	0.07	0.26	0.16	0.20
NH_3 nitrogen	2.06	0.73	1.36	0.93	7.35	0.33	5.27	1.47	1.51
Organic nitrogen	2.34	0.69	1.25	0.13	6.58	0.28	0.26	1.21	1.14
NO_3	0.41	0.29	0.26	0.02	0.33	2.32	0.25	0.14	0.27
NO_2	0.11	0.02	0.10	0.00	0.11	0.26	0.17	0.10	0.19
Cr	0.01	0.01	0.05	0.01	0.09	1.46	0.01	0.06	0.02
Cu	0.09	0.04	1.38	0.01	0.17	0.16	0.01	0.05	0.05
Fe	0.70	1.50	2.24	0.26	0.79	10.51	0.38	0.20	0.14
Temp. (°C)	11.56	10.27	12.80	1.90	14.60	6.40	12.81	12.74	12.17

[a] Source: Onondaga Lake Monitoring Report (1973).

[b] Lake water concentration correction factor not applied.

TABLE A-15

Average, 1973 (lb/day)[a]

Parameter	Harbor	Onondaga	Ley Creek	Ninemile Creek	Metro plant	Crucible	East Flume	Outlet (In)
BOD	3,181	2,924	3,201	2,090	59,153	187	707	457
Dissolved oxygen	1,436	9,624	3,313	26,683	2,714	220	1,880	967
Alkalinity	45,238	189,705	87,497	392,405	110,888	2,915	45,616	12,341
Conductivity	189,606	1,087,603	415,271	NA[b]	1,202,655	19,180	472,386	197,864
Ca	25,107	89,884	14,596	NA[b]	100,070	2,354	838	14,756
Mg	5,284	11,244	3,189	NA[b]	12,889	800	2,827	2,775
Na	8,126	109,490	13,910	NA[b]	177,012	1,251	59,491	16,365
K	423	2,109	742	6,424,345	8,921	90	157	495
Cl	33,988	202,585	85,913	12,705	297,872	2,981	265,823	65,998
SiO_2	1,358	5,825	3,515	245	6,553	221	950	153
Total phosphate	143	148	195	169	1,198	5	199	7
Ortho-P	86	85	135	5,136	704	4	173	3
NH_3 nitrogen	372	621	717	759	4,244	18	3,519	8
Organic nitrogen	487	814	435	205	3,884	15	172	110
NO_3	92	391	194	12	198	126	168	14
NO_2	20	25	39	41	65	15	111	5
Cr	1	10	11	92	55	80	5	1
Cu	20	35	64	1,877	99	9	5	5
Fe	143	2,248	1,955		461	571	251	19

[a] Source: Onondaga Lake Monitoring Report (1973).

[b] Flow data not available for dates when parameter was measured.

TABLE A-16

Percentage Contribution, 1973[a]

Parameter	Harbor	Onondaga	Ley Creek	Ninemile Creek	Metro plant	Crucible	East Flume	Outlet (In)	Total
BOD	4.4	4.0	4.5	3.0	82.2	0.3	1.0	0.6	71,900
Dissolved oxygen	3.1	20.5	7.1	57.0	5.8	0.4	4.1	2.0	46,837
Alkalinity	5.1	21.4	9.9	44.3	12.5	0.3	5.1	1.4	886,603
Conductivity	1.9	11.0	4.2	63.9	12.1	0.2	4.1	2.0	9,917,834
Ca	10.1	36.3	5.9	N/A[b]	40.4	1.0	0.3	6.0	247,605
Mg	13.5	28.8	8.2	N/A[b]	33.0	2.1	7.3	7.1	39,008
Na	2.2	28.4	3.7	N/A[b]	—	0.3	15.4	4.3	385,648
K	3.3	16.3	5.7	N/A[b]	69.0	0.7	1.2	3.8	12,937
Cl	0.5	2.7	1.2	87.0	4.0	0.1	3.6	0.9	7,379,511
SiO_2	4.3	18.7	11.2	40.6	21.0	0.7	3.0	0.5	31,281
Total phosphate	6.7	6.9	9.1	11.5	56.0	0.2	9.3	0.3	2,140
Ortho-P	6.3	6.3	9.9	12.4	51.8	0.3	12.7	0.3	1,359
NH_3 nitrogen	2.5	4.3	4.9	35.1	29.0	0.1	24.0	0.1	14,635
Organic nitrogen	7.3	12.2	6.5	11.4	58.2	0.2	2.6	1.6	6,676
NO_3	6.6	28.2	14.0	14.7	14.3	9.1	12.1	1.0	1,388
NO_2	6.8	8.6	13.4	4.1	22.3	5.1	38.0	1.7	292
Cr	0.5	4.9	5.4	20.1	27.0	39.2	2.4	0.5	204
Cu	6.1	10.6	19.5	28.0	30.1	2.7	1.5	1.5	329
Fe	1.9	29.9	26.0	24.9	6.1	7.6	3.3	0.3	7,525

[a] Source: Onondaga Lake Monitoring Report (1973).
[b] Flow data not available for dates when parameter was measured.

TABLE A-17

Lake Residence Equivalents, 1973 (mg/liter)[a]

Parameter	Harbor	Onondaga	Ley Creek	Ninemile Creek	Metro plant	Crucible	East Flume	Outlet (In)
BOD	0.88	0.81	0.89	0.58	16.39	0.05	0.20	0.13
Dissolved oxygen	0.40	2.67	0.92	7.39	0.75	0.06	0.52	0.29
pH	0.41	1.90	1.08	4.75	1.15	0.02	0.03	0.24
Alkalinity	12.54	52.57	24.25	108.74	30.73	0.81	12.64	3.42
Conductivity	52.54	301.37	115.07	4,228.21	333.26	5.31	130.90	54.83
Ca	6.95	24.91	4.04	NA[b]	27.73	0.65	0.23	4.09
Mg	1.46	3.12	0.88	NA[b]	3.57	0.22	0.78	0.77
Na	2.25	30.34	3.85	NA[b]	49.05	0.35	16.49	4.5
K	0.12	0.58	0.21	NA[b]	2.47	0.02	0.04	0.14
Cl	9.42	57.14	23.81	1,780.20	82.54	0.83	73.66	18.29
SiO_2	0.38	1.61	0.97	3.52	1.82	0.06	0.26	0.04
Total phosphate	0.04	0.04	0.05	0.07	0.33	0.00	0.05	0.00
Ortho-P	0.02	0.02	0.04	0.05	0.20	0.00	0.05	0.00
NH_3 nitrogen	0.10	0.17	0.20	1.42	1.18	0.00	0.98	0.00
Organic nitrogen	0.13	0.23	0.12	0.21	1.02	0.00	0.05	0.03
NO_3	0.03	0.11	0.05	0.05	0.05	0.04	0.05	0.00
NO_2	0.01	0.01	0.01	0.00	0.02	0.00	0.03	0.00
Cr	0.00	0.00	0.00	0.01	0.02	0.02	0.00	0.00
Cu	0.00	0.01	0.02	0.03	0.03	0.00	0.00	0.00
Fe	0.04	0.62	0.54	0.52	0.13	0.15	0.07	0.01
Temp. (°C)	0.59	2.20	1.75	3.98	2.29	0.10	2.37	0.64

[a] Source: Onondaga Lake Monitoring Report (1973).
[b] Flow data not available for dates when parameter was measured.

TABLE A-18

Average Concentrations, 1974 (mg/liter)[a]

Parameter	Harbor	Onondaga	Ley Creek	Ninemile Creek	Metro plant	Crucible	East Flume	Outlet (In)
BOD_5	21.103	2.924	6.435	1.151	124.880	9.608	2.119	4.273
Dissolved oxygen	8.303	11.015	6.879	7.732	4.359	4.444	3.895	8.895
pH	7.791	7.879	7.611	7.7	7.359	7.8	7.8	7.804
Alk (as $CaCO_2$)	240.160	224.560	216.280	137.529	211.600	58.163	77.660	151.652
Conductivity (μmhos)	1,219.363	944.318	965.318	—	1,964.091	—	—	2,182.954
Ca	156.000	90.428	86.714	690.737	111.357	45.638	37.042	174.928
Mg	40.000	24.500	24.714	29.245	23.785	14.252	2.261	19.785
Na	46.014	108.071	90.821	1,385.503	249.642	73.296	137.542	229.571
K	4.307	3.285	3.814	12.401	11.214	6.517	4.645	5.576
Cl	204.360	193.800	217.600	2,862.172	538.600	33.730	240.960	923.913
SiO_2	6.115	5.663	6.662	4.275	8.121	2.732	0.880	2.280
Total inorg PO_4 (as P)	0.359	0.087	0.310	0.018	1.902	0.052	0.087	0.154
Orthophosphate (as P)	0.285	0.053	0.258	0.015	1.185	0.022	0.056	0.106
NH_3 (as N)	1.423	0.334	1.444	0.717	9.261	0.371	5.542	1.487
Organic nitrogen	1.619	0.469	0.442	0.599	6.785	0.480	0.442	0.708
NO_3 (as N)	0.921	0.530	0.297	0.170	0.267	1.545	0.130	0.250
NO_2 (as N)	0.071	0.022	0.067	0.008	0.134	0.313	0.600	0.067
Cr	0.011	0.034	0.017	0.033	0.104	1.699	0.007	0.024
Cu	0.038	0.035	0.035	0.007	0.123	0.000	0.000	0.049
Fe	0.749	1.622	0.949	0.705	0.972	5.234	0.290	0.254
Temp. (°C)	8.740	8.333	9.133	11.8	12.840	16.7	29.0	11.459
Total coliform (colonies/100 ml)	1,184,000	260,000	113,000	39	10,200	6,800	141	
Fecal coliform (colonies/100 ml)	190,000	44,000	12,200	70	2,557	2,400	113	
Fecal streptococci (colonies/100 ml)	39,000	27,100	3,460	717	7,400	250	62	
Flow (MGD)	19.079	133.190	73.299	124.125[b]	77.167	6.500	80.000	

[a] Source: Onondaga Lake Monitoring Report (1974).

[b] Ninemile Creek discharge calculated from values recorded by the USGS gauge on Ninemile Creek at Camillus using factor of 1.36, the ratio of contributing drainage areas (see p. 15, Onondaga Lake Monitoring Program: Jan. 1973–Dec. 1973).

TABLE A-19

Average, 1974 (lb/day)[a]

Parameter	Harbor	Onondaga	Ley Creek	Ninemile Creek	Metro plant	Crucible	East Flume	Total
BOD_5	5,022.635	3,365.037	3,477.716	1,735.726	79,031.187	520.879	1,414.463	94,567
Dissolved oxygen	1,320.397	12,800.835	5,311.361	9,522.974	2,908.267	240.928	2,599.407	34,702
Alkalinity (as $CaCO_3$)	37,687.007	228,888.281	111,709.828	159,806.312	136,458.656	3,153.016	51,814.703	729,515
Ca	21,863.218	111,152.609	46,372.281	567,478.126	71,600.687	2,474.072	24,714.992	835,655
Mg	5,763.352	29,144.996	13,901.414	29,490.574	15,384.126	772.622	1,508.824	95,965
Na	6,542.511	95,153.890	40,972.265	899,418.251	160,201.187	3,973.403	91,768.562	1,298,030
K	470.671	10,416.986	2,540.983	11,079.855	7,185.392	353.340	3,099.617	35,174
Cl	21,001.660	151,418.094	88,094.406	2,180,546.005	345,539.875	1,828.529	160,768.406	2,949,197
Total inorganic PO_4	59.455	144.412	145.477	53.358	1,202.529	2.853	58.660	1,666
Orthophosphate (as P)	47.825	110.934	127.489	44.101	742.106	1.205	37.790	1,111
NH_3 (as N)	247.536	887.500	673.028	465.300	5,754.475	20.162	3,698.103	11,746
Organic N	302.908	342.808	187.720	543.767	4,267.906	26.054	295.569	5,968
NO_3 (as N)	123.553	542.878	157.497	198.321	176.871	83.780	87.053	1,370
NO_2 (as N)	10.009	63.050	50.106	11.356	85.466	16.990	400.622	637
Cr	0.608	100.246	29.890	84.090	66.223	92.140	4.966	378
Cu	7.005	28.189	14.927	37.738	77.645	0.000	0.000	166
Fe	40.023	9,879.058	632.432	1,712.968	632.074	283.762	194.059	13,374
SiO_2	949.522	5,260.573	3,123.333	4,349.267	5,186.727	148.114	587.425	19,605

[a] Source: Onondaga Lake Monitoring Report (1974).

TABLE A-20

Lake Residence Equivalents, 1974 (mg/liter)[a,b]

Parameter	Harbor	Onondaga	Ley Creek	Ninemile Creek	Metro plant	Crucible	East Flume
BOD_5	1.461	0.979	1.012	0.505	22.998	0.151	0.411
Dissolved oxygen	0.384	3.725	1.545	2.771	0.846	0.070	0.756
Alkalinity (as $CaCO_3$)	10.967	66.607	32.508	46.504	39.710	0.917	15.078
Ca	6.362	32.346	13.494	165.139	20.836	0.719	7.192
Mg	1.677	8.481	4.045	8.581	4.476	0.224	0.439
Na	1.903	27.690	11.923	261.735	46.619	1.156	26.705
K	0.136	3.031	0.739	3.224	2.090	0.102	0.902
Cl	6.111	44.063	25.635	634.551	100.554	0.532	46.784
SiO_2	0.276	1.530	0.908	1.265	1.509	0.043	0.170
Total inorganic PO_4 (as P)	0.017	0.042	0.042	0.015	0.349	0.000	0.017
Orthophosphate (as P)	0.013	0.032	0.037	0.012	0.215	0.000	0.010
NH_3 (as N)	0.072	0.258	0.195	0.135	1.674	0.005	1.076
Organic nitrogen	0.088	0.099	0.054	0.158	1.241	0.007	0.086
NO_3 (as N)	0.035	0.157	0.045	0.057	0.051	0.024	0.025
NO_2 (as N)	0.002	0.018	0.014	0.003	0.024	0.004	0.116
Cr	0.000	0.029	0.008	0.024	0.019	0.026	0.001
Cu	0.002	0.008	0.004	0.010	0.022	0.000	0.000
Fe	0.011	2.874	0.184	0.498	0.183	0.082	0.056

[a] Source: Onondaga Lake Monitoring Report (1974).

[b] Based on a volume of 37,083 million gallons and a total flow rate of 90 MGD.

TABLE A-21

Average Concentrations, 1975 (mg/liter)[a]

Parameter	Harbor	Onondaga	Ley Creek	Ninemile Creek	Metro plant	Crucible	East Flume	Outlet (In)[b]	Outlet (O)[c]
BOD_5	15.962	7.654	8.796	1.383	102.271	9.017	1.308	6.514	6.000
Dissolved oxygen	7.671	8.508	5.783	7.654	4.758	7.387	4.500	9.455	6.082
pH	7.942	7.892	7.687	7.476	7.433	7.967	7.532	7.818	7.764
Alk (as $CaCO_3$)	255.542	231.542	218.667	159.980	230.125	140.792	73.652	160.591	158.500
Conductivity (μmhos)	1,123.542	925.417	976.042	6,820.000	1,930.417	1,926.042	3,695.200	2,425.909	2,771.304
Ca	180.246	96.754	102.308	935.028	132.154	235.615	134.308	330.846	356.923
Mg	40.000	23.077	23.538	23.721	22.000	18.154	2.031	27.154	9.986
Na	56.977	78.438	111.115	654.410	291.169	241.769	111.692	322.692	1,147.650
K	2.554	3.869	3.846	13.443	12.338	16.292	5.546	9.715	2.723
Cl	129.810	157.905	199.095	2,702.178	523.905	606.905	149.714	713.100	
SiO_2	5.652	5.639	6.574	5.399	8.200	7.057	.165	2.577	
Total coliform (cells/100 ml)	364,405.0	292,570.0	82,915.0	1,090.0	110,577.0	6,421.0	2,578.0	5,118.0	5,714.0
Total inorganic phosphorus[d]	0.290	0.196	0.223	0.042	1.669	0.123	0.020	0.210	0.210
Orthophosphorus	0.212	0.112	0.162	0.025	0.927	0.071	0.025	0.131	0.159
Fecal coliform (cells/100 ml)	221,500.0	108,118.0	13,860.0	14.0	138,880.0	1,262.0	859.0	264.0	230.0
NH_3 nitrogen	1.372	0.634	1.640	0.540	7.763	1.771	5.210	1.613	1.836
Organic nitrogen	1.204	0.430	0.861	0.286	4.700	0.770	2.170	1.167	0.683
NO_3 nitrogen	0.225	0.198	0.117	0.120	0.205	0.842	0.514	0.178	0.182
NO_2 nitrogen	0.123	0.054	0.111	0.008	0.125	0.279	1.091	0.216	0.256
Cr	0.906	0.002	0.008	0.014	0.079	0.217	0.015	0.009	0.013
Cu	0.075	0.035	0.038	0.061	0.075	0.070	0.068	0.065	0.043
Fe	0.867	1.553	1.208	1.192	0.772	1.322	0.561	0.163	0.120
Fecal streptococci (cells/100 ml)	80,545.0	28,431.0	6,757.0	741.0	116,715.0	440.0	340.0	302.0	202.0
Temp. (°C)	10.049	10.689	15.494	13.000	13.602	16.027	24.700	13.020	12.662
Flow (MGD)	4.697	136.769	40.721	121.021[e]	72.187	0.912[e]	47.079		

[a] Source: Onondaga Lake Monitoring Report (1975).
[b] Sampled at 2-ft depth in the lake outlet.
[c] Sampled at 12-ft depth in the lake outlet.
[d] Total inorganic phosphorus represents the sum of ortho and acid-hydrolyzable phosphorus.
[e] Flow in "Crucible" stream was calculated using mass balance considerations and Crucible Steel's internal monitoring program data. Ninemile Creek calculated from USGS gauge at Camillus using a drainage area correction of 1.36.

TABLE A-22

Average Mass Loading Rates, 1975 (lb/day)[a]

Parameter	Harbor	Onondaga	Ley Creek	Ninemile Creek	Metro plant	Crucible	East Flume	Total
BOD_5	822.991	7,562.184	3,087.008	1,515.547	61,469.625	66.887	501.042	75,001
Dissolved oxygen	307.958	10,016.039	2,159.790	8,431.234	2,878,716	57.676	1,781.266	25,633
Alkalinity (as $CaCO_3$)	3,555.367	261,670.500	73,001.875	176,734.875	138,884.000	1,060.015	29,395.016	689,299
Ca	6,314.859	113,952.688	36,938.148	754,404.500	83,057.875	1,805.736	56,044.137	1,052,521
Mg	1,482.776	28,807.137	7,383.578	21,709.652	13,665.063	136.673	750.882	73,936
Na	2,473.748	95,960.750	40,698.391	535,345.687	183,661.813	1,980.074	43,821.078	903,250
K	171.488	6,287.387	1,789.261	12,669.172	7,741.941	115.120	2,151.211	30,626
Cl	5,290.746	168,448.500	76,124.500	2,550,828.000	312,391.375	4,600.117	57,185.332	3,174,852
SiO_2	171.656	6,398.730	2,065.435	6,138.148	4,881.887	52.129	62.416	19,770
Total inorganic phosphorus	17.068	286.408	84.309	61.538	995.688	.966	8.057	1,454
Orthophosphorus	12.029	167.719	61.338	38.759	552.431	.558	10.046	843
NH_3 nitrogen	52.448	829.135	509.916	489.897	4,653.461	15.973	2,045.250	8,596
Organic nitrogen	31.504	245.775	195.542	257.270	2,783.867	5.750	938.527	4,456
NO_3 nitrogen	10.925	257.246	52.000	98.922	129.865	6.255	207.303	763
NO_2 nitrogen	3.944	78.723	32.688	7.475	79.397	2.239	435.236	638
Cr	.349	3.503	1.960	8.769	48.776	1.658	5.901	71
Cu	3.481	64.615	23.251	70.113	47.548	.502	26.640	232
Fe	168.689	2,833.817	773.255	1,371.067	494.844	8.258	216.563	5,867

[a] Source: Onondaga Lake Monitoring Report (1975).

TABLE A-23

Mass Loading Percent Contributions, 1975[a]

	Harbor	Onondaga	Ley Creek	Ninemile Creek	Metro Plant	Crucible	East Flume
BOD_5	1.097	10.077	4.113	2.019	81.909	0.089	0.668
Dissolved oxygen	1.197	38.938	8.396	32.777	11.191	0.224	6.925
Alkalinity (as $CaCO_3$)	1.240	37.925	10.580	25.615	20.129	0.154	4.260
Ca	0.600	10.822	3.508	71.645	7.888	0.171	5.322
Mg	2.004	38.936	9.980	29.343	18.470	0.185	1.015
Na	0.274	10.611	4.500	59.198	20.309	0.219	4.846
K	0.560	20.524	5.841	40.376	25.272	0.376	7.022
Cl	0.167	5.304	2.397	80.322	9.837	0.145	1.801
SiO_2	0.868	32.361	10.446	31.043	24.689	0.264	0.316
Total inorganic phosphorus	1.174	19.693	5.797	4.231	68.463	0.066	0.554
Orthophosphorus	1.427	19.897	7.277	4.598	65.535	0.066	1.192
NH_3 nitrogen	0.610	9.645	5.931	5.699	54.130	0.186	23.791
Organic nitrogen	0.706	5.508	4.337	5.766	62.389	0.129	21.033
NO_3 nitrogen	1.433	33.731	6.818	12.971	17.028	0.820	27.182
NO_2 nitrogen	0.618	12.027	6.124	1.172	12.446	0.351	68.225
Cr	0.490	4.928	2.758	12.338	68.627	2.332	8.302
Cu	1.499	26.103	10.012	30.193	20.476	0.216	11.472
Fe	2.875	48.304	13.181	23.371	8.435	0.141	3.691

[a] Source: Onondaga Lake Monitoring Report (1975).

APPENDIX B: ONONDAGA LAKE WATER COLUMN PHYSICAL AND CHEMICAL DATA

TABLE B-1

Water Column Physical and Chemical Data (1969)

Line plot symbol	Station no. 1				Station no. 2				Mean			
	Epilimnion		Hypolimnion		Epilimnion		Hypolimnion		Epilimnion		Hypolimnion	
	Mean	Geological mean	Mean	Geological mean	Mean	Geological mean	Mean	Geological mean	Diffusion	Percent diffusion	Diffusion	Percent diffusion
T	13.91	10.27	8.13	7.12	13.78	9.96	7.98	6.77	0.13	0.9	0.15	1.8
Alkalinity	170.39	168.20	198.55	196.23	165.90	164.33	196.37	194.52	449.00	2.6	2.68	1.1
BOD	6.21	4.85	12.40	9.02	4.85	3.89	13.44	9.28	1.36	21.9	−1.04	−8.4
Cl	1458.41	1403.63	1930.23	1914.56	1474.95	1437.03	1886.88	1864.64	−16.54	−1.1	43.35	2.0
CO_2	5.16	4.05	10.87	10.03	4.26	3.48	10.26	9.56	0.90	17.4	0.61	5.6
Organic nitrogen	1.96	1.72	1.48	1.11	1.76	1.50	1.42	1.06	0.20	10.2	0.06	4.1
NH_3 nitrogen	2.14	1.62	4.31	3.79	1.90	1.47	4.08	3.50	0.24	11.2	0.23	—
NO_3	0.39	0.22	0.13	0.04	0.41	0.26	0.13	0.05	−0.02	−5.1	0.00	—
NO_2	0.06	0.02	0.08	0.00	0.08	0.02	0.14	0.01	−0.02	−33.3	−0.06	−75.0
Total phosphate	2.34	1.75	3.17	2.51	2.60	1.71	2.80	2.31	−0.26	−11.1	0.37	11.7
O-P	0.94	0.83	1.57	1.46	0.97	0.80	1.49	1.36	−0.03	−3.2	0.08	5.1
pH	7.64	7.63	7.39	7.39	7.69	7.68	7.42	7.42	−0.05	−0.6	−0.03	−0.4
SO_4	182.35	180.43	184.41	182.60	177.56	175.09	186.78	184.77	4.79	2.6	−2.37	−1.3
S	6.63	5.26	3.50	3.18	6.41	5.61	3.46	3.30	0.22	3.3	0.04	1.1
Dissolved oxygen	4.79	3.79	1.53	1.20	5.78	4.60	1.56	1.22	−0.99	−2.1	−0.03	−2.0
Ca	639.37	605.10	849.05	827.57	651.41	618.23	841.00	821.41	−12.04	1.9	8.05	0.9
Na	554.54	528.05	669.75	658.58	558.61	530.74	660.32	650.67	−4.07	0.7	9.43	1.4
K	17.17	15.18	17.79	16.42	16.72	15.03	18.75	16.88	0.45	2.6	−0.96	5.4
Mg	30.38	29.80	30.99	30.51	29.89	29.17	31.48	30.99	0.49	1.6	−0.49	1.6
Conductivity	4625.83	4533.71	5810.36	5745.89	4715.11	4624.93	5869.31	5808.83	−89.98	−3.8	−141.05	2.4
Cu	0.05	0.04	0.05	0.04	0.05	0.04	0.04	0.04	0.00	—	0.01	20.0
Cr	0.02	0.02	0.02	0.02	0.02	0.02	0.03	0.02	0.00	—	−0.01	50.0
Fe	0.02	0.17	0.26	0.22	0.20	0.18	0.27	0.24	0.00	—	−0.01	3.8
Mn	0.07	0.06	0.20	0.17	0.08	0.06	0.20	0.17	−0.01	14.3	0.00	—
Zn	0.07	0.06	0.07	0.06	0.07	0.06		0.07	0.00	—	0.00	—
F	4.43	4.43	4.56	4.56	4.46	4.46	4.65	4.65	−0.03	−0.7	−0.09	−2.0
SiO_2	4.89	4.45	8.16	7.92	4.73	4.31	7.60	7.42	0.16	3.3	0.56	6.9
Ss	17.75	13.77	15.90	12.13	16.23	12.10	15.94	11.53	1.52	8.6	−0.04	−0.2
Ds	3200.22	3087.96	3896.25	3818.21	3305.49	3174.52	3910.62	3807.28	−105.27	−3.3	−24.37	−0.6
Secchi disk	1.09	0.98	—	—	1.22	1.07	—	—	−0.13	−11.8	—	—
E. Coli	1199.61	391.57	250.86	95.65	406.58	88.97	119.92	82.27	793.03	66.1	130.94	5.2

TABLE B-2

Onondaga Lake Water Column Physical and Chemical Data (1970)[a]

	Station I		Station II	
Parameter	Epilimnion	Hypolimnion	Epilimnion	Hypolimnion
Alkalinity	168.15	178.84	159.28	175.83
BOD	4.09	5.17	3.56	7.61
Cl	1505.21	2064.29	1513.64	2078.26
Dissolved oxygen	3.99	1.29	5.15	1.30
Total phosphate	1.43	1.77	1.95	2.06
Ortho PO_4^{3-}	.70	0.96	0.72	1.05
pH	7.64	7.43	7.57	7.39
SO_4^{2-}	186.83	197.88	182.22	200.87
S^{2-}	—	—	—	—
SiO_2	4.89	6.97	5.00	6.14
CO_2	—	—	—	—
F^-	—	—	—	—
T.S.	3627.67	4670.55	3163.92	4655.63
S.S.	10.61	7.22	6.62	5.44
D.S.	3617.05	4663.33	3157.29	4650.19
Organic nitrogen	1.90	1.34	2.27	1.21
NH_3 (N)	3.05	4.50	2.25	4.66
NO_2	0.04	0.00	0.08	0.00
NO_3	0.36	0.01	0.43	0.16
Secchi disk (m)	3.26	—	2.66	—
Temp. (°C)	15.55	8.86	13.79	9.08
Conductivity (μohm)	4577.07	5934.25	4653.61	5996.00
Co	4.25	1.98	3.68	3.56
Cr	0.05	0.06	0.05	0.05
Cu	0.06	0.04	0.06	0.05
Fe	0.22	0.25	0.17	0.20
Mg	41.07	41.64	36.67	43.53
Mn	0.07	0.15	0.08	0.12
K	17.29	19.02	12.76	17.81
Na	832.79	982.26	687.62	967.25
Zn	0.07	0.05	0.07	0.05
Ca	815.81	1084.24	752.65	1142.25
E. coli (counts)/100 ml	1643.33	1024.00	1588.00	279.00

[a] All of the above parameters are recorded as mg/liter except where noted otherwise.

TABLE B-3

Onondaga Lake Water Column Physical and Chemical Data (1971)[a]

Parameter	Epilimnion	Hypolimnion
Alkalinity	189.09	199.29
BOD	6.20	7.72
Cl	1321.46	1761.38
Dissolved oxygen	5.04	0.81
Total phosphate	1.03	1.95
Orthophosphate	0.70	1.20
pH	7.65	7.34
SO_4^{-2}	173.30	171.80
SiO_2	3.36	5.63
Organic nitrogen	3.03	3.13
NH_3 nitrogen	2.48	4.09
NO_3^-	0.46	0.34
Secchi disk (m)	1.11	—
Temp. (°C)	12.19	6.78
Conductivity (mmhos)	4601.92	5578.68
Cr	0.05	0.05
Cu	0.04	0.05
Fe	0.38	0.31
Mg	66.47	69.36
K	14.69	16.35
Na	677.75	728.42
Ca	706.38	978.24
TOC[b]	11.2	11.5

[a] The parameter concentrations are reported in mg/liter unless shown otherwise.

[b] Measured only at the last seven sampling dates.

TABLE B-4

Onondaga Lake Water Column Physical and Chemical Data (1972)

	Station I	
Parameter	Epilimnion	Hypolimnion
Alkalinity	169.54	195.11
BOD	4.63	5.90
Cl^-	1386.18	2075.28
Dissolved oxygen	5.95	0.97
Total phosphate	0.50	1.12
Orthophosphate	0.36	0.94
pH	7.69	7.41
SO_4^{-2}	155.00	172.25
Organic nitrogen	1.84	1.33
NH_3 nitrogen	2.06	4.74
NO_3^-	0.42	0.18
Secchi disk (m)	1.08	—
Temp. (°C)	11.74	5.23
Conductivity (mmhos)	3939.40	5371.79
Cr	0.036	0.036
Cu	0.050	0.038
Fe	0.305	0.237
Mg	27.50	28.42
K	11.97	15.13
Na	490.63	647.28
Cu	514.43	634.02
TOC	12.99	13.42
Chlorophyll *a*	31.79	9.18

TABLE B-5

Onondaga Lake Water Column Physical and Chemical Data (1973)[a]

	Station I	
Parameter	Epilimnion	Hypolimnion
Alkalinity	144.63	177.98
BOD	6.88	5.90
Cl^-	1,294.90	1,566.47
Dissolved oxygen	5.14	1.51
Total phosphate	0.28	0.60
Orthophosphate	0.19	0.48
pH	7.60	7.45
SO_4^{-2}	157.12	155.04
Organic nitrogen	1.11	0.54
NH_3 nitrogen	1.74	3.45
NO_3^- nitrogen	0.16	0.07
NO_2^- nitrogen	0.26	0.06
Secchi disk (m)	1.16	—
Temp. (°C)	14.22	9.51
Conductivity (mmhos)	3,262.65	3,834.71
Cr	0.00	0.00
Cu	0.07	0.09
Fe	0.17	0.20
Mg	26.66	25.08
K	21.35	22.07
Na	512.71	499.38
Ca	471.21	477.79
TIC	34.65	38.57
TOC	16.69	15.08
Chlorophyll *a* (μg/liter)	24.95	8.93
Biomass	5.64	5.22
Total coliform	16,110.49	2,675.35
S^{2-}	0.01	0.40
SiO_2	3.65	5.37

[a] All parameter values are reported in mg/liter unless otherwise noted.

TABLE B-6

Onondaga Lake Water Column Physical and Chemical Data (1974)[a]

Parameter	Station 1 Epilimnion	Station 1 Hypolimnion
pH	7.6	7.6
Temp. (°C)	13.4	10.4
Dissolved oxygen	6.3	2.2
BOD_5	8.3	8.5
Total coliform (cells/100 ml)	19079	7415
Fecal coliform (cells/100 ml)	341	249
Fecal Streptococci (cells/100 ml)	270	357
Chlorophyll *a* (μg/liter)	30.5	9.7
Total pigment (μg/liter)[b]	37.5	11.2
TKN	3.5	5.1
NH_3 (as N)	2.8	4.8
Organic nitrogen	0.69	0.34
NO_3 (as N)	0.28	0.15
NO_2 (as N)	0.12	0.056
Total inorganic PO_4 (as P)	0.20	0.47
Orthophosphate (as P)	0.14	0.39
SO_4	164.7	169.5
Cl	1445	1587
SiO_2	2.49	4.46
Alkalinity (as $CaCO_3$)	1169	206
Na	486	556
K	11.2	11.0
TOC	14.4	13.9
CO_2	9.8	14.1
TIC	45.2	52.1
TC	59.6	65.9
S	0.008	1.53
Conductivity (μmhos)	3715	3858
Fe	0.137	0.129
Cr	0.0087	0.0043
Cu	0.047	0.046
Ca	399	426
Mg	25.7	27.5
Secchi disk	0.78	

[a] All parameters reported as mg/liter of the substance unless otherwise noted.

[b] Total pigment is the sum of chlorophyll *a* and phaeophytin measurements.

TABLE B-7

Onondaga Lake Water Column Physical and Chemical Data (1975)[a]

Parameter	Station 1	
	Epilimnion[b]	Hypolimnion[c]
pH	7.7	7.5
Temp. (°C)	13.3	8.0
Dissolved oxygen	7.0	2.2
BOD_5	7.2	8.3
Total coliform (counts/100 ml)	17900	3300
Fecal coliform (counts/100 ml)	2100	180
Fecal *Streptococcus* (counts/100 ml)	160	97
Chlorophyll *a* (μg/liter)	58.6	10.7
Total pigment (μg/liter)[d]	63.7	10.3
Total Kjeldahl nitrogen	2.6	4.6
NH_3 (as N)	1.8	4.3
Organic nitrogen	0.98	0.40
Nitrate (as N)	0.47	0.13
Nitrite (as N)	0.26	0.14
Total inorganic phosphorus (as P)[e]	0.24	0.54
Orthophosphorus (as P)	0.15	0.43
Sulfate	150	160
Chloride	1250	1500
Silica (SiO_2)	2.4	4.9
Alkalinity (as $CaCO_3$)	182	216
Sodium	373	446
Potassium	10.5	11.5
Total organic carbon	14.7	13.3
Total inorganic carbon	33.9	41.5
Total carbon	49.1	54.8
Carbon dioxide	13.5	25.9
Sulfide	0.0	1.8
Conductivity (μmhos)	3290	3920
Iron	0.09	0.11
Chromium	0.0047	0.0032
Copper	0.023	0.026
Calcium	345	426
Magnesium	23.3	25.9
Secchi disk (m)	0.77	
Mercury (μg/liter)	4.5	6.1
Adenosine triphosphate (μg/liter)	1.7	0.95
Standard plate count (counts/ml)	10900	500

[a] All parameters reported as mg/liter of the substance unless otherwise noted.

[b] Epilimnion values are defined as the average of the values measured at the 0-, 3-, and 6-m depths at the south deep station.

[c] Hypolimnion values are the average of measurements from the 12-, 15-, and 18-m depths.

[d] Total pigment is the sum of measured chlorophyll *a* and phaeophytin *a*.

[e] Total inorganic phosphorus is the sum of the orthophosphorus and acid hydrolyzable phosphorus species defined in Standard Methods, thirteenth edition.

APPENDIX C: ONONDAGA LAKE DEPTH SYNOPTIC PLOTS

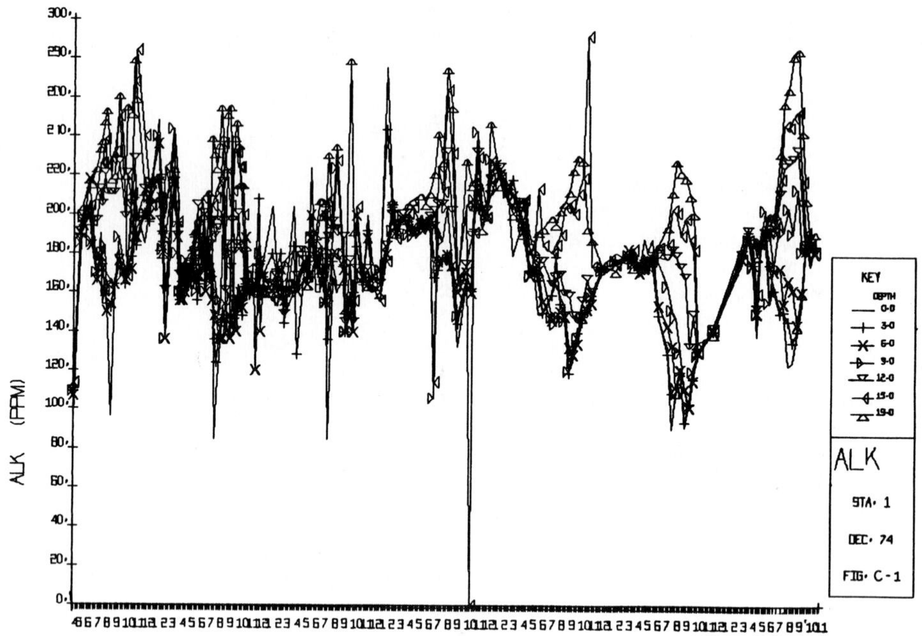

Fig. C-1.

Fig. C-2.

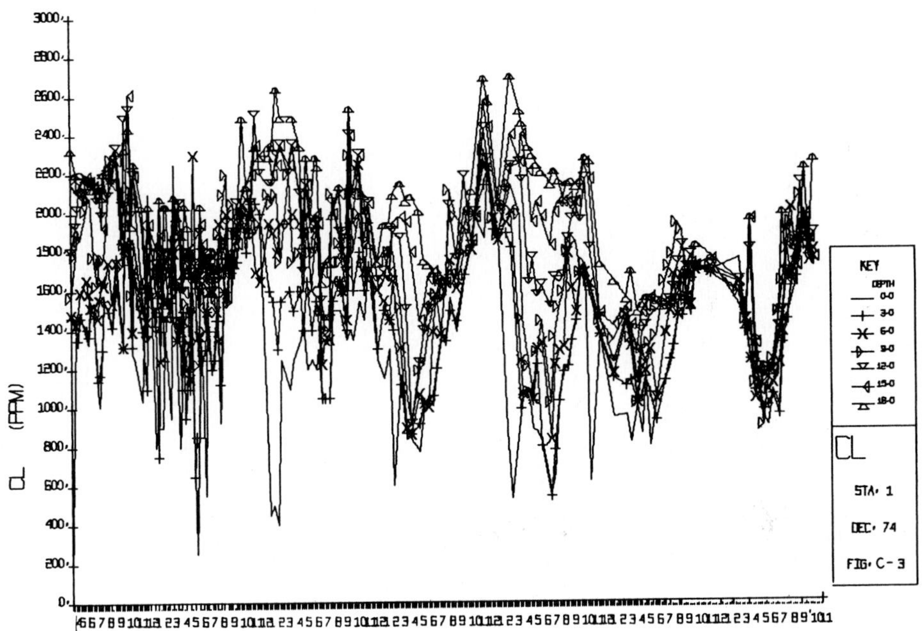
CL (PPM)
KEY
DEPTH
0.0
3.0
6.0
9.0
12.0
15.0
18.0
CL
STA. 1
DEC. 74
FIG. C-3

Fig. C-3.

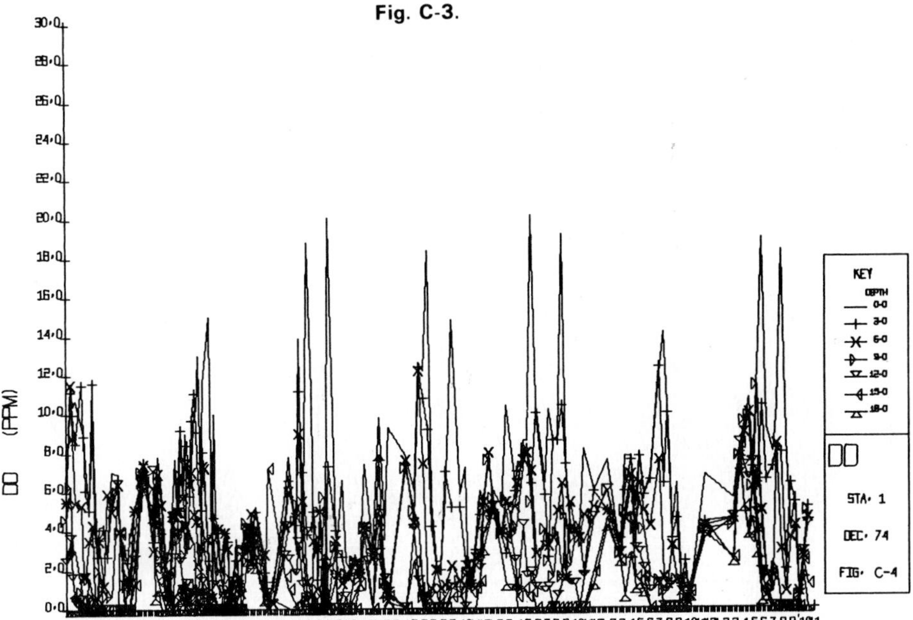
DO (PPM)
KEY
DEPTH
0.0
3.0
6.0
9.0
12.0
15.0
18.0
DO
STA. 1
DEC. 74
FIG. C-4

Fig. C-4.

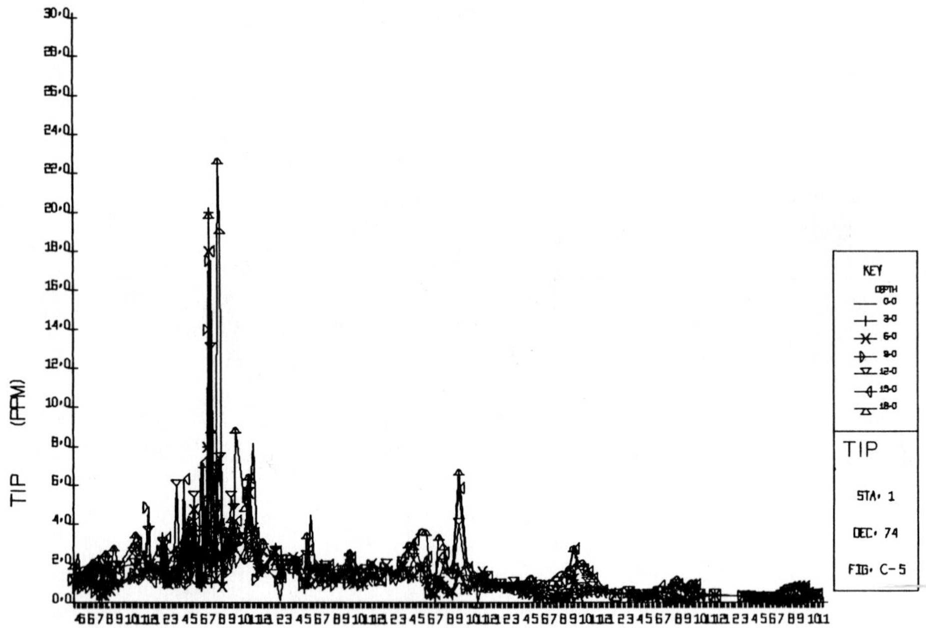
TIP (PPM)
KEY
DEPTH
0.0
3.0
6.0
9.0
12.0
15.0
18.0
TIP
STA. 1
DEC. 74
FIG. C-5

Fig. C-5.

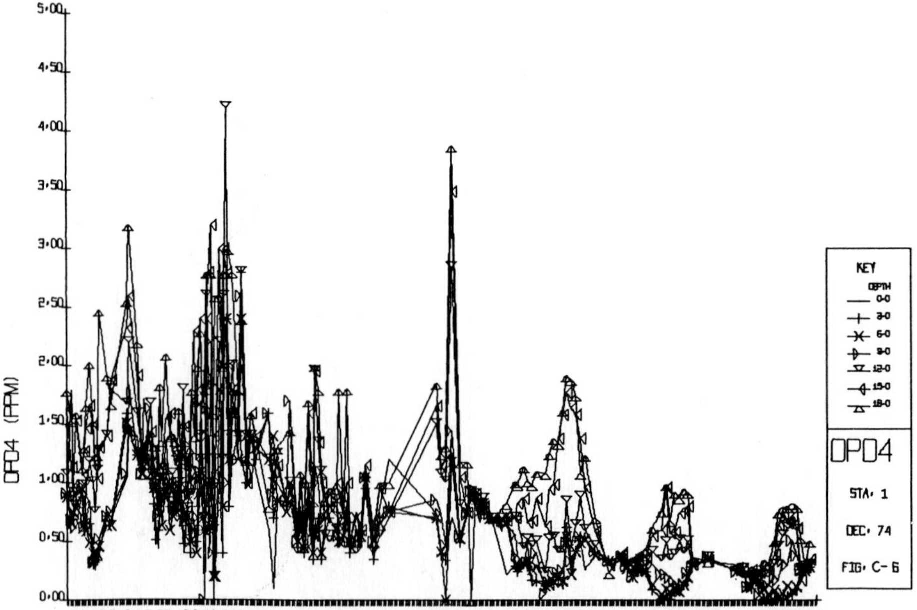
OPO4 (PPM)
KEY
DEPTH
0.0
3.0
6.0
9.0
12.0
15.0
18.0
OPO4
STA. 1
DEC. 74
FIG. C-6

Fig. C-6.

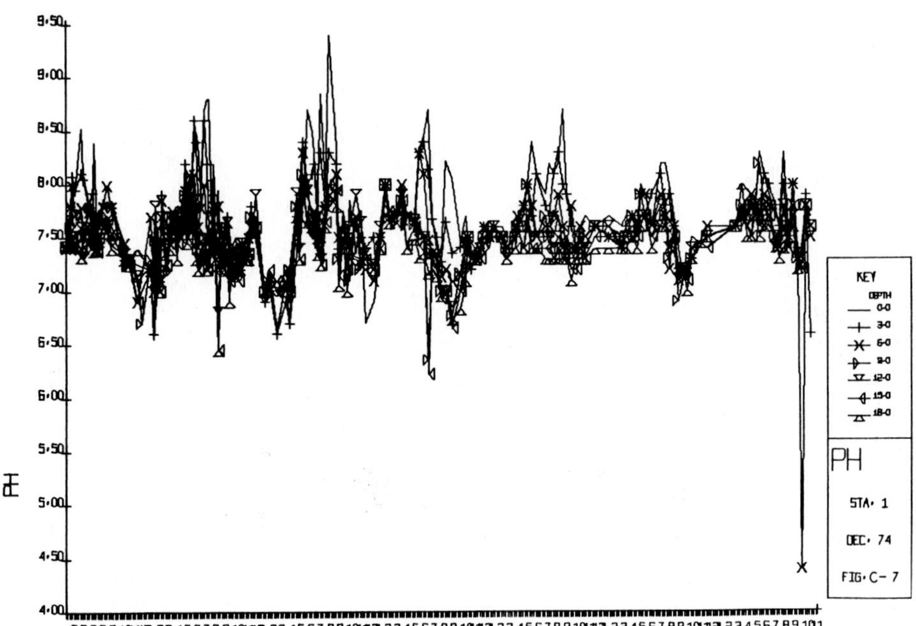

9.50
9.00
8.50
8.00
7.50
7.00
6.50
6.00
5.50
5.00
4.50
4.00
PH
KEY
DEPTH
0.0
3.0
6.0
9.0
12.0
15.0
18.0
PH
STA. 1
DEC. 74
FIG. C- 7

Fig. C-7.

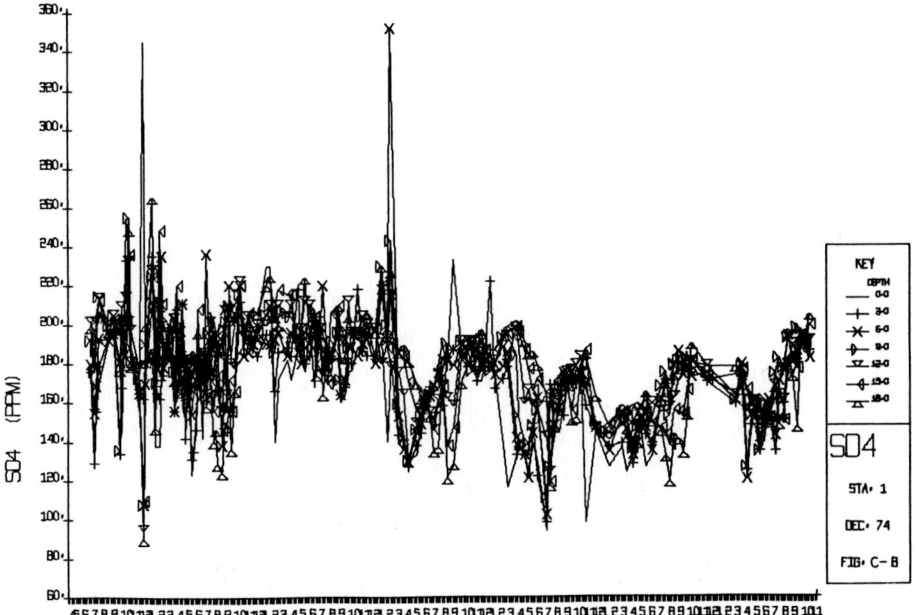

360.
340.
320.
300.
280.
260.
240.
220.
200.
180.
160.
140.
120.
100.
80.
60.
SO4 (PPM)
KEY
DEPTH
0.0
3.0
6.0
9.0
12.0
15.0
18.0
SO4
STA. 1
DEC. 74
FIG. C- 8

Fig. C-8.

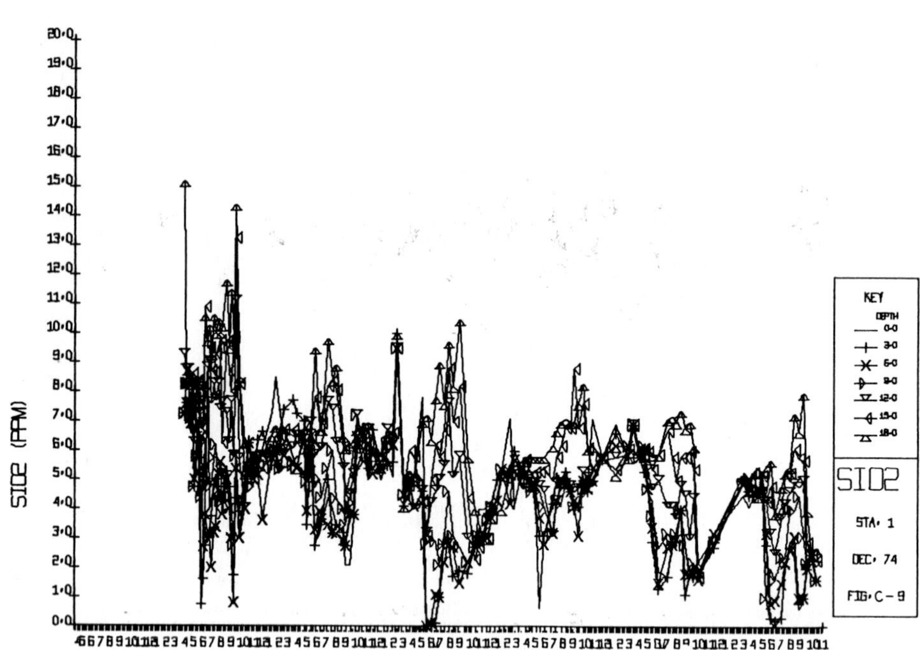

Fig. C-9.

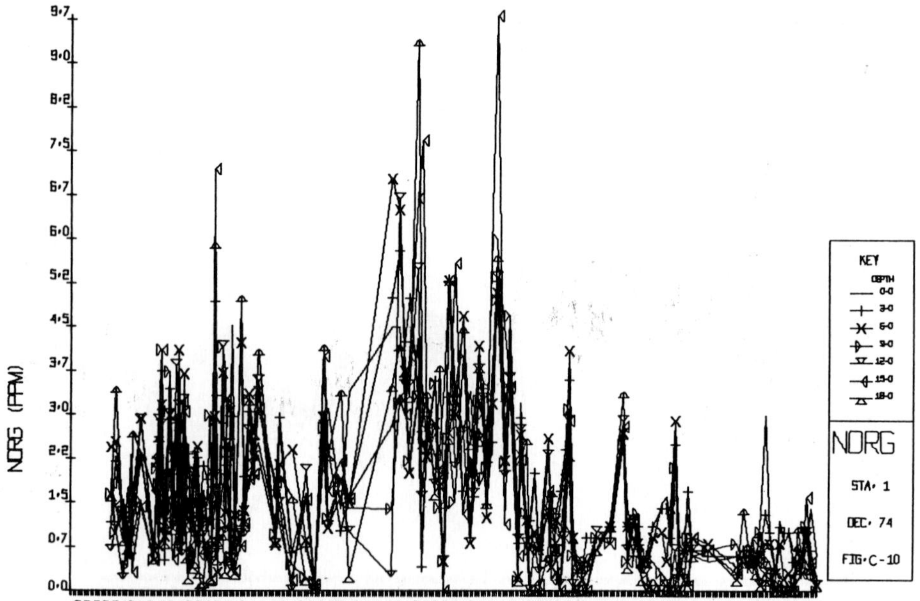

Fig. C-10.

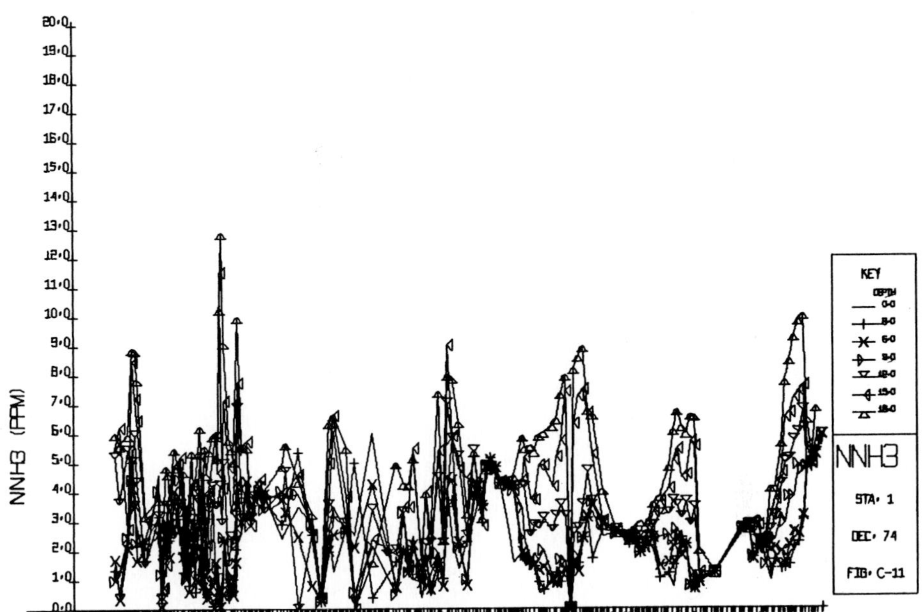

Fig. C-11.

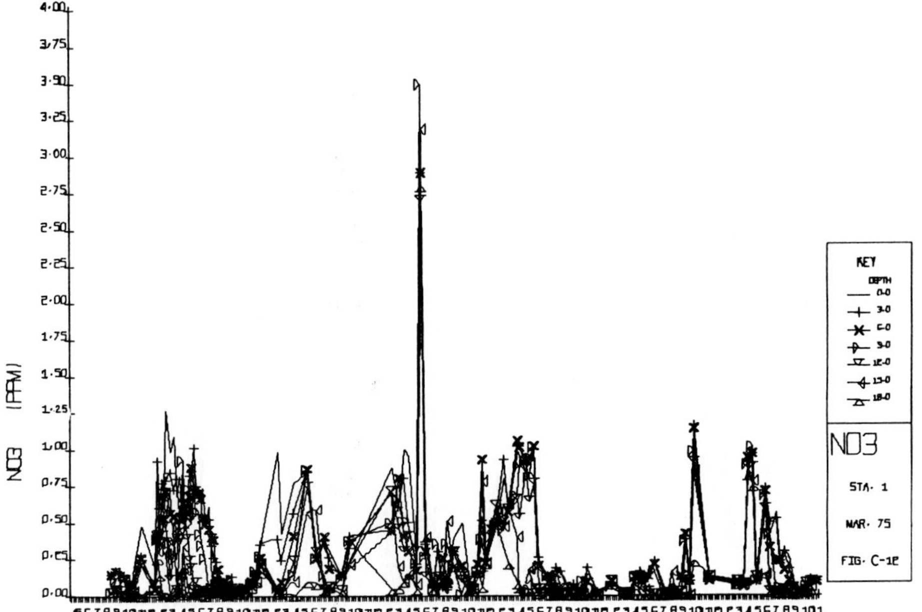

Fig. C-12.

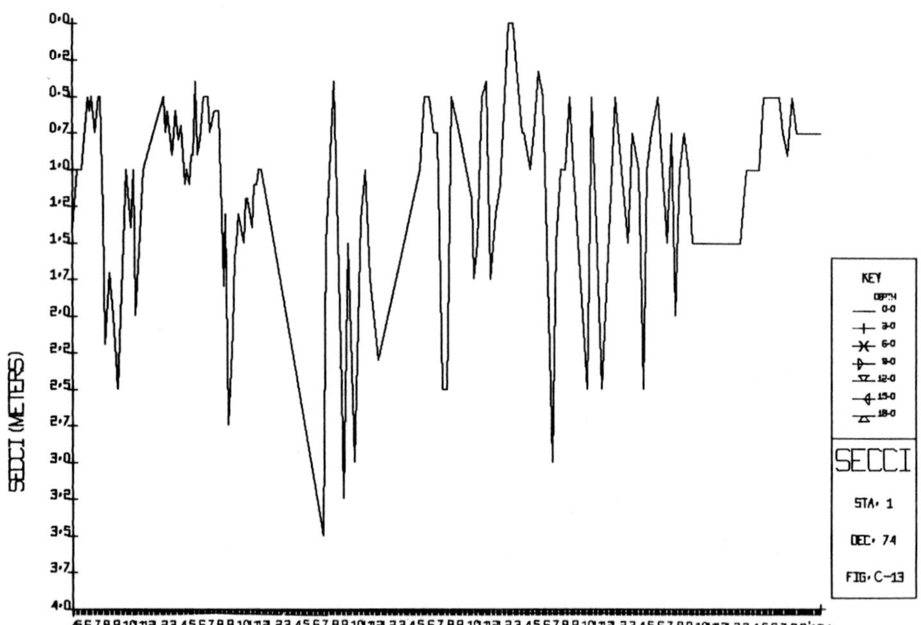

Fig. C-13.

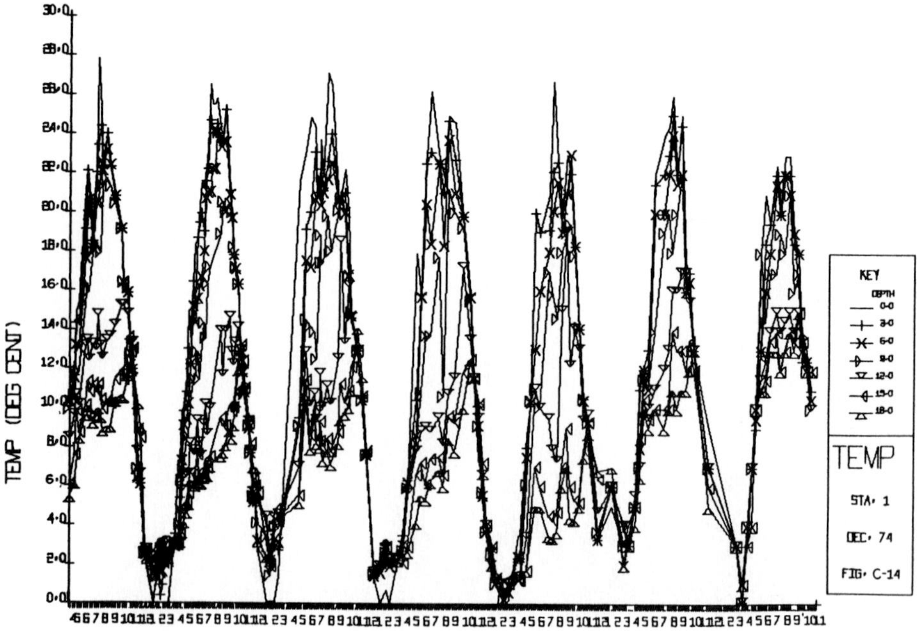

Fig. C-14.

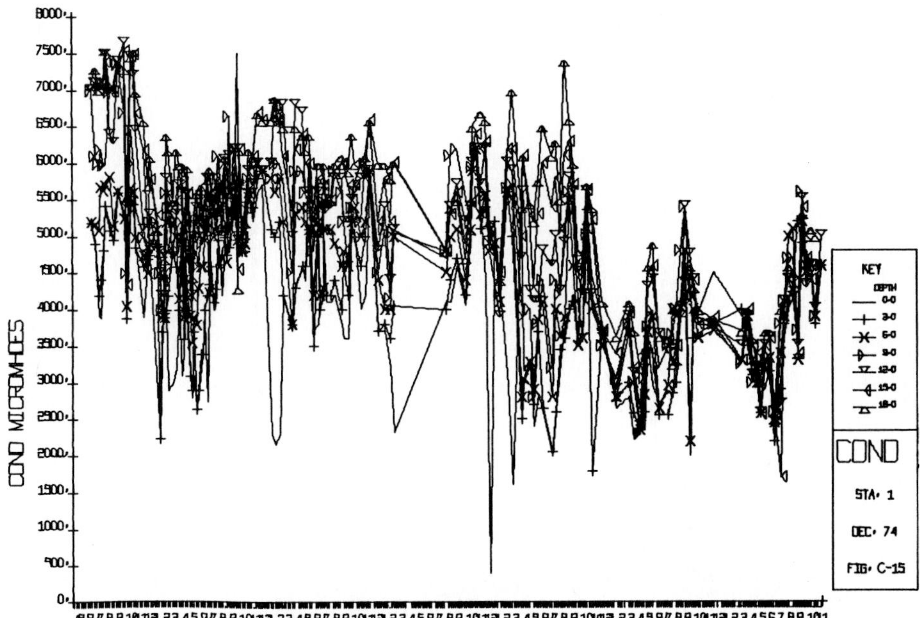

Fig. C-15.

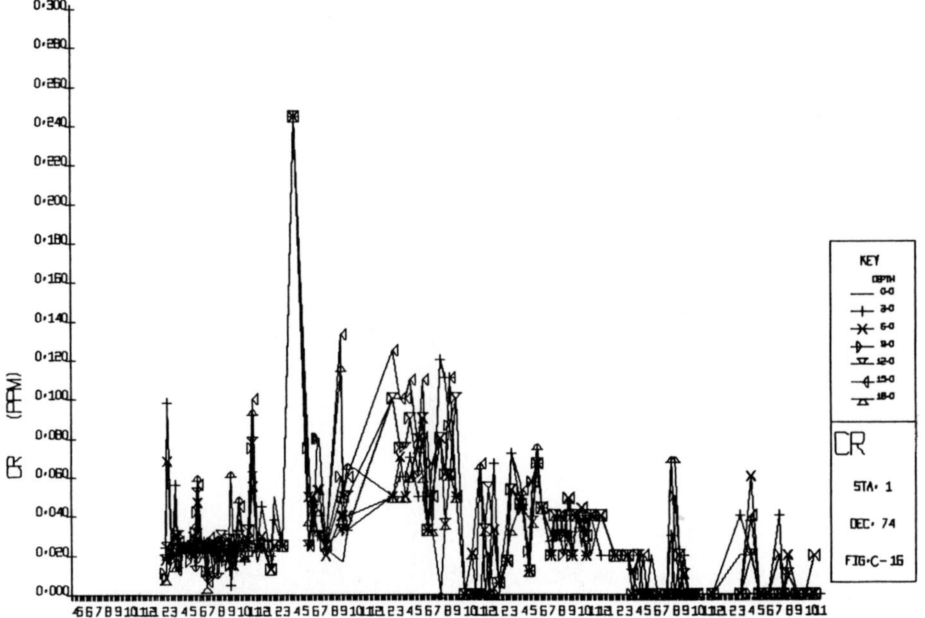

Fig. C-16.

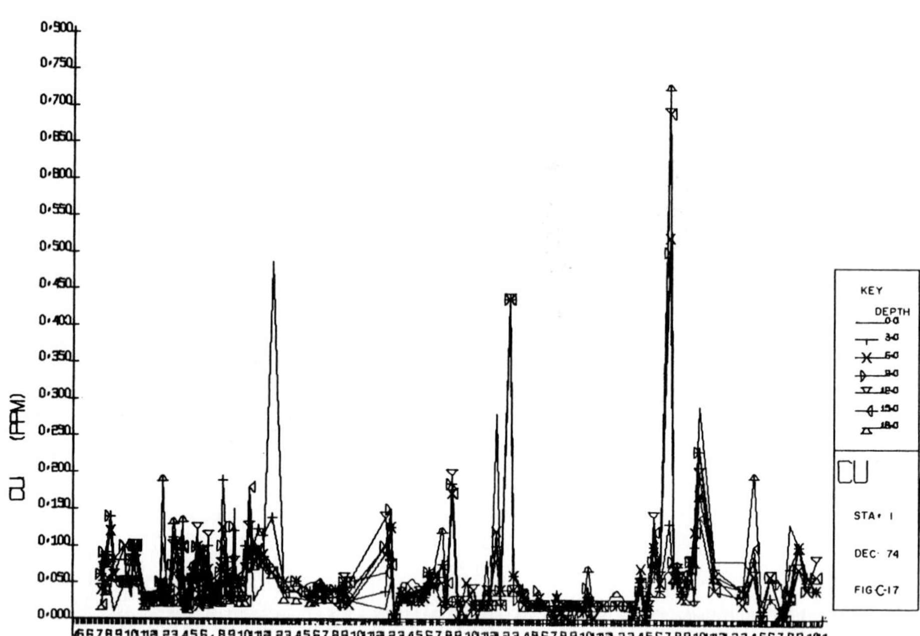

Fig. C-17.

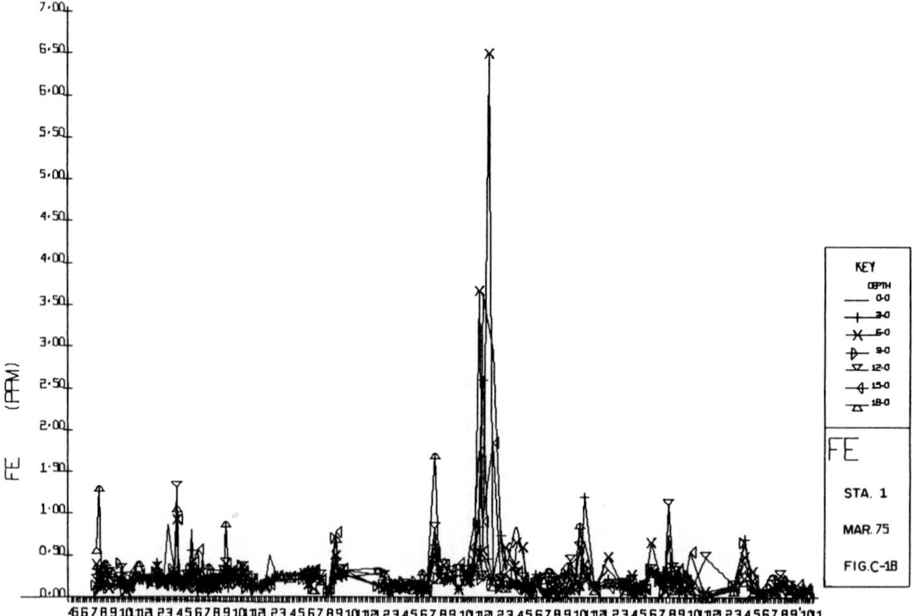

Fig. C-18.

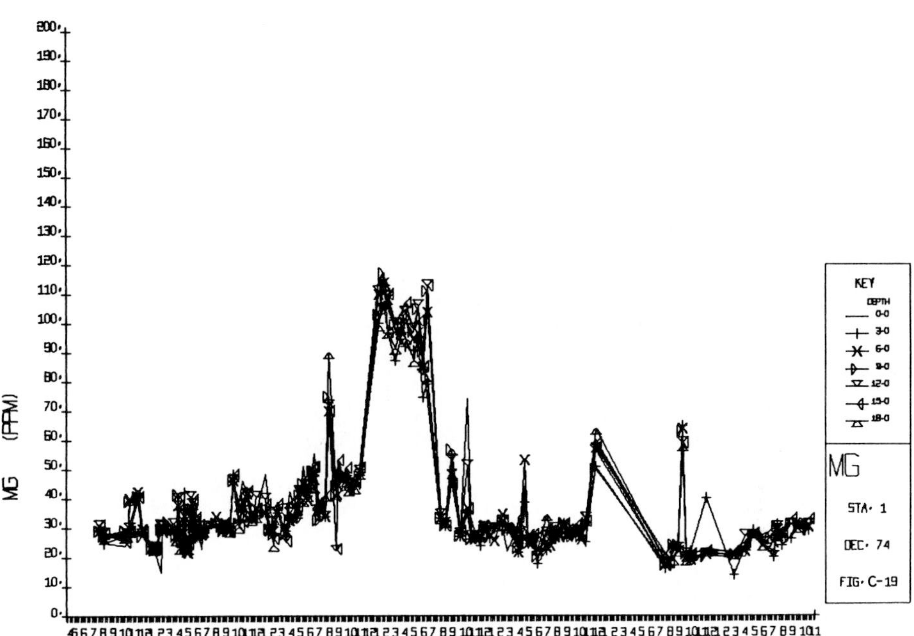

Fig. C-19.

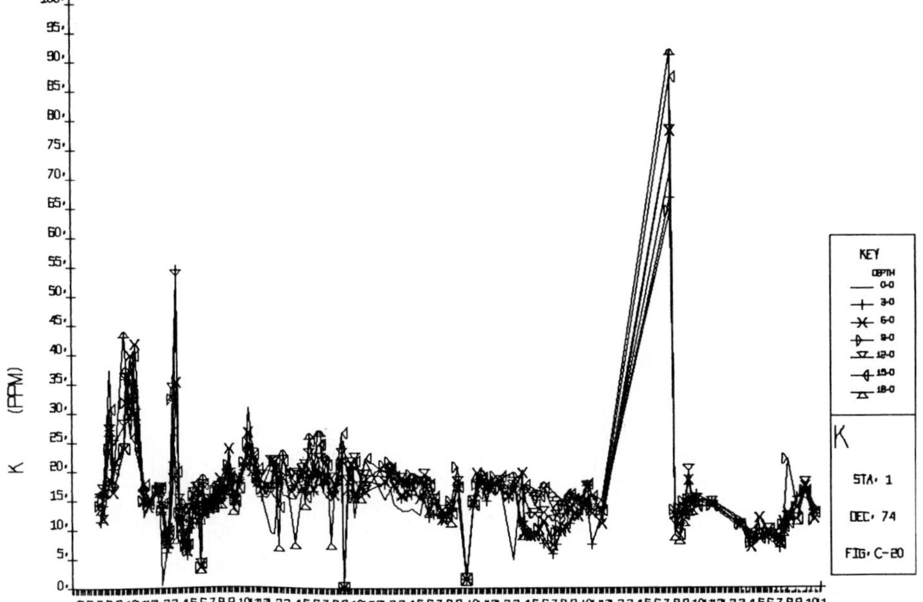

Fig. C-20.

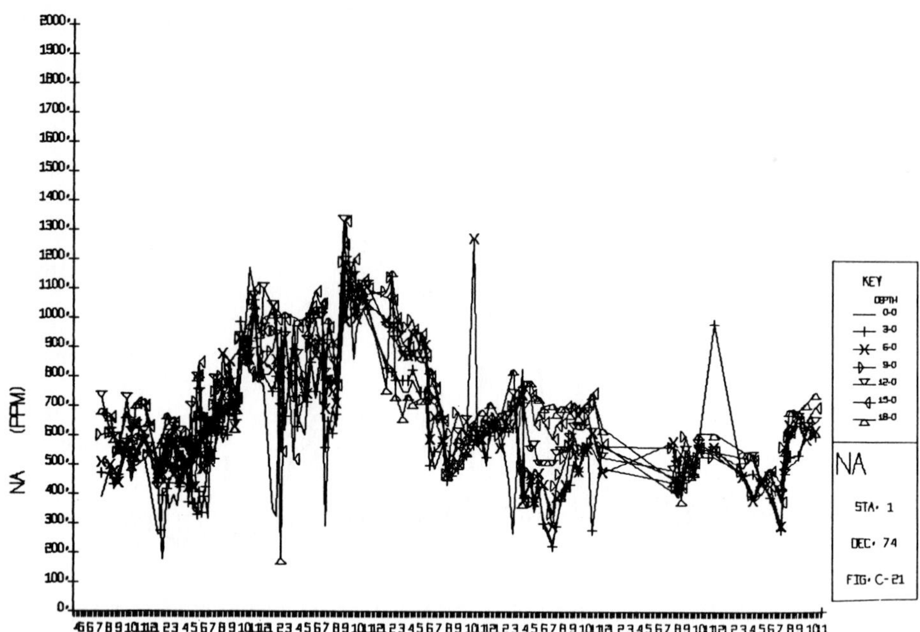

Fig. C-21.

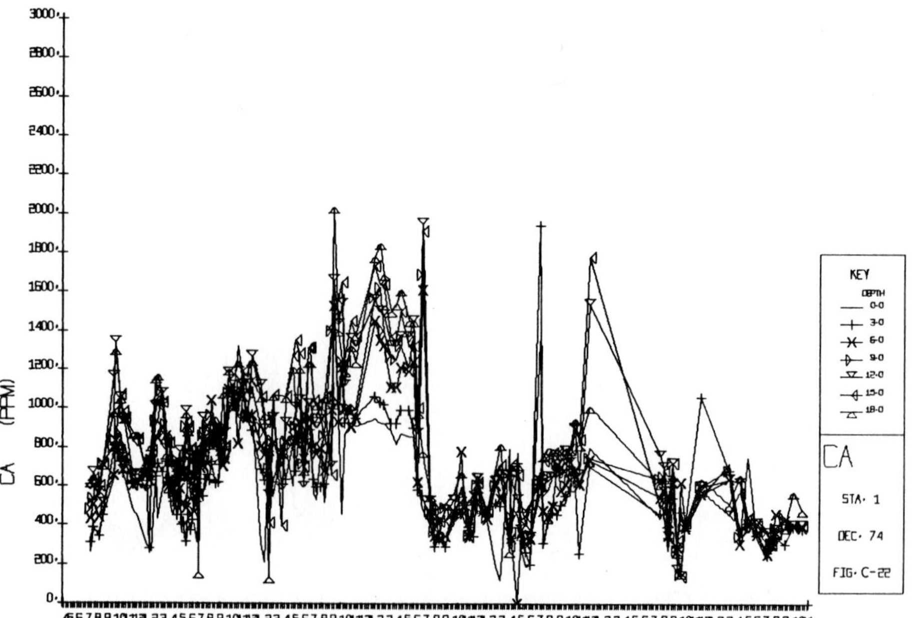

Fig. C-22.

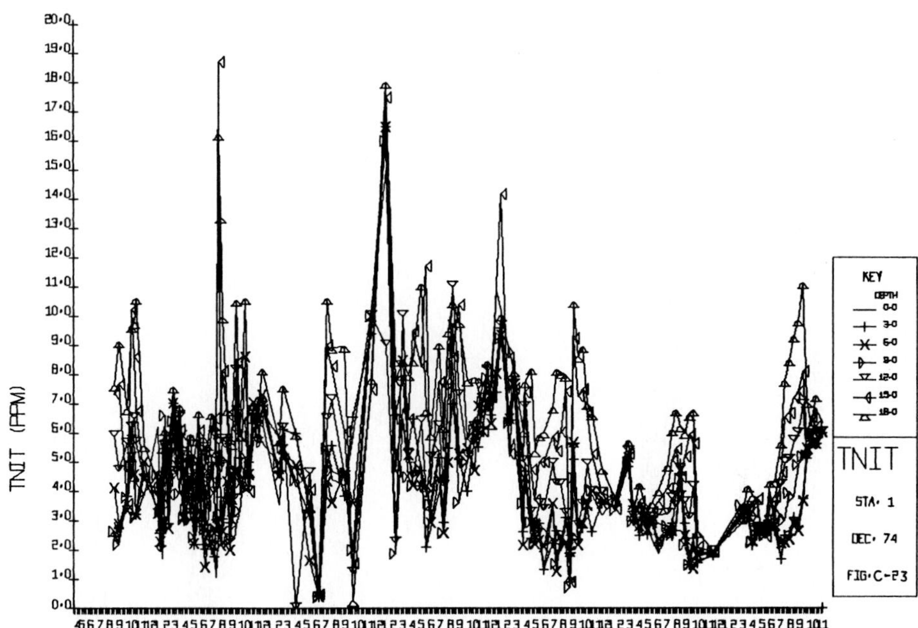

Fig. C-23.

Fig. C-24.

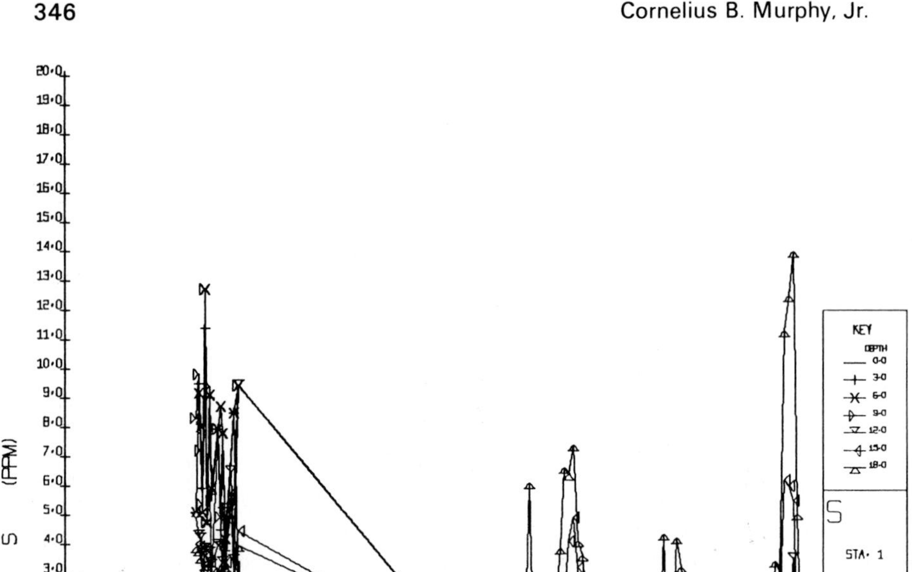

Fig. C-25.

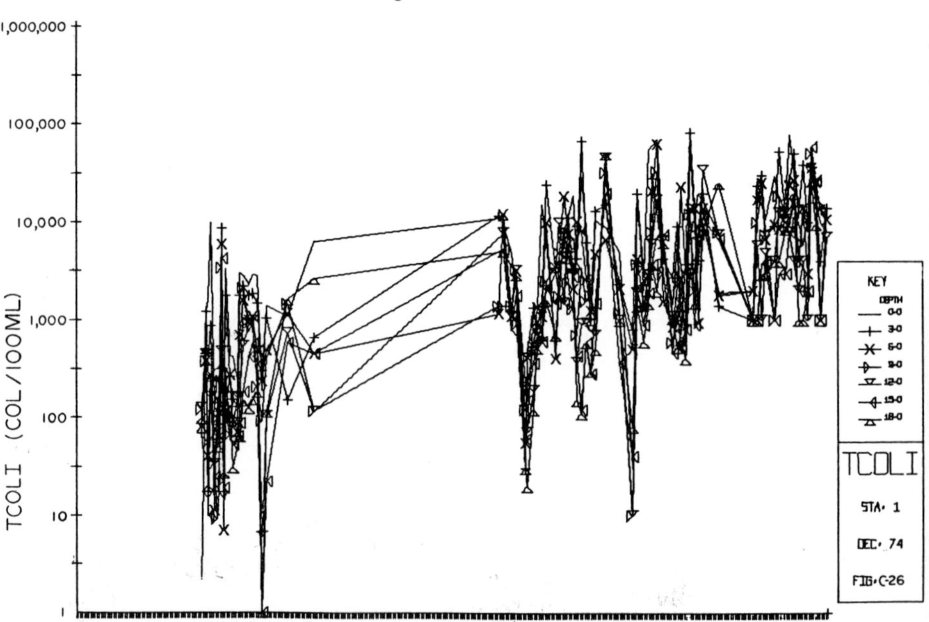

Fig. C-26.

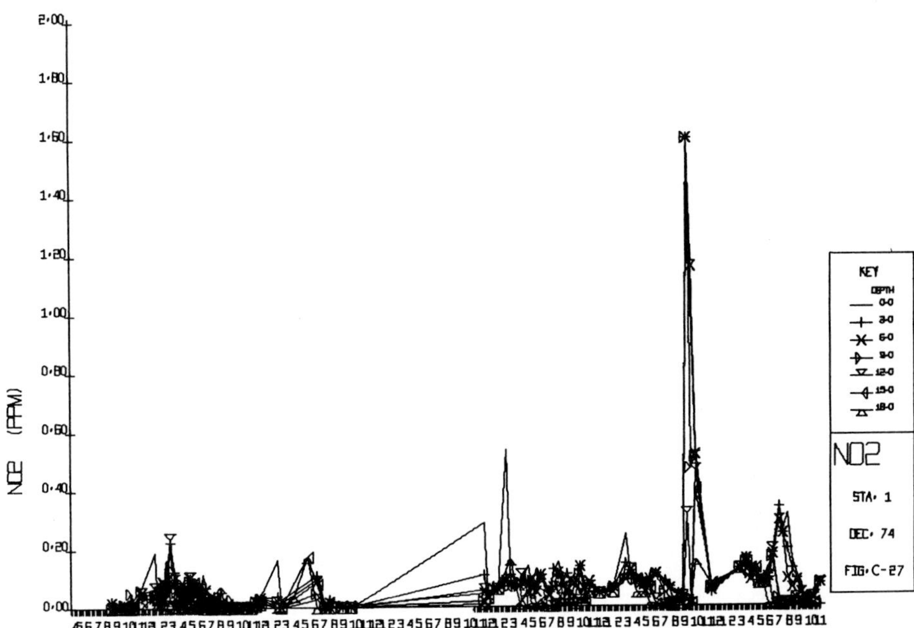

Fig. C-27.

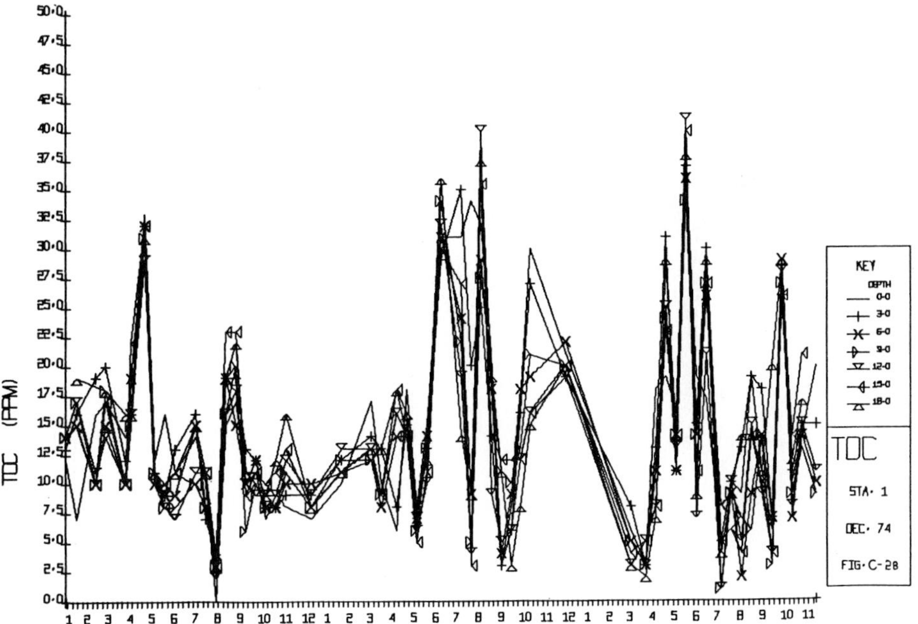

Fig. C-28.

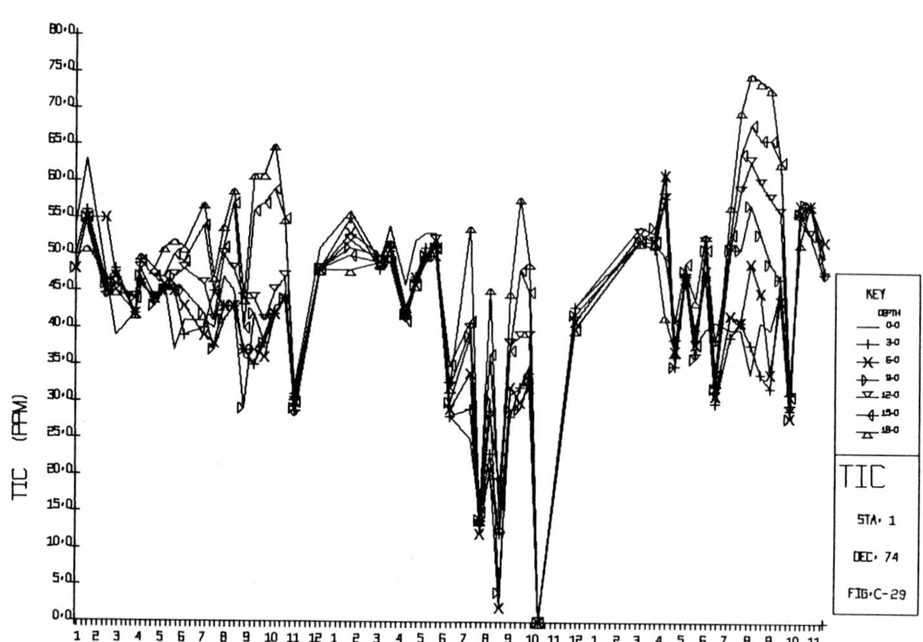

Fig. C-29.

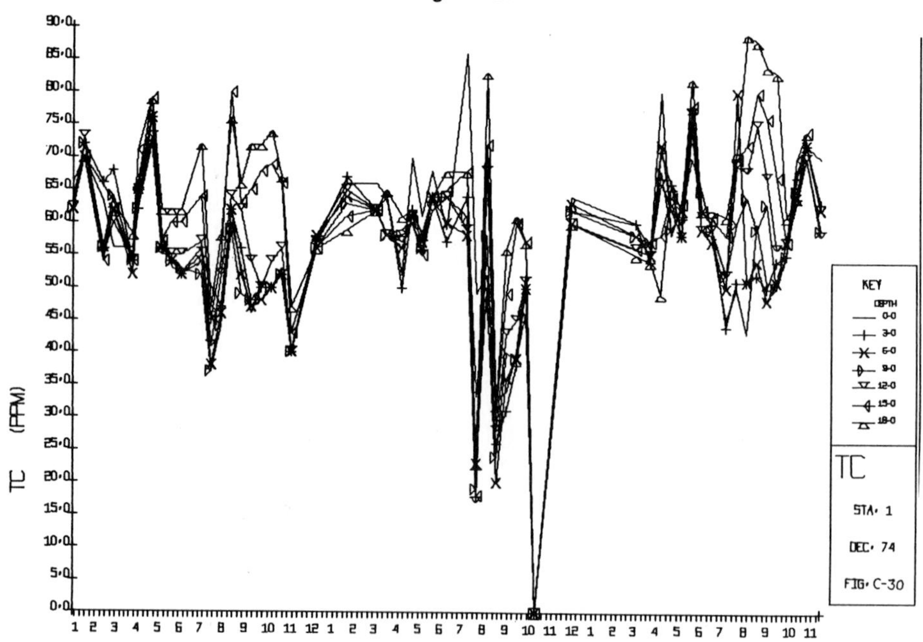

Fig. C-30.

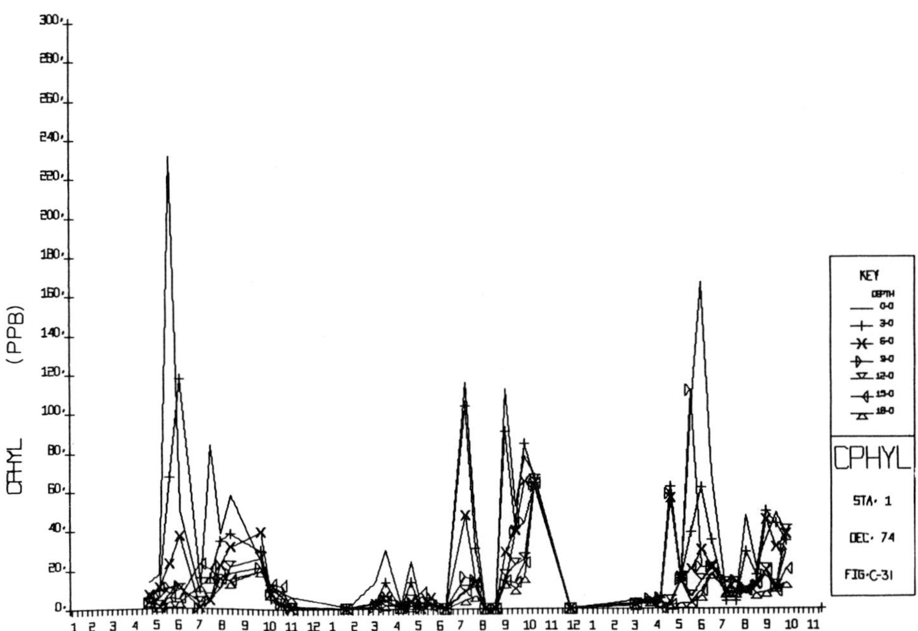

Fig. C-31.

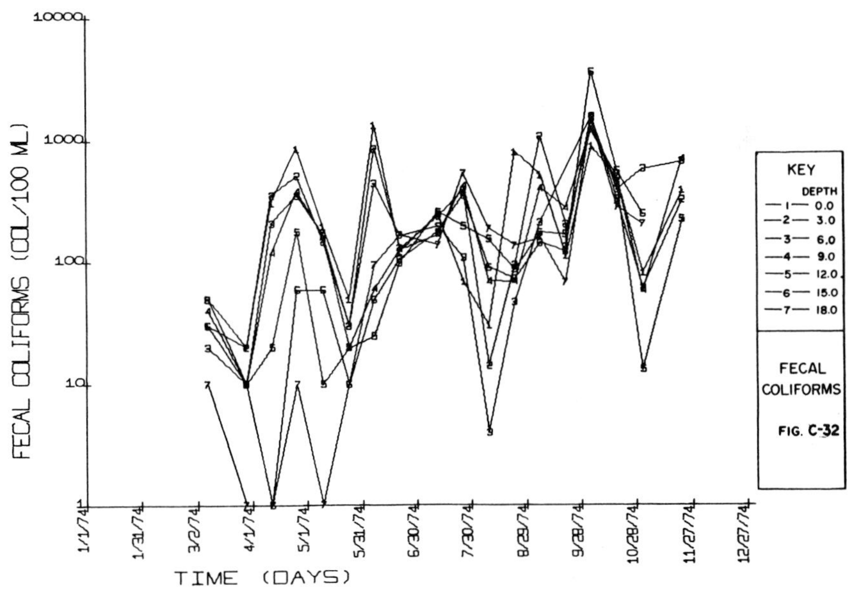

Fig. C-32.

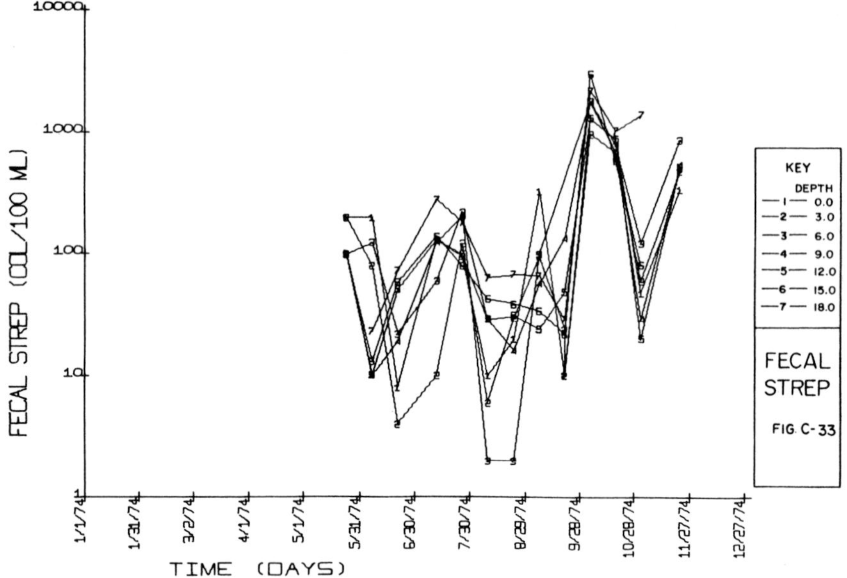

Fig. C-33.

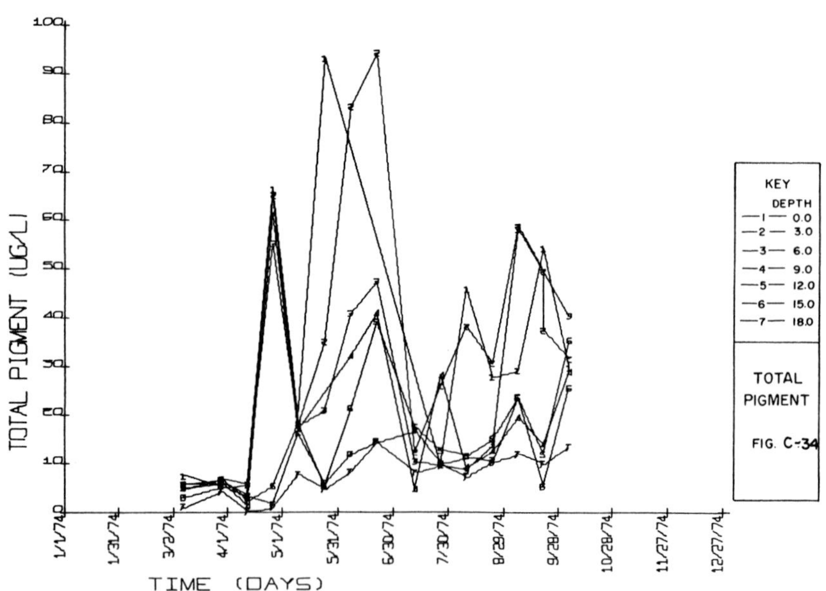

Fig. C-34.

APPENDIX D: ONONDAGA LAKE PHYTOPLANKTON DATA

TABLE D-1

Seasonal Succession of Dominant Phytoplankton (1969)

Species	Year	Outlet	Station: North Deep	South Deep	Ninemile Creek	Ley Creek
Pediastrum boryanum	1969	6/4–10/29	5/27–7/2	(June 25)	(7/2–9/18)	never
Pediastrum duplex	1969	6/4–11/13	6/11–11/20	(6/4–10/8)	(6/11–9/18)	(7/23)
Pediastrum simplex	1969	7/2–11/13	8/6–11/20	8/27–11/20	(8/27–10/29)	never
Closterium gracile	1968	never	never	never	never	never
	1969	9/18–10/29	10/2–10/29	10/8–11/5	10/29–11/13	never
Staurastrum paradoxum	1968	(8/8–11/14)	never	never	never	never
	1969	8/27–11/13	9/18–10/29	10/2	never	never
Cyclotella bodanica	1969	6/4–11, 7/9	5/27–6/11 (7/2–16)	5/27–6/11 (7/2–16)	5/27–6/18	(6/4–11) (7/16)
Cyclotella comta	1969	(9/25–12/4)	10/2–11/20	9/25–12/18	10/25	9/25–11/20
Cyclotella glomerata	1969	5/7–7/2 10/15[a]–12/4	5/7–6/25 10/8–12/4	5/7–7/2 10/8–12/18	5/7–6/25 10/25[a]–11/20	5/7–6/18 11/13–20[b]
Coscinodiscus subtilis	1969	7/2–12/4	6/25–11/13	(7/2–9, 10/2–11/20)	(7/2–11/20)	(7/2–10/8)
Fragilaria crotonensis	1969	8/27–12/4	7/28–11/20	9/25–11/13	9/25–12/4	never
Glenodinium pulvisculus	1968	never	never	never	never	never
	1969	7/23–10/2	8/6–9/18–10/8	10/8	never	never
Ceratium hirundinella	1968	8/8–27 9/30–10/7	8/14–27 9/30	8/20 never	8/14–27	never
	1969	7/23–8/13 9/18–25	8/22 9/18–25		8/27	never
Chroomonas nordstetii	1969	99/11–10/2	7/2–10/15	7/2–10/15	7/2–10/15	(7/2–8/27)
Cryptomonas ovata	1969	5/21–12/14	5/21–12/18	5/21–12/18	5/27–12/18	5/27–12/18

[a] No samples at Outlet and Ninemile Creek on Oct. 3.
[b] No samples at Ley Creek on Oct. 23 and 29, and on Nov. 5.

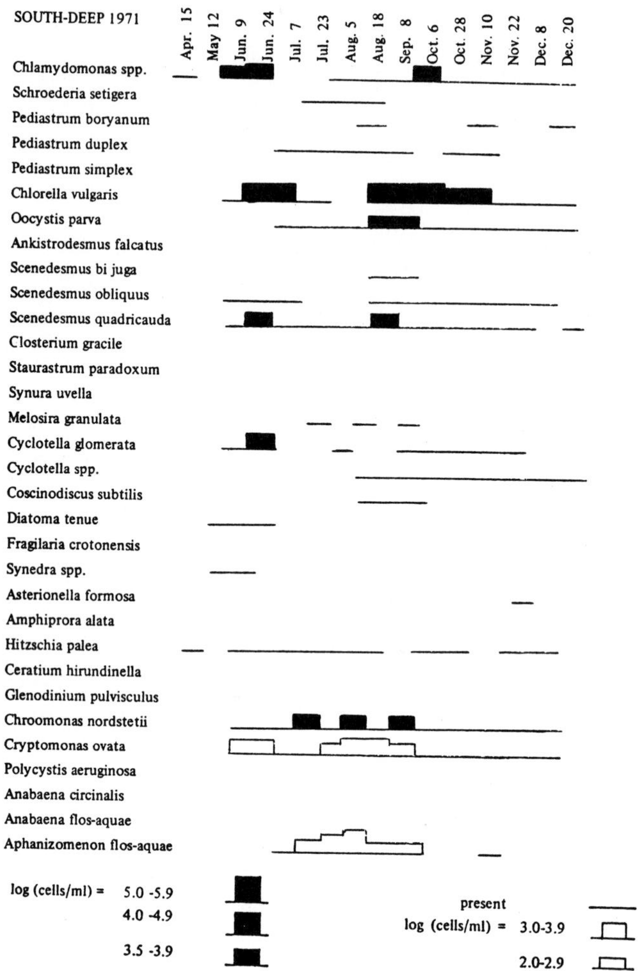

Fig. D-1. Seasonal succession of dominant phytoplankton (1971).

TABLE D-2

Seasonal Succession of Dominant Phytoplankton (1970)

Feb. 12, 25	Practically no algae
Mar. 9, 25, April 8	*Chlamydomonas* sp. moderately common
Apr. 24	Phytoplankton reduced
May 8	*Cyclotella glomerata* dominant; *Chlamydomonas* sp. common
May 21	*Cyclotella glomerata* dominant, but reduced from May 8
Jun. 4	*Chlamydomonas* sp. dominant; *Cryptomonas ovata* common
Jun. 18	*Chlamydomonas* sp., *Scenedesmus obliquus* and *Scenedesmus quadricauda* dominant; *Chlorella vulgaris* and *Cyclotella glomerata* common.
(Jul. 6—net sample only)	*Aphanizomonon flos-aquae* common
Jul. 22	*Aphanizomonon flos-aquae* dominant; *Chroomonas* sp. common
Aug. 12	*Chlorella vulgaris* dominant; *Aphanizomenon flos-aquae* common at the North-Deep
Sep. 10	*Aphanizomenon flos-aquae* dominant; *Cryptomonas ovata* and *Chroomonas* sp. common
Oct. 7	*Aphanizomenon flos-aquae* dominant
Oct. 21	Algae much reduced
Nov. 5	Algae reduced

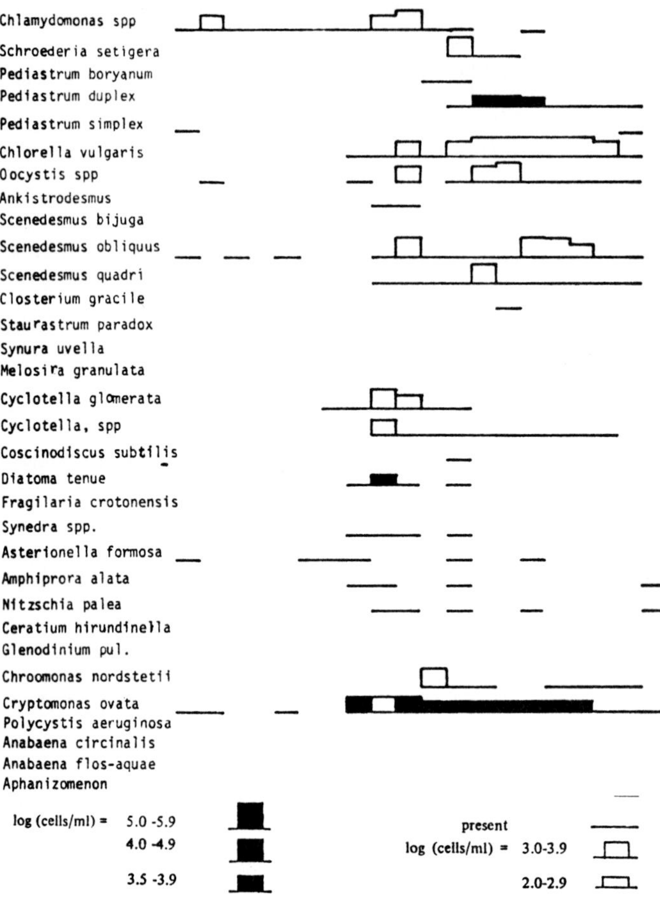

Fig. D-2. Seasonal succession of dominant phytoplankton (1972).

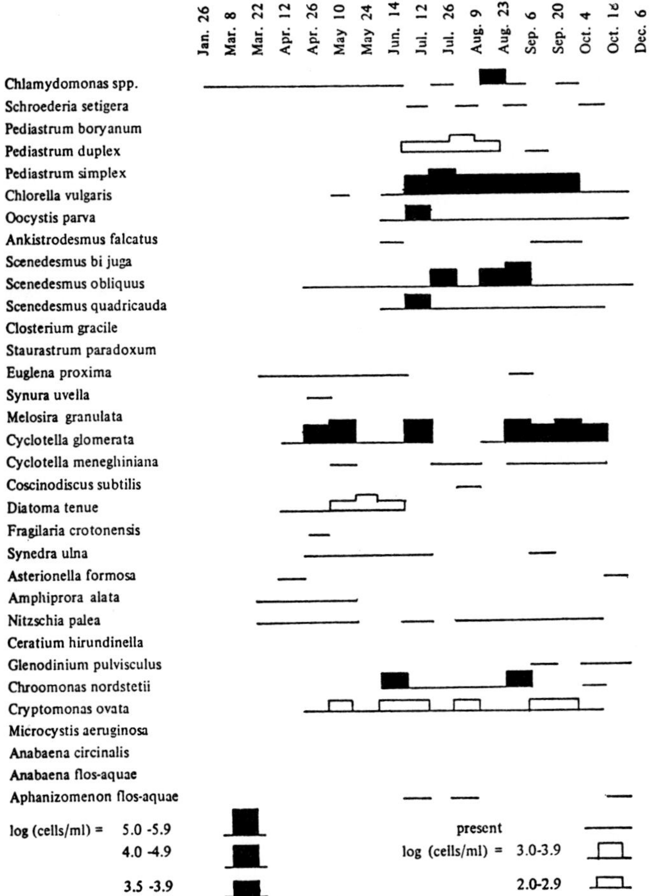

Fig. D-3. Seasonal succession of dominant phytoplankton (1973).

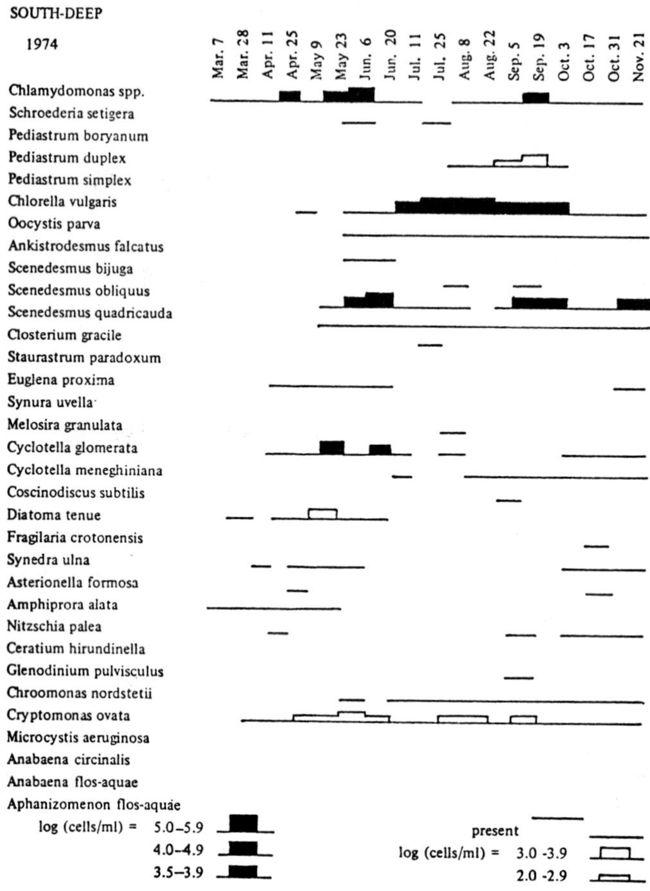

Fig. D-4. Seasonal succession of dominant phytoplankton (1974).

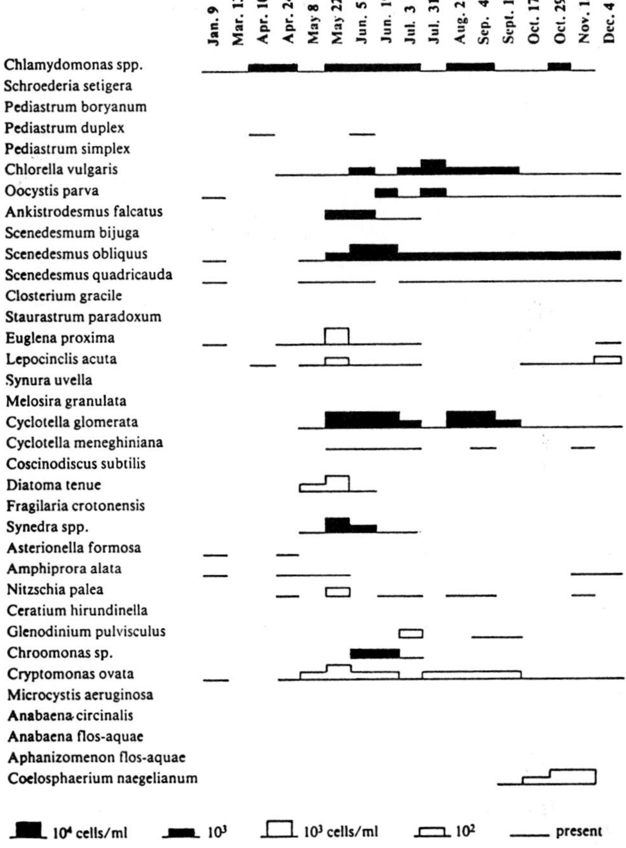

Fig. D-5. Seasonal succession of dominant phytoplankton (1975).

APPENDIX E: ONONDAGA LAKE ZOOPLANKTON DATA

TABLE E-1

Zooplankton (animals/100 liters)—1969

	Jan.		Feb.	Mar.	
	14	30	14	5	19
Cladocera					
Daphnia similis: total	1[a]	—[b]	—	—	—
with eggs	1	—	—	—	—
Daphnia pulex: total	8	2	1	tr[c]	—
with eggs	1	1	—	—	—
Ceriodaphnia quadrangula					
Copedoda					
all copepod nauplii	1	tr	—	—	—
cyclopoid copepodids	2	5	1	tr	1
Cyclops vernalis	1	1	—	—	—
Rotifera					
Keratella hiemalis	—	—	1	—	—
Brachionus sp.	—	—	—	—	—
Polyarthra sp.	—	—	—	—	—
Rare spp					
Asplanchna sp.	—	—	—	—	—
Keratella cochlearis	—	—	—	—	—
Filinia longiseta	—	—	—	—	—
Rotifer B	—	—	—	—	—
Chydorus spaericus	—	—	—	—	—
Bosmina longirostris	—	—	—	—	—
Cyclops bicuspidatus thomasi	—	—	—	— tr	—
Mesocylops edax	—	—	—	—	—
Diaptomid copepodids	—	—	—	—	—
Diaptomus sicilis	—	—	—	—	—

[a] Arithmetic mean of two station means (North and South basin near center), rounded to integers.

[b] Blank space means none present in aliquots counted.

[c] tr, Trace.

TABLE E-2

Zooplankton (animals/100 liters)—1970

Date	Rotifers	Cladocerans	Copepods
South-Deep			
Feb. 12	—	—	—
Mar. 25	—	—	—
Apr. 8	—	—	—
Apr. 24	—	—	—
May 7	6	—	3
May 21	1292	—	347
Jul. 22	1845	464	1391
Aug. 12	—	927	829
Sep. 10	2723	689	1584
Sep. 23	5437	1415	958
Oct. 7	393	212	482
Oct. 21	43	40	270
North-Deep			
Mar. 9	—	—	—
Apr. 8	—	—	—
May 21	608	—	101
Jul. 22	1664	1142	2901
Aug. 12	—	967	771
Sep. 10	1652	1394	789
Oct. 7	1299	166	1572

TABLE E-3

Zooplankton (animals/100 liters)—1971

Date	Rotifers	Copepods	Cladocerans
Jan. 21	183	—	—
Feb. 4	12	—	6
Feb. 17	15	—	—
Mar. 12	3	—	—
Jun. 9	—	5964	—
Jun. 24	3	5520	60
Jul. 7	3	2196	984
Jul. 23	405	1848	804
Aug. 5	1234	249	126
Aug. 18	675	7380	2996
Sep. 8	2382	3463	1523
Oct. 6	1203	1081	332
Oct. 28	—	9	—
Nov. 10	114	68	3
Nov. 22	40	12	9
Dec. 8	37	—	3
Dec. 20	86	3	12

TABLE E-4

Zooplankton (animals/100 liters)—1972

Date	Rotifers	Copepods	Cladocerans
Jan. 6	21	—	—
Feb. 17	3	—	—
Mar. 2	—	—	—
Mar. 31	3	—	—
May 11	—	28	—
May 25	52	1133	—
Jun. 8	3	4703	12
Jul. 6	—	356	71
Jul. 20	1437	2038	838
Aug. 3	725	5059	3095
Aug. 17	2726	4421	5944
Aug. 31	1658	227	387
Sep. 14	17167	8842	1731
Sep. 28	4151	393	2345
Oct. 13	1032	21	1756
Oct. 26	3266	86	3168
Nov. 10	405	86	540
Dec. 14	15	18	58

TABLE E-5

Zooplankton (animals/100 liters)—1973

Date	Rotifers	Copepods	Cladocerans	*Vorticella*
Jan. 26	—	—	tr	37
Mar. 8	tr[a]	tr	—	1056
Mar. 22	tr	—	—	319
Apr. 12	tr	—	25	—
Apr. 26	21	15	—	—
May 10	427	—	—	—
May 24	111	18	—	—
Jun. 14	—	4175	353	—
Jul. 12	921	6410	1339	—
Jul. 26	126	712	182	—
Aug. 9	292	181	18	—
Aug. 23	651	731	49	—
Sep. 6	420	3045	141	—
Sep. 20	7024	4519	tr	—
Dec. 6	24855	—	tr	—

[a] tr, Trace.

TABLE E-6

Zooplankton (animals/100 liters)—1974

Date	Rotifers	Copepods	Cladocerans
Mar. 7	28	—	—
Mar. 28	31	tr[a]	—
Apr. 11	28	—	—
Apr. 25	25	tr	—
May 9	28	12	—
Jun. 6	46	494	—
Jun. 20	995	3119	68
Jul. 11	37	7957	2898
Jul. 25	25	3414	5452
Aug. 8	5108	5158	7466
Aug. 22	2947	1793	626
Sep. 5	196	675	150
Sep. 19	405	4716	1056
Oct. 3	147	64	—
Oct. 17	28	356	—
Oct. 31	64	294	—
Nov. 21	1412	774	—

[a] tr, Trace.

TABLE E-7

Zooplankton (animals/100 liters)—1975

Date	Rotifers	Copepods	Cladocerans
Jan. 9	101	25	—
Apr. 10	—	3	—
Apr. 24	—	—	—
May 8	—	43	—
May 22	18	761	—
Jun. 5	15	6165	—
Jul. 3	17241	15669	68
Jul. 31	141	5501	9529
Aug. 21	2260	5944	239
Sep. 4	422	1867	6
Sep. 18	203	384	25
Dec. 4	1093	135	3

ACKNOWLEDGMENTS

The development of this contribution would not have been possible without two principal previous studies. The "Onondaga Lake Study" and the "Environmental Impact Statement on Wastewater Treatment Facilities Construction Grants for the Onondaga Lake Drainage Basin" provided the majority of the information in the Onondaga Lake study. The author wishes to express his thanks to all who have contributed to the above sources. Additionally, those utilizing the material contained within this contribution should realize that the information exists only because of the forward thinking of Onondaga County and in particular, the Department of Drainage and Sanitation.

Particular thanks is extended to Ralph McClurg and Gregory J. Welter, research engineers with O'Brien and Gere Engineers, Inc., who assisted in the preparation and editing of materials presented in this manuscript.

REFERENCES

Agar, C., and Sanderson, W. (1947). "Investigation of Onondaga Lake." Onondaga County, New York State Health Department, Albany.

American Public Health Association. (1971). "Standard Methods for the Examination of Water and Wastewater." American Water Works Association, Water Pollution Control Federation, Washington, D.C.

Anonymous. (1968). Data from Allied Chemical Corporation dating from 1919 to present, Syracuse, New York (unpublished).

Baker, D. G., and Haines, D. A. (1969). Solar radiation and sunshine duration relationships in the North-Central region and Alaska. *Minn., Agric. Exp. Stn., Tech. Bull.* 262.

Bannister, T. T. (1974). Production equations in terms of chlorophyll concentration, quantum yield and upper limit to production. *Limnol. Oceanogr.* **19,** 1.

Beeton, A. M. (1958). *Trans. Am. Fish. Soc.* **87,** 000.

Berg, C. O. (1963). Middle Atlantic states. *In* "Limnology in North America" (D. G. Frey, ed.), pp. 191–237. Univ. of Wisconsin Press, Madison.

Brennan, P. J., Grove, C. S., Jr., Jackson, D. F., Katz, M., Nemerow, N. L., and Rand, M. C. (1968). "Investigations Leading to the Restoration of Urban Lakes with Application to Onondaga Lake, New York," Report of the Department of Civil Engineering. Syracuse University, Syracuse, New York (unpublished).

Brylinsky, M. and Mann, K. H. (1973). An Analysis of Factors Governing Productivity in Lakes and Reservoir. *Limnol. Oceangr.* **18,** 1.

Burdick, G., and Lipschuetz, M. (1947). "A Supplementary Report to the Survey of Onondaga Lake and Tributary Streams in 1946 and 1947." New York State Conservation Department, Albany.

Chappelle, E. W., and Piccolo, G. L. (1975). "Laboratory Procedures Manual for the Firefly Luciferase Assay for Adenosine Triphosphate," S-726-75-1. Goddard Space Flight Center, Greenbelt, Maryland.

Clark, J. V. H. (1849). "Onondaga," Vol. II, Chapter XII, pp. 7–44. Stoddard & Babcock, Syracuse, New York.

Clarke, J. M. (1902). The squids from Onondaga Lake, New York. *Science* **16,** 1947.

Cressy, G. B. (1966). Land forms. *In* "Geography of New York State" (J. H. T. Thompson, ed.), pp. 19–53. Syracuse Univ. Press, Syracuse, New York.

Dawson, E. Y. (1966). "Marine Botany," p. 371. Holt, New York.

De Noyelles, F., Jr. (1968). A stained-organism filter technique for concentrating phytoplankton. *Limnol. Oceanogr.* **13,** 562.

DiToro, D. M., O'Connor, D. J., and Thomann, R. V. (1971). A dynamic model of the phytoplankton population in the Sacramento-San Joaquin Delta. *Adv. Chem. Ser.* **106,** 131.

Dornbush, J. M., Andersen, J. R., and Harms, L. L. (1974). "Quantification of Pollutants in Agricultural Runoff," USEPA Publ. EPA-660/2-74-005. Civil Engineering Department, South Dakota State University, Brookings.

Doudoroff, P., and Katz, M. (1953). Critical review of literature on the toxicity of industrial wastes and their components to fish. II. The metals, as salts. *Sewage Ind. Wastes* **25,** 802.

Eppley, R. W. (1972). Temperature and phytoplankton growth in the sea. *Fish. Bull.,* **70,** 1063.

Faro, R., and Nemerow, N. L. (1969). "Measurement of Benefits of Water Pollution Control." Syracuse University Department of Civil Engineering, Syracuse, New York.

Garrey, W. C. (1916). The resistence of fresh-water fish to changes of osmotic and chemical conditions. *Am. J. Phys.* **39,** 313.

Holmes, C. D. (1922). Sixteenth Annual Report of the Syracuse Intercepting Sewer Board, Syracuse, New York.

Hutchinson, G. E. (1957). "A Treatise on Limnology," Vol. I, p. 1115. Wiley, New York.

Jackson, D. F. (1968). Onondaga Lake, New York—An unusual algal environment. *In* "Algae, Man and the Environment" (D. F. Jackson, ed.), pp. 514–524. Syracuse Univ. Press, Syracuse, New York.

Jackson, D. F. (1969). Primary productivity studies in Onondaga Lake, N.Y. *Verh. Int. Ver. Theor. Angew. Limnol.* **17,** 86.

Jones, J. E. (1938). The relative toxicity of salts of lead, zinc, and copper to the stickleback. *J. Exp. Biol.* **15,** 394.

Keenan, J. D. (1970). The algal growth bioassay as a comparative sanitary and limnological parameter applied to Onondaga Lake, New York. M.S. Thesis, Syracuse University, Syracuse, New York.

Lorenzen, C. J. (1967). Determination of chlorophyll and peho-pigments: Spectorphotometric equations. *Limnol. Oceanogr.* **12,** 343.

McKee, J. E., and Wolf, H. W. (1963). "Water Quality Criteria," Publ. No. 3-A. Resources Agency of California, State Water Quality Control Board, Sacramento.

Metcalf, L., and Eddy, H. P. (1920). "Report to Syracuse Intercepting Sewer Board upon the Treatment and Disposal at Syracuse." Metcalf & Eddy, Consulting Engineers, Boston, Massachusetts.

Murphy, C. B. (1973a). Effects of nutrient reductions on Onondaga Lake. *46th Ann. Conf., Water Pollut. Control Fed., October, 1973* p. 000.

Murphy, C. B. (1973b). Effect of restricted use of phosphate based detergents on Onondaga Lake. *Science* **182,** 379.

Murphy, C. B., and Welter, G. J. (1976). Indices of algal biomass and primary production in Onondaga Lake. *Nat. Conf. Environ. Res. Dev. Des., ASCE Environ. Eng. Div., Univ. of Washington.*

Murphy, C. B., Moffa, P. E., and Karanik, J. M. (1973). Effects of nutrient reductions on Onondaga Lake. *46th Annu. Conf. Water Pollut. Control Fed., 1973.*

New York State Department of Health. (1951). "Oswego River Drainage Basin," Surv. Ser. Rep. No. 1. NYSDH, Albany.

National Oceanic and Atmospheric Administration (NOAA). (1974). "Local Climatological Data, 1974 Annual Summary." Environ. Data Serv., Nat. Climatic Cent., Asheville, North Carolina.

Noble, R. L., and Forney, J. L. (1971). Fish survey of Onondaga Lake-Summer, 1969. *In* "Onondaga Lake Study, April 1971 by Onondaga County, New York." USEPA, Water Quality Office Publ. No. 11060 FAE 4/71. US Govt. Printing Office, Washington, D.C.

O'Brien & Gere Engineers, Inc. (1952). "Onondaga Lake Watershed." Onondaga County Department of Public Works, Syracuse, New York.

O'Brien & Gere Engineers, Inc. (1972a). "Onondaga Lake Monitoring Program Jan. 1970–Dec. 1970." Onondaga County, Syracuse, New York.

O'Brien & Gere Engineers, Inc. (1972b). "Onondaga Lake Monitoring Program Jan. 1971–Dec. 1971." Onondaga County, Syracuse, New York.

O'Brien & Gere Engineers, Inc. (1973a). "Environmental Assessment Statement." Syracuse Metropolitan Sewage Treatment Plant and the West Side Pump Station and Force Main, Onondaga County, Syracuse, New York.

O'Brien & Gere Engineers, Inc. (1973b). "Onondaga Lake Monitoring Program Jan. 1972–Dec. 1972." Onondaga County, Syracuse, New York.

O'Brien & Gene Engineers, Inc. (1974). "Onondaga Lake Monitoring Program Jan. 1973–Dec. 1973." Onondaga County, Syracuse, New York.

O'Brien & Gere Engineers, Inc. (1975). "Onondaga Lake Monitoring Program Jan. 1974–Dec. 1974." Onondaga County, Syracuse, New York.

O'Brien, & Gere Engineers, Inc. (1976). "Onondaga Lake Monitoring Program Jan. 1975–Dec. 1975." Onondaga County, Syracuse, New York.

Odum, E. P. (1971). "Fundamentals of Ecology." Saunders, Philadelphia, Pennsylvania.

Onondaga County. (1971). "Onondaga Lake Study," Project No. 11060, FAE 4/71. Quality Office, Environmental Protection Agency, Onondaga County, Syracuse, New York.

Pitts, N. (1949). "Pollution Survey of Onondaga Lake and Tributaries Within the City of Syracuse, N.Y." Department of Engineering of the City of Syracuse.

Sawyer, C. N. (1957). Some new aspects of phosphates in relation to lake fertilization. *Sewage Ind. Wastes* **24,** 768.

Schultz, O. (1810). "Travels on an Inland Voyage Through the States of New York, Pennsylvania, Virginia, Kentucky, and Tennessee, and Through the Territories of Indiana, Louisiana, Mississippi and New Orleans—performed in the years 1807 and 1808; Including a Tour of Nearly Six Thousand Miles," Vol, I, pp. 29–34. Printed by Issac Riley, New York.

Shapiro, J. (1973). Blue-green algae: Why they become dominant. *Science* **179,** 382.

Shattuck, J. H. (1968). "Annual Report of the Division of Parks and Conservation for the Year 1967," Department of Public Works, Onondaga County, Liverpool, New York.

Steele, J. H. (1962). Environmental control of photosynthesis in the sean. *Limnol. Oceanogr.* **7,** 137.

Sweet. (1874). Maps of the Towns of Geddes, Salina and City of Syracuse, Scale 150 Ruds to the inch. Part of the Township of Manlius and Onondaga Salt Spring Reservation.

Syracuse-Onondaga County Planning Agency (1972). Onondaga County Population Projections.

Syracuse, University Research Corporation. (1966). "Onondaga Lake Survey 1964–1965," Final Rep., Contract No. 153. Prepared for Onondaga County Department of Public Works, Syracuse, New York.

Sze, P. (1975). Possible effect of lower phosphorus concentrations on the phytoplankton in Onondaga Lake, New York. *Phycologia* **14,** 197.

Sze, P., and Kingsbury, J. M. (1972). Distribution of phytoplankton in a polluted saline lake, Onondaga Lake, New York. *J. Phycol.* **8,** 25.

Thomann, R. V., DiToro, D. M., Winfield, R. P., and O'Connor, D. J. (1975). "Mathematical

Modeling of Phytoplankton in Lake Ontario," EPA-660/3-75-005. National Environmental Research Center, Environmental Protection Agency.

U.S. Bureau of the Census. (1962). 1970 Census of Population and Housing, New York.

U.S. Bureau of the Census. (1972). 1970 Census of Population and Housing, New York.

U.S. Environmental Protection Agency. (1974). "Environmental Impact Statement on Wastewater Treatment Facilities Construction Grants for the Onondaga Lake Drainage Basin," Region II, New York.

Vollenweider, R. A. (1974). "A Manual on Methods for Measuring Primary Production in Aquatic Environments," Int. Biol. Program., Blackwell, Oxford. p. 225.

Walker, W. (1976). "Exploring the Onondaga Lake Data Base," Tech. Pap. No. 760609. Environ. Syst. Programs, Harvard University, Cambridge, Massachusetts.

Weber, P. J. (1958). An analysis of the factors affecting the dissolved oxygen in Onondaga Lake, New York. Master's Thesis, Civil Engineering Department, Syracuse University, Syracuse, New York.

Wright, R. (1969). Director of Onondaga Historical Society (private correspondence).

Oneida Lake

E. L. Mills, J. L. Forney, M. D. Clady, and W. R. Schaffner

INTRODUCTION

Oneida Lake, the largest lake wholly within New York State, is located on the low Ontario Lake Plain of central New York (Fig. 1). Unusual limnological conditions result in high productivity and the lake provides a

ISBN 0-12-107302-5

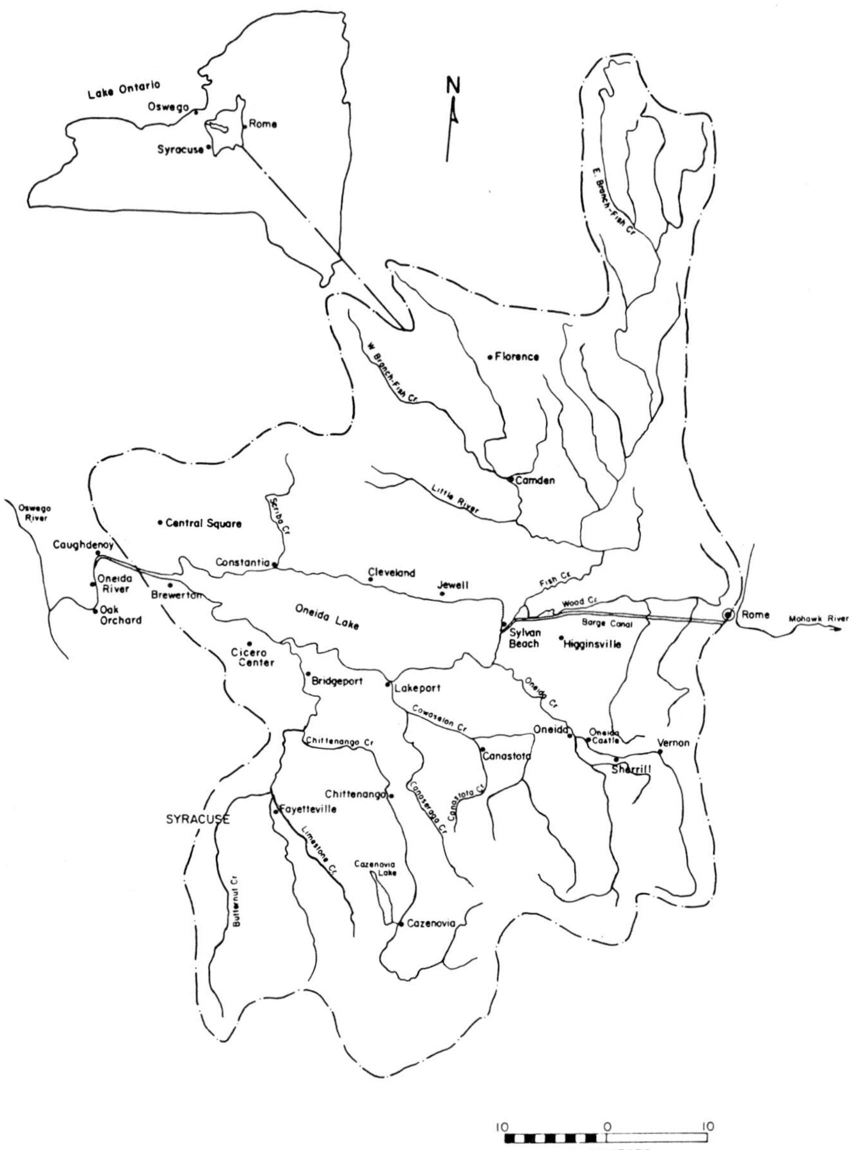

Fig. 1. The Oneida Lake drainage basin.

substantial recreational and economic resource for the region. A reservoir of scientific knowledge on both the lake and its drainage basin, accumulated since the early 1900's, has provided historical documentation of a number of environmental changes within this aquatic ecosystem. We hope that the following synthesis of important studies of Oneida Lake will provide a base for assessing future ecological change and stimulus for further research.

HISTORY OF THE ONEIDA LAKE BASIN

Oneida Lake is a remnant of Lake Iroquois, which was impounded by a glacier at the end of the pleistocene era approximately 12,500 years ago (Karrow *et al.,* 1961). While the ice was melting, the level of Lake Iroquois was maintained by an outlet east of present-day Rome, New York (Muller, 1965) and outflow to the north via the St. Lawrence River was blocked by the glacier. Lake Iroquois subsequently drained into the St. Lawrence Valley (approximately 10,000 years ago) as the glacier receded from the Adirondack Mountains (MacClintock and Terasmae, 1960) leaving behind numerous undrained depressions, one of which was Oneida Lake.

The wetlands at Cicero and Canastota on the southern shore of the lake were part of the original depression, but it has not been established when these areas became isolated from the lake. A col was formed at Lakeport, and another at Bridgeport. The regions south of the two cols, now drained by Chittenango and Canaseraga Creeks, consist of extensive muck lands and/or peat deposits which overlie well-preserved, charophytically deposited marl. At one time a bay probably extended into these areas, and upon being separated from the lake the bay gradually filled (Greeson, 1971).

The first inhabitants of the region were probably nomadic paleo-Indian hunters. These were followed by the Archaic Indians who existed from about 3500 to 1000 B.C. The final period of native culture reached its culmination with the Iroquois (Rayback, 1966). Two of the Five Nations of the Iroquois Confederacy, the Oneidas and Onondagas, lived in the Oneida Lake drainage basin; they were joined by a third, the Tuskaroras, in 1715 (Landgraff, 1926). The Oneidas, who called Oneida Lake Tsrioqui—"white water"—first settled on the north side of Oneida Creek near the lake, later moving to the southern shore. During the sixteenth century they gave up this village and built the fortified village of Oneida Castle. The Oneidas also had a fishing village at the mouth of Wood Creek at what is now Sylvan Beach. This latter site was the location of an annual salmon festival that initiated the fishing season for that species. Another important fishing village, belonging to the Onondagas, was at the outlet of the lake on the

Oneida River. In the eighteenth century the Tuskaroras had a village next to the lake at what is now Canaseraga Flats. These four locations were the only regular or permanent habitations in the area (Landgraff, 1926), although evidence for the existence of other sites has been found. The actual number of Iroquois that lived around Oneida Lake is unknown, but during the sixteenth century the population of the whole confederacy was estimated at 20,000, and except for small peripheral settlements, was congregated at about a dozen locations mostly outside the Oneida Lake basin (Rayback, 1966).

After the Revolutionary War, the area around Oneida Lake was purchased from the natives by the United States government and opened for settlement in 1789, although much of the land was subsequently sold without ever being settled (Landgraff, 1926). The first permanent settler arrived in 1789 and settled at Fort Brewerton on the north bank of the Oneida River. Shortly thereafter, settlements were established at Brewerton, Constantia, and Cleveland (Landgraff, 1926). In general, the land to the south and east of Oneida Lake was first colonized in 1790, immediately north of the lake from 1800 to 1809, and north of this after 1810. Tug Hill was never extensively colonized (Meining, 1966).

The Erie Canal was built in the early 1800's but did not pass through the lake. The construction of a side cut in the canal from Higginsville to Wood Creek in 1835 reopened the lake as a trade route (Landgraff, 1926). A lock and dam were built on the Oneida River at Oak Orchard in 1840, and the lock at Caughdenoy in 1841. Steamboats, which appeared on the lake in 1846, eventually replaced vessels propelled by sail and poles (Landgraff, 1926). Through dredging was completed in 1848. The first plank road was laid along the north shore in 1849, permitting stagecoach travel between Syracuse and the lakeside villages. A rail line was built along the north shore in 1868 (Greeson, 1971).

With the completion of the dam at Caughdenoy in 1910, and of the New York Barge Canal which passes through the lake in 1916, Oneida Lake became an important link in the route from the Great Lakes to the eastern seaboard. Over 13.5 million metric tons of freight were transported on the lake in 1962 (Barber, 1963), and although the canal is less extensively used today, traffic is still heavy between Oswego and Albany.

GEOGRAPHY OF THE DRAINAGE BASIN

Physiography

The Oneida Lake drainage basin is underlain by sedimentary rocks that range in age from Middle Ordovician to Upper Devonian (Broughton *et al.*,

1962; Fig. 2 and Table 1). The material is undeformed, and is made up of rock units that differ widely in their resistance to erosion (Greeson, 1971). The basin consists of three physiographic regions (Fig. 3): The Appalachian Upland, Erie–Ontario Lowland, and Tug Hill Upland (Muller, 1965).

The Appalachian Upland is delineated by a north-facing escarpment that drops sharply into the Erie–Ontario Lowland (Greeson, 1971), and is

TABLE 1

Description of Bedrock Geology (Key to Fig. 2)

Period	Symbol	Characteristics
Middle Ordovician	Ou	Utica shale
Upper Ordovician	Opw	Pulaski and Whetstone Gulf formations—siltstone, shale
	Oo	Oswego sandstone
Lower Silurian	SmOq	Undifferentiated Medina group and Queenstone formation—sandstone, shale, siltstone
	Scl	Herkimer sandstone; Kirkland hematite; Willowvale shale; Westmoreland hematite; Sauquoit formation—sandstone, shale; Oneida conglomerate
Upper Silurian	Sl	Oak orchard and Penfield dolostones, both replaced eastwardly by Sconondoa formation—limestone, dolostone
	Sv	Vernon formation—shale, dolostone
	Ssy	Syracuse formation—dolostone, shale, gypsum, salt
	Scc	Cobleskill limestone; Bertie and Camillus formations—dolostone, shale
Lower Devonian	Dhg	Coeymans and Manlius limestones; Rondout dolostone
	Don	Onondaga limestone—Seneca, Morehouse (cherty), and Nedrow limestone members, Edgecliff cherty limestone member, local bioherms
	Dhmr	Marcellus formation—In west: Oakta Creek shale member; In east: Cardiff and Chittenango shale members; Cherry Valley limestone and Union Springs shale members
Middle Devonian	Dhsk	Skaneateles formation—In west: Levanna shale and Stafford limestone members; In east: Butternut, Pompey, and Delphi Station shale members; Mottville sandstone member
	Dhld	Ludlowville formation—In west: Deep Run shale, Tichenor limestone, Wanakah and Ledyard shale members; Centerfield limestone member; In east: King Ferry shale and other members; Stone Mill sandstone member
	Dhmo	Moscow formation—In west: Windom and Kashong shales, Menteth limestone members; In east: Cooperstown shale member; Portland Point limestone member
	Dt	Tully limestone
Upper Devonian	Dg	West River shale; Genundewa limestone; Pen Yan and Geneseo shales; all except Geneseo replaced eastward by Ithaca formation—shale, siltstone, and Sherburne siltstone

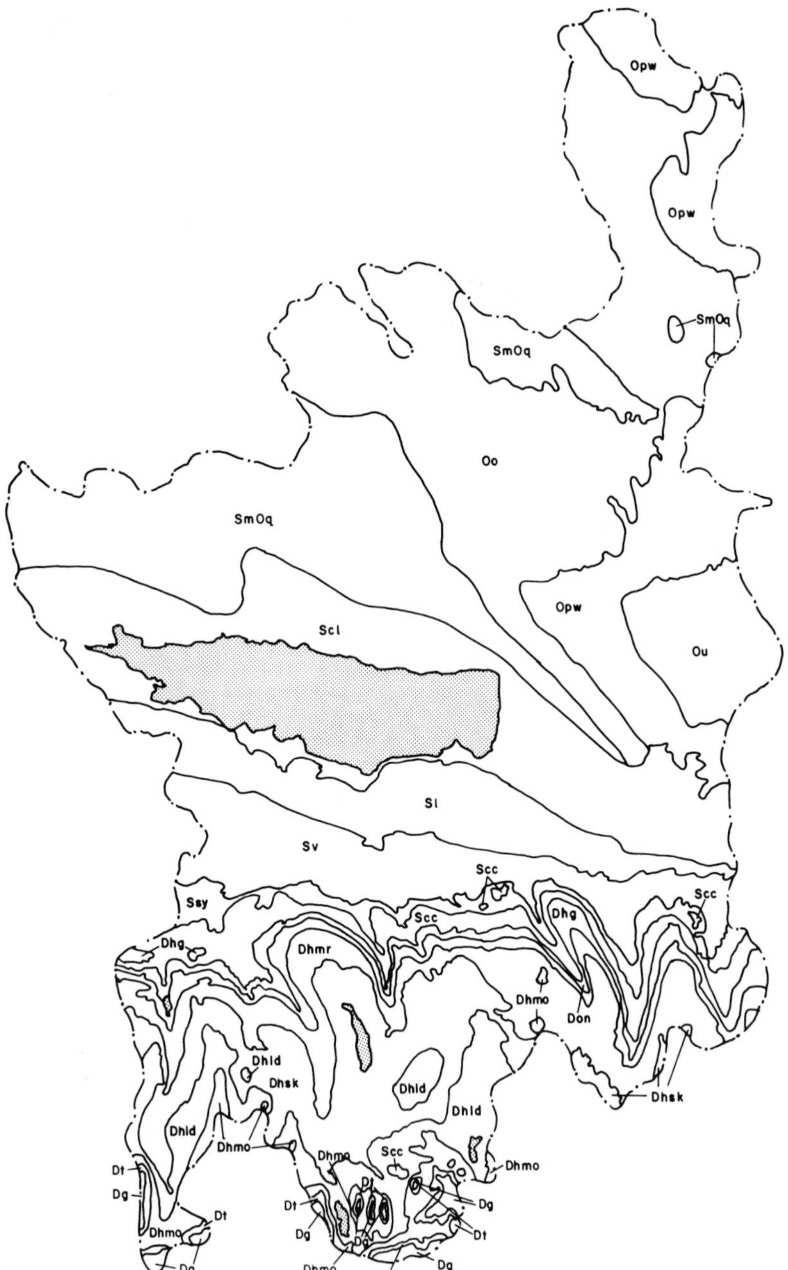

Fig. 2. Bedrock geology of the Oneida Lake drainage basin. Redrawn from Rickard and Fisher (1970a,b). (See Table 2.)

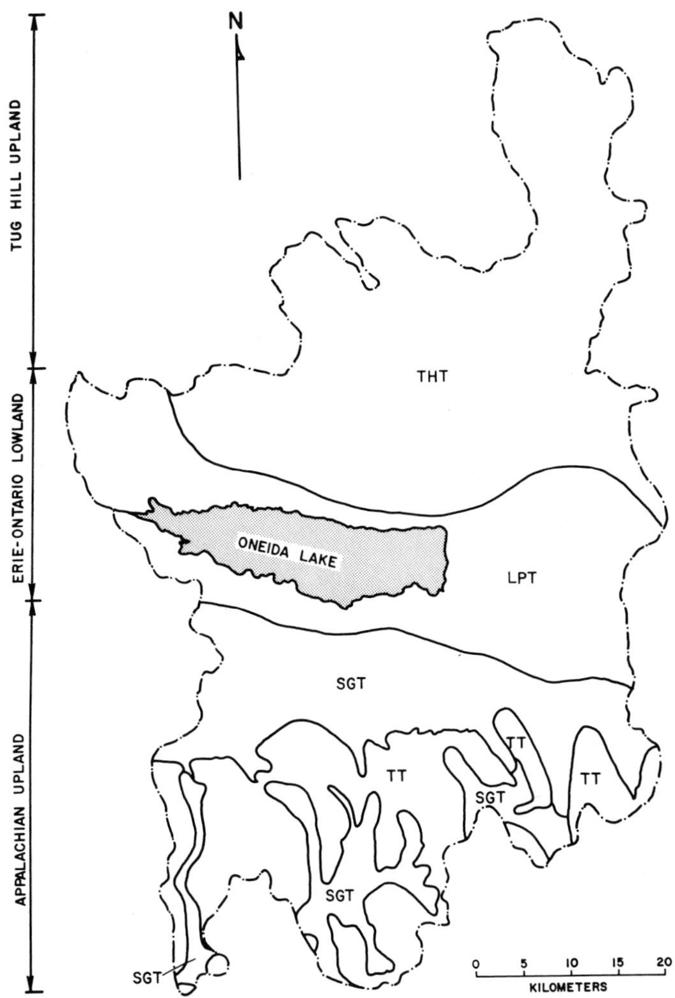

Fig. 3. Physiographic regions and terranes of the Oneida Lake drainage basin. (THT, Tug Hill terrane; LPT, Lake Plain terrane; TT, Till terrane; S and G, sand and gravel terrane; and SGT , Saline group terrane). After Greeson (1971).

characterized by a series of hills and valleys that trend in a north–south direction (Shampine, 1973). Many of the streams draining the southern portion of the basin arise in this region from small, steeply graded drainage systems (Greeson, 1971). The underlying bedrock is made up of upper shale and limestone units. The latter, cropping out at the crest of the escarpment, has an altitude of about 610 m (Shampine, 1973).

The Erie–Ontario Lowland is characterized by an expanse of low, flat land lying between the Appalachian Upland and the Tug Hill Upland. Its elevation is entirely below 250 m, with a mean altitude of 120 m (Greeson, 1971). The bedrock, which is susceptible to erosion (Muller, 1965), is made up of middle shale, dolomite, sandstone and shale, and sandstone units (Shampine, 1973). The limestone unit forms low ridges (cuestas) trending from east to west. The topography of the area is also strongly drumlinized through glacial action. In places the bedrock is overlain by unconsolidated glacial deposits up to 30 m in thickness. Much of the region is characterized by marsh, swampy areas, and sluggish streams (Greeson, 1971).

The Tug Hill Upland is a plateau remnant consisting of broad uplands and valleys, the structure of which has been preserved by the resistance of the Ordovician Oswego Sandstone to erosion (Muller, 1965). This westward-dipping sandstone unit rests on a series of sandy shales (Shampine, 1973). The altitude of the region approaches 600 m (Greeson, 1971).

Shampine (1973) points out that the chemistry of surface waters is related to the basin's geological framework, primarily in terms of its water-bearing bedrock (Kantrowitz, 1970; Fig. 4 and Table 2). Since Pleistocene and Holocene deposits that cover the basin are derived from material eroded from the bedrock, the water in these deposits is similar to that found in the rock strata. An exception occurs in the southern part of the basin. In this region the unconsolidated deposits are underlain by upper shale, but the deposits themselves contain limestone that was removed from outcroppings to the north by glacial action (Kantrowitz, 1970), and surface waters resemble groundwater in the limestone unit rather than those in the upper shale (Shampine, 1973).

Greeson (1971) has divided the drainage basin into five hydrochemically distinct and internally homogenous terranes (see Fig. 3). The subbasins of the Oneida Lake drainage basin (Fig. 5) and percentage of area contributing to each terrane is presented in Table 3.

(1) The Tug Hill Terrane includes most of the drainage basin north of the lake. The region is primarily underlain by chemically unreactive materials of which the major component is sandstone. The streams draining this region carry small chemical loads. Fish Creek contributes 44% of the water entering the lake, but only 10% of the total dissolved solids. Other creeks such as Scriba and Wood contain similar chemical loads (Greeson, 1971).

(2) The Lake Plain Terrane is that region surrounding the lake (the Erie–Ontario Lowland); 72% of which is the Wood Creek subbasin. Its water-bearing rock units are primarily dolomite, sandstone, and shale (see Table 2).

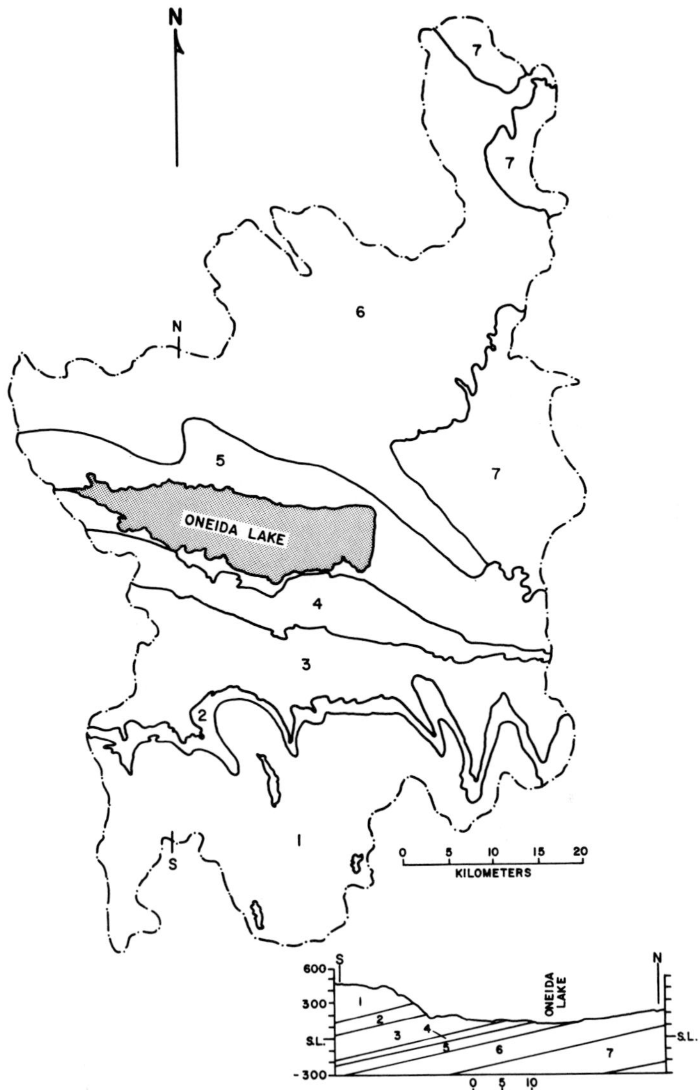

Fig. 4. Water-bearing bedrock units of the Oneida Lake drainage basin. 1, Upper shale; 2, limestone; 3, middle shale; 4, dolomite; 5, sandstone and shale; 6, sandstone; and 7, lower shale. After Shampine (1973) and Kantrowitz (1970). (See Table 2.)

TABLE 2

Description of Water-Bearing Bedrock Units (Key to Fig. 4)

Water-bearing unit	Thickness (m)	Geologic rock unit	Description
Upper shale (1)	585	Genesee formation	Black and gray shales; some thin sandstone layers
		Tully limestone	Black limestone, 7.5 m thick
		Hamilton group	Black shales; calcareous shales; thin limestone layers
Limestone (2)	105	Onondaga limestone	Blue-gray massive limestone
		Manlius limestone	Dark blue thin-bedded limestone
		Rondout limestone	Gray shaly dolomite
		Cobleskill limestone	Gray limestone and dolomite
		Bertie limestone	Gray dolomite; some thin shale partings; layers of gypsum
Middle shale (3)	260	Camilus shale	Gray thin-bedded shale; beds of gypsum, salt and dolomite
		Syracuse salt	Dolomite; gray shale; gypsum; salt
		Vernon shale	Red soft shale; beds of green shale; gypsum; dolomite
Dolomite (4)	45	Lockport dolomite	Dark-gray dolomite
Sandstone and shale (5)	75	Clinton group	Alternating layers of red and green shale and sandstone; some thin beds of limestone
Sandstone (6)	150	Albion group[a]	Red fine- to coarse-grained massive sandstone
		Oswego sandstone	Gray fine-grained sandstone
Lower shale (7)	245	Lorraine shale	Black and gray shale
		Utica shale	Black and gray shale

[a] Medina Group of Broughton *et al.* (1962).

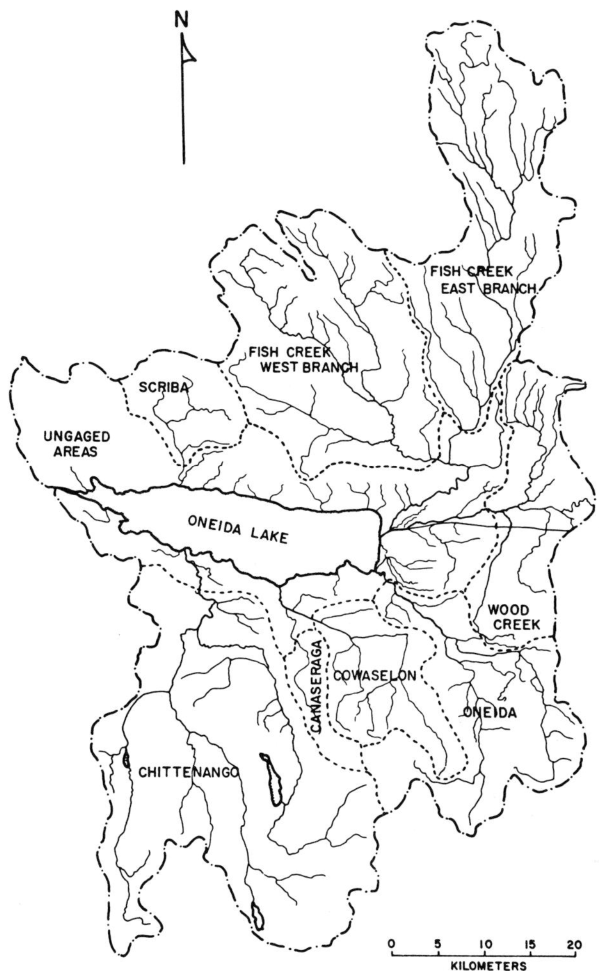

Fig. 5. Subbasins of the Oneida Lake drainage basin. After Greeson (1971).

(3) The Till Terrane is found in the Appalachian Upland, and consists of poorly sorted glacial deposits and small areas of sand and gravel. The deposits are essentially unreactive chemically. More than 25% of the Oneida, Canaseraga, Cowaselon, and Chittenango subbasins are in this terrane.

(4) The Sand and Gravel Terrane, also found in the Appalachian Upland, has areas of sand and gravel interspersed with till terrane. The material is well sorted and contains silt and clay. Relatively large amounts of groundwater are yielded to area streams, but the mineral content is low.

TABLE 3

Relative Sizes of the Subbasins in the Oneida Lake Drainage Basin and the Areas of Contributing Terranes[a]

			Terrane[b]									
	Drainage area		THT		LPT		TT		S and G		SGT	
Subbasin	km^2	Percent	km^2	Percent	km^2	Percent	km^2	Percent	km^2	Percent	km^2	Percent
Scriba Creek	99	3	96	97	3	2	—	—	—	—	—	—
West Branch Fish Creek	528	16	528	100	—	—	—	—	—	—	—	—
East Branch Fish Creek	487	14	487	100	—	—	—	—	—	—	—	—
Wood Creek	236	7	65	28	171	72	—	—	—	—	—	—
Oneida Creek	344	10	—	—	67	20	114	33	41	12	122	35
Canaseraga Creek	52	2	—	—	—	—	21	42	—	—	31	60
Cowaselon Creek	179	5	—	—	16	9	47	26	8	4	109	61
Chittenango Creek	813	24	—	—	10	1	430	53	163	20	210	26
Ungaged areas	635	19	142	22	466	73	—	—	—	—	26	4
Total drainage basin	3373	100	1318	39	733	22	612	18	212	6	498	15

[a] Source: Greeson (1971).

[b] Values rounded to nearest square kilometer. THT, Tug Hill terrane; LPT, Lake Plain terrane; TT, Till terrane; S and G, sand and gravel terrane; and SGT, Salina group terrane.

(5) The Salina Group is the only one of the five terranes that has any great effect on stream water chemistry. It occurs in that part of the Appalachian Upland underlain by Vernon Shale, Syracuse Salt, Camillus Shale, and Bertie Limestone. The first three units make up the Middle Shale unit while Bertie Limestone is part of the Limestone unit (see Table 2). This terrane ranges from 213 to 274 m in thickness, is exposed along the lower edge of the escarpment that borders the Appalachian Upland, and comprises 60% of the Canaseraga and Cowaselon Creek drainage systems. It also crops out along the sides and bottoms of the deeper valleys in the upland, and extends under the unconsolidated material of the Erie–Ontario Lowland. The units are easily erodable, and contribute considerable amounts of dissolved minerals to streams draining the region, primarily as calcium bicarbonate and sulfate. The Syracuse Salt unit contains sodium chloride beds that are in contact with the groundwater.

The Oneida Lake drainage basin has been divided into three distinct base-flow, water quality regions (Fig. 6) based on geological information (Shampine, 1973). Region I, located to the north of the lake, is made up of the Sandstone and Shale, Sandstone, and Lower Shale Units described in Table 2. Surface water leaving this region is of the calcium bicarbonate type, but is low in total dissolved solids (mean 80 mg/liter). Minor amounts of calcareous material supply most of the dissolved solids, the content of which is higher in streams in the western part of the region due mainly to greater amounts of debris in the lowland glacial deposits (Kantrowitz, 1970). Chemically, the water is acceptable for public supply, but since it probably contains some domestic waste, treatment would be required.

Region IIA contains most of the Appalachian Upland and terminates with the Bertie Limestone. Water is of the calcium bicarbonate type, but dissolved solids are higher than in Region I (mean 200 mg/liter). Streams draining the Upper Shale unit of this region are low in dissolved solids whereas those from the Limestone unit are moderately mineralized. The water has potential for public supply, but would require treatment.

Region III is made up of the low, flat land lying to the south of the lake, the southern limit of which is the outcrop of Vernon Shale. This region contains the most highly mineralized base flow in the entire basin. The Camillus and Vernon Shales (Middle Shale unit, see Table 2) contain calcium sulfate and sodium chloride. Concentrations of sodium chloride are localized, but high levels of calcium sulfate is widespread. Other water-bearing units are present, i.e., dolomite, sandstone, and shale, but water quality is primarily influenced by the Middle Shales. Thus, surface water is of the calcium sulfate type, high in total dissolved solids (mean 700 mg/liter). The water of this region is not suitable for public use because of

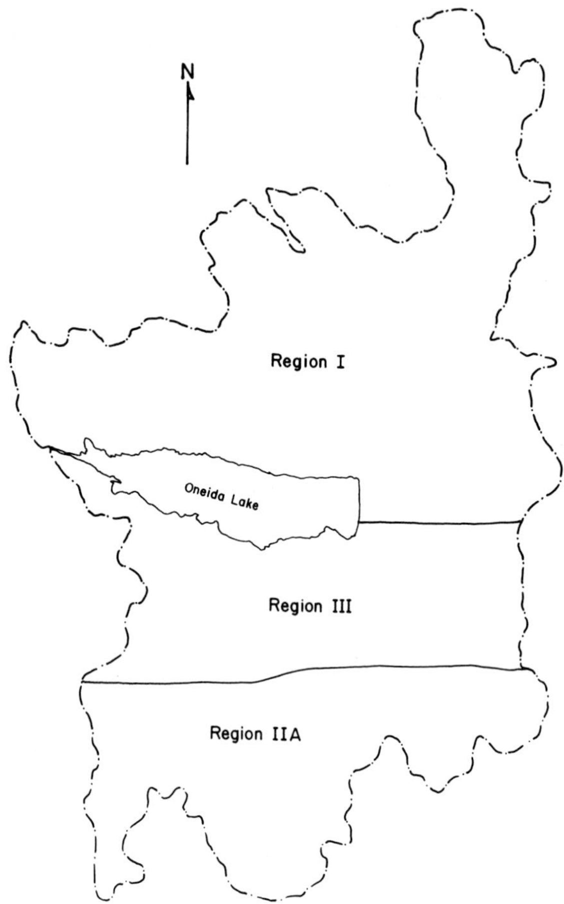

Fig. 6. Water quality regions of the Oneida Lake drainage basin. After Shampine (1973).

the high concentration of dissolved solids and extensive domestic waste pollution.

Seasonal variations in stream water are found in all regions, and reflect the influence of climatic patterns on the ratio of overland to base flow. Maximum concentrations of dissolved solids are usually observed during the summer months when base flow is primarily made up of groundwater. Minimum concentrations occur during winter and spring when snow melt is the largest contributor to stream flow.

The soils of the Oneida Lake basin are young since the region was glaciated only a few thousand years ago. The parent material lies relatively close to the surface, and is usually not seriously depleted of soluble nutrients. The better soils in the area are likely to have developed on well-drained, relatively level sites where the parent material is derived from limestone and glacial drift. Such conditions are often found on the alluvial valley bottoms, the broader, glaciated plains, or the more level uplands (DeLaubenfels, 1966b). These soils are likely to be less acid than most humid climate soils, and may even be neutral in character. The subsoil is usually nonacid.

The soils in the Oneida Lake drainage basin can be broken down into six general categories or regions (Fig. 7) and have been summarized by DeLaubenfels (1966b). A more detailed description is found in Fig. 8 and Table 4 (Cline *et al.*, 1975). Cline and Marshall (1977) have recently described the soil resources of New York State, including the Oneida Lake basin, and appraised their usefulness for agriculture.

First, there are limey soils on glacial till over undulating or rolling terrain. Limestone outcrops provided material that was ground up as the glaciers moved southward. Such outcroppings are found along the northern and eastern escarpment of the Appalachian Upland and westward along the Erie–Ontario Lowlands. The glaciers spread the broken and pulverized limestone mixed with till in significant amounts for 20 miles or so beyond the outcrops. This has resulted in the deposition of a broad belt of parent material rich in lime that extends from Lake Erie to the Hudson Valley. New York State's most productive soils have developed in this area.

Second, there are limey soils on glacial lake sediments over level or undulating terrain. These soils are similar to those in the first region, but the parent material is lake sediment and has a finer texture, fewer rocks, and a lower lime content. Such soils are found in the Erie–Ontario Lowlands, and are in general of good texture, with ample supplies of lime in the subsoil. These soils vary greatly from place to place in their drainage characteristics, depending upon the slope of the surrounding landscape. The generally level terrain and the near proximity of these low-lying sites to lakes make these soils highly suitable for fruit and vegetable crops.

The third category consists of alluvial soils in valley bottoms. These soils were laid down initially by the melt waters of the retreating glaciers. During the postglacial period streams have added alluvium during flood stage. Thus, a substantial portion of the principal valleys have developed narrow, relatively level, alluvial bottoms, examples of which are to be found in the Appalachian Uplands. The soils that have developed in this fashion are immature and lack well-developed profiles. They are free of stones and have a high production potential, although many areas suffer from either poor or excessive drainage.

The fourth category consists of deep, acid soils on glacial till over hilly

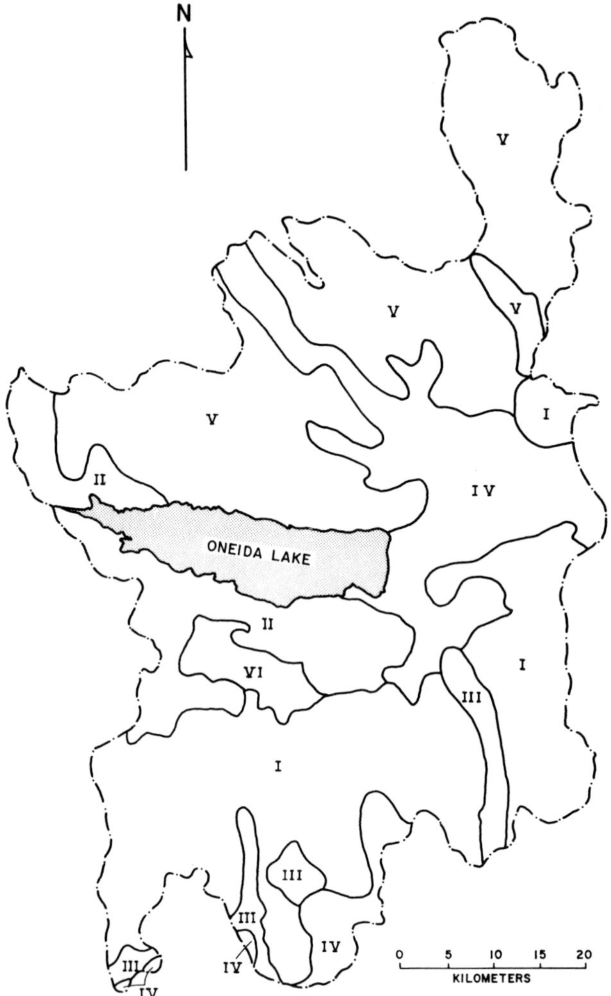

Fig. 7. Soil regions in the Oneida Lake drainage basin. After DeLaubenfels (1966b).

terrain. The glacial till in this region is derived less from limestone, and more from the shale, sandstone, and slate found in the Appalachian Uplands. There are some areas that are steep with thin soils, and poorly drained pockets can be found, but characteristically the soils are deep and moderately well drained. Fragipan development is common, creating drainage problems where the terrain suggests they should not exist. (This condition exists in acid soils derived from glacial till, and is a portion of the B horizon that is both tightly packed and impermeable.) The acidity of the soil necessitates heavy fertilization and, although all of the region has been cultivated at some time, a great deal of the land has been abandoned. As might be expected, the region also suffers from erosion.

The fifth category is made up of shallow, acid soils on glacial till over steep terrain. Such soils are found in the Tug Hill area where thin, glacial till overlies resistance sandstones. Interruption of the preglacial drainage pattern has resulted in poor drainage. Glaciation has produced a stony condition similar to that found in New England. The cooler climate, heavy

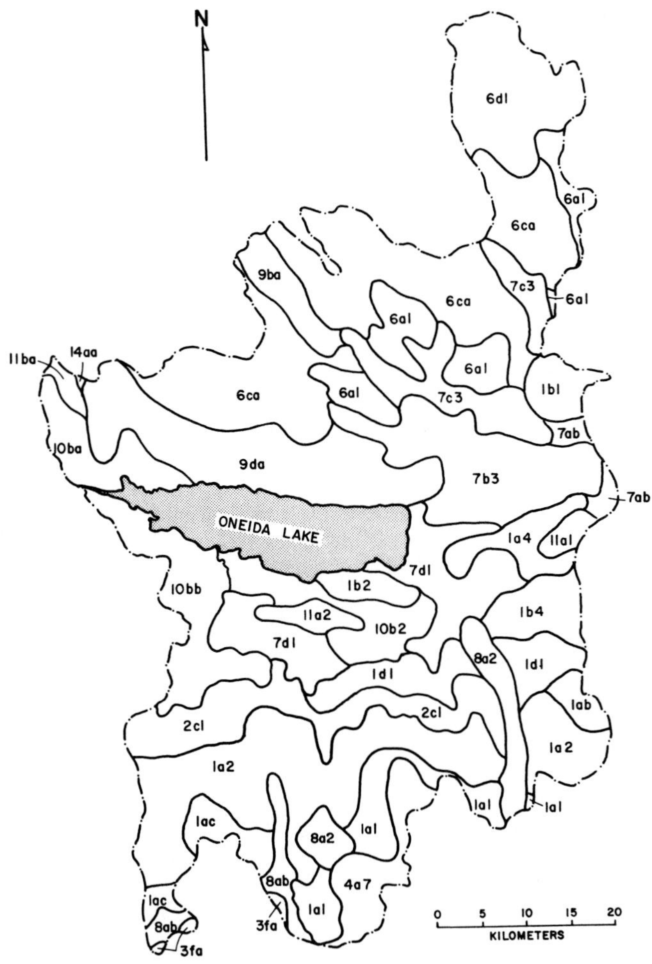

Fig. 8. Soils of the Oneida Lake drainage basin. After Cline *et al.* (1975). (See Table 4.)

TABLE 4

General Soil Areas of the Oneida Lake Basin (Key to Fig. 8)

Symbol	Most extensive series	Percent of area that is			
		>15% slope	Somewhat poor or wetter	Very stoney	>0.5 m to bedrock
Glacial till—calcareous substrata; clay accumulation in subsoil					
			Driest		
1ab	Honeoye-Lima	2-20	8-30	0	0-10
1a2	Honeoye-Lima	2-20	8-30	0	0-10
1a1	Lansing-Conesus	4-12	2-30	0	0-5
1ac	Conesus-Lansing	4-12	2-30	0	0-5
1a4	Hilton-Appleton	1-10	20-40	0	1-10
			Wet		
1b1	Burdett-Ilion	0-8	55-85	0-1	0-10
1b2	Appleton-Lansing	0-15	48-50	0	0-2
1b4	Ovid-Cazenovia	0	40-50	0	0-2
1d1	Lairdsville-Cazenovia	2	15-35	0	0
Shallow to limestone					
2c1	Benson-Wassaic	0-30	2-5	20-70	3
Till; acid soils without fragipan—hilly areas—dominantly 0.5-1.0 m to rock					
3fa	Lordstown-Volusia	30-45	15-35	0	5-20
Acid soils with fragipan in glacial till					
4a7	Marden-Volusia-Lordstown	5-30	15-55	0-8	1-5
Very acid soils with spodic horizons (podzol) in glacial till					
6a1	Worth-Empyville	7-20	5-20	5-30	0
6ca	Worth-Empyville, very stoney	3-25	1-25	50-90	0-6
6b1	Camroden-Penchney, wet	5	65-70	0	0
Acid soils in glacial outwash gravel and sand					
7ab	Alton-Altmar (gravel)	5	2	0	0
7b3	Windsor-Deerfield (sand)	0-15	3-20	0	0
7c3	Colonie (sand, hilly)	90	3	0	0
7d1	Minoa-Lawson (sand, wet)	0	50-80	0	0
Soils in glacial outwash; calcareous substrata					
8ab	Palmyra-Teel	2-21	3-18	0	0
8a2	Palmyra-Phelps	0-20	5-30	0	0
Very acid soils with spodic horizons (podsols) in outwash sand and gravels					
9ba	Colton (gravel)	5-10	10-25	0-12	0-25
9da	Adams-Colton (wet sand)	0-10	35-55	5-25	0
Soils in lake silts and clays—mainly wet					
10ba	Rhinebeck-Madelin (clays)	0-5	65-75	0-5	0
10bb	Niagara-Collamer (silts)	0	50	0	0
10b2	Niagara-Canandaigua (silts)	0	50-90	0	0
Organic soils					
11a1	Carlisle (deep, decomposed)	0	80-95	0	0
11a2	Edwards (shallow over marl)	0	100	0	0
11ba	Carlisle-Alton-Canandaigua (intermingled with mineral sod)	2	60	5	0
Intermingled soil with fragipan in silt of lakes and soil with fragipan in gravelly till					
14aa	Williamson-Ira	2-8	15-45	0-10	0

snow falls, and coniferous vegetation have all adversely affected soil development, resulting in extremely acid soils of shallow profile. The region is very poor for agriculture, and is almost exclusively forest. As was pointed out earlier, the Tug Hill area was avoided by both the Iroquois and settlers.

Finally, there are coarse textured soils on sand and gravel. The sands and gravels of this region were originally beaches, deltas, and marginal lake-bottom deposits. Soils derived from such parent material are very poor. They lack organic material and are shallow, coarse, acid, and excessively drained. They are poor for agriculture, and clearing of the natural vegetation may result in the complete destruction of the existing soil through the action of wind and water erosion.

Demography

The population in the Oneida Lake watershed has increased rapidly in recent years, particularly through urbanization of the Syracuse metropolitan area. The total population of the basin is about 91,200 (Greeson, 1971). Approximately 85,000 live south of the lake, with the remainder in the sparsely populated northern part of the basin. The shoreline population of Oneida Lake increased from 1600 in 1900 to 14,100 in 1960.

Land Use

Generally, the basin north of the lake is undeveloped, with 76% of the land being forest and wetland (Greeson, 1971; Table 5). Agriculture makes up 15% of the land use, and residential and commercial less than 1%. The southern portion of the basin is more extensively developed, with 56% of the land in agriculture. Dairy farms are numerous in the Appalachian Uplands, as are truck farms on the lake plain. Forest and wetland occupy 33% of the southern basin, and residential and commercial areas 5%. The total drainage basin is primarily rural in that 1933 km^2 (59% of the total) are in forest and wetland, and 1,253 km^2 (38% of the total) are in agriculture (Oglesby and Schaffner, 1977).

Greeson (1971) estimated that the number of shoreline dwellings within 320 m of the lake increased ninefold over a 60-year period. In 1900 there were 476 dwellings, and by 1960 the number had increased to 4298. About 17% of the lake was bordered by undeveloped forest and wetland in 1900. This has decreased to less than 50%, with the greatest reduction along the southern shore (Table 6).

The forests in the basin have developed in response to the following three variables: climate, soil, and degree and kind of past disturbance. Climatic differences are responsible for the broader vegetative patterns, of which

TABLE 5

Percentage Land Uses of the Oneida Lake Drainage Basin[a]

Drainage basin	Residential and commercial	Farmland	Forestland	Wetland	Other uses[b]
Scriba Creek	<1	15	53	31	<1
West Branch Fish Creek	<1	16	68	13	2
East Branch Fish Creek	<1	7	78	13	1
Wood Creek	2	34	46	13	5
Oneida Creek	3	62	30	2	3
Canaseraga and Cowaselon Creeks	3	61	28	6	2
Chittenango Creek	6	50	28	7	9
Ungauged areas	3	38	40	16	3
Lake surface	0	0	0	0	100
Total basin above Caughdenoy Dam	3	35	43	11	8

[a] Source: Greeson (1971).
[b] Includes water surfaces, transportation surfaces, industrial and extractive areas, and natural sand and rock surfaces.

there are five in New York (DeLaubenfels, 1966a). Four of these major forest zones occur in the Oneida Lake drainage basin (Fig. 9). In general, the northern hardwoods occupy most of the state except for the extreme southeast and high Adirondacks, and the other zones are best described by their relation to this group.

Northern Hardwood Zone

This assemblage is made up of large groups of trees dominated by the beech (*Fagus grandifolia* Ehrh.) and sugar maple (*Acer saccharum* Marsh.). Basswood (*Tilia americana* L.), white ash (*Fraxinus americana* L.), and black cherry (*Prunus serotina* Ehrh.) are found in the warmer areas. Yellow birch (*Betula lutea* Michx. f.) becomes a dominant in cooler areas. Hemlock (*Tsuga canadensis* (L.) Carr.), white pine (*Pinus strobus* L.), and white cedar (*Thuja occidentalis* L.) are abundant but not evenly distributed. Hemlock is found on moist, shady slopes and in ravines and may form nearly pure stands. This species was once more abundant, but was severely cut over in the early days for its wood and bark (a source of tannin). There are occasional oaks, but these appear to be energy limited during the summer. The climate is too warm for spruce and fir. In this zone, there are four frost-free months with the growing season seldom longer than 150 days.

Oak–Northern Hardwood Zone

This is a transition zone where oak (*Quercus* spp.) intermingles or alternates with the northern hardwoods. Oak requires more heat than the northern hardwoods, and flourishes in the moister, deeper soils at low and intermediate levels in the Appalachian Uplands, and at lower elevations to the north. Slope direction can produce local variations. South- and southwest-facing slopes at times may support pure stands of oak or oak and hickory (*Carya* spp.). White pine can be found on dry and sandy soils and in abandoned fields. The zone is well suited for American elm if the vegetative cover is open. Red cedar (*Juniperus virginiana* L.), white ash, hawthorn (*Crataegus* spp.), and locust (*Robinia pseudoacacia* L.) tend to invade the abandoned pastures in this zone.

Elm–Red Maple–Northern Hardwood Zone

This group is widespread on the poorly drained areas of the Ontario plain where the natural forest has been removed. It is distinguished by the fre-

TABLE 6

Development of Shoreline of Oneida Lake between 1900 and 1960[a]

	About 1900			About 1960		
	Area (km^2)	Percent	Number	Area (km^2)	Percent	Number
Land use						
North shore						
Residential or municipal	4.38	33	—	5.15	39	—
Farmland	—	—	—	0.34	2	—
Undeveloped forestland	6.11	46	—	5.67	43	—
Wetland (swamp, marsh)	2.80	21	—	2.12	16	—
South shore						
Residential or municipal	3.08	20	—	5.95	39	—
Farmland	0.78	5	—	3.05	20	—
Undeveloped forestland	8.13	53	—	3.91	25	—
Wetland (swamp, marsh)	3.39	22	—	2.46	16	—
Total shore						
Residential or municipal	7.46	26	—	11.11	39	—
Farmland	0.78	3	—	3.39	12	—
Undeveloped forestland	14.24	50	—	9.58	33	—
Wetland (swamp, marsh)	6.19	21	—	4.58	16	—
Residential housing						
North shore	—	—	308	—	—	1716
South shore	—	—	168	—	—	2582
Total shore	—	—	476	—	—	4298

[a] Source: Greeson (1971).

Fig. 9. Forest zones in the Oneida Lake drainage basin. (N Hw, Northern hardwood; O–N Hw, oak–northern hardwood; E–RM–N Hw, elm–red maple–northern hardwood; and S–F, spruce–fir.) After DeLaubenfels, (1966a) and Stout (1958).

quency of American elm (*Ulmus americana* L.), where it has not been killed by disease, and red maple (*Acer rubrum* L.). The presence of these trees is due to disturbed edaphic conditions. Oak is not abundant because the well-drained soils in the region are still used for agriculture.

Spruce–Fir Zone

The coolest part of the drainage basin, Tug Hill, contains this group. Northern hardwoods are present, but the growing season is usually less than

125 days, and the mean July temperature is less than 18.9°C. Spruce, fir [*Abies balsamea* (L.) Mill.], and larch [*Larix laricina* (Du Roi) K. Koch.] are found in this poorly drained area. Red spruce (*Picea rubens* Sarg., synonym *P. rubra* Link) dominates the assemblage. These softwoods have been greatly reduced in number by lumbering operations.

Today the heaviest forests in the basin are found in the Tug Hill region. Areas where forests and woodlots occupy less than one-third of the land surface are usually low, relatively level, with good, well-drained soils such as the agriculturalized Erie–Ontario Lowlands.

Point-Source Pollutants

Changes in the Oneida Lake watershed have been complex and modifications have occurred in domestic sewage treatment, demographic patterns, and agricultural activities. For many years, the city of Rome discharged its raw sewage into Wood Creek and Wagner (1928) described this stream as a "veritible open sewer." In 1932, Rome's effluent was diverted out of the Oneida Lake basin into the Mohawk River (Frank Clark, City of Rome Engineer, personal communication). In addition, wastes from milk and cheese plants, canneries, and paper mills were discharged into Oneida, Chittenango, and Fish Creeks in the early 1900's. By 1965, acreage in agriculture declined and direct contribution of wastes from this activity and related industries was greatly reduced.

The Great Lakes Basin Framework Study sponsored by the Environmental Protection Agency (Anonymous, 1975) has summarized present point-source pollutants entering Oneida Lake. Chittenango Creek receives municipal wastes and residual wastes from Limestone and Butternut creeks totaling more than 3.8 mld (million liters per day), of which only one-half receives minimal primary treatment. Butternut Creek is polluted by grease and oil wastes from a railroad diesel repair facility. Municipal wastes are also a problem in Oneida Creek and the Canaseraga–Cowaselon–Canastota Creek network. Oneida Creek receives primary effluents totaling more than 7.6 mld from the cities of Oneida and Sherrill. In addition, the Kenwood plant of the Oneida Limited silver plating plant discharges industrial wastes into Oneida Creek after treating the effluent for removal of toxic metals, oil, grease, and other materials. For many years, the Canaseraga–Cowaselon–Canastota Creek system was degraded by untreated sewage from the village of Canastota; however, Canastota recently introduced a secondary treatment facility to abate this problem. The Fish Creek basin contains only a few municipal and industrial waste sources, the most significant of which is the primarily treated effluent from Camden. Plans are now underway to construct a secondary treatment plant there.

The number of shoreline cottages has increased in recent years but the increase in domestic sewage has been partly countered by more rigorous regulation and inspection of cottage disposal systems. The Oneida Lake Shore Sewer District, which is completed, will serve the area from Chittenango Creek on the south shore west to Brewerton. Additional sewage treatment for Oneida's eastern and southeastern shores are presently under consideration.

The estimated sewered population discharging into stream tributaries of Oneida Lake in 1956 was 37,000 (New York State Department of Health, 1957). J. L. Forney (written communication) estimated that the contribution of phosphorus from the sewered population at this time was 53,182 kg. The other 50,000 or more people in the Oneida Lake basin generally used septic tanks. Between 1967–1969, Greeson (1971) estimated that individuals living in 17 major communities located in the Chittenango, Canaseraga, and Oneida watersheds and seven major communities in the northern portion of the Oneida Lake basin would contribute a maximum of 53,290 kg total nitrogen/year and 34,675 kg total phosphorus/year to the lake. Oglesby and Schaffner (1977) estimate the total phosphorus contribution from this component of the population to be 22,350 kg/year. Input of nitrogen was not estimated.

MORPHOMETRY AND HYDROLOGY OF ONEIDA LAKE

Hydrography and Morphology

Oneida Lake, the largest water body within New York, has a surface area of 206.7 km² and is located approximately 18 km northeast of Syracuse. The lake, 33.6 km long, averages 8.8 km wide. Its long axis is oriented west–northwest to east–southeast and is fully exposed to the prevailing winds. The lake is a shallow (maximum depth 16.8 m, mean depth 6.8 m), spoon-shaped depression which deepens toward the eastern end (Fig. 10). Numerous shoal areas exist with about 26% of the lake bottom consisting of areas shallower than 4.3 m. Other hydrographic information taken from Greeson (1971) is presented in Table 7.

The water level was maintained by a natural sill at the outlet until 1909 when a concrete dam was built in connection with the New York Barge Canal. Currently, the water level is regulated by a Taintor-gate dam constructed in 1952 on the Oneida River at Caughdenoy, New York, and operated by the New York State Department of Transportation. The lake is drawn down in late fall to minimize ice damage to shoreline structures and

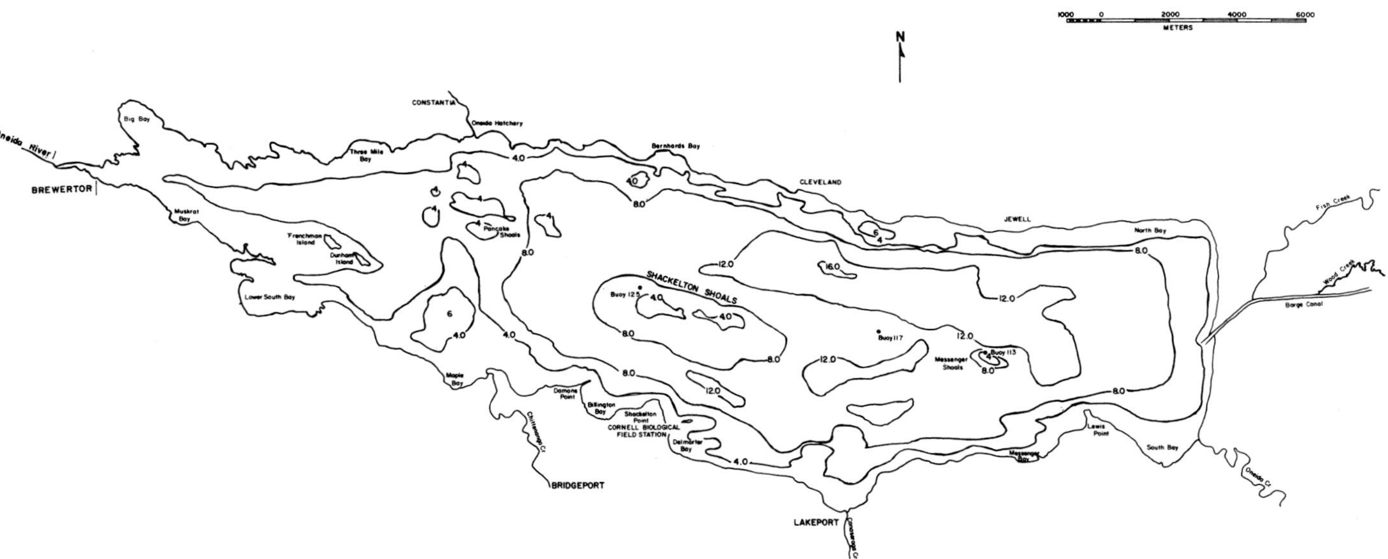

Fig. 10. Bathymetric map of Oneida Lake.

TABLE 7

Hydrographic and Geographic Description of Oneida Lake[a]

Latitude	43° 12.5′ N
Longitude	75° 55′ W
Elevation above sea level	112 m
Drainage area	3579 km^2
Surface area	206.7 km^2
Shoal area	53.1 km^2
Length	33.6 km
Width	
Maximum	8.8 km
Mean	6.1 km
Depth	
Maximum	16.8 m
Mean	6.8 m
Shoal	4.3 m
Volume	140×10^7 m^3
Development of volume	1.2
Length of shoreline	88.0 km
Development of shoreline	1.7
Hydraulic retention time	235 days

[a] Source: Greeson (1971).

to maximize storage capacity during heavy spring runoff. Normal elevation above sea level is 111 m during the winter and 112 m during the summer. The lake volume at an elevation of 112 m is 140×10^7 m^3. An average hydraulic retention is 235 days.

Hydrology and Stream Flow

The total annual discharge of water from the Oneida Lake basin via the Oneida River is approximately 213×10^7 m^3/year, an estimate based on discharge for the standard period 1931–1960 (Greeson, 1971). Greeson estimated precipitation to be 379×10^7 m^3/year in the basin. If evapotranspiration is assumed to be the difference between precipitation and outflow from the basin, it would then be equal to 166×10^7 m^3/year or 48.8 cm/year. Thus 56% of the precipitation that falls in the basin eventually reaches the lake. Input of groundwater is not significant, and probably does not exceed 1% of the total hydrologic flow into the lake (Kantrowtiz, 1970).

Fish, Oneida, and Chittenango Creeks contribute most of the inflow to the lake (Table 8). The east and west branches of Fish Creek, which drain the northern part of the basin, have the highest discharge rates (15.7 and

14.1 m^3/sec), and make up 23% and 21% of the total discharge to the lake, respectively, while Chittenango and Oneida Creeks, which drain the southern portions of the basin, contribute 25% of the total inflow to the lake (Greeson, 1971).

During Greeson's (1971) 3-year study discharge from the lake varied slightly from the long-term average of 67.7 m^3/sec (Table 8). In 1967 the outflow was 23.9% lower (51.5 m^3/sec). Higher rates of 75.0 and 82.0 m^3/sec were measured in 1968 and 1969, respectively (U.S. Department of the Interior, 1968, 1969, 1970).

Climatology

Three of New York's six climatic regions (Fig. 11) are found in the Oneida Lake drainage basin (Carter, 1966):

Region I—Extremely cold, snowy winters with very cool, wet summers. This is a region of high elevation and latitude. Energy supplies are low and the growing season short. This is the wettest part of the basin, with snowfall being very high. The Tug Hill Upland, which lies wholly within the region, receives from 3.3 to 5.7 m of snow each winter. The cold, damp summers do not lend themselves to either agriculture or tourism.

Region III—Cold, snowy winters with cool, wet summers. The Appalachian Upland is found in this region. Low energy supplies and deep snows in

TABLE 8

Discharge for the Subbasins in the Oneida Lake Drainage Basin

Drainage basin	Mean annual discharge[a] (m^3/sec)	Percentage of total discharge
Scriba Creek	2.3	4
West Branch Fish Creek	14.1	21
East Branch Fish Creek	15.7	23
Wood Creek	4.0	6
Oneida Creek	4.9	7
Canaseraga Creek	0.8	1
Cowaselon Creek	2.7	4
Chittenango Creek	12.0	18
Ungauged areas	11.2	16
Total basin (at Caughdenoy Dam)	67.7	—

[a] Values represent discharge for standard period, 1931–1960.

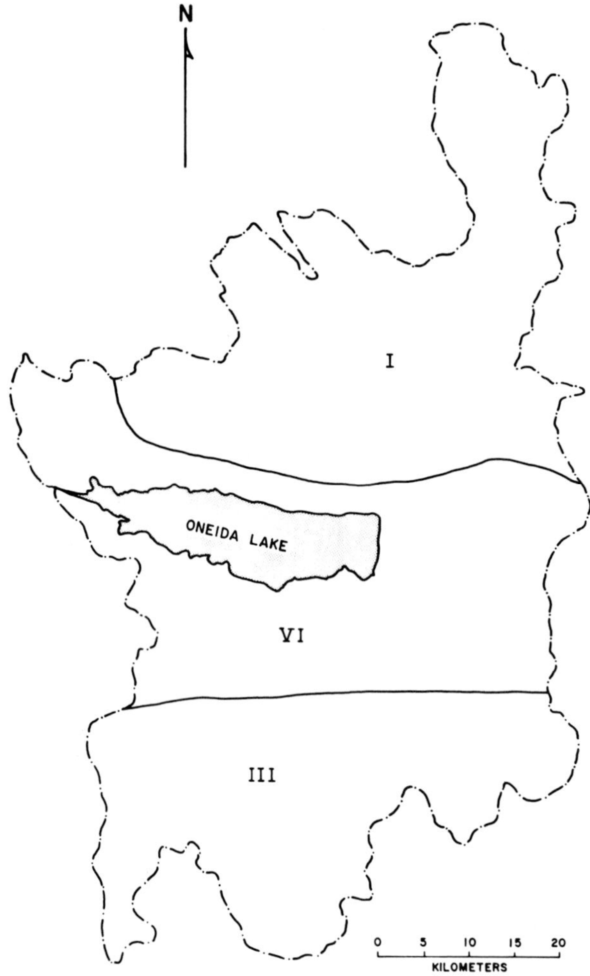

Fig. 11. Climatic regions of the Oneida Lake drainage basin. After Carter (1966).

conjunction with the poor soil conditions mentioned earlier have created an area with a high rate of agricultural abandonment.

Region IV—Cold, snowy winters with warm, dry summers. This region is warmer in both summer and winter than the previous two. Climatic conditions are well suited for crop production, even though this is one of the driest areas in the state. Moisture supply is usually at a maximum during the summer, but crop irrigation is still practiced in many instances. The Erie–Ontario Plain, which is at the eastern extent of the region, receives about 2.2 m of snowfall a year.

Precipitation

Precipitation based on long-term averages (Dethier, 1966) is presented in Fig. 12. Maximum precipitation occurs in the Tug Hill Upland, and results from the cooling of moisture-laden air coming off Lake Ontario as it rises into the higher altitudes of the region. Annual average precipitation decreases to the north and south, with the least amount falling in the southwestern corner of the basin. These precipitation patterns are in general agreement with those presented by Carter (1966), but not with those shown in Greeson (1971), who indicated that precipitation steadily increases north of the lake to the northern limits of the basin, rather than having a maximum over Tug Hill. The long-term station averages (Fig. 12) indicate that such a pattern does not occur.

Temperature

Mean monthly temperatures expressed as deviations from long-term averages (Carter, 1966) are in agreement with the regional descriptions in this section. The highest average temperatures are at Oswego (Region IV), and the lowest at Lowville (Region I) (Fig. 13).

Groundwater

Oneida Lake receives negligible inputs of groundwater because groundwater tends to flow into the valley containing the tributaries to the lake. Here, it enters the streams via springs or seepage. During extended dry periods stream flow is made up almost entirely of water entering in this manner (Greeson, 1971).

LIMNOLOGY

Physical Properties

Temperature, Waves, Seiches, and Currents

Unlike most north-temperate lakes, Oneida Lake is usually homothermal, but thermal stratification may develop during prolonged calm periods. Early investigators made temperature measurements, but most were of insufficient number or detail to provide a general basis for historical comparison. As part of their comprehensive survey of Oneida Lake, the United States Geological Survey made temperature measurements from 1967–1969. They found that during the growing season there is a gradual decline in water temperature from the surface to deeper waters; homothermal conditions prevail during the fall and spring; and slight

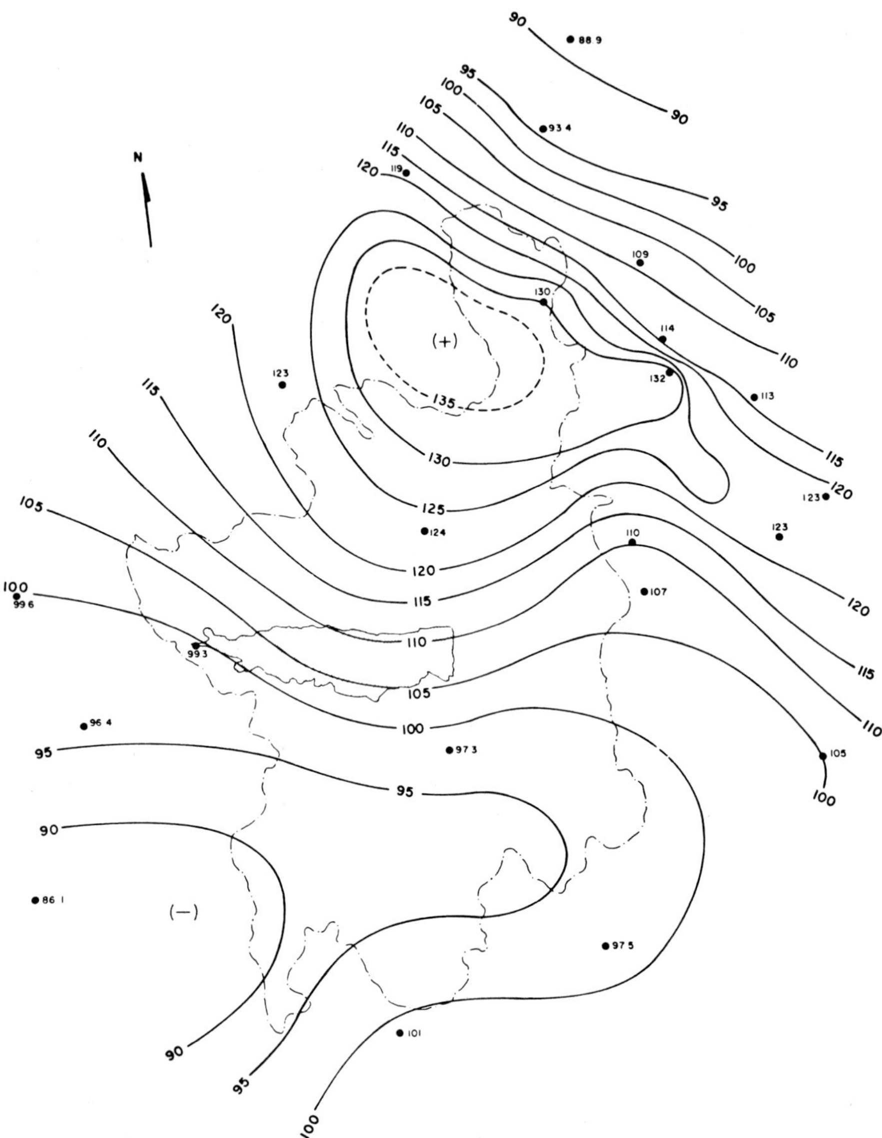

Fig. 12. Precipitation based on long-term averages in the Oneida Lake drainage basin.

thermal stratification exists under the ice cover. The temperature has been continuously monitored at one station off Shackelton Point from May through October since 1970. Typically, the water temperature of Oneida Lake increases rapidly during May and June with a maximum of approximately 25°C in mid-summer (Fig. 14). However, maximum temperatures during summer 1974 were less than 24°C and no significant thermal stratification occurred in the deeper waters. As the air temperature decreases there is a rapid decline in water temperatures during September and October.

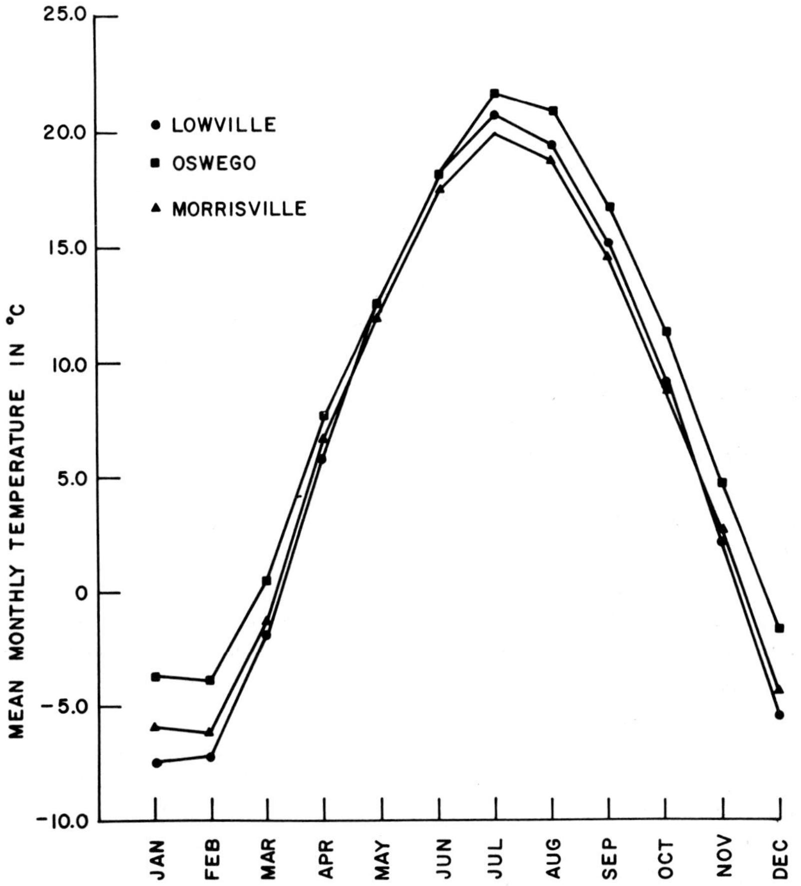

Fig. 13. Mean monthly temperatures in °C at Lowville, Oswego, and Morrisville, New York.

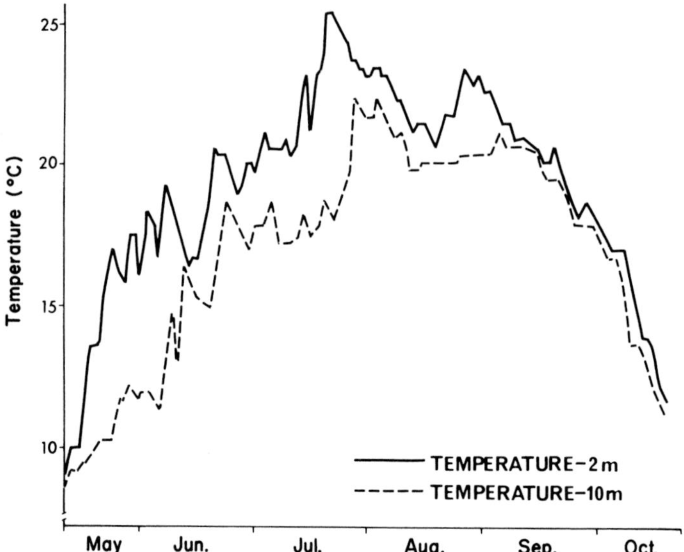

Fig. 14. Water temperatures at 2 m and 10 m in Oneida Lake, 1972.

The principal mechanism preventing permanent thermal stratification in Oneida Lake during the summer months is wind-generated wave action. Since the lake is oriented in an east–west direction and is shallow, it is constantly mixed by the prevailing westerly and north-westerly winds. According to Greeson (1971), northwest winds occurred about 57% of the time with an average velocity of 3.5 m/sec during the growing season of 1967–1969. Maximum wave heights of 0.9 m were recorded on 21.3, 22.1, and 18.0% of the days in the growing of 1967, 1968, and 1969, respectively.

Additional water mixing and partial horizontal transfer of materials in Oneida Lake are commonly produced by seiches, the largest of which result from persistent west, northwest, and east winds. Greeson (1971) reported that displacement of water can exceed 0.5 m over a 2.4-hour period given the proper velocity and wind direction.

Water currents, which can influence the transport of materials and indirectly hinder prolonged thermal stratification in a shallow lake such as Oneida, have been examined by Houde (1968). He found that surface currents usually moved in the direction of the prevailing winds, while strong subsurface currents moving in a direction opposite the wind were often detected in the open lake at depth more than half the distance to the bottom.

Ice Cover

Ice forms in mid- to late December and persists for 3 to 4 months, depending on the severity of the winter. Ice thicknesses of 0.9 m or more have occurred during late March (Greeson, 1971). However, in recent years, winters have been relatively mild and ice thicknesses of 0.3 m or less have been common. Records made for the past 17 years at the New York State Hatchery in Constantia, New York (Herbert Lake, personal communication) indicate that ice breakup generally occurs between mid-March to mid-April (Table 9).

Secchi Disk Transparency

Muenscher (1928) measured transparency on four occasions during June through August 1927 and readings ranged from 2.9 to 4.1 m with a mean of 3.5 m. With the exception of 1966, transparency in the 1960's was much lower (Table 10). Water clarity as observed by Hall (1967) in 1964 and 1965 was extremely low and similar readings were obtained in 1967–1969 by Greeson (1971). Greeson noted prolonged periods of high phytoplankton density in 1967 while algal blooms were, seemingly, of shorter duration in 1968. Secchi disk transparencies recorded in the 1970's bracketed the mean value reported by Muenscher in 1927 and suggest a marked decrease in algal density.

Changes in water transparency for the summer months (1970–1974) are shown graphically in Fig. 15. In general, a period of maximum light penetration is common in early June. Noble (1971, 1972) observed that high transparency in June of 1970 and 1971 correspond with the spring zoo-

TABLE 9

Date of Ice-Out on Oneida Lake, 1958–1975[a]

Year	Date	Year	Date
1958	Apr. 6	1967	Apr. 9
1959	Apr. 18	1968	Apr. 1
1960	Apr. 13	1969	Apr. 14
1961	Apr. 11	1970	Apr. 14.
1962	Apr. 3	1971	Apr. 19
1963	Apr. 9	1972	Apr. 19
1964	Apr. 7	1973	Mar. 17
1965	Apr. 17	1974	Mar. 17
1966	Mar. 29	1975	Mar. 28

[a] Herbert Lake, Constantia Hatchery.

TABLE 10

Mean and Range of Secchi Disk Transparency from June through September in Oneida Lake, 1927–1974

Year	Number of observations	Mean Secchi disk transparency (m)	Range	Source	Comments
1927	4	3.5	2.9–4.1	Muenscher (1928)	Measurements taken on 4 days from one station
1964	9	1.8	1.2–2.7	Hall (1967)	Measurements taken on 9 or more days each year at one or more stations
1965	14	2.3	1.8–5.1		
1966	12	3.8	1.1–5.3		
1967		1.9	0.4 m	Greeson (1971)	Measurements taken at 1–2-week intervals at two stations each year
1968	120	1.9	to	(Aug. 14, 1969)	
1969		1.6	4.4 m	(June 14, 1968)	
1970	61	3.6	1.1–7.0	Cornell Biological Field Station (unpublished data)	Measurements taken on 8–10 days at seven stations
1971	58	3.1	1.8–8.5		
1972	54	2.8	1.4–5.3		
1973	75	3.9	1.8–6.6		
1974	69	3.0	1.4–6.4		

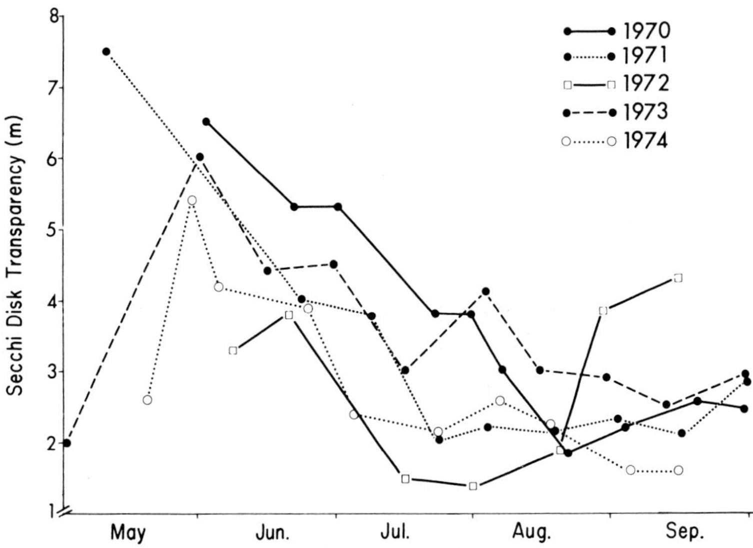

Fig. 15. Seasonal changes in Secchi disk transparency in Oneida Lake, 1970–1974.

plankton pulse, and suggested that zooplankton grazing was responsible for the increased transparency. A dramatic decline in Secchi disk transparency is evident in mid- to late summer presumably due to increased algal production and a predominance of blue-green phytoplankton blooms.

Lake Sediments

The nature of the bottom material in the limnetic portion of Oneida Lake has been described by Greeson (1971). Bottom sediments have been categorized into five groups on the basis of composition and size of materials (Fig. 16). Silt and clay, the dominant sedimentary material, covers 40.3% of the bottom of Oneida Lake while mud (mixture of silt, clay, and organic material), sand, cobble and rubble, and gravel constitute 27.5, 17.8, 10.2, and 4.2%, respectively. The cobble and rubble were deposited in Oneida Lake in the form of drumlins (Baker, 1916b). Four major drumlins exist in the lake—Shackelton Shoals (the largest), Pancake Shoals, Frenchman and Dunham Islands, and Messenger Shoals.

Iron–manganese concretions, which are abundant on the bottom in some areas, have been termed "Oneida Lake pancakes" by local residents (Gillette, 1961). These concretions are similar to those found in the oceans and are, presumably, the result of precipitation of iron and manganese when chemically reduced sediment pore water comes in contact with well-oxygenated water (Dean, 1970; Dean *et al.*, 1972; Dean and Ghosh, 1977).

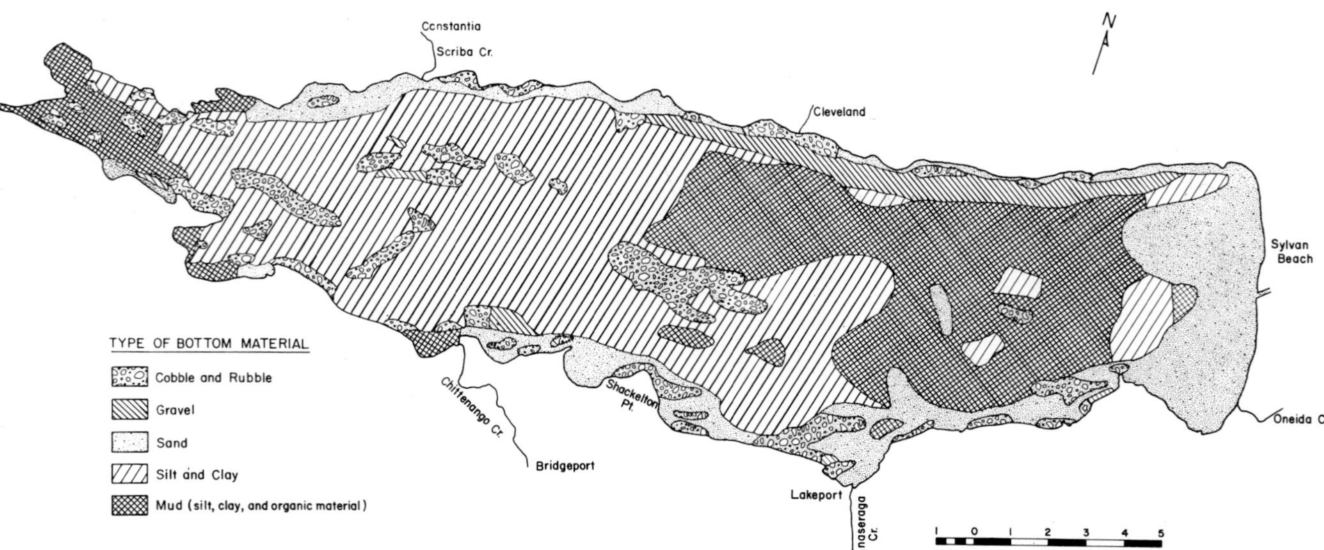

Fig. 16. Bottom sediments categorized on the basis of composition and size of materials in Oneida Lake. After Greeson (1971).

Chemical-Lake

Alkalinity and Hydrogen Ion Concentration (pH)

Wagner (1928) reported 9.7 to 31.7 mg/liter as $CaCO_3$ in Oneida Lake while total alkalinity at the outlet was 78 mg/liter. Later, Burdick and Lipschuetz (1946) and Dence and Jackson (1959) reported mean alkalinity values of 74 mg/liter and 84 mg/liter and suggested that the substantial increase since 1927 reflected an increase in productivity and change of the dominant phytoplanktors from diatoms to blue-greens. However, the lake values reported by Wagner were probably in error since subsequent surveys in 1956, 1961, and 1967–1969 found negligible differences between alkalinity in the lake and outlet. Greeson (1971) observed a mean alkalinity of 81 mg/liter as $CaCO_3$ but found this parameter to be highly variable, ranging from 31 to 221 mg/liter over a 3-year period. Summer and autumn values were typically higher than the mean while an annual minimum usually occurred during the colder winter–spring months. In general, Oneida Lake is a well-buffered, carbonate-bicarbonate system with total alkalinity on the order of 80 mg/liter as $CaCO_3$ and no significant changes have occurred since 1927 (Table 11).

Long-term averages and ranges of pH indicate that, although variability occurs within years, no dramatic changes have taken place between 1927 and 1969 (Table 12). The pH is commonly basic and has been as low as 6.8 and as high as 9.3. Seasonally, Greeson (1971) found pH to be highest in summer and lowest in winter.

TABLE 11

Alkalinity in Oneida Lake, 1927–69

Year	Total alkalinity (as mg/liter of $CaCO_3$)	Source
1927	78	Wagner (1928)
1946	74	Burdick and Lipschuetz (1946)
1954	84	Dence and Jackson (1959)
1956	77	New York State Department of Health (1957)
1961	86	Mt. Pleasant *et al.* (1962)
1963–1965	83	Lehne (1963, 1964, 1965)
1967–1969	81	Greeson (1971)

TABLE 12

Mean and Range of pH in Oneida Lake, 1927–1969

Year	Mean	Range	Source	Comments
1927	8.1	—	Wagner (1928)	One sample taken at Oneida Outlet at Brewerton, depth 12 ft.
1946	7.8	6.7–8.5	Burdick and Lipschuetz (1946)	Surface and bottom samples taken from three stations
1948	8.0	7.6–8.3	Burdick and Lipschuetz (1948)	Surface and bottom sample off Cleveland
1955	7.8	—	Shampine (1973)	One sample taken during mid-winter, depth not indicated
1960	7.0	—	Shampine (1973)	One sample taken during early November, depth not indicated
1961	7.7	6.8–8.5	Mt. Pleasant *et al.* (1962)	110 measurements taken just below surface, Apr.-Oct. from 20 stations
1963	7.5	7.0–8.1	Lehne (1963)	59 measurements at surface, Sep.–Nov. from 12 stations
1964	8.1	7.0–9.3	Lehne (1964)	110 measurements at surface, Jan.-Oct. 1964 from 12 stations
1965	8.1	6.9–8.6	Lehne (1965)	109 measurements at surface, May–Oct. 1965 from 12 stations
1967–1969	8.1	7.2–9.3	Greeson (1971)	15 stations, May 1967–Oct. 1969

Dissolved Oxygen

A clinograde oxygen curve is characteristic during the summer months (Fig. 17). Daytime oxygen concentrations from 1972 to 1975 in the upper 5 m of the water column were usually 7.1 mg/liter to 13.0 mg/liter with supersaturation most common during this period. Dissolved oxygen levels decreased markedly, as did percent saturation, in the bottom waters from June through August. However, oxygen concentrations less than 1 mg/liter are temporary and limited to brief periods of calm weather. Dissolved oxygen is usually uniform throughout the water column when autumnal and vernal mixing occurs.

Principal sources of information on dissolved oxygen are provided in Table 13. Prior to the 1950's, the depletion of dissolved oxygen in deep waters of Oneida Lake was probably uncommon (Forney, 1973) although

Burdick and Lipschuetz (1948) did report a value of 2.3 mg/liter at 40 m off Cleveland in mid-July of 1948. Low levels of dissolved oxygen were observed in 1959 when concentrations on the bottom declined to 0.5 mg/liter in some areas and below 2.0 mg/liter over 50% of the lake area (Table 14). Oxygen concentrations near the bottom in deeper waters were low in the early 1960's and in recent years. Periods of high primary production as well as meteorological conditions, particularly extended periods of calm weather, are important factors contributing to the apparent decline in dissolved oxygen since 1959.

Jacobsen (1966) reported severe oxygen deficits in 1959 and a subsequent decline in abundance of the mayfly, *Hexagenia limbata,* through 1964.

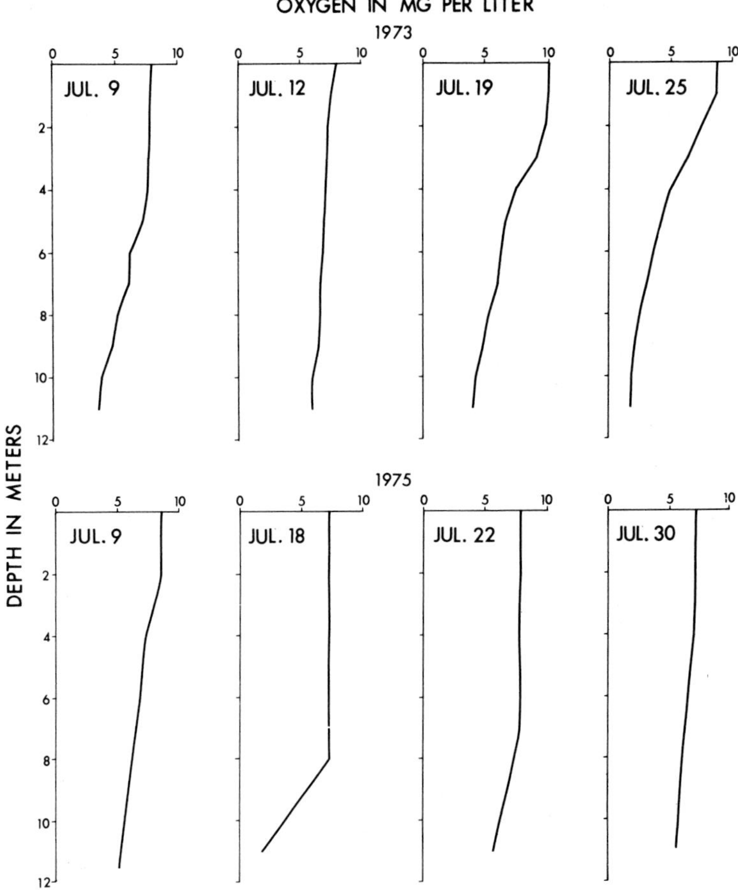

Fig. 17. Oxygen profiles in Oneida Lake, 1973 and 1975.

TABLE 13

Studies Providing Dissolved Oxygen Information on Oneida Lake, 1927–1975

Investigator	Station(s) sampled	Sampling frequency	Comments
Wagner (1928)	Off Cleveland	Jun. 29–Aug. 24, 1927	High, low, and average in profiles to 15 m
Burdick and Lipschuetz (1946)	Big, Bay, Upper South Bay, off Cleveland	4 sampling dates only, Jan., Mar., Jun., and Aug., 1946	Surface and bottom values only
Burdick and Lipschuetz (1948)	Off Cleveland	Jul. 14, 1948	Samples at 0.3 m, 6.1 m, and 12.2 m
Mt. Pleasant *et al.* (1962)	20 lake stations	May 8–Oct. 17, 1961	Samples collected at 0.6 m
Lehne (1963, 1964, 1965)	12 stations	Sep. 1963–Oct. 1964	Surface values
Jacobsen (1966)	2 stations		Vertical profiles at 5-m intervals
Greeson (1971)	5 mid-lake stations located at navigation buoys of the New York State Department of Transportation	May 1967 through Oct. 1969	Vertical profile at 10-m intervals to 40 m maximum depth
Forney (1973)	1 to 4 stations from 1957 through 1972	Usually Jun., Jul., and Aug., occasional measurements in Sep.	Summary of oxygen concentrations at depths greater than 11 m for 1957–1972
Cornell Biological Field Station (unpublished data)	1 station off Shackelton Point	Jun., Jul., and Aug., 1971–1974	Vertical profiles, raw data

TABLE 14

Mean and Range of Dissolved Oxygen (mg/liter) in the Bottom Waters of Oneida Lake during June through August, 1957–1974

Year	Mean	Number of observations	Range
1957	7.4	3	6.5–8.5
1958	7.4	3	5.9–8.7
1959	4.9	5	0.5–8.8
1960	6.6	4	6.0–7.2
1961[a]	5.9	3	2.5–8.3
1962	8.2	6	7.4–10.0
1963	6.8	8	2.6–9.0
1964	6.4	4	2.1–8.3
1965[b]	7.6	1	—
1966	4.1	5	0.8–7.3
1967[c]	3.3	10	0.8–6.8
1968[c]	5.9	7	4.1–10.0
1969	—	—	—
1970	4.0	4	3.3–4.8
1971	6.1	4	4.1–7.9
1972	4.2	8	0.4–6.6
1973	4.2	9	2.4–6.8
1974	5.3	4	2.3–7.4

[a] Partially from Mt. Pleasant *et al.* (1962).
[b] Lehne (1965).
[c] Greeson (1971).

Mayfly larvae are extremely sensitive to low oxygen and concentrations less than 1 mg/liter are lethal within 30 to 48 hours (Hunt, 1953). Although oxygen levels may have been low intermittently before 1959 (as in 1948) resulting in less pronounced losses of mayflies, the depletion of oxygen in 1959 could have been the first occurrence severe enough to kill large numbers of this benthic organism. Frequent depletions of oxygen since 1959 probably contributed to the decline of the mayfly. Forney (1973) speculated that burrowing activities of mayflies may have facilitated aeration of sediments and their loss was partly responsible for the increased oxygen demand of benthic sediments and persistent oxygen stratification in recent years.

Phosphorus

Nuisance blooms of algae, particularly common during the 1940's and 1960's, prompted several investigators to focus on phosphorus and its relation to phytoplankton production in Oneida Lake. Burdick and Lipschuetz

(1946) examined surface and bottom samples from four sites during March and August 1946 and observed a mean of 0.11 mg/liter of total phosphate. The first water-quality survey examining temporal and spatial patterns of phosphorus concentrations in Oneida Lake was conducted in 1961 by Mt. Pleasant *et al.* (1962). They found that total phosphorus was almost double that observed in 1946 and there were marked seasonal and spatial variations. Total phosphorus was usually higher in the shallow inshore stations, presumably influenced by inflowing streams, and maximum concentrations were observed during spring with a dramatic decline during the summer months when biological activity was greatest. The Onondaga County Department of Public Works collected surface samples from 12 stations at monthly intervals from 1963 to 1965 (Lehne, 1963, 1964, 1965). Total phosphorus concentrations during this period were the highest recorded in Oneida Lake (Table 15) and yearly averages were as much as four times other published data. Changes in total dissolved phosphate (TDP) were

TABLE 15

Total Phosphate as PO_4^- in Oneida Lake, 1946–1975

Year	Total phosphate as PO_4^- (mg/liter)	Source	Methods	Comments
1946	0.11	Burdick and Lipschuetz (1946)	Not given	Surface and bottom samples during Mar. and Aug., 1946, 2 stations
1961	0.21	Mt. Pleasant *et al.* (1962)	(1)	March through Oct. 1961, 20 stations. Samples collected at 0.6 m below surface
1963	0.33	Lehne (1963)	(1)	
1964	0.40	Lehne (1964)	(1)	Jul. 1963 through Oct. 1965, 12 stations, surface samples
1965	0.48	Lehne (1965)	(1)	
1967	0.16	Greeson (1971)	(1)	May 1967–Sep. 1969, 15 stations, surface samples
1968	0.14	Greeson (1971)	(1)	
1969	0.13	Greeson (1971)	(1)	
1975	0.14	Cornell Biological Field Station (unpublished data)	(2)	May through Sep. 1975, 5 stations at weekly intervals, integrated surface to bottom samples

(1) American Public Health Association (1971).
(2) Menzel and Corwin (1965).

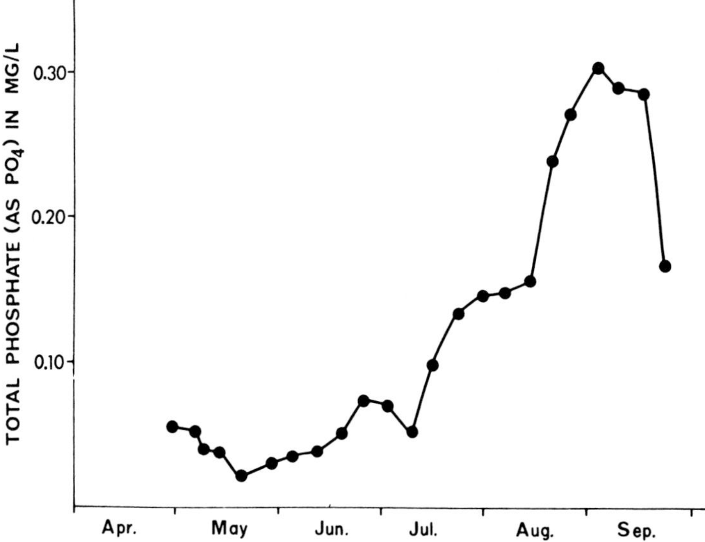

Fig. 18. Seasonal variation of total phosphate (as PO_4) in Oneida Lake, April through September 1975. Data are integrated over the entire water column and are an average of five stations.

examined spatially and temporally by Greeson (1971) from 1967–1969. Yearly average concentrations of TDP were 0.16, 0.14, and 0.13 mg/liter for 1967, 1968, and 1969, respectively. However, the inclusion of the particulate fraction would have undoubtedly increased total phosphate concentrations to within the range of previous investigators since phytoplankton production was high during this period. Total phosphate (as PO_4) concentrations for May through September 1975 (Cornell Biological Field Station, unpublished data) averaged 0.141 mg/liter, indicating a possible reduction in phosphorus concentration. Temporal variation of total phosphate (as PO_4^-) is shown graphically in Fig. 18.

Greeson (1971) did not consider phosphorus the controlling factor limiting phytoplankton production during 1967–1969 and Mt. Pleasant *et al.* (1962) suggested phosphorus was present in excess of metabolic needs of phytoplankton in 1961. In contrast, W. Fuhs (written communication to Dr. L. Hetling), conducted bioassay tests of samples collected in mid-July 1973 and concluded that phosphorus and nitrogen were in short supply.

Nitrogen

The first measurements of nitrogen in Oneida Lake were made by Burdick and Lipschuetz (1946). Later, Shampine (1973) and Mt. Pleasant *et*

TABLE 16

Concentrations (mg/liter) of Various Forms of Nitrogen, Averaged for All Samples, 1946–1969

Year	NO_3–N	NO_2–N	Ammonia NH_4^+	Organic N	Total N	Source
1946	0.09 (14)[a]	—	0.05 (14)	0.37 (10)	0.52 (14)	Burdick and Lipschuetz (1946)
1955	2.10 (1)	—	—	—	—	Shampine (1973)
1961	0.29 (112)	—	—	—	—	Mt. Pleasant *et al.* (1962)
1965	1.08 (36)	—	0.26 (36)	0.31 (25)	—	Shampine (1973)
1966	0.37 (3)	—	0.07 (3)	0.18 (3)	—	Shampine (1973)
1967–1969	0.40 (200)	0.01 (200)	0.12 (200)	0.16 (134)	—	Greeson (1971)

[a] Number of samples given in parentheses.

al. (1962) reported nitrogen in the form of nitrate for the years 1955 and 1961. Surveys by the U.S. Geological Survey (Shampine, 1973) and Greeson (1971) provide the most comprehensive information on the various forms of nitrogen (Table 16). Seasonal variations in Greeson's study were marked. Nitrate nitrogen exhibited pronounced summer minima and winter maxima. Although seasonal shift were variable from year to year, extremely low concentrations of nitrite (as NO_2^-) were observed during April–October of 1967 and October–December of 1968. Mean concentrations of total nitrogen during the growing seasons of 1967, 1968, and 1969 were 0.24, 0.27, and 0.42 mg/liter, respectively. Greeson attributed the higher concentration in 1969 to nitrogen-fixation resulting from an abundance of blue-green algae. For the lake as a whole, there appeared to be a definite increase in the organic form of nitrogen when levels of dissolved inorganic nitrogen were low and ammonium ions tended to increase with depth. Temporal variations of nitrate nitrogen are shown graphically in Fig. 19.

Major Ions

Berg (1963) listed the major ionic constituents in Oneida Lake based on collections made in the mid-1950's by the U.S. Geological Survey. A second, more extensive survey of Oneida Lake's ionic composition by the U.S. Geological Survey covered the period 1961–1967 (Shampine, 1973).

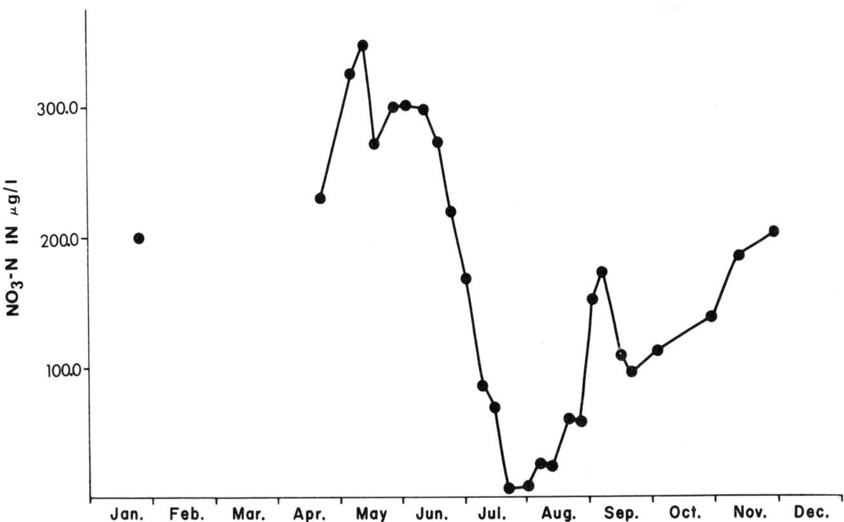

Fig. 19. Seasonal variation of nitrate nitrogen in Oneida Lake, 1975. Data are integrated over the entire water column and are an average of five stations.

Similar analyses were made on samples taken from 1967–1969 by Greeson (1971) who reported the following mean concentrations of major ions (mg/liter): calcium, 38; magnesium, 8.6; sodium, 4.9; potassium, 0.9; bicarbonate, 99; sulfate, 48; and chloride, 9.0.

Positive and negative ions expressed as milliequivalents appear to be reasonably well-balanced although anions are lower by about 0.3 mEq for 1965 and 1973 (Table 17). Of the cations, calcium is by far the dominant as well as the most variable while bicarbonate is the dominant among the anions. The most pronounced changes have occurred in sodium and chloride concentrations. Chloride ions have more than doubled since measurements were first made in 1885 and it appears that an abrupt increase in this ion occurred during the early 1960's (Table 18). Although less data are available, sodium also appears to have increased during the early 1960's. Most sodium chloride in Oneida Lake comes from tributary inflow but the origin is uncertain. It is known that groundwater can dissolve sodium chloride at shallow depths in the western portion of the Oneida Lake basin (F. J. Pearson, Jr. and G. S. Meyers, unpublished manuscript). Although speculative, runoff from salt used to deice roads during the winter may also be an important factor. In addition, it would appear that the hardness of Oneida Lake water has also increased (Table 18).

Dissolved Solids and Other Chemicals

A comprehensive examination of the content of total dissolved solids of Oneida Lake was made in 1967–1969 (Greeson, 1971). Dissolved solids showed little vertical stratification but some variation was detected. Concentrations in the southern and western portions of the lake were higher reflecting inflow from streams in the southern portion of the basin. Temporal variations indicated a winter–spring minimum and summer maximum. Greeson calculated that 409×10^6 kg of dissolved solids entered Oneida Lake from the drainage basin and over 10% of this was retained in the lake. On a lakewide basis, the mean concentration over the 3-year period was 163 mg/liter with concentrations ranging from 57 to 283 mg/liter. Similar values of total dissolved solids were obtained by Shampine (1973).

Mean concentrations of trace elements, specific conductance, chlorinated hydrocarbon insecticides, and herbicides were measured from May 1967 through October 1969 (Table 19). No attempt was made to concentrate the filtered samples for trace element analysis and many elements were below the level of detection by emission spectroscopy analysis. Concentrations of most pesticides were also below the detection limit. Seasonal variations of trace elements were minimal and no correlation was found with changes in phytoplankton.

TABLE 17

Milliequivalent Ionic Balances for Oneida Lake, 1955–1975

	Cations					Anions				
Year	Ca^{2+}	Mg^{2+}	Na^{+}	K^{+}	Total	HCO_3^{-}	SO_4^{2-}	Cl^{-}	Total	Source
1955	1.54	0.59	0.08[a]	—	2.21	1.31	0.88	0.13	2.32	Berg (1963)
1960	2.04	0.70	0.15	0.02	2.91	1.74	1.06	0.17	2.97	Shampine (1973)
1961	1.84	0.72	0.14	0.03	2.73	1.54	0.98	0.15	2.67	Shampine (1973)
1965	2.17	0.88	0.23	0.03	3.31	1.54	1.11	0.24	2.89	Shampine (1973)
1966	2.01	0.72	0.20	0.03	2.96	1.66	1.08	0.25	2.99	Shampine (1973)
1967–1969	1.89	0.71	0.21	0.02	2.83	1.62	0.99	0.25	2.85	Greeson (1971)
1973	2.04	0.82	0.21	0.04	3.11	1.60	0.86	0.35	2.81	R. T. Oglesby (unpublished data)
1975	1.79	0.89	0.22	0.04	2.94	1.59	0.89	0.25	2.73	Cornell Biological Field Station (unpublished data)

[a] Determined as sodium plus potassium.

TABLE 18

Chlorides, Sodium, and Hardness in Oneida Lake, 1885–1975

Year	Chloride (mg/liter)	Sodium (mg/liter)	Hardness as $CaCO_3$ (mg/liter)	Number of observations	Source
1885	4.2	—	—	8	New York State Board of Health (1888)
1886	3.0	—	84	2	New York State Board of Health (1888)
1955	4.5	—	—	1	Berg (1963)
1955	—	—	107	1	Shampine (1973)
1956	5.5	—	128	9	New York State Department of Health (1957)
1960	5.9	3.5	—	1	Shampine (1973)
1961	12.9	—	145	116	Mt. Pleasant *et al.* (1962)
1961	—	3.2	—	1	Shampine (1973)
1965	8.8	5.3	161	20	Shampine (1973)
1966	8.8	4.7	136	4	Shampine (1973)
1967–1969	9.0	4.0	124	>425	Greeson (1971)
1973	12.7	4.8	141	2	R. T. Oglesby (unpublished data)
1975	8.9	5.1	126	5	Cornell Biological Field Station (unpublished data)

TABLE 19

Mean Concentrations of Dissolved Chemicals and Specific Conductance in Oneida Lake, 1967–1969[a,b]

Parameter	Concentration[b]	Parameter	Concentration[b]
Aldrin	ND[c]	Lindane	Trace
Aluminum	0.030	Lithium	0.003
Barium	0.032	Magnesium	8.6
Beryllium	<0.0008	Manganese	0.016
Bismuth	<0.004	Mercury	~0.0001
Boron	0.024	Molybdenum	<0.001
Cadmium	<0.030	Nickel	<0.001
Chromium	0.007	Rubidium	<0.001
Cobalt	<0.008	Silica	2.1
Copper	0.007	Silver	<0.0004
DDD	<0.00001	Silvex	ND
DDE	ND	Specific conductance	275 μmhos/cm at 25°C
DDT	ND		
Dieldrin	ND	Tin	<0.008
Endrin	ND	Titanium	<0.004
Fluoride	0.100	2,4-D	ND
Gallium	<0.002	2,4,5-T	ND
Germanium	<0.004	Vanadium	<0.004
Heptachlor	Trace	Zinc	<0.02
Iron	0.037	Zirconium	ND
Lead	<0.006		

[a] From Greeson (1971).
[b] Concentrations given in mg/liter unless otherwise indicated.
[c] Not detectable.

Chemical–Tributaries

Wagner (1928) first recognized the importance of point-source pollutants of Oneida Lake and examined the pH, alkalinity, and dissolved oxygen in Wood Creek, Oneida Creek, and the outlet in 1927. Later, Burdick and Lipschuetz (1946) analyzed influent streams on three occasions in 1946 for various fractions of nitrogen and phosphorus. They reported an abundant supply of nutrients entering Oneida Lake, especially from the southern streams. Comprehensive surveys inventorying the chemical constituents of Oneida Lake streams were done by Mt. Pleasant *et al.* (1962) and Lehne (1963, 1964, 1965). Mt. Pleasant *et al.* concluded that the most important contributors of nutrients to the lake were the Barge canal, and Chittenango, Canaseraga, and Oneida Creeks. Increases in nutrient levels in tributaries have reflected, in part, the urbanization of the Oneida Lake watershed. Chittenango Creek, which flows through urban areas, was enriched between

1946 and 1961, and Lehne (1965) noted that the creek had deteriorated significantly from 1956 to 1965. Greeson (1971) measured selected nutrients and trace metals in large tributaries to Oneida Lake. As might be expected, streams draining the southern portion of the basin had higher chemical loads of these materials than those in the north (Table 20). Similar data reported by Shampine (1973) are in general agreement.

Greeson calculated a nutrient budget for Oneida Lake from the major tributaries and ungauged streams (Table 21). The fertile southern basin, which forms 51% of the total basin, contributed 82% of the total dissolved solids content. Some 17.6% and 61.9% of inflowing nitrogen and phosphorus was retained by the lake. In contrast, Oglesby *et al.* (1973) estimated that 45% of the nitrate nitrogen and 87% of the molybdate reactive phosphorus was retained annually in Owasco Lake, one of New York's Finger Lakes.

Biological Properties

Phytoplankton

Although early studies of phytoplankton were cursory and largely descriptive, they established that Oneida Lake was a naturally productive body of water characterized by nuisance quantities of algae. The first phytoplankton list for Oneida Lake was made by Dr. G. M. Smith (Adams and Hankinson, 1928), who visited the lake in August 1918. He reported the predominance of species of blue-greens and greens and noted the periodic abundance of *Gloeotrichia echinulata* in the surface waters during the summer. Later, the phytoplankton were surveyed by Muenscher (1928), who reported that diatoms dominated during August and September of 1927. The genera *Asterionella* and *Stephanodiscus* were common throughout the survey, while *Fragilaria* and *Tabellaria* predominated only during August. In addition, he found in his net plankton counts that blue-greens were common, represented by the genera *Anabaena* and *Microcystis* with occasional traces of *Gloeotrichia*.

Increased algal concentrations of nuisance proportions and fish kills in the early 1940's prompted another limnological investigation (Burdick and Lipschuetz, 1946). *Microcystis* and *Rivularia* (probably *Gloeotrichia*) were reported to be extremely abundant along the north shore of Oneida Lake. Two years later, these same workers reported that strong south winds had concentrated *Microcystis* along the north shore and subsequent decomposition resulted in a fish kill. Stone and Pasko (1946) sampled the north shore of Oneida Lake at Jewell during August 1946 and found that blue-greens, *Microcystis*, *Anabaena*, and *Gloeotrichia*, to be dominant. During the sum-

TABLE 20

Mean Annual Loads of Selected Nutrients from Major Tributaries of Oneida Lake[a,b]

Drainage basin	Silica (SiO_2)	Calcium (Ca)	Magnesium (Mg)	Sodium (Na)	Potassium (K)	Bicarbonate (HCO_3)	Sulfate (SO_4)	Chloride (Cl)	Total nitrogen (N)	Total phosphate (PO_4)	Total dissolved solids
Scriba	7.3	22.7	7.3	3.6	0.9	72.7	27.3	6.4	0.5	0.1	109.1
	(3.3)	(0.8)	(1.2)	(1.4)	(1.5)	(1.2)	(0.8)	(1.3)	(1.8)	(0.4)	(1.0)
West Branch	42.7	227.3	37.3	15.5	3.6	390.9	118.2	29.1	2.6	1.1	590.9
Fish Creek	(19.5)	(8.5)	(6.3)	(6.2)	(6.1)	(6.4)	(3.3)	(5.8)	(10.3)	(4.8)	(5.2)
East Branch	40.9	109.1	34.5	14.5	3.6	354.5	109.1	26.4	3.1	1.2	545.5
Fish Creek	(18.7)	(4.1)	(5.9)	(5.8)	(6.1)	(5.8)	(3.0)	(5.2)	(12.1)	(5.2)	(4.8)
Wood	12.7	90.9	21.8	27.3	3.6	272.7	72.7	55.5	1.7	2.2	445.5
Creek	(5.8)	(3.4)	(3.7)	(10.9)	(6.1)	(4.5)	(2.0)	(11.0)	(6.8)	(9.5)	(4.0)
Oneida	16.4	472.7	100.9	30.9	7.3	963.6	745.5	66.4	2.2	3.3	2000.0
Creek	(7.5)	(17.6)	(17.2)	(12.3)	(12.1)	(15.9)	(20.7)	(13.2)	(8.5)	(14.3)	(17.8)
Cowaselon	10.9	327.3	72.7	16.4	4.5	581.8	627.3	34.5	1.1	1.1	1445.5
Creek	(5.0)	(12.2)	(12.4)	(6.5)	(7.6)	(9.7)	(17.4)	(6.8)	(4.3)	(4.8)	(12.8)
Canaseraga	9.1	100.0	20.9	4.5	0.9	181.8	181.8	9.1	0.5	0.3	427.3
Creek	(4.1)	(3.7)	(3.6)	(1.8)	(1.5)	(3.1)	(5.0)	(1.8)	(2.1)	(1.2)	(3.8)
Chittenango	41.8	1018.2	218.2	60.0	23.6	2418.2	1363.7	120.0	5.5	5.7	4163.7
Creek	(19.1)	(38.0)	(37.2)	(23.9)	(39.3)	(39.9)	(37.9)	(23.8)	(21.4)	(25.0)	(37.1)
Ungauged	37.3	309.1	72.7	76.4	10.9	818.2	327.3	155.5	4.9	6.0	1472.7
Areas	(17.0)	(11.5)	(12.4)	(30.5)	(18.2)	(13.5)	(9.1)	(30.7)	(19.2)	(26.1)	(13.1)

[a] Source: Greeson (1971).

[b] Upper values represent kg $\times$ 10^2/day; lower values (in parentheses) represent percent of total.

TABLE 21

Chemical Budgets of Major Nutrients in Oneida Lake[a,b]

Inflows and outflows	Silica (SiO_2)	Calcium (Ca)	Magnesium (Mg)	Sodium (Na)	Potassium (K)	Bicarbonate (HCO_3)	Sulfate (SO_4)	Chloride (Cl)	Total nitrogen (N)	Total phosphate (PO_4)	Total dissolved solids
Northern drainage basin	110.0 (50.2)	504.6 (18.8)	122.7 (19.3)	74.5 (29.7)	13.6 (22.7)	1236.4 (20.4)	390.9 (10.9)	145.5 (28.8)	8.8 (34.5)	5.6 (24.6)	1954.6 (17.4)
Southern drainage basin	109.1 (49.8)	2172.7 (81.0)	472.7 (80.5)	174.5 (69.2)	45.5 (75.8)	4818.2 (96.1)	3181.9 (79.6)	357.3 (88.3)	13.3 (70.8)	15.3 (52.0)	9245.5 (82.2)
Other sources[c]	—	4.5 (0.2)	0.9 (0.2)	1.8 (0.7)	0.9 (1.5)	—	27.3 (0.8)	1.8 (0.4)	3.5 (13.5)	2.0 (8.7)	4.5 (0.4)
Total inflow to lake	219.1	2681.8	596.4	250.9	60.0	6054.6	3600.0	504.6	25.5	22.9	11,245.6
Total outflow to lake	218.2	2363.7	527.3	245.5	59.1	6000.1	3000.0	536.4	20.5	8.7	10,000.1
Retention by lake	0.9 (0.4)	318.2 (11.9)	69.1 (10.2)	5.5 (2.2)	0.9 (1.5)	54.5 (0.9)	600.0 (16.7)	— —	5.0 (19.6)	14.2 (61.9)	1245.5 (11.1)

[a] Source: Greeson (1971).
[b] Upper values represent kg × 10^2/day; lower values (in parentheses) represent percent of total.
[c] Precipitation on lake and lake-shore population.

mers of 1954 and 1955, Dence and Jackson (1959) found that the great bulk of net plankton consisted of the blue-greens *Gloeotrichia* and *Aphanizomenon* as dominants and *Microcystis* and *Anabaena* as subdominants. The greens and diatoms often were common but never dominant.

Mt. Pleasant *et al.* (1962) compiled a qualitative list of the dominant genera of phytoplankton between June 27 and September 10, 1961 which indicated a preponderance of genera representing the blue-greens, greens, and diatoms. In 1964 and 1965 phytoplankton samples were taken on eight different occasions by Hall (1967) who indicated that *Anabaena flos-aquae, Coelosphaerium naegelianum,* and *Microcystis incerta* were the dominant blue-greens and *Fragilaria crotonensis, Melosira granulata,* and *Stephanodiscus niagarae* were the most abundant during the summer and fall.

Later, Greeson (1971) conducted a lakewide study of temporal and spatial variations of phytoplankton composition and abundance from May 1967 through June 1969. Temporally, greens dominated during late spring and early summer, but were replaced by blue-greens during the summer months. Diatoms became dominant with the decline of the blue-green algae and remained dominant during the fall, winter, and spring (Fig. 20). Spatial distribution was seldom uniform and the highest concentrations of phytoplankton were generally found in bays and shoal areas in the western end of the lake and along the southern shore. The four most common species of blue-green algae were *Microcystis aeruginosa, Anabaena circinalis, Anabaena flos-aquae,* and *Aphanizomenon holsaticum,* with the latter two constituting 89% of the blue-green community (Table 22). Diatoms became a large portion of the total standing crop during the colder months, dominated by such species as *Melosira granulata, Asterionella formosa, Fragilaria brevistriata,* and *Fragilaria crotonensis.* Common among the greens were *Pediastrum duplex, Pandorina morum, Sphaerocystis schroeteri, Microspora willeana,* and *Staurastrum paradoxum.*

Zooplankton

Composition. In 1927, Muenscher (1928) found pulses in crustaceans in late June and early September and most of these organisms were concentrated in the upper 3 m. Changes in the Rotifera were less marked with a maximum abundance in mid-August. *Daphnia* was the most common cladoceran; other cladocerans included *Leptodora, Sida,* and *Pseudosida.* Among the copepods, *Diaptomus* and *Cyclops* were the most abundant.

Studies during the summer months in the mid-1960's by Hall (1967) were concerned with the structure and seasonal changes of the zooplankton community in Oneida Lake. Hall found 14 zooplanktors common and five rare (Table 23) and most of these species have been found in recent years

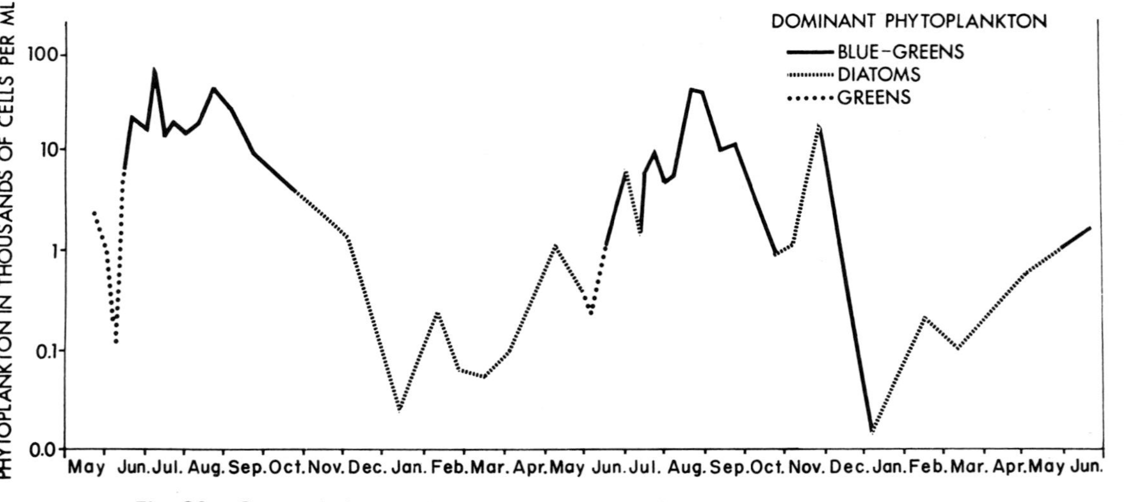

Fig. 20. Seasonal changes in relative abundance of major phytoplankton groups in Oneida Lake, 1967–1969. After Greeson (1971).

TABLE 22

Common or Abundant Phytoplankton Species Found in Oneida Lake 1967–1969[a]

- Chlorophyta
  - *Actinastrum hantzschii* Lagerheim
  - *Chlamydomonas sp.*
  - *Cladophora glomerata* (Linneaus) Kützing
  - *Dictyosphaerium pulchellum* Wood
  - *Eudorina elegans* Ehrenberg
  - *Microspora willeana* Lagerheim
  - *Mougeotia sp.*
  - *Pandorina morum* (Müller) Bary
  - *Pediastrum boryanum* (Turpin) Meneghini
  - *Pediastrum duplex* Meyen
  - *Sphaeriocystis schroeteri* Chodat
  - *Staurastrum paradoxum* Meyen
  - *Volvox aureua* Ehrenberg
- Chrysophyta (including Bacillariophyceae)
  - *Asterionella formosa* Hassall
  - *Cocconeis pediculus* Ehrenberg
  - *Dinobryon cylindricum* Imhof
  - *Fragilaria brevistriata* Grunow
  - *Fragilaria capucina* Desmazieres
  - *Fragilaria crotonensis* Kitton
  - *Melosira ambigua* (Grunow) Müller
  - *Melosira granulata* (Ehrenberg) Rolfs
  - *Melosira varians* Agardh
  - *Navicula sp.*
  - *Nitzschia signoidea* (Mitzsch) Smith
  - *Stephanodiscus niagarae* Ehrenberg
  - *Synedra delicatissima* Smith
  - *Synedra ulna* (Ehrenberg) Grunow
  - *Tabellaria fenestrata* (Lyngbye) Kützing
- Pyrrophyta
  - *Ceratium hirundinella* (Müller) Schrank
- Cyanophyta
  - *Anabaena circinalis* Rabenherst
  - *Anabaena flos-aquae* (Lyngbye) DeBrefisson
  - *Anabaena spiroides* Klebahn
  - *Aphanizomenon holsaticum* Richter
  - *Chroococcus limneticus* Lemmermann
  - *Gleotrichia echinulata* (Smith) Richter
  - *Lyngbya birgei* Smith
  - *Microcystis aeruginosa* Elenkin
  - *Microcystis incerta* Lemmermann
  - *Oscillatoria subbrevis* Schmidle
  - *Synechococcus aeruginosa* Näglelü

[a] By Greeson (1971).

TABLE 23

Zooplankton Species Found in Oneida Lake[a]

Species	Common	Rare
Calanoid copepods		
Diaptomus minutus	X	
Diaptomus oregonensis	X	
Diaptomus sicilis	X	
Epischura lacustris	X	
Cyclopoid copepods		
Cyclops bicuspidatus thomasi	X	
Cyclops vernalis	X	
Eucyclops agilis		X
Ergasilis chautauquaensis		X
Mesocyclops edax	X	
Tropocyclops prasinus	X	
Cladocera		
Bosmina longirostris		X
Ceriodaphnia quadrangula		X
Chydorus sphaericus	X	
Daphnia galeata mendotae	X	
Daphnia pulex	X	
Daphnia retrocurva	X	
Diaphanosoma leuchtenbergianum	X	
Leptodora kindtii	X	
Sida crystallina	X	

[a] Source: Hall (1967).

(Cornell Biological Field Station, unpublished data). Generally, copepods have predominated in numbers while the cladocerans have constituted most of the biomass.

Pronounced annual differences in composition and density of the major zooplanktors during the summer has occurred since 1964. Prior to 1968, *Daphnia pulex* was seldom observed in Oneida Lake and *D. galeata* nd *D. retrocurva* were the dominant species. As shown in Fig. 21, *D. pulex* was the dominant species in 1970, 1973, and 1974 and was common in 1971 and 1972. *Diaphanosoma* appears to have decreased since 1964, while *Chydorus* has fluctuated greatly in abundance. Other cladocerans such as *Leptodora* and *Sida* have remained typically low. The calanoid copepod, *Diaptomus,* has been consistently more abundant than *Epischura* and the cyclopoids *Cyclops* and *Mesocyclops*. *Diaptomus* was a subdominant only in 1973 when *Mesocyclops* was most abundant. *Ergasilis* has occurred sporadically, and usually in negligible numbers.

Seasonal variation in zooplankton populations for 1970–1974 is illustrated in Fig. 22. During 1970–1972, zooplanktors exhibited spring and late summer pulses with minimum density occurring in mid-summer. The spring pulse consisted of *Daphnia* and calanoid copepods and the late summer pulse consisted of *Chydorus*. Peak numbers of *Daphnia* in 1973–1974 were markedly reduced from those in 1970 and 1972. *Chydorus*, which was

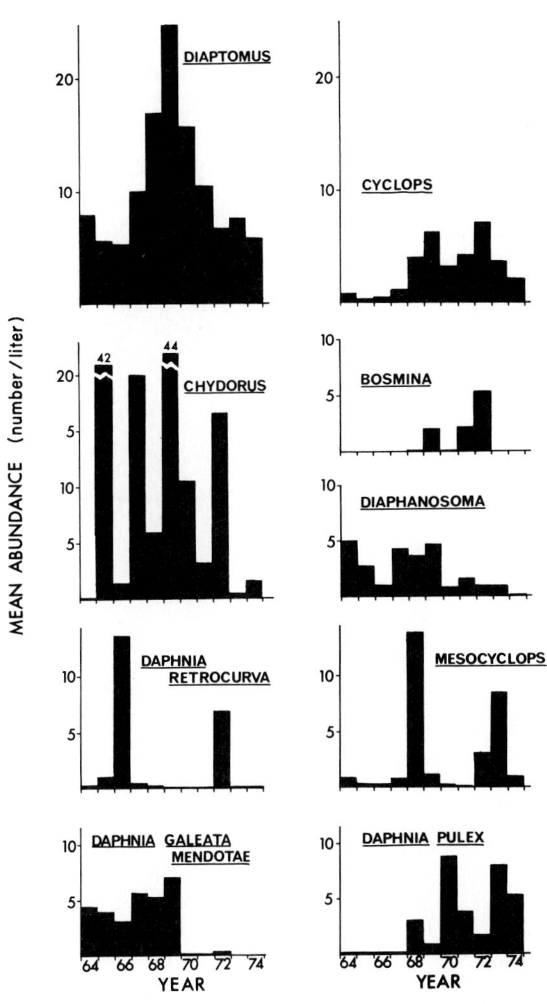

Fig. 21. Annual changes of dominant zooplankton in Oneida Lake, 1964–1974.

abundant in earlier years, was virtually absent in 1973, and calanoid and cyclopoid copepods reached a peak of 16 organisms/liter in late June to early July, about one-half their peak abundance in previous years.

Utilization by Fish. Zooplankton is important as food for Oneida Lake fish. In 1916–1927, Adams and Hankinson (1928) described the composition of zooplankton from a variety of habitats and noted the predominance during the summer of these organisms in the diet of fish. Zooplankton eaten included the copepods *Cyclops, Diaptomus,* and *Epischura* and the cladocerans *Bosmina, Chydorus, Daphnia, Diaphanosoma,* and *Leptodora.* Much later, Tarby (1974) examined the frequency of occurrence of food items in stomachs of the large yellow perch, *Perca flavescens,* from 1965–1971. Zooplankton, particularly *Daphnia,* and chironomids, a benthic insect, predominated in the diet (Table 24) in June of all years. The occurrence of *Daphnia* in the stomachs of perch declined in July and August, paralleling trends observed in mean monthly densities of these food organisms (Fig. 23). Noble (1975) examined the food habits of young yellow perch during 1968–1971. *Daphnia* was the principal food from mid- to late summer, and cyclopoid copepods were usually of secondary importance. Consumption of *Chydorus* corresponded to their population pulse. Noble

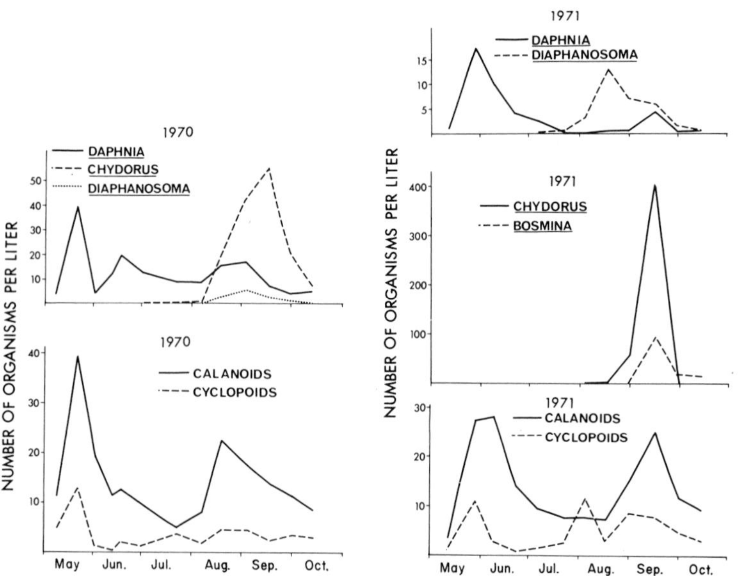

Fig. 22. Seasonal changes of dominant zooplankton in Oneida Lake, 1970–1974.

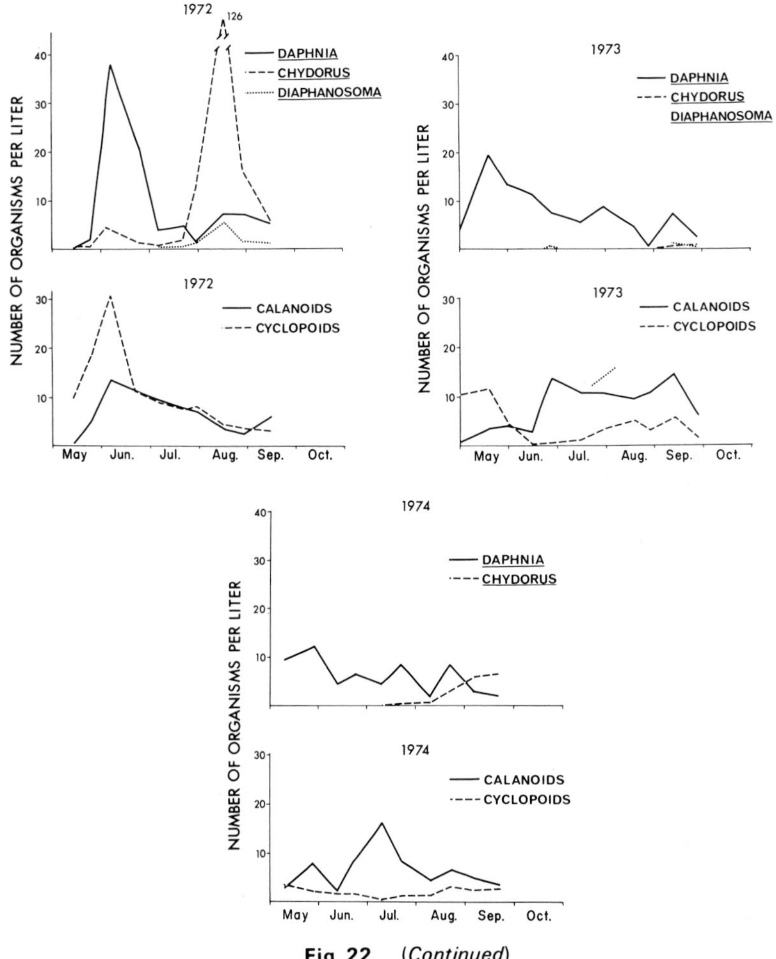

Fig. 22. (*Continued*)

demonstrated that production of young fish per day and mean biomass per day were highly correlated with zooplankton abundance. A decline and near extinction of *Daphnia pulex* occurred in 1968 and 1971 when the biomass of yellow perch was high. Consumption of this cladoceran by the strong-year classes of yellow perch probably contributed to its marked decline. In 1973 and 1974, years of weak-year classes of perch, predation by planktivorous fish was not a significant cause of mortality of *D. pulex* but rather physiological mortality from thermal stress may have strongly influenced the population (Clark, 1975).

Benthic Invertebrates

Populations. Early studies of the bottom fauna of Oneida Lake were concerned primarily with the classification of snails and clams living in the western end of the lake and with their qualitative relationships to fish, vegetation, bottom types, and other major components of the aquatic ecosystem (Baker, 1916a–c, 1918a–c; Pilsbry, 1917, 1918). Aspects of molluscan biology investigated included taxonomy, food habits, natural enemies, microhabitat, and the associated plants and animals in general life zones. Working in water less than 3 m deep, Baker found the molluscan fauna was very rich in both number of individuals and of species and

TABLE 24

Percent Monthly Mean Frequency of Occurrence of Food Items in Stomachs of Yellow Perch Taken in Gillnets, June–August, 1965–1971[a]

Year	Number of stomachs examined	Food items		
		Zooplankton	Chironomidae	Amphipoda
June				
1965	199	6.6	87.6	1.7
1966	184	19.6	59.8	11.1
1967	203	22.4	65.8	10.5
1968	200	62.8	11.6	14.9
1969	139	45.6	29.1	25.3
1970	257	50.8	44.8	2.7
1971	221	88.3	9.2	9.9
July				
1965	146	1.2	54.3	17.3
1966	201	0.0	46.6	31.8
1967	203	2.8	43.9	28.0
1968	193	9.4	21.7	37.0
1969	180	5.3	40.9	54.5
1970	168	57.8	41.3	30.2
1971	98	11.7	20.0	10.0
August				
1965	258	0.0	10.0	1.8
1966	205	3.0	17.9	47.7
1967	187	7.6	30.3	10.6
1968	228	3.9	5.2	7.8
1969	229	3.6	7.3	12.7
1970	144	17.6	0.0	11.8
1971	143	0.0	10.4	10.4

[a] Source: Tarby (1974).

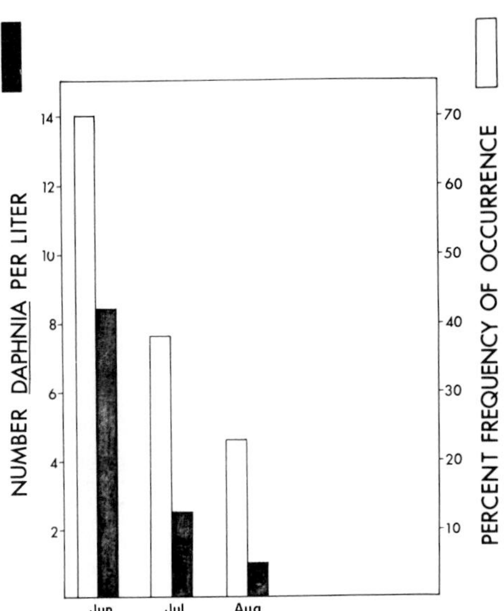

Fig. 23. Average monthly density and percentage frequency of occurrence of *Daphnia* in yellow perch stomachs, June–August 1965–1971.

varieties (over 60), especially in the zone characterized by sandy bottom and moderate densities of rooted aquatic plants.

After Baker's studies, additional species of mollusks invaded Oneida Lake through the Erie Canal (Harman, 1968). Recent studies show that species diversity has declined and the exotic gastropod *Bithynia tentaculata* has become predominant (Harman and Forney, 1970). Displacement of native forms was attributed to a combination of competitive interaction and increasing eutrophy of the lake.

Baker (1916b) recognized the potential value of the quantitative study of benthic organisms as a tool of fish and lake management. Subsequently, in one of the earliest studies of its kind in North America, he determined the density and relative abundance of benthic animals living in Lower South Bay, a 375-ha area on the southwest shore of Oneida Lake (Baker, 1918c; Fig. 24). Considering the entire benthic community, Baker again concluded that sand-bottomed areas of Lower South Bay were the richest and boulder areas the poorest, a finding heavily influenced by the dense populations of mollusks, which were more abundant than all other invertebrates combined.

Baker's detailed descriptions of his sampling methods and sites in Lower South Bay allowed replication of his studies 51 years later in 1967. As in the lake proper, species diversity of gastropods had been reduced in Lower

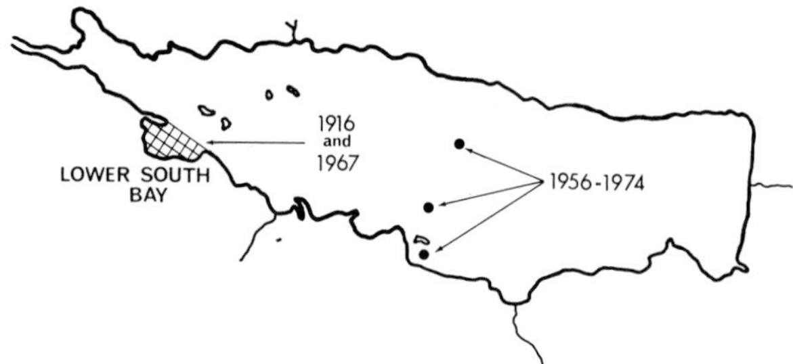

Fig. 24. Map of Oneida Lake showing areas sampled for benthic invertebrates in 1916 and 1967 and 1956–1974.

South Bay (from 23 to 10 species), *Bithynia* had become very abundant, and the total number of snails had declined (Harman and Forney, 1970; Fig. 25). Populations of other invertebrates also changed markedly between 1916 and 1967 on different bottom types in Lower South Bay (Clady, 1975; Fig. 25). The density and frequency of occurrence of Tubificidae and Amphipoda were significantly higher in 1967, while Ephemeroptera, Trichoptera, and Isopoda decreased. Although total numbers of Chironomidae did not change significantly (Fig. 25), there was an apparent shift in species composition to forms more tolerant of adverse environmental conditions (i.e., *Procladius*). Sensitive genera of mayflies also disappeared, while the pre-

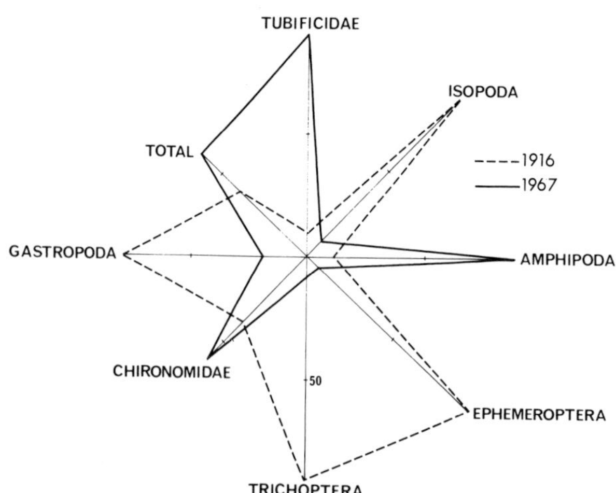

Fig. 25. Percentage of the total number of individuals per site of each benthic invertebrate that occurred in 1916 and 1967 in Lower South Bay.

TABLE 25

Yearly Mean Density (individuals/m²) of Benthic Invertebrates in Oneida and Other Lakes

	Oneida Lake	Great Lakes[a]	Inland Lakes[b]
Amphipoda	4–184	123–300	290–1162
Oligochaeta	0–292	677–5949	9–34
Sphaeriidae	12–130	Tr[c]–438	—
Gastropoda	7–146	40–211	39
Chironomidae	516–1493	73–424	1045–1210
Hexagenia	0–329	1–139	10–343

[a] From Carr and Hiltunen (1965) and Schneider *et al.* (1969).
[b] From Craven and Brown (1969), Schneider (1965), and Tebo (1955).
[c] Tr, trace.

dominant amphipod changed from *Hyalella* to *Gammarus*. Changes in the nonmolluscan populations in Lower South Bay were generally consistent with changes in the populations of snails and clams and were probably due to eutrophication and shoreline development (Clady, 1975).

Benthic animals have been quantified since 1956 in Oneida Lake at three sites which vary from 5.5 to 12.2 m deep (see Fig. 24). These populations

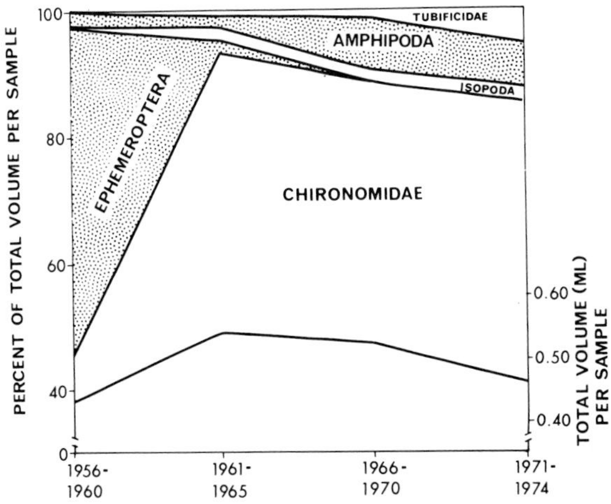

Fig. 26. Percentage of total volume per sample (5-year means based on a mean volume per individual of 0.01 ml for Isopoda and Amphipoda and 0.005 ml for Tubificidae) made up of various organisms at three sites in Oneida Lake, 1956–1974.

appear to be intermediate between productive areas of the Great Lakes and other eutrophic inland lakes. In 1956–1958, there was a yearly average of 256 *Hexagenia limbata* nymphs/m^2, densities which were similar to populations in inland lakes and the open waters of Lake Erie in 1930 (Carr and Hiltunen, 1965). Moderate densities of amphipods, sphaerid clams, and snails also were similar to those found in shallow areas of the Great Lakes (Table 25). Sparse numbers of oligochaetes (primarily Tubificidae) and extremely dense populations of chironomids were more typical of inland lakes, however. In 1959, the first detected acute oxygen depletion occurred throughout most of the deeper area of the lake (Jacobsen, 1966). This and subsequent depletions of oxygen have probably affected the composition of the bottom fauna of Oneida Lake more than any other factor, including introductions of exotic forms. Density of once-predominant *Hexagenia limbata* nymphs began to decline in 1959 (Jacobsen, 1966). The species is now apparently extinct since the last nymph was seen in 1968 (Clady and Hutchinson, 1976; Fig. 26). Chironomids increased from 1956 through 1964 (Jacobsen, 1966) and have stabilized in subsequent years as the predominant organism in terms of numbers and biomass. Abundance of amphipods (*Gammarus*) and isopods (*Asellus*) increased significantly after 1962 (Clady and Hutchinson, 1976). Low-oxygen-tolerant tubificids, which were not present in the bottom samples at the three sites in 1956, were common by 1974 when they constituted nearly 5% of the total volume of the intertebrates (Fig. 26). *Hexagenia* nymphs have apparently been replaced by other organisms (primarily chironomids) since volume per sample has not declined (Fig. 26). Except for isopoda, changes from 1956 through 1974 in the bottom fauna at more open-water areas were similar to those which occurred earlier in the century and perhaps over a longer period of time in Lower South Bay.

Utilization by Fish. Benthic invertebrates have been reported in the stomachs of 20 species of fish and undoubtedly are eaten by most of the other fish in the lake (Table 26). *Hexagenia* nymphs, chironomids, crayfish, and amphipods apparently are most utilized. Consumption is heavy at times, especially by the yellow perch. For example, in 1959 *Hexagenia* occurred in 63% of 277 perch stomachs containing food (Clady and Hutchinson, 1976) and in June, 1965 chironomids were found in 88% of the 121 perch that had fed (Tarby, 1974; see Table 24).

The effect of predation on benthos populations is largely unknown. Clady and Hutchinson (1976) found that the yellow perch slowly altered its feeding habits in response to changes in bottom fauna over 15 years. They concluded, however, that the perches' preferences for *Hexagenia* nymphs coupled with lags in switching to consumption of other organisms may have

TABLE 26

Occurrence of Benthic Invertebrates in the Stomachs of Fish from Oneida Lake

Species of fish	Ref.[a]	Mollusca	Chironomidae	Trichoptera	Ephemeroptera	Odonata	Crayfish	Amphipoda	Isopoda
Northern pike	1			X					
Carp	10	X	X	X			X	X	
Golden shiner	3	X	X	X					
Spottail shiner	1				X				
Fallfish	3						X		
White sucker	3	X	X	X	X	X			
Shorthead redhorse	3		X		X				
Yellow bullhead	3	X	X		X				
Brown bullhead	3	X				X		X	
Burbot	3						X		
Brook silverside	3		X					X	
White perch	2		X		X			X	
White bass	7,11		X		X		X	X	
Rock bass	3,11	X			X	X	X	X	
Pumpkinseed	1,3	X	X	X		X	X	X	X
Smallmouth bass	6,11	X					X	X	
Largemouth bass	1						X		
Tessellated darter	3	X	X					X	X
Yellow perch	3,4,11,12	X	X	X	X	X	X	X	X
Logperch	3	X	X		X	X	X	X	
Walleye	5,8,9,11	X	X	X	X			X	
Total occurrences		12	13	7	10	6	10	12	3

[a] Key to references: 1, Adams and Hankinson (1928); 2, Alsop and Forney (1962); 3, Baker (1916b); 4, Clady and Hutchinson (1976); 5, Forney (1966); 6, Forney (1972); 7, Forney and Taylor (1963); 8, Houde (1967); 9, Raney and Lachner (1943); 10, Smallwood and Struthers (1928); 11, Stone and Pasko (1946); 12, Tarby (1974).

contributed to the failure of the mayfly population to recover from reduction by intermittent oxygen depletion. Utilization of post-*Hexagenia* populations of benthos by walleyes and yellow perch appears to be largely opportunistic (as when a species is exceptionally abundant or vulnerable, such as during emergence of chironomids; Tarby, 1974) or as an alternate source when preferred foods (fish and zooplankton) are scarce.

Fish

History of the Fisheries. The fish of Oneida Lake have long been important as a source of food and recreation. Before 1800, native Americans took large numbers of Atlantic salmon (*Salmo salar*) and American eels (*Anguilla rostrata*) with the use of spears and weirs (Adams and Hankinson, 1928). Little is known of the commercial fishery which supplanted the native fishing during the 1800's. DeWitt Clinton in July, 1810 observed settlers seining salmon which were shipped to the New York City market (Campbell, 1849) but the salmon population declined rapidly (DeKay, 1842). Eels remained abundant and were the most important commercial fish in 1915 (Adams and Hankinson, 1916). Cisco (*Coregonus artedii*), channel catfish (*Ictalurus punctatus*), and several percids, catostomids, esocids and centrarchids contributed to the early commercial fishery but their relative importance in the catch is uncertain.

Cobb (1905) estimated the commercial harvest was 22,700 kg in 1894 and 99,500 kg in 1895, excluding eels taken in weirs. Harvest may have increased in subsequent years since the number of trapnets licensed for Oneida Lake rose from 69 in 1895 (New York State Commissioners of Fisheries, Game and Forest, 1896) to 123 in 1899 (New York State Commissioners of Fisheries, Game and Forest, 1899). Netting was prohibited after 1900, although the sale of most species caught by hook and line was permitted. Despite legislation, netting continued and Cobb (1905) noted that the value of illegal fishing gear seized in 1901 exceeded the value of the whole investment in the legal commercial fisheries of inland New York. This would suggest a much larger fishery than indicated by the number of nets licensed in earlier years. C. F. Davison, fish dealer and former commercial fisherman, estimated the total catch including illegal fishing in 1928 was 227,000–273,000 kg or 7.8–9.4 kg/ha (Adams and Hankinson, 1928). Although conflicts between anglers and commercial fishermen continued into the 1930's, sport fishing gradually supplanted commercial fishing.

Smallmouth bass (*Micropterus dolomieui*), largemouth bass (*Micropterus salmoides*), walleye (*Stizostedion vitreum*), yellow perch (*Perca flavescens*), and chain pickerel (*Esox niger*) were considered the most valuable sport fish by Adams and Hankinson (1928) and, with the exception of the chain pickerel, have remained the most important species. During the 1957–1959 fishing seasons, the average harvest was 205,000 walleye, 198,000 yellow perch, and 22,000 smallmouth bass per year (Grosslein, 1961). These three species made up 77–84% of the total catch. The walleye and bass harvested in these years averaged 0.57 kg, the perch 0.23 kg (Cornell Biological Field Station, unpublished data). Assuming other species averaged 0.23 kg each, the annual sport fishing catch was 197,000 kg or 9.5 kg/ha in 1957–1959.

Changes in the Fish Populations. Beginning with the extinction of the salmon the native fish population has been substantially altered. The eel, walleye, pickerel, and northern pike (*Esox lucius*) were important piscivorous species in the early 1900's (Adams and Hankinson, 1928). Of these only the walleye has remained abundant. Reported catch of eels in weirs set in the Oneida River was 19,000 kg in 1913 (Adams and Hankinson, 1928) but the population declined in subsequent years and the fishery was abandoned (Dence, 1964). Cobb (1905) noted chain pickerel and walleyes were caught in about equal numbers by ice fishermen in 1896. Although the chain pickerel occupied the inshore area its loss had little detectable effect on the structure of the nearshore population.

Yellow perch was the predominant species taken by shore seining in 1916–1927 (Adams and Hankinson, 1928), 1946 (Stone and Pasko, 1946), and 1960–1970 (Clady, 1976a). The frequency of capture of 56 other species suggested that populations had remained fairly stable except for the silvery minnow (*Hybognathus nuchalis*), spottail shiner (*Notropis hudsonius*), and trout perch (*Percopsis omiscomaycus*) which all exhibited marked declines. Failure of prey species to respond to a decrease in chain pickerel, eel, and northern pike may indicate a proportional increase in biomass of the walleye. Young walleyes were caught in seines more frequently in 1960–1970 than in 1916–1927 which suggests an increase in adult stocks. The number of female walleyes netted and processed for eggs by the Oneida Hatchery increased from an average of 1380 in 1912–1920 to 7186 in 1932–1940, while the number of persons employed during the spawning run decreased from 27 to 19 (New York State Department of Environment and Conservation, unpublished records). Expansion of the walleye population during a period when abundance of other piscivores diminished is consistent with evidence that prey populations remained fairly stable.

Invaders have fared rather poorly. Freshwater drum (*Aplodinotus grunniens*) were first observed in the early 1950's (Dence and Jackson, 1959). A few large drum were caught by anglers in subsequent years but young did not appear in seine or trawl catches until 1973. White perch (*Morone americana*) probably entered Oneida Lake prior to 1950 (Scott and Christie, 1963) and a strong-year class was produced in 1954 (Alsop and Forney, 1962). The catch of age 1 and older white perch peaked in 1960 and declined substantially in later years (Cornell Biological Field Station, unpublished data). The presence of gizzard shad (*Dorosoma cepedianum*) in Oneida Lake had not been reported prior to 1954 when vast schools of young attracted public attention (Dence, 1964). The explosion was brief and few gizzard shad were detected after 1957. However, the high density of gizzard shad in 1954 may have briefly disrupted predator–prey relations. Exceptionally large year classes of white perch, yellow perch white bass, *Morone chrysops,* smallmouth bass, and walleye developed in 1954 (Alsop and Forney, 1962; Forney, 1967, 1972; Forney and Taylor, 1963).

Cultural Influences. Agriculture and industrialization both played an important role in the decline of several native species. Increased siltation of streams following deforestation and cultivation probably destroyed many salmon spawning grounds. Later, dams built for water power blocked spawning runs and contributed to the early disappearance of salmon in Oneida and the entire Lake Ontario basin (Christie, 1972). The eel population dwindled soon after completion of the Barge Canal. Whether dams and locks on the Oswego and Oneida Rivers were an effective barrier to migration elvers is unknown, but the sequence of events is suggestive.

Changes in water levels following construction of the Barge Canal probably contributed to the demise of the chain pickerel and led to a reduction in northern pike. Manipulation of water levels in the spring which were designed to reduce property damage also reduced flooding of marshes where both species spawned. The final blow was the drainage of vast marshes for agriculture and the filling of low-lying areas of lake shore for cottage development.

Low oxygen during periods of temporary thermal stratification has had a destabilizing effect on fish populations since 1959. Distribution of trout perch shifted from depths of over 10 m to a zone 6–10 m deep, apparently in response to reoccurring periods of low oxygen (Clady, 1977). Displacement may have exposed trout perch to more intense competition or predation since the catch in trawls has decreased substantially. Trawl catches of tessellated darters (*Etheostoma olmstedi*) and logperch (*Percina caprodes*) also decreased while the catch in shore seining remained stable, indicating their range was restricted. Low oxygen in deeper, slightly cooler water has also imposed additional stress on the remaining cisco population which is living at temperatures near their upper tolerance level (Smith, 1972).

Population Dynamics. The walleye has been the most abundant piscivorous species in Oneida Lake during the past two decades and the standing crop has averaged about 22 kg/ha (Forney, 1967). Walleyes spawn along much of the shoreline and in several tributaries, but the distribution of larvae following hatching in 1960–1967 suggested most young originated from stocking (Houde and Forney, 1970). Subsequent measurements of larval abundance during a period of alternate-year stocking confirmed the fact that most larvae were of hatchery origin (Forney, 1975). Larvae are limnetic for 4 to 6 weeks after hatching and concentrate in bays (Houde, 1969). Copepods are important in the diet of larval walleye (Houde, 1967) but fish predominate after young become demersal at a length of 35 to 45 mm (Forney, 1966).

Young-of-the-year yellow perch are the most important item in the diet of yearling and older walleyes and annual growth increments are directly correlated with perch abundance (Forney, 1965). Walleyes begin feeding on young perch in late June when the year class averages 20 mm and predation continues into the following summer (Forney, 1974). Consequently, production attained by each year class of yellow perch during their first summer largely determines the amount of forage available to the walleye population.

Factors regulating production of young perch were examined in several studies conducted between 1961 and 1974. Differences in number of eggs spawned and in the survival of eggs and prolarvae due to meteorological conditions cause a nine-fold variation in size of year classes by the time young attain a length of 88 mm (Clady, 1976b; Noble, 1968a). Year classes which were initially abundant usually remain dominant through the first year and attain the highest biomass and production (Forney, 1971). Annual differences in the growth rates of young perch increased variability in biomass and production. Young perch selected *Daphnia pulex* in preference to smaller daphnids and growth was rapid in years when *D. pulex* remained abundant through the summer (Lin, 1975). Yellow perch consumed 17% of the *Daphnia pulex* standing crop per day in 1968 when perch density was high but less than 2% of the daphnids in 1969 and 1970 when the perch biomass was low.

The catch of young perch in trawls usually declines from 1000–3000 per haul in August to 1–100 by the following May, which suggests a high mortality rate (Forney, 1971). Most mortality of young perch is probably attributable to predation by walleyes although cannibalism by adult perch is significant in some years (Tarby, 1974). As the density of young perch declines in late summer, walleyes prey on other species including young walleyes.

In years when young perch are scarce, adult walleyes can seriously deplete year classes of young walleyes (Chevalier, 1973). Relative survival of the 1961–1970 year classes of walleyes between August of the first and second year of life was closely correlated with the biomass of young perch in late summer (Forney, 1974). This suggests recruitment of walleyes is a function of perch production and interaction between prey abundance and cannibalism may adjust the biomass of walleyes to some mean level of forage production.

Aquatic Macrophytes

Abundance and composition of aquatic macrophytes has changed markedly since the early 1900's when dense stands of emergent vegetation occupied much of the shoreline (Baker, 1916b, 1918c; Adams and Hankinson, 1928). In most bays, bands of bullrush (*Scirpus*) and water willow (*Dianthera*) extended 100 to 300 m from shore while scattered stands lined the more exposed shoreline. Only remnants of the once vast bed of emergents remain and these are largely restricted to the littoral area bordering the islands. In 1916, the major emergent species in Lower South Bay were *Justica americana, Scirpus acutus, S. americana, S. smithii,* and *Pontederia cordata* (Baker, 1916b). Noble (1968b) found these were still the dominants in 1967 but stands were sparse and limited to a few relatively undisturbed sections of Lower South Bay. Shoreline development and manipulation of water levels probably contributed to the demise of emergents but the exact cause is obscure.

Rooted submergent macrophytes have remained moderately abundant although the density has varied. In the 1940's, Burdick and Lipschuetz (1946) noted a sharp decline in the abundance of higher aquatic plants which they attributed, in part, to increased algal concentrations. Lehne (1965) observed that rooted aquatics became more abundant in 1965 and extensive mats of *Heteranthera dubia* reached "nuisance proportions" in some areas, athough this macrophyte had not been previously reported among the dominants in Oneida Lake. Noble (1968b, 1969) examined the distribution of aquatic macrophytes of Lower South Bay in 1967 and compared his findings with those of Baker (1916b) (Table 27). The diversity of submergents had increased since 1916 and the species composition of submergent vegetation had changed from principally *Potamogeton* spp. to an abundance of *Heteranthera dubia* and *Ceratophyllum demersum.*

The dynamic nature of the aquatic macrophyte community is illustrated by recent changes in distribution and abundance. Beds of aquatic plants which reached the surface and were sufficiently dense to restrict or prevent the operation of an electrofishing boat in September–October 1965, 1967, and 1972 were qualitatively outlined by J. Forney (unpublished data) (Fig. 27). Plants were somewhat more abundant and widely distributed in 1967 than in 1965 but a dramatic decline in density and extent of aquatic macrophytes was apparent in 1972. Inflow of allochthonous materials following Tropical Storm Agnes increased turbidity and may have

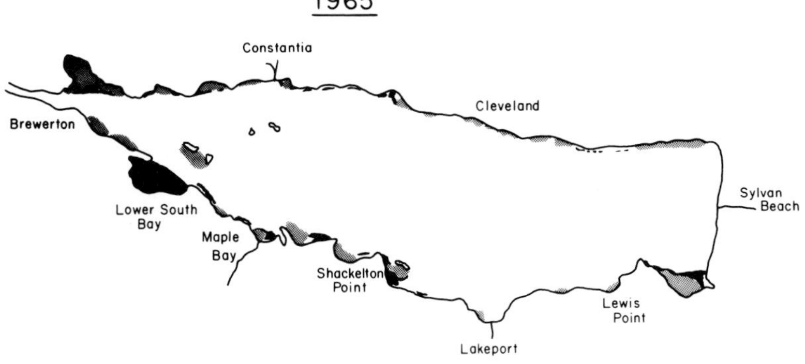

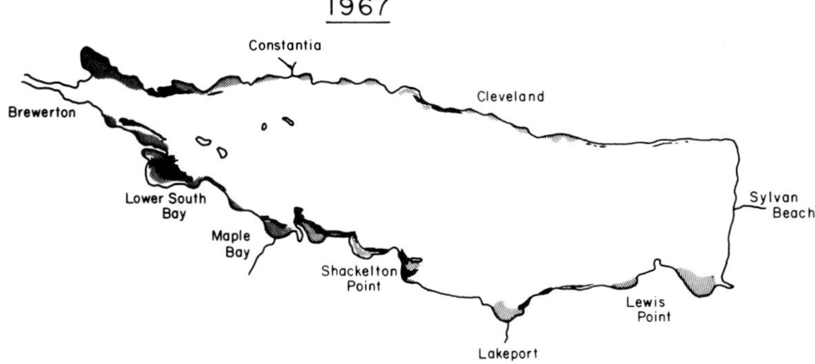

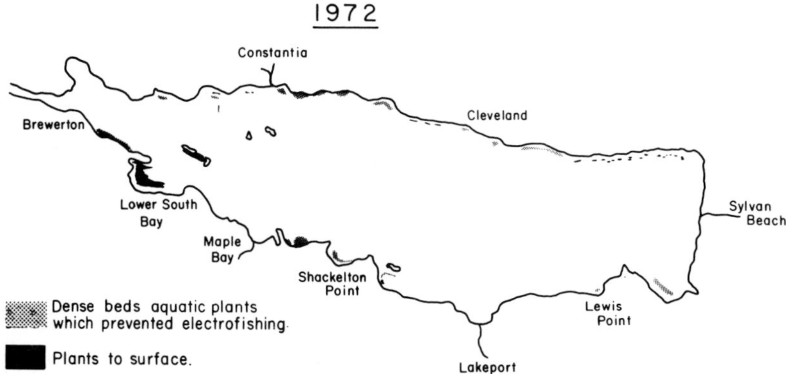

Fig. 27. Distribution of aquatic macrophytes in Oneida Lake, 1965, 1967, and 1972.

TABLE 27

Frequency of Occurrence of Submerged Plant Species in Lower South Bay at 134 Sampling Stations in 1916 and 1967

Species	1916	1967
Vallisneria americana	30 (22.4)[a]	81 (60.4)
Heteranthera dubia	0 (0.0)	76 (56.7)
Anacharis canadensis	16 (11.9)	67 (50.0)
Myriophyllum verticillatum	28 (20.9)	53 (39.6)
Chara sp.	1 (0.7)	47 (35.1)
Potamogeton spp.[b]	61 (45.5)	44 (32.8)
Najas flexilis	22 (16.4)	43 (32.1)
Ceratophyllum demersum	0 (0.0)	38 (28.4)
Lemna trisulca	19 (14.2)	10 (7.5)
Nitella sp.	3 (2.2)	0 (0.0)
Myriophyllum scabratum	1 (0.7)	0 (0.0)
Utricularia vulgaris	1 (0.7)	0 (0.0)

[a] Percent occurrences in parentheses.

[b] In 1916, ten species were recorded—*P. natans, P. perfoliatus, P. foliosus, P. friesii, P. interruptus, P. lucens, P. praelongus, P. richardsoni, P. robbinsii,* and *P. zosteriformis.* In 1967, seven species were recorded—*P. natans, P. perfoliatus, P. foliosus, P. richardsoni, P. crispis, P. pectinatus,* and *P. zosteriformis.*

restricted the degree of rooted plant growth. Recovery was apparently rapid since complaints of excessive weed growth in 1974 led to a state-supported program of aquatic plant control in 1975.

DISCUSSION

Trophic Status

During its recorded history, Oneida Lake has apparently been a highly productive body of water. In 1809, James Fenimore Cooper described the lake as a "broad dark colored body of water, unwholesome to drink and strangely blended with dark particles the boatmen called lake blossoms." Present-day descriptions by casual observers are often similar (Kooyoomjian and Clesceri, 1972). However, the objectionable blooms of algae, die-offs of fish, and associated problems are usually, and probably erroneously, attributed to "pollution" rather than the natural productivity of the lake.

Species composition of the phytoplankton and benthic community are characteristic of a productive lake. Nuisance blooms of *Anabaena,*

Aphanizomenon, Microcystis, and *Gloeotrichia* frequently develop during late summer and fall. These genera are typical of Norrviken (Ahlgren, 1970), the western basin of Lake Erie (Vollenweider *et al.*, 1974), Gravenhurst Bay (Michalski *et al.*, 1973), and other enriched waters. *Melosira,* a diatom characteristic of eutrophic lakes (Davis, 1964) is often abundant. The benthos is dominated by species which tolerate low oxygen. Chironomids are numerous while oligochaetes and the fingernail clam (*Pisidium*) are common in deeper waters. Inshore, the pulmonate snail (*Bithynia tentaculata*) is abundant and is widely distributed in highly fertile waters. In general, biological parameters establish Oneida Lake as a productive system but its position in the continuum from oligotrophy to eutrophy can be best examined by comparison with other lakes.

Secchi disk transparency is one of the most meaningful parameters used in defining water quality and has been employed by Rawson (1960) and others as an index of trophic state. Figure 28 describes the relationship between chlorophyll *a* and transparency for the Muskoka Lakes, several recreational lakes located in southern Ontario, Canada, four Finger Lakes

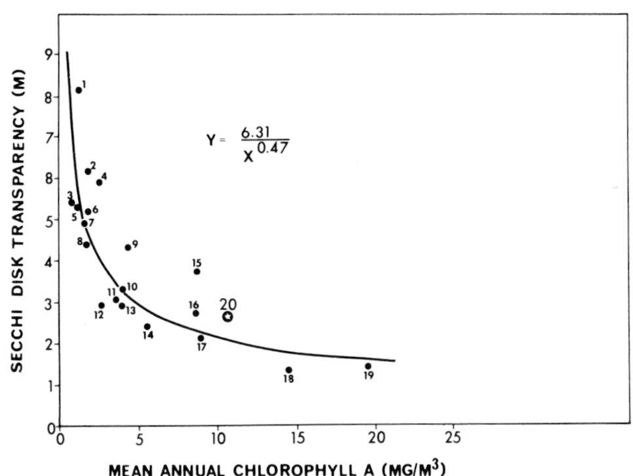

Fig. 28. The relationship between Secchi disk transparency and chlorophyll *a* for Oneida Lake and other north-temperate lakes. All data are yearly means. (1) Joseph Lake, 1969–1970; (2) Rosseau Lake, 1969–1970; (3) Skaneateles Lake, 1973; (4) Little Joseph Lake, 1969–1970; (5) Skaneateles Lake, 1972; (6) Skeleton Bay, 1969–1970; (7) Dudley Bay, 1969–1970; (8) Muskoka Lake, 1969–1970; (9) Conesus Lake, 1973; (10) Sammamish Lake, 1973; (11) Owasco Lake, 1973; (12) Owasco Lake, 1972; (13) Hemlock Lake, 1973; (14) Hemlock Lake, 1972; (15) Conesus Lake, 1972; (16) Gravenhurst Bay, 1969–1970; (17) Sturgeon Lake, 1971; (18) Lake Erie—Western Basin, 1970–1971; (19) Bay of Quinte, 1967; (20) Oneida Lake, 1975.

of New York State, and Oneida Lake. Oneida, positioned in the eutrophic section of the curve, is characterized by high chlorophyll *a* concentrations and moderate to low transparency.

The eutrophic character of Oneida Lake reflects both morphometric and edaphic features. The lake basin is shallow and aligned with the prevailing westerly winds which results in effective regeneration of nutrients from bottom sediments. The lake plain immediately surrounding the lake and soils of the southern half of the basin are fertile and rich in lime. Streams draining these soils provide the nutrients needed to support high primary production. In Fig. 29, mean depth plotted against standing crop as measured by chlorophyll *a* forms an inverse hyperbolic relationship; Rawson (1955) would have considered Oneida Lake as morphometrically eutrophic due to its shallowness and high standing crop of phytoplankton.

Vollenweider (1968) and Brezonik *et al.* (1969) have established that increased nutrient loading characteristically results in an increased phytoplankton standing crop. Oglesby and Schaffner (1977) have calculated phosphorus loading for Oneida Lake to be 0.72 gm P/m^2/year. Later, Vollenweider (1973) and Dillon (1974) modified this relationship and found that the trophic condition of a lake could be broadly delineated into areas of eutrophy, mesotrophy, and oligotrophy by characterizing a lake according to phosphorus loading, mean depth, and water renewal time. According

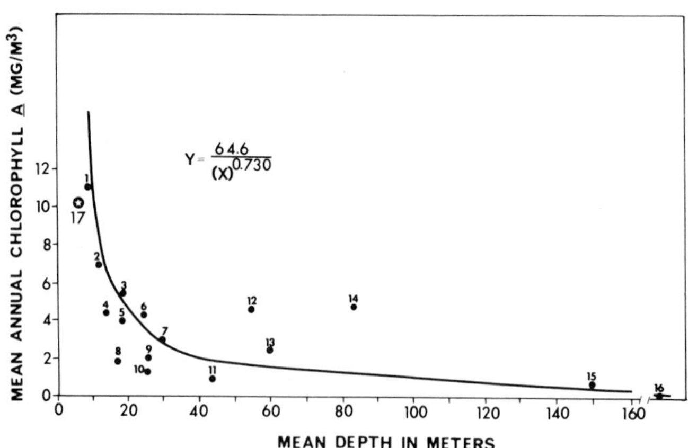

Fig. 29. Relationship between mean depth and mean annual chlorophyll *a* for Oneida Lake and a number of north-temperate lakes. (1) Lake Erie—Western Basin; (2) Conesus Lake; (3) Lake Erie—Central Basin; (4) Hemlock Lake; (5) Lake Sammamish; (6) Lake Erie—Eastern Basin; (7) Owasco Lake; (8) Lake Muskoka; (9) Lake Rosseau; (10) Lake Joseph; (11) Skaneateles Lake; (12) Cayuga Lake; (13) Lake Huron; (14) Lake Ontario; (15) Lake Superior; (16) Crater Lake; (17) Oneida Lake.

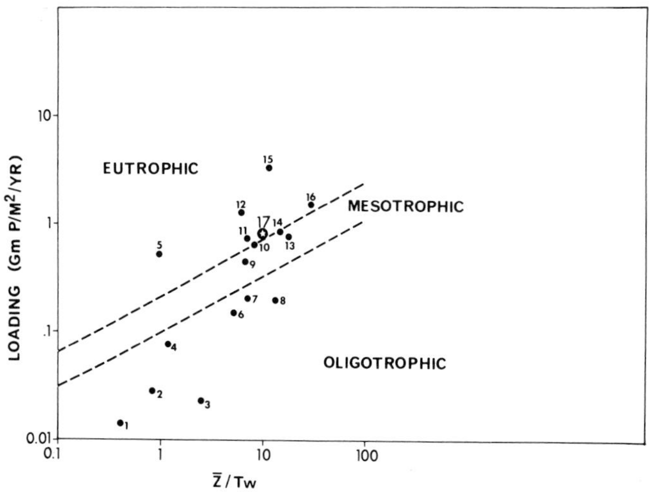

Fig. 30. Relationship of total phosphorus specific loadings (gm P/m²/year) to mean depth divided by the water retention time for Oneida Lake and other north-temperate lakes. (1) Lake Tahoe; (2) Lake Superior; (3) Skaneateles Lake; (4) Lake Joseph; (5) Lake Mendota; (6) Lake Rosseau; (7) Little Joseph Bay; (8) Dudley Bay; (9) Hemlock Lake; (10) Conesus Lake; (11) Lake Ontario; (12) Lake Erie; (13) Muskoka Lake; (14) Owasco Lake; (15) Lake Washington; (16) Lake Zurich; (17) Oneida Lake.

to this scheme, Oneida Lake would lie on the borderline between mesotrophy and eutrophy (Fig. 30).

Edmondson (1961) found that the potential concentration of phosphorus, calculated by dividing the areal income of phosphorus (gm P/m²) by the mean depth or gm P/m³, was closely related to algal standing crop. Using this relationship, Oneida Lake was compared with several Finger Lakes and lakes plotted by Bachmann and Jones (1974) (Fig. 31). Oneida Lake falls close to other lakes which have been considered eutrophic; however, with increased nutrient reduction Oneida Lake could easily be considered as meso-eutrophic.

The Oneida Lake basin has been extensively altered by agriculture and urbanization. The impact of these cultural changes on the trophic status of the lake is uncertain since the most significant shifts in land use largely predate quantitative studies of nutrient input. Investigations of the benthic community which span a period of 50 years tend to support the view that the lake has become more eutrophic. Ephemeroptera and Trichoptera were well represented in the near-shore benthic community of Lower South Bay in 1916. By 1967, few representatives of these families were present but tubificids exhibited a significant increase. Replacement of a diverse mollus-

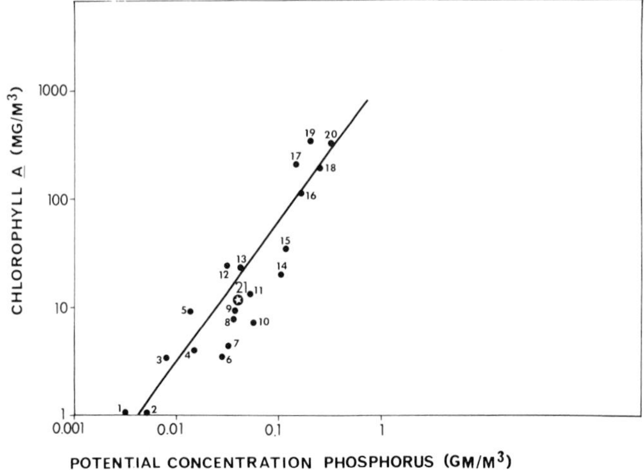

Fig. 31. Relationship between chlorophyll *a* (mg/m^3) and potential phosphorus concentrations (gm/m^3) for Oneida lake and other lakes (Bachman and Jones, 1974). (1) Clear Lake, Canada; (2) Skaneateles Lake; (3) Lake West Okoboji, Iowa, 1972; (4) Lake West Okoboji, Iowa, 1973; (5) Lake Washington, Washington, 1967–70; (6) Owasco Lake; (7) Hemlock Lake; (8) Lake Sammamish, Washington, 1970; (9) Lake Sabasticook, Maine; (10) Conesus Lake; (11) Lake Sammamish, Washington, 1965; (12) Lake Washington, Washington, 1957–1966; (13) Lower Lake Minnetonka, Minnesota; (14) Lake Mendota, Wisconsin; (15) Lake 227, Canada; (16) Lake East Okoboji, Iowa, 1973; (17) Lower Gar Lake, Iowa, 1973; (18) Lake East Okoboji, Iowa, 1971; (19) Lower Gar Lake, Iowa, 1971; (20) Lake Norrviken, Sweden; (21) Oneida Lake.

can fauna in 1916 by a dense population of *Bithynia tentaculata* is also suggestive of enrichment. The collapse of the abundant burrowing (*Hexagenia limbata*) mayfly population in deep water and its replacement by chironomids after 1959 reinforces the concept of progressive eutrophication. However, when viewed individually these events are less convincing.

Photographs show that the shoreline of Lower South Bay was largely undeveloped and bordered by a wide band of emergent vegetation (Baker, 1916b; Adams and Hankinson, 1928). By 1967 most of the low-lying shore areas had been filled and emergent vegetation was sparse. These physical changes alone may have generated perceptible changes in the benthic community. Eutrophication of the inshore area may have contributed to the extinction of many native molluscans between 1916 and 1967 but competition with an exploding *Bithynia* population seems equally plausible. This European species probably entered Oneida Lake from Lake Ontario. Its restricted distribution near the outlet of Oneida Lake in 1916 suggests that the full effect of the invasion was not evident at this time. The disappearance of *Hexagenia* in 1959 as a result of reoccurring oxygen depletions in the deeper waters of the lake is the most compelling evidence supporting cultural eutrophication. However, the sewered population was higher in 1959 than in 1931 before the city of Rome diverted its sewage to the Mohawk River so the apparent abrupt increase in productivity was not paralleled by a sudden increase in population or agricultural activity—the events usually associated with cultural enrichment. Finally, there is no evidence that changes in macrophytes, bottom fauna, or water quality have influenced the community structure of the fish, although low oxygen during extended calm periods has had a destabilizing effect on fish populations in recent years.

Special Problems

Oneida Lake is a large, shallow, productive body of water which has undergone a number of chemical and biological changes. Cultural influences particularly have had considerable impact on changes in Oneida Lake. An examination of the nutrient and chlorophyll content of the lake sediments deposited over the past 300 years would provide insights into human impacts on changes in lake productivity.

A substantial body of knowledge on both the dynamics of the Oneida Lake ecosystem and its drainage basin has been accumulated during the present century. Considerable data are available on the higher trophic levels and on such descriptive properties as geology, morphology, and major ion concentrations. A thorough description of the phytoplankton and further studies of nutrients are needed to determine if phosphorus is the primary factor controlling phytoplankton production.

Nutrient control has become a prime goal of water quality management in the Oneida Lake basin but further reduction in nutrient loading may alter the food web since the lake approaches mesotrophy in the type of fish community and potential phosphorus concentration. Presently, little information exists on the energy transfer in complex aquatic communities and on causes and effects relating to food-chain dynamics in large eutrophic lakes. A quantitative assessment of the primary food chain of Oneida Lake would be valuable in assessing the consequences of eutrophication or nutrient reduction since this food chain also characterizes Lake Erie and many other inland lakes of the Northeast and findings would have general application to water management problems. Manipulation of food webs in eutrophic lakes may be a viable alternative to nutrient control. These and other alternatives should be established before management schemes advocating nutrient reduction are finalized.

SUMMARY

Oneida Lake, the largest water body wholly within New York State, with a surface area of 206.7 km², is located 18 km northeast of Syracuse on the low Ontario Lake Plain of central New York. Oriented in a roughly west–northwest pattern, Oneida is one of the remnants of the once more extensive Lake Iroquois. This large, glacially impounded lake eventually drained into the St. Lawrence Valley about 10,000 years ago leaving behind many undrained depressions, one of which was Oneida Lake.

Nearly 100,000 people inhabit the Oneida Lake Basin with nearly 85% of these in the southern half. Oneida Lake is a major recreational resource for the region and is utilized extensively for boating and sport fishing. Seventy-six percent of the drainage basin north of the lake is forest and wetland. Only 15% is used for agriculture and less than 1% for residential and commercial purposes. In Oneida's southern basin about 56% of the land is farmed with an additional 33% in forests and wetlands and 5% for commercial usages.

The bedrock of the basin consists mostly of shale and limestone to the south of the lake and erosion-resistant sandstone to the north. The lake plain surrounding the lake consists of shale and sandstone overlain by unconsolidated glacial deposits. The soils of the southern half of the basin are highly buffered and consist of parent material rich in lime. These soils to the north contain less organic material, are shallow, and more acid.

The chemistry of Oneida Lake is strongly influenced by nutrient-rich streams that flow from the south over the Onondaga limestone and through the fertile and highly populated Ontario Lake Plain. In marked contrast, Oneida is fed from the north by nutrient-poor streams from the Tug Hill Upland, a plateau composed of sandstone highly resistant to erosion. Forty-three percent of the drainage basin north of the lake contributes 67% of the water and only 18% of the total dissolved solids entering the lake. On the other hand, 51% of the drainage basin south of the lake contributes 33% of the water and 82% of the total dissolved solids.

The climate is of the humid continental type with warm summers and cold, snowy winters. Maximum precipitation occurs in the Tug Hill Upland and results from cooling of moisture-laden air coming off Lake Ontario. The least amount of precipitation falls in the southwestern corner of the basin.

Oneida Lake is shallow, with a mean depth of 6.8 m and a maximum depth of 16.8 m. Numerous shoals are present and about 26% of the lake bottom is shallower than 4.3 m. The lake is 33.6 km long, 8.8 km wide, and has a surface area of 206.7 km². The volume is 140×10^7 m³, and the lake has a hydraulic retention time of 235 days.

Oneida Lake is usually homothermal but, during the summer, brief periods of temperature and oxygen stratification may develop. Permanent thermal stratification is prevented by wind-generated wave action since the lake is fully exposed to the prevailing winds. Typically, the water temperature increases rapidly during May and June and reaches a maximum of approximately 25°C in mid-summer. A clinograde oxygen curve is characteristic during the summer months; dissolved oxygen levels decrease markedly, as does the percent saturation in the bottom waters, from June through August.

Seasonal changes in Secchi disk transparency would indicate that maximum light penetration occurs in early June. A dramatic decline in transparency is evident in mid- to late summer presumably due to increased algal production.

The lake is a well-buffered, carbonate–bicarbonate system with total alkalinity on the order of 80 mg/liter as $CaCO_3$. The pH is usually above 8 and has been reported as low as 6.8 and as high as 9.3; the pH is highest in summer and lowest in winter. Calcium is the dominant as well as the most variable cation while bicarbonate is the most common anion. Sodium and chloride concentrations have nearly doubled in the last 20 years.

Phosphorus was present in excess of metabolic needs of phytoplankton in 1961 and was not considered the controlling factor limiting algal production during 1967–1969. However, recent evidence would suggest that this element may be scarce enought to limit primary production. Seasonal and spatial variations in total phosphorus are marked. Maximum concentrations occur during the spring with a decline during the summer months and in-shore sites are usually higher in total phosphorus than offshore sites. Concentrations of nitrogen compounds appear to vary from year to year and nitrate nitrogen exhibits a summer minimum and a winter maximum.

Blue-green algae have been a dominant component of the summer phytoplankton of Oneida Lake since early in the present century. The four most common genera of blue-green algae in Oneida Lake are *Anabaena, Aphanizomenon, Microcystis,* and *Gleotrichia.* Diatoms are the second most abundant algal group and occur mainly during the fall, winter, and spring.

Pronounced annual differences in the composition and density of the major zooplanktors during the summer has occurred since 1964. Prior to 1968, *Daphnia pulex* was seldom observed in Oneida lake and *D. galeata* and *D. retrocurva* were the dominant species, but in recent years *D. pulex* has been the dominant daphnid. Copepods have usually predominated in numbers while the cladocerans have constituted most of the biomass. The latter group is an important food resource for Oneida Lake fish.

Since the early 1900's, changes have occurred in the benthic fauna. The mayfly (*Hexagenia limbata*) previously was the predominant benthic organism and at times emerged from Oneida Lake in such great numbers as to be considered a nuisance. Between 1954 and 1966, there was a dramatic decline in *Hexagenia* which was accompanied and followed by increases in chironomids, tubificids, and amphipods. The diversity of the molluscan fauna has decreased since 1915, suggestive of Oneida's enriched conditions.

The fish of Oneida Lake have long been important as a source of food and recreation. Cultural influences have played an important role in changes in native fish populations. The lake is managed to maximize warm-water fish production, particularly walleye and adult perch which are the most abundant piscivorous species and make up a high proportion of the total production. The principal food chain in Oneida Lake is:

phytoplankton → zooplankton → yellow perch → walleye

The abundance and composition of aquatic macrophytes has changed since the early 1900's. The diversity of submergents has increased since 1916 and species composition has changed from principally *Potamogeton* spp. to an abundance of *Heteranthera dubia* and *Ceratophyllum demersum*. Recent changes in distribution and abundance indicate the dynamic nature of the aquatic macrophyte community. Dense growth of rooted macrophytes in 1974 has led to a state-sponsored program of aquatic plant control in Oneida Lake.

During its recorded history, Oneida Lake has apparently been highly productive. A comparison of several trophic state indicators in Oneida Lake with other north-temperate lakes would categorize the lake as eutrophic. Diagnostic variables indicating eutrophy are low mean depth, decreases in oxygen in the bottom waters during summer, increasing predominance of tolerant chironomids and tubificids in the benthic fauna, and nuisance blooms of blue-green algae during the summer.

REFERENCES

Adams, C. C., and Hankinson, T. L. (1916). Notes on Oneida Lake fish and fisheries. *Trans. Am. Fish. Soc.* **45,** 155–169.

Adams, C. C., and Hankinson, T. L. (1928). The ecology and economics of Oneida Lake fish. *N.Y. State Coll. For., Roosevelt Wildl. Ann.* **1,** 235–548.

Ahlgren, G. (1970). Limnological studies of Lake Norrviken, an eutrophied Swedish lake. 2. Phytoplankton and its production. *Schweiz. Z. Hydrol.* **32,** 353–396.

Alsop, R. G., and Forney, J. L. (1962). Growth and food of white perch in Oneida Lake. *N.Y. Fish Game J.* **9,** 133–136.

American Public Health Association. (1971). "Standard Methods for the Examination of Water and Wastewater," 13th ed. APHA, Chicago, Illinois.

Anonymous. (1975). "Great Lakes Basin Framework Study—Water Quality," Appendix 7. US Environ. Prot. Agency, Washington, D.C.

Bachmann, R. W., and Jones, J. R. (1974). Phosphorus inputs and algal blooms in lakes. *Iowa State J. Res.* **49,** 155–160.

Bailey, R. M., ed. (1970). "A List of Common and Scientific Names of Fishes from the United States and Canada," 3rd ed., Spec. Publ. No. 6. Am. Fish. Soc.

Baker, F. C. (1916a). The fresh-water Mollusca of Oneida Lake, New York. *Nautilus* **30,** 5–9.

Baker, F. C. (1916b). The relation of mollusks to fish in Oneida Lake, New York. *N.Y. State Coll. For., Tech. Publ.* **4,** 1–366.

Baker, F. C. (1916c). A new variety of *Lampsilis* from Oneida Lake with notes on the *luteola* group. *Nautilus* **30,** 74–77.

Baker, F. C. (1918a). Further notes on the Mollusca of Oneida Lake, N.Y.: The mollusks of lower South Bay. *Nautilus* **31,** 81–93.

Baker, F. C. (1918b). Descriptions of a new variety of fresh-water mussel from Oneida Lake, New York. *N.Y. State Coll. For., Tech. Publ.* **9,** 247–248 (append.).

Baker, F. C. (1918c). The productivity of invertebrate fish food, on the bottom of Oneida Lake, with special reference to mollusks. *N.Y. State Coll. For., Tech. Publ.* **9,** 1–264.

Baker, F. C. (1918d). The relation of shellfish to fish in Oneida Lake, New York. *N.Y. State Coll. For., Circ.* **21,** 1–34.

Barber, W. D. (1963). "The New York State Canal System Evaluation of Current Status as a Transportation Facility and Proposals for Future Use," Planning Rep. No. 5. Department of Civil Engineering, Syracuse University, Syracuse, New York.

Berg, C. O. (1963). Middle Atlantic States. *In* "Limnology in North America" (D. G. Frey, ed.), pp. 191–237. Univ. of Wisconsin Press, Madison.

Brezonik, P. L., Morgan, W. H., Shannon, E. E., and Putman, H. D. (1969). Eutrophication factors in North Central Florida lakes. *Bull., Water Resour. Res. Cent., Univ. Fla., Publ.* No. 5, Ser. No. 134.

Broughton, J. G., Fisher, D. W., Isachson, Y. W., and Rickard, L. V. (1962). The geology of New York State. *N.Y. State Mus. Sci. Serv. Geol. Surv., Map Chart Ser.* No. 5.

Burdick, G. E., and Lipschuetz, M. (1946). "A Report on the Investigation of Oneida Lake," mimeo. New York State Dept. of Conservation, Albany.

Burdick, G. E., and Lipschuetz, M. (1948). "Stream Pollution Report," Mimeo. New York State Conserv. Dept., Albany.

Campbell, W. W. (ed.). (1849). "Life and Writings of DeWitt Clinton." Baker and Scribner, New York.

Carr, J. F., and Hiltunen, J. K. (1965). Changes in the bottom fauna of western Lake Erie from 1930 to 1961. *Limnol. Oceanogr.* **10,** 551–569.

Carter, D. B. (1966). Climate. *In* "Geography of New York State" (J. H. Thompson, ed.), Syracuse Univ. Press, Syracuse, New York.

Chevalier, J. R. (1973). Cannibalism as a factor in the first year of survival of Walleye in Oneida Lake. *Trans. Am. Fish. Soc.* **102,** 739–744.

Christie, W. J. (1972). Lake Ontario: Effects of exploitation, introductions, and eutrophication on the salmonid community. *J. Fish. Res. Board Can.* **29,** 913–929.

Clady, M. D. (1975). Comparison of the bottom fauna in 1916 and 1967 in a bay of Oneida Lake, New York. *N.Y. Fish Game J.* **22,** 114–121.

Clady, M. D. (1976a). Changes in abundance of inshore fishes in Oneida Lake, 1916–1970. *N.Y. Fish Game J.* **23,** 73–81.

Clady, M. D. (1976b). Influence of temperature and wind on the survival of early stages of yellow perch, *Perca flavescens*. *J. Fish. Res. Board Can.* **33**(9), 1887–1893.

Clady, M. D. (1977). Decline in abundance and survival of three benthic fishes in relation to reduced oxygen levels in a eutrophic lake. *Am. Midl. Nat.* **97**(2), 419–432.

Clady, M. D., and Hutchinson, B. (1976). Food of the yellow perch, *Perca flavescens,* following a decline of the burrowing mayfly, *Hexagenia limbata. Ohio J. Sci.* **76**(3), 133–138.

Clark, H. L. (1975). The population dynamics and production of *Daphnia* in Oneida Lake, New York, with reference to predation by young yellow perch. M.S. Thesis, Cornell University, Ithaca, New York.

Cline, M. G., Arnold, R. W., and Olson, G. W. (1975). Working draft soil association map of New York. Unpublished.

Cline, M. G., and Marshall, R. L. (1977). Soils of New York Landscapes. *Cornell Information Bulletin 119.* Cornell University, Ithaca, New York.

Cobb, J. N. (1905). "The Commercial Fisheries of the Interior Lakes and Rivers of New York and Vermont," U.S. Fish Comm. Rep. for 1903, pp. 225–246.

Craven, R. E., and Brown, B. E. (1969). Ecology of *Hexagenia* naiads (Insecta-Ephemeridae) in an Oklahoma reservoir. *Am. Midl. Nat.* **82,** 346–358.

Davis, C. C. (1964). Evidence for the eutrophication of Lake Erie from phytoplankton records. *Limnol. Oceanogr.* **9,** 275–283.

Dean, W. E. (1970). Fe-Mn oxidate crusts in Oneida Lake, New York. *Proc., Conf. Great Lakes Res.* **13,** 217–226.

Dean, W. E., and Ghosh, S. K. (1977). Factors contributing to the formation of ferromanganese nodules in Oneida Lake, New York. (In press.)

Dean, W. E., Ghosh, S. K., Krishnaswami, S., and Moore, W. S. (1972). Geochemistry and accretion rates of freshwater ferromanganese nodules. *NSF-IDOE* pp. 13–19.

DeKay, J. E. (1842). "Zoology of New York. The New York Fauna." Part 4, "Fishes". Appleton and Co., and Wilby and Putnam, Albany.

DeLaubenfels, D. J. (1966a). Vegetation. *In* "Geography of New York State" (J. H. Thompson, ed.), pp. 90–103. Syracuse Univ. Press, Syracuse, New York.

DeLaubenfels, D. J. (1966b). Soil. *In* "Geography of New York State" (J. H. Thompson, ed.), pp. 104–110. Syracuse Univ. Press, Syracuse, New York.

Dence, W. A. (1964). Oneida Lake—a lake in transition. *In* "Some Aquatic Resources of Onondaga County" (D. F. Jackson, ed.), pp. 13–21. Onondaga County Dept. of Public Works, Div. Parks Conserv., Onondaga County, New York.

Dence, W. A., and Jackson, D. F. (1959). Changing chemical and biological conditions in Oneida Lake, New York. *School Sci. Math.* **59,** 317–325.

Dethier, B. E. (1966). Precipitation in New York State. *N.Y. Agric. Exp. Sta., Ithaca, Bull.* **1009,** 1–78.

Dillon, P. J., and Rigler, F. H. (1975). A simple method for predicting the capacity of a lake for development based on lake trophic status. *J. Fish. Res. Board Can.* **32,** 1519–1531.

Edmondson, W. T. (1961). Changes in Lake Washington following an increase in the nutrient income. *Verh. Int. Ver. Limnol.* **14,** 167–175.

Forney, J. L. (1965). Factors affecting growth and maturity in a walleye population. *N.Y. Fish Game J.* **12,** 217–232.

Forney, J. L. (1966). Factors affecting first-year growth of walleyes in Oneida Lake, New York. *N.Y. Fish Game J.* **13,** 146–167.

Forney, J. L. (1967). Estimates of biomass and mortality rates in a walleye population. *N.Y. Fish Game J.* **14,** 176–192.

Forney, J. L. (1971). Development of dominant year classes in a yellow perch population. *Trans. Am. Fish. Soc.* **100,** 739–749.

Forney, J. L. (1972). Biology and management of smallmouth bass in Oneida Lake, New York. *N.Y. Fish Game J.* **19,** 132–154.

Forney, J. L. (1973). "Monitoring the Ecosystem," N.Y. Fed. Aid Project No. F-17-R-17, Job No. I-c (mimeo). N.Y. Dep. Environ. Conserv., Albany, New York.

Forney, J. L. (1974). Interactions between yellow perch abundance, walleye predation, and survival of alternative prey in Oneida Lake, New York. *Trans. Am. Fish. Soc.* **103,** 15–24.

Forney, J. L. (1975). Contribution of stocked fry to walleye fry populations in New York lakes. *Prog. Fish Cult.* **37,** 20–24.

Forney, J. L., and Taylor, C. B., (1963). Age and growth of white bass in Oneida Lake, New York. *N.Y. Fish Game J.* **10,** 194–200.

Gillette, M. J. (1961). Oneida Lake Pancakes. *N.Y. State Conserv.* **18,** 41.

Greeson, P. E. (1971). "Limnology of Oneida Lake with Emphasis on Factors Contributing to Algal Blooms," Open-file Rep. U.S. Geol. Surv., N.Y. Dep. Environ. Conserv., Albany, New York.

Grosslein, M. D. (1961). Estimation of angler harvest on Oneida Lake, New York. Ph.D. Thesis, Cornell University, Ithaca, New York.

Hall, D. J. (1967). "Limnology of Oneida Lake," Prog. Rep. Hatch 479 (mimeo). Cornell University Biol. Field Station, Bridgeport, New York.

Harman, W. N. (1968). Replacement of pleurocerids by *Bithynia* in polluted waters of central New York. *Nautilus* **81,** 77–83.

Harman, W. N., and Forney, J. L. (1970). Fifty years of change in the molluscan fauna of Oneida Lake, New York. *Limnol. Oceanogr.* **15,** 454–460.

Houde, E. D. (1967). Food of pelagic young of the walleye, *Stizostedion vitreum vitreum,* and yellow perch, *Perca flavescens. J. Fish. Res. Board Can.* **26,** 1647–1659.

Houde, E. D. (1968). The relation of water currents and zooplankton abundance to distribution of larval walleyes, *Stizostedion vitreum,* in Oneida Lake, New York. Ph.D. Thesis, Cornell University, Ithaca, New York.

Houde, E. D. (1969). Distribution of larval walleyes and yellow perch in a bay of Oneida Lake and its relation to water currents and zooplankton. *N.Y. Fish Game J.* **16,** 185–205.

Houde, E. D., and Forney, J. L. (1970). Effects of water currents on distribution of walleye larvae in Oneida Lake, New York. *J. Fish. Res. Board Can.* **27,** 445–456.

Hunt, B. P. (1953). The life history and economic importance of a burrowing mayfly, *Hexagenia limbata* in southern Michigan lakes. *Mich., Dep. Conserv. Inst. Fish. Res., Bull.* **4,** 1–151.

Jacobsen, T. L. (1966). Trends in abundance of the mayfly, *Hexagenia limbata,* in Oneida Lake, New York, 1956–1964. *N.Y. Fish Game J.* **13,** 168–175.

Kantrowtiz, I. H. (1970). "Ground-water Resources in the Eastern Oswego River Basin, New York," Basin Planning Rep. ORB-2. New York State Water Resour. Comm., Albany.

Karrow, P. F., Clarke, J. P., and Terasmae, J. (1961). The age of Lake Iroquois and Lake Ontario. *J. Geol.* **69,** 659–667.

Kooyoomjian, K. J., and Clesceri, N. L. (1972). Perception of water quality by select respondent groupings in inland water-based recreational environments. *Rensselaer Fresh Water Inst. Rep.* **73–7,** 1–34.

Landgraff, H. C. (1926). "Oneida Lake; Past and Present." Lakeside Press.

Lehne, M. C. (1963). Annual Report, Oneida Lake Pollution Control Engineer, Onondaga County Dep. Drainage and Sanitation, Syracuse, New York (unpublished).

Lehne, M. C. (1964). Annual Report, Oneida Lake Pollution Control Engineer, Onondaga County Dep. Drainage and Sanitation, Syracuse, New York (unpublished).

Lehne, M. C. (1965). Annual Report, Oneida Lake Pollution Control Engineer, Onondaga County Dep. Drainage and Sanitation, Syracuse, New York (unpublished).

Lin, Y.-S. (1975). Food and growth of young yellow perch during the pelagic and demersal stages in Oneida Lake. Ph.D. Thesis, Cornell University, Ithaca, New York.

MacClintock, P., and Terasmae, J. (1960). The galcial history of Covey Hill. *J. Geol.* **68,** 232–241.

Meining, D. W. (1966). Geography of expansion, 1785–1855. *In* "Geography of New York State" (J. H. Thompson, ed.), pp. 140–171. Syracuse Univ. Press, Syracuse, New York.

Menzel, D., and Corwin, N. (1965). The measurement of total phosphorus on the liberation of organically bound fractions by persulfate oxidation. *Limnol. Oceanogr.* **10,** 280–282.

Michalski, M. F. P., Johnson, M. G., and Veal, D. M. (1973). Muskoka Lakes Water Quality Evaluation," Rep. No. 3. Ontario Ministry of the Environment.

Mt. Pleasant, R. C., Rand, M. C., and Nemerow, N. L. (1962). "Chemical and Microbiological Aspects of Oneida Lake, New York," Res. Rep. No. 6. Civil Engineering Dept., Syracuse University, Syracuse, New York.

Muenscher, W. C. (1928). Plankton studies of Cayuga, Seneca, and Oneida lakes. *In* "A Biological Survey of the Oswego River System," Suppl. to 17th Annu. Rep., pp. 140–157. New York State Dept. of Conservation, Albany.

Muller, E. H. (1965). Quaternary geology of New york. *In* "Quaternary of the United States" (H. E. Wright and D. G. Frey, eds.), pp. 99–112. Princeton Univ. Press, Princeton, New Jersey.

New York State Board of Health. (1888). "Annual Report of the New York State Board of Health," pp. 194–202 and 233–235. NYSBH, Albany.

New York State Commissioners of Fisheries, Game and Forest. (1896). "First Annual Report." Wynkogs Hallenbeck Crawford Co., Albany, New York.

New York State Commissioners of Fisheries, Game and Forest. (1899). "Fourth Annual Report." Wynkogs Hallenbeck Crawford Co., Albany, New York.

New York State Department of Health. (1957). "Oneida River Drainage Area," Oswego River Drainage Basin Surv. Rep. No. 5. NYSDH, Albany.

Noble, R. L. (1968a). Mortality rates of pelagic fry of the yellow perch, *Perca flavescens* (Mitchill), in Oneida Lake, New York, and an analysis of the sampling problem. Ph.D. Thesis, Cornell University, Ithaca, New York.

Noble, R. L. (1968b). "Effect of Limnological Changes on Survival of Young Fish in Oneida Lake," N.Y. Fed Aid Project No. F-17-R-12, Job No. I-e (mimeo). N.Y. Dep. Environ. Conserv. Albany, New York.

Noble, R. L. (1969). "Effect of Limnological Changes on Survival of Young Fish in Oneida Lake," N.Y. Fed Aid Project No. F-17-R-13, Job I-e (mimeo). N.Y. Dep. Environ. Conserv. Albany, New Yorl.

Noble, R. L. (1971). "Monitoring the Ecosystem," N.Y. Fed Aid Project No. F-17-R-15, Job No. I-c (mimeo). N.Y. Dep. Environ. Conserv. Albany, New York.

Noble, R. L. (1972). "Monitoring the Ecosystem," N.Y. Fed Aid Project No. F-17-R-16, Job No. I-c (mimeo). N.Y. Dep. Environ. Conserv. Albany, New York.

Noble, R. L. (1975). Growth of young yellow perch (*Perch flavescens*) in relation to zooplankton populations. *Trans. Am. Fish. Soc.* **104,** 731–741.

Oglesby, R. T., and Schaffner, W. R. (1977). The response of lakes to phosphorus. *In* "Nitrogen and Phosphorus: Agriculture, Wastes, and the Environment." (K. H. Porter, ed.), pp. 25–118. Ann Arbor Sci. Publ., Ann Arbor, Michigan.

Oglesby, R. T., Hamilton, L. S., Mills, E., and Willing, P. (1973). "Owasco Lake and its Watershed." Report to Cayuga County Planning Board and the Cayuga County Environmental Management Council.

Pearson, F. J., Jr., and Meyers, G. S. (1970) Hydrochemistry of the Oneida Lake Basin, New York, N.Y., Dep. Environ, Conserv. Dept. of Inv. RI-12, Albany, New York.

Pilsbry, H. A. (1917). Ammicolidae from Oneida Lake, New York. *Nautilus* **31,** 44–46.

Pilsbry, H. A. (1918). New species of Ammicolidae from Oneida Lake, New York. *N.Y. State Coll. For., Tech. Publ.* **9,** 244–246 (append.).

Raney, E. C., and Lachner, E. A. (1942). Studies of the summer food, growth, and movements of young yellow pike-perch. *Stizostedion v. vitreum,* in Oneida Lake, New York. *J. Wildl. Manage.* **6,** 1–16.

Rawson, D. S. (1955). Morphometry as a dominant factor in the productivity of large lakes. *Verh. Inst. Ver. Limnol.* **12,** 164–175.

Rawson, D. S. (1960). A limnological comparison of twelve large lakes in northern Saskatchewan. *Limnol. Oceanogr.* **5,** 195–211.

Rayback, R. J. (1966). The Indian. *In* "Geography of New York State" pp. 113–120. Syracuse Univ. Press, Syracuse, New York.

Rickard, L. V., and Fisher, D. W. (1970a). "Geological Map of New York," Adirondack Sheet. New York State Museum and Science Service, Albany, New York.

Rickard, L. V., and Fisher, D. W. (1970b). "Geological Map of New York," Finger Lakes Sheet. New York State Museum and Science Service, Albany, New York.

Schneider, J. C. (1965). Further studies of the benthic ecology of Sugarloaf Leke, Washtenaw County, Michigan. *Pap. Mich. Acad. Sci., Arts Lett.* **50,** 11–29.

Schneider, J. C., Hooper, F. F., and Beeton, A. M. (1969). The distribution and abundance of benthic fauna in Saginaw Bay, Lake Huron. *Proc. Conf. Great Lakes Res.* **12,** 80–90.

Scott, W. B., and Christie, W. J. (1963). The invasion of the lower Great Lakes by the white perch, *Roccus americanus* (Gmelin). *J. Fish. Res. Board Can.* **20,** 1189–1195.

Shampine, W. J. (1973). "Chemical Quality of the surface Water in the Eastern Oswego River Basin, New York," Basin Planning Rep. ORB-6 New York State Dept. of Environmental Conservation, Albany.

Smallwood, W. M., and Struthers, P. H. (1928). Carp control studies in Oneida Lake. *In* "A Biological Survey of the Oswego River System," Suppl. to 17th Annual Report, pp. 67–83. New York State Dept. of Conservation, Albany.

Smith, D. B. (1972). Age and growth of the cisco in Oneida Lake, New York. *N.Y. Fish Game J.* **19,** 83–91.

Stone, U. B., and Pasko, D. (1946). Oneida lake investigations. mimeo *N.Y. State Conserv. Dep.,* Albany, New York.

Stout, N. J. (1958). Atlas of forestry in New York. *State Univ. Coll. For., Bull.* **41.**

Tarby, M. J. (1974). Characteristics of yellow perch cannibalism in Oneida Lake and the relation to first year survival. *Trans. Am. Fish. Soc.* **103,** 462–471.

Tebo, L. B., Jr. (1955). Bottom fauna of a shallow eutrophic lake, Lizard Lake, Pocahontas County, Iowa. *Am. Midl. Nat.* **54,** 89–103.

United States Department of the Interior. (1968). "Water Resources Data for New York, 1967," Part 1. Water Resour. Div., Washington, D.C.

United States Department of the Interior. (1969). "Water Resources Data for New York, 1968," Part 1. Water Resour. Div., Washington, D.C.

United States Department of the Interior. (1970). "Water Resources Data for New York, 1969," Part 1. Water Resour. Div., Washington, D.C.

Vollenweider, R. A. (1968). "Scientific Fundamentals of the Eutrophication of Lakes and Flowing Waters," Tech. Rep. DAS/CSI 68.27. Organ. Econ. Coop. Dev., Paris.

Vollenweider, R. A. (1973). Input-output models. *Schweiz. Z. Hydrol.*

Vollenweider, R. A., Munawar, M., and Stadelmann, P. (1974). A comparative review of phytoplankton and primary production in the Laurentian Great Lakes. *J. Fish. Res. Board Can.* **31,** 739–762.

Wagner, F. E. (1928). Chemical investigation of the Oswego Watershed. *In* "A Biological Survey of the Oswego River System," Suppl. 17th Annu. Rep., pp. 1–23. New York State Dept. of Conservation, Albany, New York.

Index

F

I

J

K

L

M

N

O

A
B 8
C 9
D 0
E 1
F 2
G 3
H 4
I 5
J 6